TRENDS IN LOGIC

TRENDS IN LOGIC
Studia Logica Library

VOLUME 21

SCOPE OF THE SERIES

Trends in Logic is a bookseries covering essentially the same area as the journal *Studia Logica* – that is, contemporary formal logic and its applications and relations to other disciplines. These include artificial intelligence, informatics, cognitive science, philosophy of science, and the philosophy of language. However, this list is not exhaustive, moreover, the range of applications, comparisons and sources of inspiration is open and evolves over time.

Volume Editor
Ryszard Wójcicki

The titles published in this series are listed at the end of this volume.

TRENDS IN LOGIC

50 Years of Studia Logica

Edited by

VINCENT F. HENDRICKS
Department of Philosophy and Science Studies,
Roskilde University, Denmark

and

JACEK MALINOWSKI
Institute of Philosophy and Sociology,
The Polish Academy of Sciences, Poland

SPRINGER-SCIENCE+BUSINESS MEDIA, B.V.

A C.I.P. Catalogue record for this book is available from the Library of Congress.

ISBN 978-90-481-6414-1 ISBN 978-94-017-3598-8 (eBook)
DOI 10.1007/978-94-017-3598-8

Printed on acid-free paper

CONTENTS

Preface

Precisely 50 years ago the first volume of *Studia Logica* appeared. The present volume celebrates the 50th Anniversay of *Studia Logica – An International Journal for Symbolic Logic.* The papers in this Anniversary volume are the results of the two conferences *Trends in Logic – 50 Years of Studia Logica.*

The first conference took place at The Polish Academy of Sciences, Jabłonna, Poland, October 4–6, 2003 and featured lectures Michael Dunn (USA), Jens Erik Fenstad (Norway), Dagfinn Føllesdal (Norway), Vincent F. Hendricks (Denmark), Joachim Lambek (Canada), Grzegorz Malinowski (Poland), Ewa Orłowska (Poland) and Krister Segerberg (Sweden). The conference was sponsored by The Institute of Philosophy and Sociology, The Polish Academy of Sciences; Faculty of Mathematics and Computer Science, Adam Mickiewicz University, Poznań, Poland; Faculty of Mathematics and Computer Science, University of Warmia and Mazury, Olsztyn, Poland; The Banach Center and Kluwer Academic Publishers. The Honorary Committee included Henryk Domański, Zbigniew Palka, Jerzy Pelc, Jan Rychlewski and Henryk Samsonowicz. The program committee consisted of Ryszard Wójcicki, Wojciech Buszkowski, Jacek Malinowski, Ewa Orłowska and Jan Zygmunt.

The second conference took place at The Carlsberg Academy, Copenhagen, Denmark, November 20–22, 2003 and featured lectures by Johan van Benthem (The Netherlands), Wojciech Buszkowski (Poland), Leo Esakia (Georgia), Melvin Fitting (USA), Joseph M. Font (Spain), Robert Goldblatt (New Zealand), Daniele Mundici (Italy), Hiroakira Ono (Japan) and Heinrich Wansing (Germany). The conference was organized by Vincent F. Hendricks, Jacek Malinowski, Stig Andur Pedersen, Ryszard Wójcicki, with the very substantial assistance of Pelle Guldborg Hansen and sponsored by ΦLOG – The Danish Network for Philosophical Logic and Its Applications.

Studia Logica would like to thank all individuals and institutions contributing so kindly to the celebration of this Anniversary. Finally, *Studia Logica* would also like to thank Kluwer Academic Publishers for the continuous fruitful cooperation.

Vincent F. Hendricks and Jacek Malinowski
July 2003

VINCENT F. HENDRICKS
JACEK MALINOWSKI

50 Years of *Studia Logica*: Editorial Introduction

In 1953, exactly 50 years ago to this day, the first volume of *Studia Logica* appeared under the auspices of The Philosophical Committee of The Polish Academy of Sciences. Now, 5 decades later the present volume is dedicated to a celebration of this 50th Anniversary of *Studia Logica*. It features a series of papers by distinguished scholars reflecting both the aim and scope of this journal for symbolic logic.

A Brief History of *Studia Logica*

Studia Logica was founded by the famous Polish logician and philosopher Kazimierz Ajdukiewicz during a time which was rather inhospitable to such initiatives. All the academic activities were controlled by the communistic party (The Polish Unified Workers Party) which intended to capture the life of Polish science under a Marxist biased heading. Despite these local political intentions, ever since the publication of the first issue *Studia Logica* aimed at close shaves with the international and more liberal world of science.

The first issues of the journal were published under the editorship of Henryk Greniewski, Tadeusz Kotarbiński, Andrzej Mostowski, Roman Suszko with Kazimierz Ajdukiewicz as the editor-in-chief. Continuously appearing since 1953, *Studia Logica* was from the beginning strongly influenced by the extraordinarily rich and fruitful tradition of logical investigations which had surfaced in Poland between the World Wars. Prominent logicians and philosophers like Jan Łukasiewicz, Alfred Tarski, Stanisław Leśniewski, Leon Chwistek, Tadeusz Kotarbiński, Kazimierz Ajdukiewicz, Mordechaj Wajsberg, Andrzej Mostowski and Adolf Lindenbaum were all central figures in establishing that tradition. After Ajdukiewicz's death in 1963 the editor-in-chief position was taken over by Jerzy Słupecki and then in 1970 by Klemens Szaniawski.*

In 1975 the editorship of *Studia Logica* was undertaken by Ryszard Wójcicki, first assisted by Jan Zygmunt and then for a very long period

*Ryszard Wójcicki's paper in this volume includes more detailed historical information on *Studia Logica*.

V. F. Hendricks and *J. Malinowski* (eds.),
Trends in Logic: 50 Years of Studia Logica

Trends in Logic **21**: 1–10, 2003.

of time by Grzegorz Malinowski. R. Wójcicki has since been leading the editorial work of the journal with the exception of the years 1991-1992 when the journal was directed by Jan Zygmunt. The mid 1970's were the beginning of significant and elaborate changes. In 1976 an international editorial board was formed and the journal was published exclusively in English. In 1979 North-Holland became the publisher of *Studia Logica* along with Ossolineum (a Polish publishing house). The journal then gained a solid international position.

Even though the initiatives of these changes were Wójcicki's, his efforts were strongly backed by a number of other people, notably Andrzej Mostowski, Helena Rasiowa, Klemens Szaniawski. In co-publishing the journal North-Holland was replaced in 1984 by Reidel; Reidel later became part of Kluwer Academic Publishers. Since 1992 *Studia Logica* have been published exclusively by Kluwer Academic Publishers, i.e. without the cooperation of Ossolineum.

The 1990's added yet more velocity to the journal. The year 1995 announced an increase in the volume and frequency of published issues to 6 yearly issues, which again was increased in 2000 to 9 issues every year. The same year *Studia Logica* also expanded its publication profile by adding the book series *Trends in Logic: Studia Logica Library.* The current volume is the twenty-first in this series. The book series has the same editorial statement as the journal:

> *Studia Logica* publishes original papers on various logical systems, which utilize methods of contemporary formal logic (those of algebra, model theory, proof theory, etc.). More specifically, *Studia Logica* invites articles on topics in general logic and on applications of logic to other branches of knowledge such as philosophy of language, cognitive science and artificial intelligence. The distinctive feature of *Studia Logica* is its series of monothematic issues edited by outstanding scholars and devoted to important topics of contemporary logic or covering significant conferences.

In 1998, notably due to the work of Graham Priest, the reviews section of *Studia Logica* became an integral part of the journal. From 2000-2002 the reviews section was directed by Edwin Mares.

Over the years *Studia Logica* has achieved an unquestionable position as one of the most important logic journals available. This success owes much to a plethora of logicians and philosophers who closely cooperate with the journal. Immediately below is a selective list of scholars who over the years either have been or are currently acting on the editorial board:

K. Ajdukiewicz, S. Artemov, A. Arruda, A. Avron, T. Batóg, N. Belnap, J. van Benthem, W. Blok, L. Borkowski, W. Buszkowski, W. Carnielli, M.L. Dalla Chiara, N. da Costa, M. Cresswell, J. Czelakowski, K. Dôsen, A. Dragalin M.J. Dunn, W. Dziobiak, L. Esakia, M. Fitting, K. Fine, V. Finn, D. Føllesdal, A. Fuhrmann, D.M. Gabbay, R. Gilles, R. Goldblatt, H. Greniewski, A. Grzegorczyk, P. Hajek, J. Hintikka, R. Hilpinen, W. van der Hoek, L. Iturrioz, S. Jaśkowski, D. Kaplan, M. Kokoszyska-Lutmanowa, T. Kotarbiński, J. Kotas, G. Kreisel, J. Lambek, D. Makinson, L. Maksimowa, G. Malinowski, W. Marciszewski, E. Mares, R.K. Meyer, D. Monk, A. Mostowski, D. Mundici, A. Nakamura, M. Nasieniewski, I. Nemeti, H. Ono, E. Orłowska, Z. Pawlak, J. Perzanowski, D. Pigozzi, W. Pogorzelski, G. Priest, T. Prucnal, M. Przełęcki, C. Rauszer, W. Rautenberg, H. Rasiowa, R. Routley, I. Ruzsa, V. Rybakov, D. Scott, A. Skowron, K. Segerberg, W. Sieg, V. Shehtman, J. Słupecki, V. Smirnov, S. Surma, R. Sylvan, P. Suppes, R. Suszko, K. Szaniawski, M. Tokarz, A. Troelstra, S. Thomason, R. Turner, A. Urquhart, D. Vakarelov, H. Wansing, B. Węglorz, P. Weingartner, T. Williamson, P. Wojtylak, A. Wroński, M. Zakharyashev, J. Zygmunt.

Studia Logica would like to take this opportunity to express its sincere gratitude to all who have contributed, in the most lucid and encouraging way, to the status of the journal today.

The journal is edited by independent managing editors. At present, the board of managing editors consists of Wojciech Buszkowski, Melvin Fitting, Andre Fuhrmann, Robert Goldblatt, Jacek Malinowski, Daniele Mundici and Heinrich Wansing. Vincent F. Hendricks manages the reviews section.

The official name of the journal is *Studia Logica – An International Journal for Symbolic Logic.* The Institute of Philosophy and Sociology of the Polish Academy of Sciences owns the journal's name, while the exclusive publisher and owner of copyrights is Kluwer Academic Publishers. At present *Studia Logica* is directed by Ryszard Wójcicki – the chairman of the editorial board and Jacek Malinowski – the coordinating editor responsible for the day-to-day operations.

The editorial office is located in the Staszic Palace, Section of Logic, Language and Action, Institute of Philosophy and Sociology, The Polish Academy of Sciences, Nowy Świat 72, Warsaw, Poland. The official *Studia Logica* website is located at `http://www.studialogica.org`.

Research Trends in *Studia Logica* – the Past and the Future

The present volume consists of papers reviewing selected topics in logic which have found their way into *Studia Logica* over the years and now have become the quint-essential trademarks of the journal. The vast development of logic and logic related areas over the last decades have forced a selection of topics to be covered in this volume which unfortunately leaves a great many logical themes unaddressed for which apologies are in order.

The volume opens with the paper by R. Wójcicki "Polish Logic in the Postwar Period". This paper includes an overview of the history of logical investigations over the last fifty years in Poland. It also contains an exposition of the history of *Studia Logica*.

Johan van Benthem's paper then presents an overview of the logical and philosophical landscape to which *Studia Logica* belongs and have been taking an important part in shaping. This landscape remains both constant and ever changing. One constant is the methods inherent in first-order logic which have been utilized in foundational mathematical studies, in natural language semantics, in studies of various forms of inference patterns and consequence operations, in the philosophy of science and artificial intelligence. The dynamics of logic is then again reflected in studies of communication and information updates, in various epistemic and doxastic logics, in game theory and logical theories for social software, in belief revision and theory change. Many of the themes touched upon in van Benthem's contribution are subsequently dealt with in greater detail in some of the other papers of this Anniversary volume.

Algebraic logics and consequence operations have for a long time been dominant topics of *Studia Logica*. These fields consist in investigating, by means of algebraic methods, the consequence operations and deductive systems defined by means of classes of abstract algebras or more generally classes of logical matrices. Logical matrices and the general theory of consequence operations stem from results obtained by Tarski and his collaborators. J. Czelakowski, G. Malinowski [85] presented an analysis of Tarski's contributions.

Studia Logica published a number of significant results in the field of consequence operations. J. Łoś [54] presented an algebraic approach to deductive systems. R. Wójcicki [73] started the modern model-theoretic treatment of logical matrices. R. Wójcicki's book [88] is the most complete treatment of the theory of logical consequences and contains the full list of references to papers in this field published in *Studia Logica*.

For a few samples of the papers demonstrating the intimate relationship between consequence operations and abstract algebraic logic published in *Studia Logica* the following should be mentioned: S. Bloom [75] studied some representations and properties of the lattice of consequence operations. This paper contained a seminal idea that allowed the introduction of first-order methods in the study of matrix models of sentential logics. J. Zygmunt [74] investigated ultraproducts of logical matrices. R. Wójcicki [74] and G. Malinowski [74] scrutinized the lattice of many-valued consequence operations. W. Dziobiak [81] published an important paper investigating the notion of degrees of maximality in the lattice of consequence operations. J. Czelakowski [80], [81] studied matrix models and reduced matrix models of given logics. W. Rautenberg [81] proposed a full characterization of two-element matrices. Rautenberg [93] also provided a first-order axiomatization and the conditions of first-order axiomatizability for a class of reduced matrices. Over the next few years Czelakowski and Dziobiak published a number of papers devoted to the structural interrelations between the deduction theorem and the properties of matrix models for given logics.

The above mentioned papers have a seminal continuation in Blok and Pigozzi's paper [86]. This particular paper initiated a rich stream of results pertaining to the Leibniz-operator. B. Herrmann [97] characterized equivalential and algebraizable logics by properties of the Leibniz-operator. In [91] Blok and Pigozzi and later in 2000 and 2003 J.M. Font, R. Jansana and D. Pigozzi published three special issues of the journal devoted exclusively to this field, now called *abstract algebraic logic.* The paper by J.M. Font in this volume systematically summarizes the results in this area.

Modal logics have been an important part of the journal since the very beginning. The special issues of H. Rasiowa [77] and K. Segerberg [81] were dedicated to modal logics. The special issue, *Provability Logic* [91], edited by L. Esakia and S. Artemov addressed the modal interpretation of the concept of proof. A number of papers on modal logics appeared in subsequent issues. To mention a few of them; J.M. Dunn [95] considered a semantics for negation-free modal systems; Jansana [95] approached some important, and still unsolved problems, related to logical modal entailment. The special issue edited by D.M. Gabbay and F. Pirri [97] comprised investigations of combining modal logics.

This Anniversary volume holds two papers on modal logics. The paper by Melvin Fitting discusses two different ways in which first-order logic may be expanded to serve as foundations for mathematics. Both approaches are modal in nature – the one in terms of higher-order intensional logic, the other in terms of intensional set theory. The second paper by Rob Goldblatt deals with modal logics from an algebraic perspective and provides an analysis of the canonical models of modal logics.

An area like many-valued logic is closely connected to results in consequence operations and logical matrices. Papers by Wójcicki [74] and Malinowski [74] are typical examples of the interrelation between many-valued logics and the theory of consequence operations. Due to spatial limitations it is impossible to list all the papers presenting new results in this area. Mundici [98] edited a special issue in this field. Fitting [95] developed the tableaux method for many-valued logics. The special issue edited by R. Dyckhoff and H. Wansing [2001] furnished more results concerning the tableaux methods. Daniele Mundici's paper in this volume presents a comprehensive survey of what has happened in many-valued logic with particular emphasis on the rapid and extensive developments in this field over the past twenty years.

Non-classical logics overlap with the subjects mentioned immediately above. *Studia Logica* received an extensive input with respect to the algebraic approach to non-classical logics by the work of by H. Rasiowa and her collaborators. The special issue edited by H. Rasiowa [77] is just one representative example of this input and of her influence on *Studia Logica* in general. The so-called BCK-logics belonging to the mainstream of investigations in non-classical logics have likewise been widely discussed in papers published in the journal. Hiroakira Ono's paper in this volume contains a review of the part of non-classical logics related to substructural logics.

Logical systems designed to study the concept of consistency have been another trademark of the journal. J. Kotas edited a special issue of *Studia Logica* [75] devoted to Jaśkowski's logics of discourse. Paraconsistent logics were also the focus of the special issue of the journal edited by G. Priest and R. Routley [84].

Two papers in this volume deals with some of the main topics of paraconsistency. The paper by Heinrich Wansing and Sergiej Odintsov reviews the constructive approach to this area. Graham Priest's paper discusses inconsistent models of arithmetic and is concluded by a number of open questions and problems.

Quantum logics have been a frequently addressed subject of the journal. To recall only some selected papers: J. Kotas [67], J. Czelakowski [74-75], R. Giuntini [88] and J. Malinowski [92] have published important algebraic results with respect to deductive systems of quantum logics. The paper by M.L. Dalla Chiara, R. Giuntini and R. Leporini in the current volume treats this subject matter.

P. Suppes published an analysis of logical entailment as it appears in the natural language of English [79]. J. Lambek then edited a special issue of the journal focussing on categorial logics [89]. In some of his numerous publications in *Studia Logica* W. Buszkowski have considered the theory of categorial grammars. Buszkowski was the editor of a special issue devoted to Lambek-calculus and its many applications in logic and linguistics [2002]. The current paper by Buszkowski is a survey of this development.

A number of important cross-fertilized subjects have found their way into the journal like the intimate connection between logic and cognitive science. van Benthem's introductory paper touches upon the extensive developments in dynamic logic. Segerberg [92] edited a special issue of the journal dedicated to this field.

The relation between logic and issues in cognitive science may also be traced through a subject like non-monotonic logic – the systematic study of the mechanisms of common-sense reasoning. D. Makinson published his seminal philosophical analysis of the notion of preference modelling common-sense reasoning in the journal [94]. Various special issues of *Studia Logica* have also been dedicated to non-monotonicity in particular Jones, Sergot, and D. Makinson [96]; H. Rasiowa and E. Orłowska [97]; M. Ojeda-Aciego, I. P. de Guzman, G. Brewka, L.M. Pereira [2002], and M-A. Williams [2001] hold a number of interesting results.

Further close encounters between logic and cognitive science may be found in belief dynamics. In the 1980's C. Alchurrón, P. Gärdenfors and D. Makinson published a series of papers outlining the now classical theory of belief change known as AGM. Recently W. van der Hoek and M. Olsson edited a special issue centered around the recent achievements in belief revision theory.

Studia Logica has proved itself to being one of the most influential journals in logic today. This is due in part to the historical and even political circumstances surrounding the journal and in part due to the visions and vigor of the collaborators and contributors determinedly working to fulfill *Studia Logica*'s ambition. Symbolic logic has turned out be a discipline so vital to both foundational studies as well as applications. *Studia Logica* will continue to perfect the means for fulfilling the ends: To serve as a fruitful forum for formal thought in philosophy as philosophy translates into the pursuit of knowledge.

References

M.E. Adams W. Dziobiak ed. [96] Priestley Duality, special issue of *Studia Logica* vol. 56, no. 1-2.

W. Blok, D. Pigozzi [86] Protoalgebraic logics, *Studia Logica* vol. 45, no. 4, p. 337–370.

W. Blok, D. Pigozzi [91] Algebraic Logic, special issue of *Studia Logica* vol. 50, no. 3-4.

S. Bloom [75] Some theorem on structural consequence operations, *Studia Logica* vol. 34 no. 1, p. 1–10.

W. Buszkowski [2002] The Lambek calculus in logic and linguistics, special issue of, *Studia Logica* vol. 71, no. 3.

J. Czelakowski [74-75], Logics based on partial Boolean σ-algebras, *Studia Logica* vol. 33 no. 4., p. 371–398 and vol. 34, no. 1, p. 69 - 86.

J. Czelakowski [80] Reduced products of logical matrices, *Studia Logica* vol. 39, no. 1, p. 19–43.

J. Czelakowski [81] Equivalential logics I and II, *Studia Logica* vol. 40, no. 3, p. 227–236 and vol. 40, no. 4, p. 355–372.

J. Czelakowski, G. Malinowski [85] Key notions of Tarski's methodology of deductive systems, *Studia Logica* vol. 44, no. 3, p. 321–251.

M. Dunn [95] Positive modal logics, *Studia Logica* vol. 55, no. 2, p. 301–345.

R. Dyckhoff, H. Wansing ed. [2001] Analytic tableaux and related methods, special issue of *Studia Logica*, vol. 69. no. 1 and 2.

W. Dziobiak [81] The lattice of strengthenings of a strongly finite consequence operation, *Studia Logica* vol. 40, no. 2, p. 177–194.

L. Esakia, S. Artemov, ed. [91], Provability Logics, special issue of *Studia Logica* vol. 50 no. 1.

D.M. Gabbay, F. Pirri [97], Combining Logics, special issue of *Studia Logica* vol. 59 no. 1 and no, 2.

M. Fitting [95] Tableaux for Many-Valued modal logics, *Studia Logica* vol. 55, no. 1, p. 63–87.

J.M. Font, R. Jansana, D. Pigozzi [2000] [2003] *Abstract Algebraic Logic*, special issue of *Studia Logica* vol. 65 no. 1 and and vol. 74 no. 1-2.

R. Giuntini [88] Quantum logics and Lindenbaum property, *Studia Logica* vol. 46, no. 1, p. 17–36.

B. Herrmann [97] Characterizing equivalential and algebraizable logics by the Leibniz operator,*Studia Logica* vol. 58, no. 2, p. 305–337.

R. Jansana [95] Abstract modal logics, *Studia Logica* vol. 55, no. 2, p. 273–299.

Jones, M. Sergot, D. Makinson [96] Papers in Deontic Logic, special issue of *Studia Logica* vol. 57, no. 1.

J.Kotas [67] An axiom system for modular logic, *Studia Logica*, vol. 21, p. 17–38.

J. Kotas [75] Stanisław Jaśkowski achievements in mathematical logics, special issue of *Studia Logica* vol. 34 no. 2.
J. Lambek ed. [89] Categorial Logic, special issue of *Studia Logica* vol. 48, no. 3.
J. Łoś [54] The algebraic treatment of the methodology of elementary deductive systems, *Studia Logica* vol. 2, p. 151 - 210.
D. Makinson [93] Five faces of minimality *Studia Logica* vol. 52, no. 4. p. 339–379.
G. Malinowski [74] S-algebras and the degrees of maximality for three and four valued logics of Łukasiewicz, *Studia Logica* vol. 33 no. 4., p. 359–370.
J.Malinowski [92] Strong versus weak quantum consequence operation, *Studia Logica* vol. 51 no. 1, p. 113–124.
D. Mundici ed. [98] Many-Valued Logics, special issue of *Studia Logica* vol. 61, no. 1.
M. Ojeda-Aciego, I. P. de Guzman, G. Brewka, L.M. Pereira ed. [2002] Logistics in Artificial Intelligence, special issue of *Studia Logica*, vol. 72, no. 1.
G. Priest, R. Routley ed, [84], Paraconsistent Logics, special issue of *Studia Logica* vol. 43, no. 1–2.
H. Rasiowa ed. [77] Algebraic and Kripke-style approaches to logic, special issue of *Studia Logica*, vol. 36, no. 1–2.
H. Rasiowa, E. Orłowska, ed. [97] Reasoning with Incomplete Information, special issue of *Studia Logica* vol. 58, no. 1.
W. Rautenberg [81] 2-element matrices, *Studia Logica* vol. 40 no. 4, p. 315–354.
W. Rautenberg [93] On reduced matrices, *Studia Logica* vol. 52, no. 1, p. 63–72.
K. Segerberg [81] Essays in modal logic, special issue of *Studia Logica* vol. 40 no. 1.
K. Segerberg [92] Logic of Action, special issue of *Studia Logica* vol. 51, no. 3-4.
P. Suppes [79] Logical inference in English: A preliminary analysis, *Studia Logica* vol. 38 no. 4, p. 375–393.
M-A. Williams ed, [2001] On commonsense reasoning, special issue of Studia Logica, vol. 67 no. 2 and no. 3.
R. Wójcicki [73] Matric approach in the methodology of sentential calculus, *Studia Logica* vol. 32, p. 7–37.
R. Wójcicki [74] The logics stronger than three valued sentential calculus. The notion of degree of maximality versus the notion of degree of completeness, *Studia Logica* vol. 33 no. 2, p. 201–214.

R. Wójcicki, *Theory of logical calculi. Basic theory of consequence operations*, vol. 199 of *Synthese Library*. Reidel, Dordrecht, 1988.
J. Zygmunt [74] A note on direct products and ultraproducts of logical matrices, *Studia Logica* vol. 33 no. 4., p. 349–358.

RYSZARD WÓJCICKI
in collaboration with
JAN ZYGMUNT

Polish Logic in Postwar Period

1. Introductory Remarks

During the 10th Congress of Logic, Methodology and Philosophy of Science (Florence, August 19-25, 1995), I took part in a panel discussion on the situation of logic in Eastern Europe during the time of Soviet domination. This essay, originally written to celebrate the 50th anniversary of the Polish Academy of Sciences[1], is primarily based on the paper I presented on that occasion[2]. The list of people with whom I consulted while preparing first the Florence paper and then the present one is rather long.[3] I appreciate the assistance of all of them. My special thanks are due to Wojciech Buszkowski, Andrzej Grzegorczyk, Witold Marciszewski, Wiktor Marek, Roman Murawski, Jerzy Tiuryn and Jan Zygmunt, who in addition to offering various suggestions, remarks and criticism provided me with brief overviews of selected areas of logical investigations carried out by Polish logicians. The postwar Polish logic is too rich and too diversified for one person to be able to present it in an adequate manner, and I would not have been able to complete this paper without these people's kind assistance. Yet, it goes without saying that the final responsibility for this paper is mine. I tried as far as I was able to evaluate critically all pieces of information I was offered and occasionally revise them. In this manner, the Polish version of this survey

[1]The Polish version "Logika polska okresu powojennego, próba rzutu oka wstecz" of this paper was published in *Nauka* 4(2002), 157-175.

[2]"The Postwar Panorama of Logic in Poland", in: *Logic and Scientific Methods*, eds. M.L. Dalla Chiara et al., Kluwer 1997, 597-608.

[3]Various suggestions regarding the earlier "Florence" version of this survey were offered by: Janusz Czelakowski, Andrzej Grzegorczyk, Jacek Malinowski, Marcin Mostowski, Roman Murawski, Ewa Orłowska, Witold A. Pogorzelski, Kazimierz Świrydowicz, Max Urchs, Jan Woleński, Andrzej Wójcik, Jan Zygmunt. While preparing this version of the survey I received assistance from: Zofia Adamowicz, Wojciech Buszkowski, Janusz Czelakowski, Witold Marciszewski, Wiktor Marek, Roman Murawski, Jan Mycielski, Mieczysław Omyła, Jerzy Pogonowski, Jerzy Tiuryn, Anita Wasilewska, Andrzej Wiśniewski, Jan Woleński, and Jan Zygmunt.

V. F. Hendricks and *J. Malinowski* (eds.),
Trends in Logic: 50 Years of Studia Logica

Trends in Logic **21**: 11–33, 2003.

was prepared in cooperation with Jan Zygmunt, who was also the author of the initial draft of its English version. The latter was accessible on the *Studia Logica* home page for quite some time. Eventually, I rewrote its various parts taking into account suggestions offered to me by its readers. Of those, I particularly appreciate the remarks and comments offered by Z. Adamowicz, J.M. Dunn, W. Marek, R.Murawski, Z. Pawlak, J. van Benthem, and H. Wansing. I also express my gratitude to Tom Brunty who took the effort to polish this translation.

This paper was, in its initial form, addressed to a rather large audience, and though this time its expected readers are logicians, its "popular" style has been preserved. Thus, in particular, various rather "loose" definitions meant to explicate various technical notions to the reader whose knowledge of logic is limited were neither removed nor replaced by definitions thought to be more accurate.

2. Definition

The term "logic" is rather ambiguous, yet the question of how it should be understood is crucial for delimiting the scope of this paper. Thus, even though no answer to this question is likely to gain approval from all whose opinion on the matter should be considered, I do not think I should avoid undertaking it.

"Logic" (in the basic meaning of the term) is the formal theory of reasoning. Thus, it is a theory whose chief concern is to state conditions that an argument should satisfy in order to be valid. One method of approaching this issue consists of reducing the notion of the validity of an argument to that of the validity of a "rule of inference". The idea here is that an argument is valid if and only if its conclusion has been derived from premises by using valid rules of inference.

The logical analysis of reasoning is carried out under the assumption that both the premises and the conclusion are sentences rather than someone's beliefs or suppositions. The sentence is a syntactic concept and quantity and thus it should be defined in terms of vocabulary and the syntactical rules that govern the language examined. On the other hand, the notion of validity is related to that of truth and thus is a semantic one. Consequently, its purely syntactical definition is not available.

As is rather clear from the remarks above, logical investigations, both in their syntactic and semantic versions, are bound to be strongly related to and often to overlap linguistic ones. Even though logicians limit their linguistic analyses to languages deprived of numerous peculiarities characteristic of "natural" languages (i.e., to languages one might call "ideal" or "idealized"), those languages are by no means "artificial", as is often claimed. The relationship between idealized languages as the objects of logical analyses and natural languages is much the same as that between theoretical models of phenomena and the phenomena they represent, e.g. between a system of mass points whose behavior is described by Newtonian Mechanics and the system of corresponding real, three-dimensional physical objects.

If sentences examined by logicians were treated as someone's beliefs or suppositions, the logical theory of reasoning would be both relevant to and dependent upon results of investigations into human mental capacities. From the point of view of cognitive psychology, reasoning is not merely a sequence of operations on sentences (as the logicians maintain), but is a mental activity of a human being. If logic is not a part of psychology, as was maintained until the beginning of the 20-th century, it is not because it treats notions like reasoning, belief or supposition as merely informal and thus dispensable, but because its laws are not experientially grounded. Logic is a formal science and, as all formal sciences, notably mathematics, it is concerned with some "arbitrarily chosen" set of assumptions that define the objects (or rather the system they form) to be investigated. The domains of logical investigations are known as "logical systems". The fact that a logical system is determined by "arbitrarily chosen" assumptions does not mean that the selection of such assumptions is not governed by restrictions. Whenever a new logical system is proposed, one expects that it will be useful in dealing with either theoretical or practical problems of considerable significance.

Most often a system of logic is defined as a set of sentences (more broadly, well-formed formulas) that are "logically true", i.e. true regardless of how one understands the "nonlogical terms" they contain. Since the chief task of logic is to state conditions for valid reasoning, besides defining the notion of logical truth, the definition of a logical system should instruct the user as to how the concept of logically valid inference might be reduced to that of logical truth. An alternative way of defining logic consists of defining the notion of a logically valid inference directly, rather than in terms of logical truth. Thus there are two alternative ways to define a logical system. One might respectively call them "sentential" (where the logical truth of a sentence is of primary substance) and "inferential" (where the key notion is that of a valid inference).

Seeking to comply with different logical intuitions, logicians arrive at different systems of logic, or different "logics" for short. The system known as classical logic is of special significance for it provides a formal basis for all standard mathematics.

3. The Outskirts of Logic

Not without good reason, mathematicians consider logic as one of the foundations of mathematics. Two other key elements of mathematics are set theory and recursion theory. The tendency to identify logic with mathematical foundations is most natural. If one adopts such a perspective, then in addition to logic in the narrow sense (which I sought to characterize in the previous section), set theory and recursion theory, one must perforce also treat many other forms of mathematical investigation as parts of logic. For instance, one cannot help but treat investigations into the foundations of arithmetic as part of logic, for their main subject is the logical analysis of various possible formalizations of arithmetic. For the same reason, one cannot resist classifying investigations into foundations of geometry, foundations of algebra, etc, as part of logic Similarly, category theory, universal algebra, and a relatively young discipline called complexity theory might also be regarded as branches of modern logic[4].

The logical foundations of computer science define another important area of investigation strongly related to logic. The relevance of logic for computer science owes to several key attributes. First, the structure of computing machines is based on the laws of logic. Second, logic provides the tools that are necessary for the design and analysis of programming languages. Thirdly, "cognitive engineering", or the design of data bases and expert systems, fruitfully uses logical tools. The language of predicate calculus combined with the laws of logical deduction provide the general framework for defining computer languages. Therefore, along with foundations of mathematics, the foundations of computer science are also rightfully regarded as part of logic in a general sense of the word.

The foundations of mathematical linguistics and the formal foundations of communication theory are two more branches of science which employ the ideas and techniques of logic. In these cases it is extremely difficult to draw a precise border between logic and issues that do not belong to it. Finally, there is another discipline related to logic in many essential ways; namely, cognitive science, including the theory of artificial intelligence. The relations between logic and cognitive science may be readily discernible , but they can-

[4]One might think that by extending the notion of logic to its limits, logicians behave like "logical imperialists" who try to invade other branches of mathematics. This is not so. Logic and its methods are both the source of inspiration and the basic research tool for mathematics. Applying the term "logic" to the foundations of mathematics enables the experts on foundational issues to establish their scientific identity. It also enables them to see their highly varied field of research as a whole that differs from the rest of mathematics.

not, however, be easily characterized. One reason for this is that, in spite of its various considerable achievements, cognitive science remains a conglomerate of concepts and ideas that split into often dramatically different areas of investigations.

In this essay I will make every effort to maintain a reasonable balance between the narrow (presented in Section 2) and the wide (presented in this Section) conceptions of logic. This will not be an easy task, and I only hope that both those who prefer to define logic in a narrow manner and those who opt for as broad a definition as possible may forgive my departures from the specific idea of logic they consider to be the right one.

4. The Interwar Period

The achievements of Polish logicians in the years 1920-1939 gained worldwide recognition. The term "Polish School of Logic" became popular within the international community of logicians. It is noteworthy that during this time the achievements of Polish logicians overlapped in a very substantial way those of another formation known as the "Lvov-Warsaw School of Philosophy"[5].

Seen from today's perspective, the activity of the Polish School of Logic resulted in achievements whose significance can hardly be overestimated. It was during this time when Kazimierz Ajdukiewicz formulated his conception of categorial grammar (inspired by some ideas of developed by Tadeusz Kotarbiński), Jan Łukasiewicz developed his many-valued logic, and Adolf Lindenbaum and Alfred Tarski invented the method of algebra of language known today as the method of Lindenbaum algebra. This was also the period when Tarski published his fundamental papers on the conception of truth and deductive systems (following Kotarbiński's reism, the term coined by Kotarbiński for his radical nominalism) and Stanisław Leśniewski attempted to form a nominalistic version of the set theory strong enough to

[5]It was formed by a group of philosophers, logicians, sociologists, and other scientists who upheld the tradition initiated by the seminars and writings of Kazimierz Twardowski, an eminent psychologist and philosopher from Lvov University. Precise analysis and lucid argument were the virtues that Twardowski considered indispensable both in scientific inquiry and philosophical analyses. Logic was regarded as the basic tool for meeting this requirement. It is not odd then that the Lvov-Warsaw School attracted logicians while at the same time logicians and their works essentially influenced the School. An impressive monograph on the School was written by Jan Woleński, *Logic and Philosophy in the Lvov-Warsaw School*, Kluwer 1989. A brief and informative article on Polish logic of the interwar period can be found in *The Routledge Encyclopaedia of Philosophy*, vol 7, Routledge, London and New York, 1998, 498-500, "Polish Logic" by J. Zygmunt.

base mathematics on it. Leon Chwistek put forward the idea of a simplified theory of types, Janina Hosiasson-Lindenbaum examined the methodological and logical aspects of probability theory, Jan Łukasiewicz and Alfred Tarski developed an algebraic treatment of logical matrices, and Stanisław Jaśkowski (independent of the German logician Gerhard Gentzen) defined the system of "natural deduction" (a system of rules of inference that make the "formalized" inference similar to the "natural" one). This list is remarkable in that it recounts a vast array of spectacular achievements, but perhaps its most notable feature lies in the fact that it is far from complete.

Within the Polish School of Logic logical investigations were carried out both in philosophical and in mathematical departments. For instance, while Leśniewski and Tarski were members of the faculty of the mathematical department of Warsaw University, Ajdukiewicz (Lvov University), Kotarbiński, and Łukasiewicz (Warsaw University) represented the "philosophical wing" of logic. Thus, the cooperation between "mathematicians" and "philosophers" was systematic and close[6].

5. World War II

The Second World War was a disaster for both Polish logic and Polish science as a whole, for three obvious reasons.First, some of the logicians, including those who made substantial contributions to the development of their discipline, did not survive. Priest Jan Salamucha, historian of logic and a close collaborator of Jan Łukasiewicz and Jacek Bocheński, was killed during the Warsaw Uprising in 1944. Because of their Jewish origin Adolf Lindenbaum, his wife Janina Hosiasson-Lindenbaum, Mojżesz Presburger, and Mordechaj Wajsberg (the author of significant results concerning many-valued logic) were murdered by the Nazis. Being abroad when the war started, Tarski avoided this tragic fate.

Second, the war destroyed the whole structure of cooperation among and between scientific institutions. The Jan Kazimierz University of Lvov and the Stefan Batory University of Vilnus ceased to exist. So did numerous research teams. The continuity of teaching, publishing, and organizational work was broken. The manuscripts of many papers being prepared for publication were destroyed. But still, during the war, the illegal "underground universities" were formed and acted on a large scale, offering

---

[6]Alfred Tarski – undoubtedly the most significant person in Polish logic during the interwar period, and one of the greatest logicians of the 20th century – dedicated the collection of his papers *Logic, Semantics, Metamathematics, papers from 1923 to 1938* (Clarendon Press, Oxford, 1956) to "my teacher Tadeusz Kotarbiński".

among others courses of logic. Among its teachers and students were Kazimierz Ajdukiewicz, Janina and Tadeusz Kotarbiński, Andrzej Mostowski, Jan Salamucha, Zygmunt Zawirski, Henryk Hiż, Zbigniew Czerwiński, Andrzej Grzegorczyk, Jerzy Pelc, Roman Suszko, Klemens Szaniawski, and many others.

Finally, when the war was over and Poland was captured by the Soviet Union, many Polish logicians who were lucky enough to leave their country earlier decided to stay abroad. Thus, for instance, Jacek M. Bocheński settled in Switzerland, Czesław Lejewski in England, Jan Łukasiewicz in Ireland, and Henryk Hiż, Bolesław Sobociński, and Alfred Tarski in the USA. Andrzej Ehrenfeucht and Jerzy Mycielski were able to emigrate later.

6. Early Postwar Period – logic in philosophy departments

Although everyone was aware that both political and social life would undergo dramatic changes in postwar Poland, the end of the war brought new hopes. Despite tremendous losses, Polish logic did not cease to exist and soon began to experience a rebirth both in philosophical and mathematical university departments.

Tadeusz Kotarbiński and his wife Janina started their academic activity in Łódź, which, since Warsaw had been completely destroyed during the war and had not yet been rebuilt, was the main scientific center of the country (the Łódź University was founded in 1945). Among the many young people grouped around them were Jerzy Pelc, Marian Przełęcki, and Klemens Szaniawski. At about this time Kazimierz Ajdukiewicz became the Rector of Poznań University. Besides his rector duties, he gathered a team of people who either used logic in their investigations or treated logic as the main subject of their scientific pursuits. Among others, Marcin Czerwiński, Jerzy Giedymin, Seweryna Łuszczewska-Romahnowa, and Roman Suszko worked under his leadership. Maria Kokoszyńska-Lutman, a student of the eminent Tarski who became recognized for her analyses of Tarski's notion of truth, settled in Wrocław. Henryk Mehlberg settled there along with her, after which he eventually succeeded in emigrating to the USA. In Cracow there was a group of logicians led by Izydora Dąmbska and Roman Ingarden. Tadeusz Czeżowski moved from Vilnius to Toruń, where one of his best known students was Leon Gumański.

Except for Roman Ingarden (one of the most prominent Polish philosophers of the 20th century) all the philosophers and logicians whom I mentioned as promoters of newly formed postwar research groups were outstanding representatives of the Lvov-Warsaw School. Besides Kazimierz Twardowski (cf. footnote v), Ajdukiewicz and Kotarbiński were perhaps the most prominent members of that formation. Not all of them (one might also mention here Izydora Dąmbska) were logicians, even in the widest sense of the term. All of them, however, appreciated the significance of logic and were ready to act in support of its development. Zygmunt Zawirski (professor at the mathematics department of the Jagiellonian University in Cracow) was an outstanding example of one displaying such an attitude. Under his supervision, Stanisław Jaśkowski (1945), Andrzej Mostowski (1945), and Jerzy Słupecki (1947) completed their habilitation theses, and R. Suszko (1945) received his master degree.

7. Early Postwar Period – logic in mathematics departments

During the early postwar period it was customary to avoid the term "logic" as either the name or part of the name given to sections of the mathematics departments in which logicians were grouped. The preferred term was "foundations of mathematics". This terminology was a form of camouflage. Even though the ideologists of the communist party considered no branch of science to be ideologically neutral, logic (viewed as part of philosophy and thus part of a discipline supposed to be the chief tool of "ideological offensive") was the subject of their special attention. At different times and in different countries, it was demanded of logicians (with varying degrees of vigor) to suppress formal logic and supplant it with the socalled "dialectical logic" implicit in the writings of Hegel and the "classics" of Marxism.

In Wrocław logic was primarily developed within two groups. One of them was led by professor Czesław Ryll-Nardzewski and the other by professor Jerzy Słupecki. Ryll-Nardzewski, at that time the author of several seminal papers on logic, grouped several talented people around himself, among them L. Pacholski, B. Węglorz, A. Wojciechowska. Among those who started their academic careers under Słupecki's supervision were Witold A. Pogorzelski, Ludwik Borkowski, Bogusław Iwanuś, Tadeusz Prucnal, and also Jerzy Loś, one of the most brilliant logicians of the postwar period (and the author of the widely known Ultraproduct Theorem).

Logic was also being developed by scholars elsewhere in Poland. Notable among them is Stanisław Jaśkowski, who began his work in the mathematics department of the Nicolas Copernicus University in Toruń in 1945. Among others his collaborators there included Jerzy Kotas and August Pieczkowski.

Polish postwar logic owes a great deal to professor Andrzej Mostowski, who pursued his academic career in Warsaw. He was such a prominent figure that his achievements deserve to be presented in a separate section.

8. Andrzej Mostowski

There are several reasons for considering A. Mostowski an exceptional scholar. First, he was an unquestionable scientific authority. Some of his results concerning the foundations of mathematics were of breakthrough significance. His contribution to set theory was immense. He was one of those who, together with A. Tarski, developed the theory of decidability (a theory whose main task is to settle whether, for a given theory, there is an algorithm that enables one to distinguish sentences that are its theorems from the remaining ones). He suggested an algebraic interpretation of the quantifiers and initiated the research of the so-called generalized quantifiers[7]. He also made a substantial contribution to model theory.

Second, he was a scientist of very well established and international reputation. Many eminent researchers in logic and foundations of mathematics visited Poland solely for the purpose of contacting him.

Third, he had a unique skill to look at logic as a discipline whose methods are mathematical and yet whose issues are philosophical in origin. His textbook *Logika Matematyczna* (*Mathematical Logic*) published in 1948 combines in an exemplary way an exposition of the "technicalities" of mathematical logic with a fairly large discussion of the philosophical aspects of its problems. Two of his other works are of a similar nature: "The Present State of Investigations on the Foundations of Mathematics"[8] and *Thirty Years of Foundational Studies*[9]. The latter will be discussed in section 14.

Among his students and close collaborators were numerous stellar representatives of the postwar movement in logic and foundations of mathematics: Zofia Adamowicz, Andrzej Ehrenfeucht, Andrzej Grzegorczyk, W. Guzicki, Wiktor Marek, Helena Rasiowa, Roman Sikorski, Paweł Zbierski, and many others[10]. At the department of mathematics of Warsaw Univer-

---

[7]This subject was examined in many Polish and foreign centers. In Poland it was examined by: A. Krawczyk, M. Krynicki, L. Szczerba, W. Szmielew, M. Zawadowski, and others. Its computational aspects were analysed by A. Pawlak, H. Rasiowa, and E. Orłowska.

[8]A. Mostowski (in collaboration with A. Grzegorczyk, S. Jaśkowski, J. Łoś, S. Mazur, H. Rasiowa, and R. Sikorski), "The Present State of Investigations on the Foundations of Mathematics", *Dissertationes Mathematicae* 9 (1955), 1-48.

[9]A. Mostowski, *Thirty Years of Foundational Studies*; *Lectures on the Development of Mathematical Logic and the Study of the Foundations of Mathematics in 1930-1964*, *Acta Philosophica Fennica* 17 (1965), 1-180.

[10]The group of close collaborators of A. Mostowski included Janusz Onyszkiewicz, Stanisław Krajewski, and Konrad Bieliński. These names are well known to all who witnessed the democratic changes in Poland (J. Onyszkiewicz was the Defense Secretary in two cab-

sity Mostowski was the Head of the Section of Algebra and the Section of the Foundations of Mathematics. He also was the director of the Institute of Mathematics at the Polish Academy of Sciences and a member of the Polish Academy of Sciences. His personal prestige and position translated into prestige and position for logic as a whole, and this surely contributed to the further development of the discipline.

9. Alfred Tarski

Speaking at the Florence Congress (see the introduction) I stated the following: "One might say, not being entirely wrong, that for a long period of time the chief Polish seminar on logic was held in Berkeley, California, in the residence of Alfred Tarski, who kept very close cooperation with his friends and colleagues from Poland, Andrzej Mostowski in particular." From the 1940's onward Berkeley was visited by Wanda Szmielew, Andrzej Mostowski, Andrzej Ehrenfeucht, Jerzy Łoś, Jerzy Mycielski, Lesław Szczerba, Leszek Pacholski, and others, who took part in various research programs initiated by Tarski[11].

Tarski was undoubtedly one of the most outstanding logicians of the 20th century. He also influenced immensely the development of research in semiotics and philosophy. Philosophical aspects of his theory of truth (Tarski was not only fully aware of them, but also put them forward in his papers published in philosophical periodicals) are even at present the subject of vivid controversies and analyses[12]. Intuitions associated with the notion of truth formed a background for his concept of the consequence operation, as well as (cf. section 15) the model theory.

The research center on foundations of mathematics created by Tarski in Berkeley was one of the most influential in the world. Despite many difficul-

---

inets, S. Krajewski remains one of the most prominent members of the Jewish community in Poland, and K. Bieliński was one of the leaders in the underground Solidarity movement). Of course there are other logicians (e.g. Jan Waszkiewicz) and scientists close to logic (e.g. Klemens Szaniawski, Rector Electus of Warsaw University) who were deeply engaged in the political movement that sought to promote democratic changes.

[11]In a letter I received as part of correspondence concerning this paper, J. Mycielski wrote: "Since there was a close cooperation (exchange of papers and ideas) between Mostowski's and Tarski's schools, one can say that there was just one Berkeley-Warsaw school and it is impossible to discuss one without discussing the other."

[12]Not only philosophical ones. Tarski's concept was the subject of numerous formal analyses and generalizations. An interesting survey of this topic can be found in S. Krajewski's essay "Prawda" in: *Logika formalna. Zarys encyklopedyczny z zastosowaniami do informatyki i lingwistyki*, edited by W. Marciszewski, PWN 1987, 144-156.

ties (among them getting permission to travel abroad), Polish logicians were able to keep in touch with him and his students (who are among the most outstanding American logicianstoday), the importance of which for postwar Polish logic is hard to overestimate[13].

10. The First Congress of Polish Science

In four previous sections I adduced some evidence that Polish logic continued to develop despite severe war losses. But even though this growth was achieved, growing ideological pressures and various decisions of a political or administrative nature posed very real threats to its continued development. Necessarily, the climate created by these factors had direct consequences for the continued development of logic.

Today hardly anybody is aware that the discussive logic put forward by Jaśkowski in 1948 was an attempt to form a logical system that admits controversies and contradictions in discussions. Jaśkowski did not care to call those controversies "dialectical" or to use the word "dialectical" on any occasion in his papers. Notably, he also never cared to quote any of the so-called "classics" of Marxism (which was still possible in Poland, though in Russia that would have been an act of desperate courage).

Jaśkowski's papers (as well as a few more published by Maria Kokoszyńska-Lutman, Ludwik. S. Rogowski, and Tadeusz Kubiński) were essentially attempts to argue, albeit implicitly, that there is a chance for cooperation between formal logic and dialectical logic; the former might serve as a useful tool for clarifying and analyzing the rather enigmatic ideas of the latter. This attempt was bound to fail, since dialectical logic was supposed to be a part of so-called "Marxist dialectic" and not reducible to any system of logic.

The postulate of submitting science to ideology, along with some institutional changes, was introduced at the First Congress of Polish Science in 1953. Logic was not the main aim of the "ideological offensive" there; rather, the Lvov-Warsaw School was. Party ideologists were committed to begin the process of full eradication of any politically incorrect philosophy (i.e., those different from "diamat" – dialectical materialism, and "hismat" – historical materialism) from academic life. Poland could not be allowed to remain an oasis, free from the rules being obeyed in other countries of the Soviet block,

---

[13] Tarski's contribution to the development of logic is discussed in J. Zygmunt's "Alfred Tarski" in: *Polska filozofia powojenna*, vol. II, edited by W. Mackiewicz, Agencja Wydawnicza Witmark, Warszawa, 2001, 342-375.

most notably the Soviet Union.

The public criticism of the Lvov-Warsaw School was directed against its most distinguished representatives: Kazimierz Ajdukiewicz, Tadeusz Kotarbiński, Maria Ossowska, and Stanislaw Ossowski. It is noteworthy , however, that this campaign did not result in the eventual expulsion of the "bourgeois" philosophers from academic life. Instead, measures were instituted to prevent them from teaching philosophy, humanities or social sciences. Some of the philosophy professors were employed as the chairs of logic. Others received jobs in the Polish Academy of Sciences.

11. Ajdukiewicz's Initiative for Expanding Logical Investigations

It is interesting to note that although the future of logic was becoming more and morein jeopardy , one of the Party's chief ideologists – Adam Schaff, whose aim as chief critic of Ajdukiewicz in the 1953 ideological debate was to eradicate bourgeois philosophy from universities – actually backed Ajdukiewicz's effort to organize the Section of Logic of the Polish Academy of Sciences. That section eventually became a part of the Institute of Philosophy and Sociology of the Academy.

Along with forming the section of Logic, Ajdukiewicz founded *Studia Logica* (SL), an academic journal intended to provide a forum for both mathematical and philosophical logicians[14]. The original editorial board consisted of the following scholars: Kazimierz Ajdukiewicz (editor in chief), Leszek Kołakowski, Tadeusz Kotarbiński, Andrzej Mostowski, and Roman Suszko (secretary). Unfortunately the presence of L. Kołakowski in this assemblage does not mean that this eminent philosopher of worldwide contemporary recognition began his career as a logician. The reason for his presence at the editorial board was rather peculiar, though certainly characteristic of that time. At least one person on the Editorial Board had to be a Party member, and no other person invited to join the Editorial Board satisfied this condition.

[14]The note "From the Editor" found in the first issue of *Studia Logica* reads that the journal will publish papers devoted to all areas of logic, including formal logic, mathematical logic, inductive logic, theory of definition and of classification, etc., and that SL invites especially works on the history of Polish logic.

12. Fifty Years of *Studia Logica*

As was intended by Ajdukiewicz, *Studia Logica* played from its very beginning the role of a bridge linking mathematical and philosophical logic. Shortly after its inception the periodical started to appear regularly, with an extended editorial board, and its papers came to be published more and more often in foreign languages, which in turn attracted authors from abroad. Just as Ajdukiewicz had intended, logicians from mathematical departments came fairly often to choose *Studia Logica* as a forum to present the results of their investigations alongside of their publications in mathematical journals. In this respect, Andrzej Mostowski had a significant effect in his work on the editorial board. After K. Ajdukiewicz's death in 1963, Jerzy Słupecki (the Department of Mathematics of Wrocław Univeristy) became the editor in chief. In 1971 Zdzisław Pawlak and then in 1978 Helena Rasiowa (both true luminaries in the field of mathematical logic) joined the editorial board. In 1987 Rasiowa went on to become one of the managing editors of the journal.

In 1976 *Studia Logica* underwent an important transformation. Thanks to the efforts of Ryszard Wójcicki (at that time the Editor-in-Chief) and other members of the editorial board (notably Stanisław Surma, Klemens Szaniawski and Jan Zygmunt) the international editorial board was established[15]. Thus began the process of transforming this periodical into a truly international publication, which was not easy, especially due to the fact that any and all international contacts were supposed to be cleared by the State Security Police[16].

Since 1953 over 150 issues of SL have been published. From the time it became an international periodical in 1976, English has been its sole language of publication, and its editorial board is of broad international constituency. Moreover, the Ossolineum Publishing Company, the original publisher of the journal, succeeded in getting North-Holland to become a co-publisher. A few years later North-Holland was replaced by Kluwer Aca-

[15]Its membership included: N. D. Belnap, Jr and J.M. Dunn (USA), B. I. Dahn (DDR), L. Esakia (USSR, Georgia), D. Follesdal (Norway), R. Gilles (Canada), J. Hintikka (Finland), L. Maksimowa and V. A. Smirnow (USSR, Russia), R. Routley (Australia), I. Ruzsa (Hungary), P.Weingartner (Austria), and P. M. Williams (UK).

[16]This "party vigilance" was not unjustified. Since the scientific periodicals were not censored, there was a danger that some of the papers might have been written by "enemies of socialism". *Studia Logica* committed this kind of crime by publishing a review of A. A. Zinoviev's (one of the leading Soviet dissidents) book *Logical Physics* (SL 35, 1976). It was a book free from ideological issues, but still the members of the Soviet Academy of Sciences protested.

demic Publishers, and in 1991 Kluwer became the sole publisher of *Studia Logica*[17]. From 1991, Ryszard Wójcicki, who at the end of the 1980's had stepped down as Editor-in-Chief to be replaced by Jan Zygmunt, resumed the responsibility of heading the journal (assisted by Jacek Malinowski). Since 2000, *Studia Logica* has been published in three volumes a year. In 1995, with the publication of SL's companion series, *Trends in Logic, the Studia Logica Library* was started.

13. Publications on Logic

Studia Logica has not been the only Polish periodical open for publishing papers on logic. Various logical papers, notably some directly related to foundations of mathematics, have been published in *Fundamenta Mathematicae* (FM), an international journal established in 1920. FM has been a vehicle for a large number of papers written by logicians from all over the world. The scope of this journal has been rather large from its very beginning, and it has covered foundational studies into mathematics of all kinds.

In 1973 *Reports on Mathematical Logic* was founded by Stanisław Surma. The profile of Reports has been much the same as *Studia Logica.* A newsletter called *Bulletin of the Section of Logic* of the Institute of Philosophy and Sociology of the Polish Academy of Sciences[18] has been active in promoting both research and international cooperation in logic. In 1993 Jerzy Perzanowski started *Logic and Logical Philosophy,* a periodical fully devoted to philosophical logic.

Since 1974, *Fundamenta Informaticae* (a journal initiated by Zdzislaw Pawlak) has been published under the editorship of Helena Rasiowa and then Andrzej Skowron, who assumed the editorship after Rasiowa's death. This journal covers applications of logic to computer science and is published by IOS Press.

This list would not be complete without mentioning the series *Biblioteka Myśli Semiotycznej* (*The Library of Semiotics Ideas*) initiated by Jerzy Pelc. It is hard to overestimate its importance, not only for semiotics but also for

[17]According to the contract, the periodical remains one of the publications of the Institute of Philosophy and Sociology of the Polish Academy of Sciences, and the main Polish libraries receive it at reduced prices

[18]The aim of this newsletter (founded by the section of logic of Inst. Phil. Soc. Pol. Ac. Sc. in 1973) was to extend international cooperation and indirectly promoting *Studia Logica.* Since 1991 it has been published by Łódź University and edited by Grzegorz Malinowski.

logic. The 46 volumes of this series give the reader some understanding of the richness and variety of the research that has occurred in those areas where logic meets linguistics[19].

14. The Foundations of Mathematics

Even the barest attempt to present only the most important achievements of the research into the foundations of mathematics would require a separate paper. Hence, such a presentation certainly cannot be provided within this section[20]. A highly competent and informative survey of this area was offered, however, by A. Mostowski's in his study, "Thirty Years of Foundational Studies" (cf . footnote 7). This work covers the developments of 1930-1964, and its list of references consists of 244 items. This reference table is truly outstanding, and some of its most remarkable citations are as follows:

A. Ehrenfeucht is mentioned both as the author of a paper on the methods of game theory applied to the problem of decidability of the first order theories and as coauthor (together with A. Mostowski) of a paper on model theory. This reference mentions six papers by A. Grzegorczyk, one of them written jointly with A. Mostowski and C. Ryll-Nardzewski. Grzegorczyk's papers, with one exception, concern the problems of decidability and computability, while the joint paper is related to some foundational problems of arithmetic. The decidability theory is also the subject of the paper by A. Janiczak. J. Łoś is credited with a list of six papers, mainly on model theory. One of them, written with C. Ryll-Nardzewski, explores the theory of representation of Boolean algebras and the problem of Stone theorem's equivalents. The paper by Łoś and Suszko is concerned with the operation of summing models. E. Marczewski's paper on abstract algebra is mentioned along with a large work by S. Mazur on computational analysis. Mostowski mentions four of his own works, and lists H. Rasiowa as the author of several papers on algebraic representation of various non-classical logics. Algebraic methods are the subject of a cited work by H. Rasiowa and R. Sikorski, who also authored the monograph *The Mathematics of Metamathematics* (PWN, 1963, position [173] on the list). Mostowski cites two papers by C.

[19] *Studia Semiotyczne* (founded by J. Pelc and published by Polskie Towarzystwo Semiotyczne) has also played an important role in this area.

[20] It requires a competence to which the author of this survey cannot aspire, nor probably can the workers in this discipline, because of its size and scope. On the other hand it would be odd not to mention the foundations of mathematics. The results of this discipline are strongly related to those of logic.

Ryll-Nardzewski;one of them presents a characterization of categorical theories, and in the other the axiom of induction is examined and it is proved that Peano arithmetic is not finitely axiomatizable. In the paper by R. Sikorski the notion of a metric space is applied to the analysis of intuitionistic logic. W. Szmielew proves in her paper the famous theorem stating the decidability of the elementary theory of abelian groups. Tarski's name appears nine times as an author and several times as a co-author, while the monograph *Undecidable Theories* (North-Holland) is jointly attributed to A. Tarski, A. Mostowski, and R. M. Robinson. This book set an important direction in the search for general methods of proving the undecidability of formalized theories in general and various mathematical theories in particular.

If one were to create an even more extensive list of the names, discoveries, and achievements that have been important since 1964, one would certainly need to mention the following. Z. Adamowicz (whose findings concerned both so-called "weak arithmetic" and arithmetic with open induction), K.R. Apt (second order arithmetic and foundations of computer science), A. Ehrenfeucht, A. Grzegorczyk and W. Guzicki (several works concerning the notion of forcing), M. Krynicki (writings related to the notion of generalized quantifiers), A. Mostowski (reflexivity of Peano arithmetic), R. Murawski (expandability of models of Peano arithmetic to models of second order arithmetic), S. Krajewski (nonstandard satisfaction classes), W. Marek, and M. Srebrny (their investigations were concerned with higher-order arithmetic and set theory, in particular with relations between Zermelo-Frankel theory and Kelley-Morse class theory), H. Kotlarski (automorphisms of nonstandard models of Peano arithmetic), J. Mycielski (infinite combinatorics, universal algebra, and the analysis of Hugo Steinhaus's axiom of determinacy), Z. Ratajczyk, C. Ryll-Nardzewski, L. Szczerba (foundations of geometry, the notion of interpretability of theories formulated in various languages), T. Traczyk (Hilbert spaces, quantuum logic), P. Zbierski (descriptive set theory), and Z. Vetulani (foundations of second and higher-order arithmetic and artificial intelligence).

The following works also should be mentioned. In the 1970's, A. Grzegorczyk published the textbook *Zarys logiki matematycznej*, whose third edition (Polish Scientific Publishers, 1973) was translated into English as *An Outline of Mathematical Logic* (Kluwer 1974). Z. Adamowicz and P. Zbierski wrote *Logika matematyczna* (PWN, 1991), translated as *Logic of Mathematics: A Modern Course of Classical Logic* (John Willey & Sons, 1997).

In retrospect, it is now obvious that even after the period covered in his landmark tome (i.e., since 1964), A. Mostowski[21] played the leading role in the development of the foundations of mathematics and its important branch – model theory (cf. section 15) – while A. Tarski played a somewhat less direct though still prominent role in this regard.

15. Model Theory

This theory has a distinguished position within the foundations of mathematics. Initiated in the 1940's - 50's by the work of L. Henkin, A. Robinson, and A. Tarski, it forms an important part of mathematical logic. It deals with the relations between language in its formalized version and its "models", i.e. the structures that may become the objects to which the language expressions refer (if they are properly interpreted).

As was pointed out in the previous section, many important results in model theory were obtained by the Poles. In the 1960's and 1970's model theory was developed by L. Pacholski and B. Węglorz, later by H. Kotlarski, and quite recently important and widely recognized results were obtained by Ludomir Newelski. An especially significant advance in this field (because of its numerous applications) was the development of Łoś's ultraproduct theorem, which was based on some of Skolem's ideas of the 1930's. The Boolean models method of H. Rasiowa and R. Sikorski[22] was another important milestone, since it enabled logicians to extend the field of model theory to non-classical logics.

16. Algebraic Logics

The history of this branch of logic (overlapping universal algebra to some extent) goes back to Lindenbaum-Tarski's idea of the algebra of language. For the sake of brevity, suffice it to state that algebraic logics deal with the relations between the laws of logic and theorems that characterize algebraic operations[23].

---

[21]The work of Andrzej Mostowski is discussed in five papers written by: A. Grzegorczyk, W. Guzicki, W. Marek, L. Pacholski, C. Rauszer, and P. Zbierski in: A. Mostowski, *Foundational Studies: Selected Works*, vol I, PWN, Warszawa, North-Holland, Amsterdam 1979. The monograph which summarizes his metamathematical research in set theory is: A. Mostowski, *Constructible Sets with Applications*, PWN, Warszawa, North-Holland, Amsterdam, 1969.

[22]H. Rasiowa and R. Sikorski, *The Mathematics of Metamathematics*, PWN, Warszawa, 1963 (3rd edition, 1970).

[23]If by laws of logic one means the laws of classical logic, then they can be characterized by the laws of Boolean algebra. Non-classical logics can be represented by "non-Boolean"

The basic tools for examining the relations between algebra and logic (or more precisely, between algebraic structures and particular systems of logic) were established by Alfred Tarski along with his collaborators and followers. The monograph by H. Rasiowa and R. Sikorski, *The Mathematics of Metamathematics*, PWN, 1963 (cf. 20), an outstanding source of information in this area, was a source of inspiration for authors all over the world to carry out research in algebraic logic. In 1974 H. Rasiowa published her monograph, *An Algebraic Approach to Non-Classical Logic* in which she extended her earlier results to the large class of so-called non-implicative logics. This class includes modal logics among others.

Besides Rasiowa and Sikorski, algebraic logic was also the research subject of Cecylia Rauszer, who was interested in logics with "constructive falsehood". Many important and difficult results were obtained by A. Wroński and his group of logicians (Section of Logic, Philosophy Department of the Jagiellonian University in Cracow). Their research, carried out partly in cooperation with Japanese logicians, made important contributions to the theory of pseudo-Boolean algebras, BCK algebras, equivalential algebras and logical systems related to these classes of algebras. In Toruń, algebraic methods were developed by J. Kotas and his group.

The issues underlying algebraic logics are closely related to those of logical matrices (cf. section 19).

17. Foundations of Computer Science

The research in this branch of logic, the importance of which is still growing, was initiated by H. Rasiowa and her group (Grażyna Mirkowska, Ewa Orłowska, Cecylia Rauszer, Andrzej Salwicki, Andrzej Skowron, Jerzy Tiuryn, Anita Wasilewska, to name but a few). The activity of these people was oriented toward developing applied logic such as algorithmic logic, logic of programming (correctness of programs), logic of information (data bases), and logic of knowledge (expert systems). An important role in the development of the foundations of computer science was played by Zdzisław Pawlak ("rough sets" theory, Pawlak machines), and also by Andrzej Blikle, Beata Konikowska, Józef Winkowski.

Research into the mathematical foundations of computer science was initiated in Poland by H. Rasiowa and Zdzisław Pawlak. Since the 1990's

---

algebras(e.g. intuitionistic logic is characterized by pseudo-Boolean algebras). H. Rasiowa, *An Algebraic Approach to Non-Classical Logic*, PWN, Warszawa, North-Holland, Amsterdam, 1974.

they have been undertaken in the Department of Mathematics, Mechanics, and Computer Science at Warsaw University by the group headed by Jerzy Tiuryn. This group consists of Damian Niwiński, Jerzy Tyszkiewicz, PawełUrzyczyn, and Igor Walukiewicz, and deals with numerous aspects of modern computer science: lambda calculus, functional programming, type theory, modal logics, and theory of finite models.

18. Theory of Consequence

The first postwar papers with reference to Tarski's consequence theory were published by J. Łoś, J. Słupecki and W. A. Pogorzelski. One of the key papers on this subject was "The Algebraic Treatment of the Methodology of Elementary Deductive Systems"[24] by J. Łoś. The research in this area was carried out also by A. Grzegorczyk and R. Suszko among others. It was Suszko who first developed the idea of "abstract logic"[25].

There are two important aspects of the theory of consequence. First is the methodological aspect. Tarski developed the consequence theory as a theory of "deductive systems". This notion is a formal equivalent to that of "theory" or even "a system of beliefs". Logical analysis of a deductive system has numerous and significant methodological implications. Second, the notions of consequence let us see the limitations of the propositional conception of the system of logic and bring into prominence the inferential concept of those systems[26].

Many profound realizations concerning the so-called structural completeness of logical calculi (which was also examined abroad) were made by W.A. Pogorzelski (who introduced this idea) and his students, including T. Prucnal and P. Wojtylak among others.

The area of consequence theory also includes the research concerned with so-called non-monotonic reasoning. This kind of research was carried out by

---

[24] *Studia Logica* 2, 1955, 151-212.

[25] This notion was also explored by Spanish logicians.

[26] These two conceptions were discussed in section 2. The extent to which they differ from each other can be illustrated by the following. The system of 3-valued Łukasiewicz logic in its inferential meaning has two "non-trivial" extensions, while the same system in its sentential meaning has only one extension. This result (obtained by R. Wójcicki) was later generalized (G. Malinowski, W. Dziobiak). It makes clearer some of the "paradoxical" results obtained earlier by H. Hiż and Rasiowa, which were concerned with the sentential meaning of the logical system. The distinction between logic understood as a set of tautologies and a logical consequence enables us to show (R. Wójcicki) that we cannot define classical logic by the use of the constants of intuitionsitic logic in the inferential case (which is possible in the sentential case).

Witold Łukasiewicz, Wiktor Marek and Mirosław Truszczyński.

19. Logical Matrices

The concept of a logical matrix, understood as a set of "logical values" (e.g. truth and falsehood in the simplest case) that can be taken by propositions, appeared in the beginning of the 20th century. J. Łukasiewicz introduced many-valued matrices while defining many-valued logics. A. Lindenbaum proved that each propositional logic (cf. section 2), i.e. set of tautologies, has an adequate logical matrix that characterizes it in a unique way. This result, along with one of the papers by A. Tarski and J. Łukasiewicz, showed that logical matrices can be used as certain generalizations of algebraic methods (the latter were discussed in section 16).

J. Łoś and R. Suszko tried to find a matrix representation of consequence operation. Their result, however, required some corrections (given by R. Wójcicki)[27]. These efforts (the former is of vital importance) gave a new direction to a very general, technically difficult, and philosophically interesting field of research.

This problem was my special concern in 1970 when I initiated a separate research project. I started to examine the representations of consequence operations together with J. Czelakowski, W. Dziobiak, J. Hawranek, M. Tokarz, T. Prucnal, R. Suszko, J. Zygmunt, A. Wroński, and P. Wojtylak. Some contributions to this project were made by logicians from abroad (e.g. W. Rautenberg), although this subject is regarded as typically "Polish"[28]. This project was continued and developed later on by J. Czelakowski and W. Dziobiak as well as by Spanish and American logicians, especially J.M. Font, W.J. Block, and D. Pigozzi[29].

---

[27]The papers addressed here are: J. Łoś, R. Suszko, "Remarks on Sentential Logic", *Indagationes Mathematicae* 20, 1958, 177-183, and R. Wójcicki "Some Remarks on the Consequence Operation in Sentential Logics", *Fundamenta Mathematicae* 68, 1970, 269-279.

[28]The results mentioned here as well as those of other logicians (W. Blok, H. Hiź, J. Łoś, D. Pigozzi, H. Rasiowa, W.A. Pogorzelski, and others) are discussed in: R. Wójcicki, *Theory of Logical Calculi; Basic Theory of Consequence Operations*, Kluwer, 1988. Some of these results were applied to the theory of automatic theorem proving (cf. Z. Stachniak, *Resolution Proof System, An Algebraic Theory*, Kluwer, 1996).

[29]This research is discussed in the monograph: J. Czelakowski, *Protoalgebraic Logics, Trends in Logic, Studia Logica Library*, Kluwer, 2001.

20. Type Logics and Categorial Grammars

These areas belong to the foundations of mathematical linguistics. They were explored in Poznań by the logicians grouped around Wojciech Buszkowski. The term "type logics" refers to the set of operations that are used while creating complex language expressions out of simple ones. "Type logics" belong to the family of substructural logics.

From the historical point of view this kind of research was initiated by Ajdukiewicz and Bar-Hillel, and refers to the calculus of Lambek (1958). Buszkowski's research (as well as that of M. Kandulski and W. Zielonka) produced a thorough understanding of some of the fundamental problems of mathematical linguistics and its applications. These researchers explored the algebraic and computational properties of various kinds of languages, especially the so-called "tree-like languages" generated by categorial grammars, the algorithms for finding minimal categorial grammars, and the relations between categorial grammars and Chomsky's generative grammars[30]. Contemporarily, research in logic and linguistics is also carried out in the Poznań center by T. Batóg and J. Pogonowski.

21. Suppositional Logic

In 1934 S. Jaśkowski published an article, "On the Rules of Suppositions in Formal Logic" (published as the first issue of the *Studia Logica* series)[31]. Ironically, Jaśkowski's and Gentzen's conceptions of natural deduction were discovered independently of each other and yet were published almost simultaneously[32]. A suppositional system of logic was the basis for the computer program "Mizar" designed by Andrzej Trybulec (Warsaw Univerity's

[30] A thorough discussion of their proceeds with a comparison to the proceeds of other centers results can be found in: W. Buszkowski "Mathematical Linguistics and Proof Theory", *Handbook of Logic and Language* (collective work edited by J. van Benthem and A. ter Meulen, Elsevier and MIT Press).

[31] S. Jaśkowski "On the Rules of Suppositions in Formal Logic", *Studia Logica. Wydawnictwo poświęcone logice i jej historji*, nr 1, Seminarium Filozoficzne Wydz. Matematyczno-Przyrodniczego Uniwersytetu Warszawskiego, Warszawa, 1934. This series was to be published under the editorship of J. Łukasiewicz. Unfortunately only one volume was published. Postwar *Studia Logica* (cf. section 12) referred to it but did not comprise its continuation.

[32] In order to make S. Jaśkowski's system better known J. Słupecki and L. Borkowski elaborated a version of it (and published it in the book *Elementy logiki i teorii mnogości*, PWN, 1963; translated as *Elements of Mathematical Logic and Set Theory*, Pergamon Press 1967) and presented the possibility of applying it to mathematical reasoning.

branch in Białystok, currently an independent university). Its function is well expressed by the title of one of the research projects coordinated by W. Marciszewski, "Systems of Logic and Algorithms for Computer Testing of the Correctness of Proofs"[33].

22. Others

A very important institution, organized under the auspices of the Polish Academy of Sciences, is the Conference on the History of Logic. It was initially chaired by T. Czeżowski, J. Słupecki, and S. Surma, and since the 1980's it has been organized exclusively by A. Wroński, Section of Logic of the Jagiellonian University in Cracow. Substantial contributions to the subject of the history of logic were made by W. Marciszewski, R. Murawski, S. Surma, J. Woleński, and J. Zygmunt among others.

The creation of logical systems was not a Polish speciality. Besides the discussive logic of Jaśkowski (which was mentioned in section 10) there are two exceptions to this, however: The "non-Fregean logic" of R. Suszko – a system of logic inspired by some intuitions taken from *Tractatus Logico-Philosophicus* of L. Wittgenstein, and the system Grz (for Grzegorczyk, its author) – one of the most widely studied systems of modal logic.

Polish logicians (J. Słupecki, H. Rasiowa, T. Traczyk, G. Malinowski, W.A. Pogorzelski, T. Prucnal, M. Tokarz, K. Trzęsicki, and R. Wójcicki, not to mention the great precursors of the subject – J. Łukasiewicz, A. Tarski, and M. Wajsberg) contributed tremendously to the research in many-valued logics[34]. Polish logicians also made important strides in the fields of intuitionistic logic (H. Rasiowa, R. Sikorski, P. Wojtylak, A. Wroński), modal logic (A. Grzegorczyk, E. Orłowska, J. Perzanowski, H. Rasiowa, C. Rauszer, J. Hawranek, and T. Skura), and relevance logic (W. Dziobiak, M. Tokarz, K. Świrydowicz).

Quite a large number of papers were concerned with deontic logic (J. Gregorowicz, L. Gumański, J. Kalinowski, T. Kubiński, W. Suchoń, K. Świrydowicz, J. Woleński, Z. Ziemba, Z. Ziembiński) and casual logic (H.

---

[33]This project is supported by international grants (from EU and NATO, e.g. Ph. D. scholarships in Germany and Japan) and its main aim is to complete an online encyclopedia of mathematics (comprising a collection of computer verified proofs of theorems). There is also a quarterly: *Formalized Mathematics - A Computer Assisted Approach* published there, edited by R. Matuszewski.

[34]A good introduction to this subject is a book by G. Malinowski, *Many-Valued Logics*, Clarendon Press, Oxford 1993.

Greniewski, A. Pieczkowski, K. Trzęsicki, and M. Urchs[35]). The latter were mostly based on some ideas initiated by S. Jaśkowski.

Polish authors have also dealt with the logic of questions (erotetic logic). This body of work was initiated by Ajdukiewicz and substantially developed by T. Kubiński, L. Koj, and A. Wiśniewski.

Scholarship conjointly embracing logic and theory of communication is being performed by M. Tokarz.

Ryszard Wójcicki
Institute of Philosophy and Sociology,
Polish Academy of Sciences

Jan Zygmunt
Wrocław University

---

[35]M. Urchs is mentioned here because of his strong connections with Polish logic. He came to Poland as a young man from the DDR and began to study in Toruń.

JOHAN VAN BENTHEM

Fifty Years: Changes and Constants in Logic

In this year of celebrations of 'First-Order Logic 75' and 'Trends in Logic – 50 Years of Studia Logica', two phenomena stand out. One is the rapid expansion of themes in modern logic, the other the remarkable constancy of the basic methods encapsulated in first-order logic. The resulting light piece may be read as a companion to Ryszard Wojcicki's survey of logic in Poland. Our main purpose is to show some patterns by means of aerial photography.

1. Agendas and Methods

Historically, the agenda of logic as the study of reasoning has changed over time. Topics like patterns of scientific explanation are now considered part of the philosophy of science, but the border-line is vague. Likewise, to take a modern example of such blurring, AI research into default rules evidently addresses a major aspect of reasoning as well. Here is one way of thinking about the subject matter. Reasoning is about handling propositions, and these lie mid-way in a hierarchy of data structures, which runs from words and phrases to complete sentences, and then upward to texts, and whole theories. Each of these levels is associated with activities that show logical structures, such as – from low to high – determining meanings of expressions, characterizing speech acts, understanding discourse and proof, or analyzing theory structure. All these levels have occurred to some extent in mathematical and philosophical logic, though the emphasis varies at times. Of course, there is a proliferation of logical systems in modern research on these topics, and logical pluralism is an undeniable fact of life. But it is also true to say that one system has shown a remarkable constancy throughout the period, viz. first-order predicate logic – so much so, that we still teach it as the gateway to the field. One obvious reason for this choice is that predicate logic sets the methodological standards. But more than that, it also serves as a sort of conceptual laboratory for implementing and testing new ideas, throughout the past 50 years. In particular, we will look at how predicate logic has fared when confronted with realities of language and reasoning,

---

V. F. Hendricks and *J. Malinowski* (eds.),
Trends in Logic: 50 Years of Studia Logica

Trends in Logic **21**: 35–56, 2003.

leading up to the second main theme of this paper. The above hierarchy of logical levels is at the same time a hierarchy of activities with their own logical structure. Our discussion will be light, pointing merely at the broad ideas at stake. Technical details are left to the references.

2. Truth and Meaning

The basic format The familiar semantics of first-order logic explains the notion of truth of a formula ϕ in a model M under assignment s. It works via obvious recursive clauses, such as $M,\ s \models \exists x\phi$ iff there exists an object d in M with $M,\ s[x := d] \models \phi$. Simple as it is, this truth definition embodies two major ideas. The first is *compositional interpretation* of formulas following their stepwise construction in an appropriate perspicuous logical syntax. The second is the selection of an appropriate *context of evaluation* making the compositional recursion possible. In first-order logic, the assignment functions do the job.

Natural language semantics Both ideas have guided all further work in the formal semantics of natural language, where they have been extended to deal with a wide spectrum of further expressions (van Benthem and ter Meulen, eds., 1997). A typical example is the semantics of generalized quantifiers such as "all", "most", "few", where compositional interpretation of their common format $Qx \bullet \phi(x)$ becomes possible as soon as one determines the appropriate type of denotation, taking a quantifier Q to denote a property of sets. At the same time, the quantifier repertoire of natural languages is much richer than that of first-order logic, and more complex 'branching' constructions such as "Ten firms owned a thousand jets (together)" raise doubts about compositionality in its straightforward form (Keenan and Westerståhl 1997). But this deviation and the ensuing discussion just show how the logical theory serves as a search-light in Popper's sense for finding interesting phenomena that high-light peculiarities of language.

Philosophical modalities The same style of semantic analysis is found in the study of various linguistic concepts for philosophical purposes. For instance, consider modalities, such as the epistemic operator 'agent j knows that ϕ': $K_j\phi$. Here the relevant contexts are possible worlds structured by an accessibility relation of epistemic indistinguishability – and the compositional semantic clause says that $K_j\phi$ is true in world s iff ϕ is true in all worlds accessible for j from s. Likewise, a standard account of conditionals $\phi \Rightarrow \psi$ says that ψ is true in all those ϕ-worlds which are closest to the current world s. In this way, the compositional methodology has led to the discovery of important general parameters in organizing logical contexts, viz. *accessibility* and *minimization* along an ordering. Incidentally, first-order logic does play a role even here, though notions of modality are not part of its repertoire of Boolean operations and standard quantifiers. Most well-behaved modal truth definitions can be *translated* into first-order terms, and this allows for a well-known transfer of insights about predicate logic to modal logic in so-called correspondence theory (cf. Blackburn, de Rijke and Venema 2001).

Dynamic predicate logic The above expansion of ideas from first-order semantics to semantics in general is well-known. What may be less known is that predicate logic itself can also serve to investigate *new mechanisms* of interpretation. One modern example was motivated by the anaphoric behaviour of natural language, which differs systematically from binding in predicate logic. E.g., an existential quantifier can bind across sentence boundaries, as in $\exists xAx.Bx$, ("A man came in. He smiled.") and also across operator boundaries as in $\exists xAx \rightarrow Bx$ ("If a man comes in, he smiles"). It turns out that such phenomena can be explained very naturally if we encode more dynamics of the evaluation process into the truth definition (Groenendijk and Stokhof 1991). This leads to a new format

> $M, s_{in}, s_{out} \models \phi$: the assignment transition (s_{in}, s_{out}) records a successful evaluation of ϕ.

In particular, an existential quantifier $\exists x$ is now an action resetting the current assignment s to any $s[x := d]$, while an atomic formula Px is a test whether the value $s(x)$ of the current assignment s has the property P^M. This can be done perfectly compositionally, with e.g., logical conjunction becoming composition of binary transition relations. In a sense, this format merely brings out the process intuitions behind ordinary first-order semantics. Predicate logic is a system for two kinds of semantic action: testing a fact, and changing a state. The latter happens either by 'random assignments' triggered by formulas $\exists x\phi$, or by definite assignments $x := t$ triggered by substitution operators $[t/x]\phi$. On this view, predicate logic and natural language are more like imperative programming languages, whose programs denote actions that change states of a computational device. Thus, *processes* become a major logical topic in their own right. This is our first logical story line which naturally runs into computer science. More are to come.

Evaluation games A more drastic process view of first-order evaluation was proposed in Hintikka 1973. In each model M, statements ϕ define an evaluation game between two players called Verifier (**V**) and Falsifier

(**F**). For atoms ϕ, **V** wins if the atom is true, otherwise **F** wins. The logical operations then become very general game operations. A disjunction amounts to a *choice* for **V** which disjunct to play, a conjunction triggers a choice for **F**. A negation indicates a *role switch* between the two players. Finally, existential quantifiers $\exists x\phi(x)$ let **V** pick an object d in M such that play continues with $\phi(d)$, and universal quantifiers do the same for **F**. The basic fact about these games is that a formula ϕ is true in model M iff Verifier has a *winning strategy.* This simple bridge links logic to game theory. E.g., logical laws now acquire game-theoretic import. E.g., Excluded Middle $\phi \vee \neg\phi$ expresses the so-called determinacy of the above evaluation games: one of the two players must have a winning strategy. This reflects an early result by Zermelo that all finite depth zero-sum games are determined. Game theory is all about *interaction* between different agents, and hence this now also becomes a logical core theme.

Conclusion The semantics of first-order logic has provided a precise modus operandi for semantics generally, highlighting ideas of compositionality, context selection, and context structure of various kinds. But also, it is still a testing ground for new ideas about semantics, leading to new logical themes of process structure and multi-agent interaction.

3. Varieties of Inference

The basic format Valid consequence in predicate logic says that every model for the premises Σ must also be a model for the conclusion ϕ: $MOD(\Sigma) \subseteq MOD(\phi)$. This notion of consequence via plain set inclusion is characterized by the familiar structural rules of Contraction, Monotonicity, Reflexivity, and Cut (van Benthem 1996). But historically, variants have occurred. E.g., Bernard Bolzano's original notion of consequence in his 1837 monograph *Wissenschaftslehre* also demanded that the premises be consistent, i.e., that $MOD(\Sigma)$ be non-empty, thereby changing the structural rules (van Benthem 2003A has the complete set). But Bolzano's work also suggests a more general viewpoint. Logic is about a variety of *styles of reasoning* for different purposes, with different structural standards.

Natural language inference We have a good grip on first-order validity, which has been captured completely by semantic tableaux, and various sorts of proof systems. But taking such ideas of deductive organization to natural language, the picture becomes much more diverse. Natural reasoning seems to fall into modules that cut the cake differently from the received logical rules for quantifiers or Boolean operations. Generalized quantifiers form an inferential family, first-order or not. A typical example is *monotonicity inference*: the ubiquitous transition from a sentence involving some predicate P to one replacing P by some predicate P' with a larger extension. This happens in contexts such as "All men are P", or "Most men are P" and many much more complex quantificational patterns. There is also a version allowing downward replacement, as in "No men are P", "Few men are P", or "If any P can do it, you can". Taken together, upward and downward monotonicity inferences account for much of what was covered in the traditional syllogistic (Sanchez Valencia 1991). But in addition, there are many other natural inferential subsystems in natural language, such as those regulating temporal perspective. The total landscape does not show the divisions familiar from standard logical proof systems. Even so, predicate logic has served as a prism for studying key phenomena. A good example is monotonicity. Predicate logic tells us that the key to this sort of reasoning lies in syntax: 'positive occurrence' of P in some precise sense licenses upward replacement, negative occurrence downward replacement. And Lyndon's Theorem even shows that this criterion is both sufficient and necessary. Thus, predicate logic can be a laboratory for understanding inferential behaviour in essentially different settings.

Philosophy of science The reference to Bolzano also suggests connections with the philosophy of science. Another aspect of reasoning are its many different manifestations: proof, refutation, consistency management, and many more. Some of the most ubiquitous are confirmation or refutation of hypotheses, or the explanation of observed phenomena in terms of relevant facts and hypotheses. This variety was still clear to a logician like Peirce, with his joint account of deduction, induction, and abduction, but it has dropped out of standard logic textbooks. Philosophers of science have a much nicer taxonomy of different manifestations of reasoning. Still, logical techniques have been used to model these, especially by neo-positivist philosophers. E.g., one formal sense of explanation is this. Proposition P explains fact F given theory T, provided that (a) $T\&P$ implies F, while (b) T by itself does not imply F. This may be viewed as a new ternary notion of consequence $E(T,P,F)$, which can be analyzed by standard logical means. In particular, it is not difficult to find a complete set of structural rules for $E(T,P,F)$, including principles like $E(T,P,F) \leftrightarrow E(T,F,F)\&\neg E(T \cap P,F,F)$.

Artificial intelligence The study of varieties of reasoning received new impetus in the 1980s when AI research started looking seriously at so-called common sense reasoning in planning and problem solving. In particular, *default reasoning* seems to deviate from ordinary logical consequence in the following sense. In making plans, we do not look at all models of the premises, but only those that seem most relevant, or plausible. E.g., if I use the default inference that I will be home on time if I catch the 5:51 train, I do not mean that every world in which I take that train has me home by 6:20, but only those without strikes, floods, and reversals of the laws of physics. These and similar observations lead to the idea of *circumscription*, where a conclusion ϕ follows from given premises Σ if ϕ is true in all *minimal* models of the premises, with minimality measured along some relevant comparison relation. For instance, McCarthy 1980 looks at models with smallest possible domains of objects, or smallest predicate extensions verifying the premises. These are both natural generalizations of logical consequence, which have been studied even for standard first-order languages. The general format behind circumscription sounds like the earlier minimization semantics of conditional logic, and indeed it is. The structural properties of default reasoning are close to the known properties of conditional logic (Veltman 1996). A final interesting idea here is that we draw inferences of various kinds. Standard logical consequence is totally safe, but it may not give us enough. Therefore, we jump to further conclusions by default reasoning, capitalizing on our ignorance. But of course, there is a price to be paid, of *non-monotonicity*: new evidence may invalidate earlier conclusions. If there is a flood after all, I will have to withdraw the conclusion that I will be home on time, and revise my further beliefs accordingly. This dynamic idea will return below.

Conclusion Predicate logic has made valid consequence precise, and thereby provided tools for charting varieties of reasoning beyond the standard format. On the other hand, in the 'conceptual lab' mode, new notions of consequence can be studied over a first-order language to highlight their deviant behaviour, for instance, in terms of structural rules.

4. Theory Architecture

The early years of foundational research saw a close co-operation between logicians and philosophers of mathematics and the natural sciences. Significantly, Tarski's early logic textbook was called *Logic and Methodology of the Exact Sciences.* Predicate logic has had much to say about the structure of theories, especially in mathematics. But again, the torch was also carried on by philosophers of science, and later on, computer scientists.

Language design Setting up a theory involves first of all choosing an appropriate language for the relevant concepts. In such a language, there is often a layering into primitive and derived notions. One important result in logic is *Beth's Theorem*, saying that the following two notions are equivalent for first-order theories $T = T(\mathbf{P}, Q)$: (a) predicate Q is implicitly definable in T: i.e., fixing the interpretation of $\mathbf{P}$ in a model for T leaves just one choice for Q, and (b) Q has an explicit first-order definition $\forall \mathbf{x} : Q\mathbf{x} \leftrightarrow D(\mathbf{P}, \mathbf{x})$ which follows from T. Such insights are relevant to issues of definability and reducibility up until today. For instance, in recent decades, the notion of 'supervenience' has been proposed as a middle ground between outright reducibility of mental predicates to physical ones and complete independence: 'when two organisms have the same physical properties, they also have the same mental ones'. But it is easy to show by Beth's Theorem that this form of implicit definition implies explicit definability after all. A looser fit between layers of vocabulary occurs in the famous distinction for empirical theories between *observational* and *theoretical* vocabulary (cf. Przełęcki 1969). Here one just wants to say that the intended empirical models $(D, \mathbf{P})$ to be described allow of at least one expansion with predicates $\mathbf{Q}$ to models $(D, \mathbf{P}, \mathbf{Q})$ for the whole theory T. This so-called Ramsey sentence view has led to further model-theoretic investigations, partly inspired by the much richer 'structuralist' framework for scientific theories (Sneed 1971).

Deductive organization Other aspects of theory structure have to do with deductive organization of principles according to relative importance. This occurs already when we axiomatize the set of theorems, putting some propositions at centre stage. But one can also have more refined architectures, placing some axiomatic principles at the core, and others at the periphery, corresponding to different degrees of willingness to give them up if the theory is to be revised by incompatible evidence. For instance, in accounts of explanation and lawlikeness in the philosophy of science, one often finds a distinction between core laws, auxiliary hypotheses, and plain facts. This theme of 'structured theories' has returned in computer science and AI, as a principle of knowledge and data base representation, and again predicate logic is used as the best platform for developing this (Ryan 1992).

Intertheory relations Finally, theories live in families, with various interesting relations between them. A typical case is (conservative) extension. A more sophisticated example is *relative interpretability*, which says that basic predicates in the vocabulary of one theory T_1 can be translated into possible complex predicates of another theory T_2 in such a way that T_2 proves all translations of the T_1-theorems relativized to some definable subdomain U of objects. This covers the way in which set theory interprets arithmetic. In other cases, we must also allow that the objects to be used are simple constructions out of those for T_2, as happens when we interpret rationals as ordered pairs of integers. There are also operations turning theories into new ones. Calculus of theories was an early theme in Polish logic, which remained alive in the philosophy of science (Kuipers 2001 has a sophisticated view of formal intertheory relations). But perhaps the most innovative technical studies of theory structure come from modern computer science – e.g., in the theory of abstract data types. Modern *module algebra* combines Polish-style calculus of theories with an account of different languages involved, leading to elegant algebraic versions of logical interpolation theorems (Maibaum and Veloso 1995). These themes are now making their way back into logic in recent work on combining logics (Gabbay 1996).

Conclusion Theory structure is not a main concern here, but we mention it because it for its relevance to our general area map. Information organization at higher aggregation levels seems a natural extension of standard logical concerns, especially when one interpolates modern accounts of structured data bases in between texts and full-fledged theories. Moreover, in the end, it seems hard to judge the performance of a logical calculus as a reasoning system without some sort of systematic account of data organization.

5. From Reasoning Phenomena to Logical Calculi

Metatheorems Perhaps the most spectacular feature of predicate logic has been the discovery of its major system properties, such as the Löwenheim-Skolem theorem, and the compactness and completeness theorems. The historical process of discovering such major results continued right into the 1950s, with Beth's definability theorem and Craig's interpolation theorem. After that, little of comparable significance has come to light, and many would view Lindström's Theorem around 1970 as an end-of-history result, saying that predicate logic is the only system satisfying the usual meta-properties as we now know them. But classical metatheory is more open-ended than that in interesting senses. E.g., Mason 1985 shows that the complete first-order theory of a two-sorted language for first-order consequence plus vocabulary sets (the minimum needed to formulate the interpolation theorem) is equivalent to True Arithmetic, and hence deeply incomplete. But whatever insights may still be around the corner, the meta-properties that we know have served as a sort of lodestone by which to measure system design elsewhere. In particular, Craig interpolation and Beth definability properties often serve as constraints on a well-designed logical calculus, and their intuitive counterparts even show up inside empirical natural language semantics (cf. ten Cate and Shan 2002 on the functioning of questions).

System pressures Nevertheless, there is also a curious twist to this very success. Predicate logic has been so successful as a calculus, and its mathematical properties so appealing, that the tool has to some extent supplanted the original object of study. Many people say nowadays that the proper object of study in logic are formal systems, rather than reasoning phenomena. This is reflected in much research practice, which is about proving completeness theorems and other meta-properties of formal systems whose designers have never used that system for any specific purpose at all, and do not even know how it works. The formal system emphasis also constrains research, in that logical key notions may get spread out over hosts of systems, rather than being topics in their full generality. But a more optimistic perspective is available. Modern logic has become the science of both natural and artificial reasoning systems, and much of our exposition so far can be interpreted as showing the fruitfulness of the interaction between the two research modes.

6. Cultural Cross-Currents: Logical Dynamics

Logic started out as a part of philosophy. Its modern phase came about through deep ties with the foundations of mathematics. But the resulting formal systems, with predicate logic as a prime example, immediately started influencing philosophy again, especially the philosophy of science and the philosophy of language. Tracing the resulting historical lines, as we have indicated above, shows a further development. Gradually, many of the above logical themes, such as compositional semantics, modality, conditionals, or language levels, have led naturally to research at interfaces with linguistics, computer science, and AI. Indeed, even a much broader survey than the few examples offered here would show that about every major topic in logic ends up in some surrounding discipline, and most often in computer science. This process of expanding horizons still continues, witness current contacts between logic, economic game theory, and even sociology. It would be highly misleading, however, to think of this as application of ready-made logical ideas in some engineering sense. In many cases, the influence is mutual, as the relevant concerns are shared with other fields. The paradigmatic example of this for the rest of our paper comes from computer science.

One of the guiding insights in studying computation and processes generally has been that representation of information and computational actions should be studied in tandem. You cannot run a process without good data structures, but you cannot design good data structures without understanding the processes they are supposed to serve. Thus, there is a tandem between 'statics' and 'dynamics'. Taking the same idea to logic results in what may called a *Dynamic Turn* (van Benthem 1996). Traditionally, logic may be about static notions of meaning and inference, but in the above, we have already seen various dynamic actions behind the scenes: evaluation, inference, or belief revision. E.g., inference is in the first place an information-generating *process*, just like asking questions, giving answers, and communication in general. These actions use or transform static propositions, but they have a logical structure of their own. In fact, even traditional terminology has this double aspect. The word 'statement' denotes both an activity and its product, and so does 'proof'. In particular, in addition to the earlier hierarchy of static logical levels: words, sentences, texts, theories, there is also a corresponding hierarchy of logical activities. In a table without any great pretensions, we might depict the playground for logicians as follows:

Expressions	Parsing, interpreting
Sentences	Speech acts: state, ask, attack, defend
Texts	Dialogue games and strategic behaviour
Theories	Long-term/large-scale activities

The table may still be of too low dimensionality, because if we take the multi-agent aspect seriously, we might also have to include different levels of social organization.

Let us now turn this general Dynamic Turn philosophy into concrete topics for logical investigation. The following three sections provide examples of some recent notions and results.

7. Communication

Simple update As we saw, existing logical practice has many dynamic features. Consider propositional logic. A person's information may be modelled as the set of all possibilities that she entertains. Now consider the following typical valid propositional inference:

from two premises $A \vee B$ and A, draw the conclusion B

Without prior knowledge, two propositions A, B span 4 candidates for the real world. The premises then trigger updates restricting this set. In the limit, one option remains, and we know the actual situation. Here is an update video of the stepwise information flow in this case. The two premises lead to an information state (in this special case, a complete one), where an update with the conclusion does not change anything:

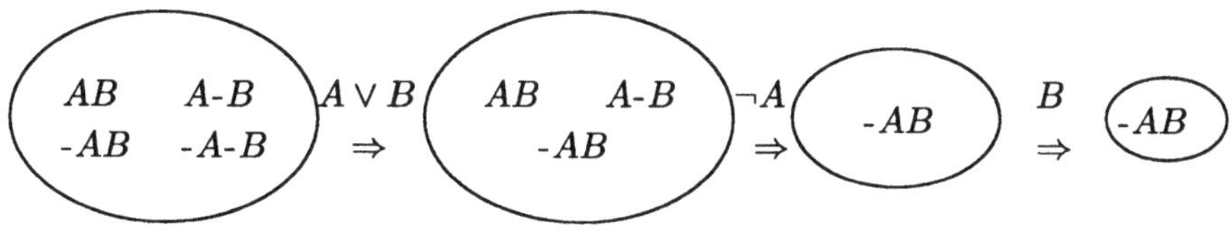

This is the hallmark of a new notion of *valid consequence* in a dynamic perspective: conclusions follow if they do not change the final information state produced by the successive premises.

Questions and answers But even the simplest episode of communication involves more than one person. Suppose I approach you with a question, and the following occurs

Q *Is it midnight?*
A *Yes.*

What information has passed? I learn that it is midnight. But there is much more logic to the situation. If my question was genuine, merely by asking it, I convey to you that I do not know if it is midnight, but I think it possible that you do. And your answer does not just convey the mere fact. It also makes sure you know that I know, I know that you know that I know, and so on to any depth of iteration: we achieve *common knowledge* of it being midnight. The latter is often a prerequisite to joint action in philosophy, linguistics or game theory. In general, our actions may even be conditioned by finer epistemic differences.

Epistemic states and actions To describe a question-answer episode precisely, we need two things in tandem, an account of the information states involved, and one of the actions transforming them. Models for the former purpose exist already in standard epistemic logic, with possible worlds standing for all relevant total states of reality, related by uncertainty relations for all agents involved. E.g., a simple epistemic model for our 2-agent group $\{\mathbf{Q}, \mathbf{A}\}$ has two states 'P', '$not - P$'. The horizontal labelled line in the following picture indicates that **Q** cannot distinguish between the two:

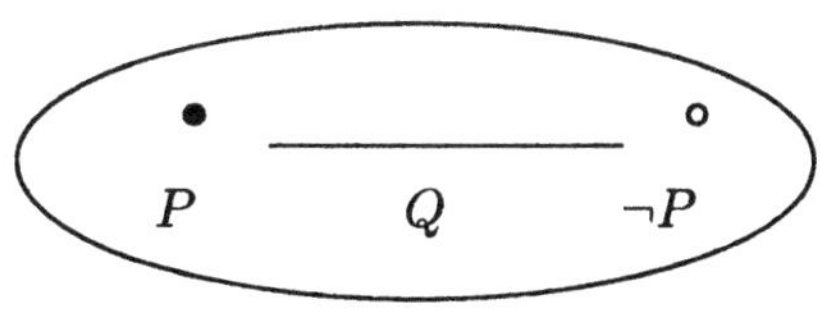

The black dot stands for the actual world. There are no uncertainty lines for **A**, who knows the real situation. In this model, evaluating epistemic

formulas in the usual way shows that, amongst other things, **Q** knows that **A** knows *whether* P is the case: $K_Q(K_A P \vee K_A \neg P)$. Next, **A**'s answer triggers an *informational update* of this information model. This update eliminates the option *not-P*, turning the model into the one-point diagram

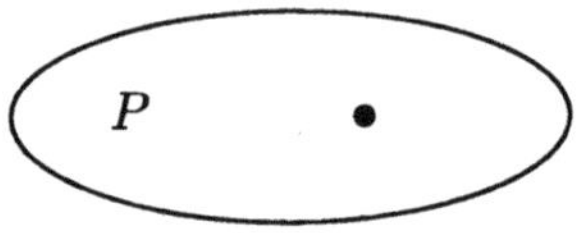

At this stage, P is common knowledge between **Q**, **A**, written $C_{\{\mathbf{Q},\mathbf{A}\}}P$. This means that P holds at each world which can be accessed by any finite sequence of the agents' uncertainty lines. The general update procedure is this. *A public announcement* ϕ! of an assertion ϕ eliminates all those worlds from the current model which fail to satisfy ϕ. With larger epistemic models, such updates may have striking effects, including accounts of the famous Muddy Children puzzle and other group scenarios (Fagin et al. 1995).

Dynamic-epistemic logic The above setting can be formalized in a joint static-dynamic epistemic logic, by using ideas from *dynamic logics* of programs in computer science. In particular, such systems for describing behaviour of programs have 'tandem expressions' describing conditions which hold in states resulting from performing actions:

$[a]\phi$ ϕ holds after every successful execution of action a

This can express statements about epistemic effects of communication, like

$[A!]K_j\phi$ after a true public announcement of A, j knows that ϕ.

Note the long intellectual history encapsulated in this single update formula. It brings together speech act theory, epistemology, and program logics from computer science. There are complete and decidable calculi for this sort of communication. These provide a grip on more general issues, such as complete specification of preconditions and effects of various speech acts. In doing so, many interesting logical phenomena have come to light (van Benthem 2002B). E.g., public assertion of a proposition does *not* always make it common knowledge, but it is an open problem to classify the syntactic propositional forms ϕ that do. Much more general systems of update logic exist by now, dealing with more private information transfer, hidden wholly or in part from other agents. E.g., van Ditmarsch 2000 gives a complete analysis of all communicative moves in the game "Cluedo". The corresponding updates produce more complex changes in epistemic models than just world elimination. The most sophisticated system to date is that of Baltag, Moss and Solecki 1998, which deals with mixtures of public and private information, and even hiding and cheating. The authors also present many results on their completeness and complexity.

Further logical aspects Epistemic-dynamic logic is a typical Dynamic Turn system. Nevertheless, it connects up with standard themes in many ways. For instance, the notion of dynamic inference can be analyzed again as to its structural rules: these now become sensitive to the order and multiplicity of occurrences of announcement actions for premises. (A complete set is found in van Benthem 2002C.) Also, in more standard predicate-logical terms, the model change involved in communication really means the following. We already noted that epistemic formulas can be translated into first-order ones over the same models. A formula of the form $[A!]\phi$ now invites us to make a cross-model evaluation: $M, s \models [A!]\phi$ iff $M|A, s \models \phi$ – with $M|A$ the *restriction* of the model M to the definable subdomain of all worlds satisfying A. But this amounts to evaluating a relativized formula $(\phi)^A$ in M, s itself. Hence, the logic of public update is an instance of the first-order logic of the operation of relativization. The latter has not been axiomatized so far – but at least, we are back with a standard logical question. Likewise, more complex epistemic model updates correspond to more complex combinations of relativization, translation, and pairing of objects, much as with relative interpretability between theories. The more general theme here is comparative evaluation of formulas across models, as has been proposed for independent reasons in Barwise and van Benthem 1999. A recent instance of this style of thinking is Löding and Rohde 2003 which shows that a cross-over modal logic describing other models with accessibility links deleted is as complex as full first-order logic. Thus, our constant theme returns. Even the most recent dynamic logics can be illuminated by viewing their key features in the light of predicate logic.

Conclusion In addition to inference or interpretation, actions of communication like telling, asking, or observing seems just as 'logical'. These can be modelled in multi-agent dynamic epistemic logics treating static propositions and dynamic actions on a par. In the end, these frameworks have some anchoring in ordinary predicate logic after all, when we look at model-crossing evaluation of first-order formulas.

8. Interaction

Update steps are single moves in a larger setting. Interpreting a statement, or taking in its prima facie informational value, do not yet amount to 'making sense of it' in a practical context. I first need to understand the purpose of the current discourse, whether just narrative, or question-oriented, or argumentative, and I need to understand your longer-term behaviour in this setting, including your responses to my actions. This is the typical setting of *game theory*, where players are engaged in strategic interaction, and stable behaviour means that they are in some form of rational *equilibrium*. A good account of modern game theory for logicians may be found in Osborne and Rubinstein 1994, and a panorama of current interactions between logic and game theory is in van Benthem 1999-. We sketch the two main prongs of this recent encounter here.

Logic games As we have already seen, the logical core activity of semantic evaluation can be cast as a two-player zero-sum game. Moreover, winning strategies for players turned out to encode truth (for Verifier) or falsity (if the winning strategy is for Falsifier). E.g., here is the evaluation game for the formula $\forall x \exists y \; x \neq y$ on a two-element domain.

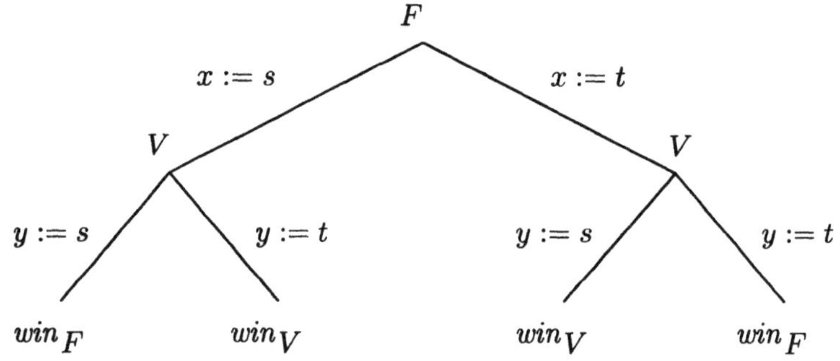

The winning strategy for **V** is choosing the object different from that of **F**. One can think of this as a specific *reason* why the given formula is true. On more general models, evaluation games can be much more interesting: see the references for illustrations. Such games have been devised for a variety of logical languages, some of the most interesting being infinite versions for languages with fixed-point operators (Stirling 1999). Other logic games have existed since the 1950's for *argumentation* (Lorenzen dialogue games) and semantic *model comparison* (Ehrenfeucht-Fraïssé games). E.g., in Lorenzen games, winning strategies for the player making the initial claim correspond to proofs, and winning strategies for the opposing player correspond to counter-models, or more generally scenarios for maintaining consistency of some set of assertions.

Predicate logic as game algebra For our remaining points, we stick with evaluation games. The latter also exemplify very general structure beyond evaluation, viz. operations of choice by one player, role reversal, and composition for arbitrary games. The latter can be studied in abstracto, and it then turns out that many logical laws express valid principles of a *game algebra*. E.g., a De Morgan law $\neg(A \vee B) \leftrightarrow \neg(A \wedge B)$ is an obvious identity thinking about the workings of a role switch. But the same is true for propositional distribution laws, or principles involving composition. The result is a decidable subsystem of predicate logic (Parikh 1985, Goranko 2000), which thereby falls into two levels: general game laws like the above, plus extra principles for the specifics of object picking and fact testing subgames, which create the undecidability. Thus, predicate logic serves as a mirror of abstract game theory, but we also learn something new about it.

Imperfect information Predicate logic can also serve to experiment with further structure from game theory, such as numerical preferences for players, or *imperfect information*. The usual logic games have perfect information: at each stage, players know exactly what was played. But e.g., card games do not give players all that information, and they have to act and communicate to find out more about the initial distribution of the cards. Evaluation games with imperfect information were proposed in Hintikka and Sandu 1997 and many subsequent publications for a variety of semantic and foundational purposes. The result is a compound of both actions which change game states and possibly deficient knowledge which players must deal with in the course of a game. This is again a mixture of epistemic and dynamic logic, which seems characteristic for much of our general activities.

Game logics Reversing perspective, we can also observe that there is logical structure to arbitrary games. Game trees are models for *dynamic logics*, and viewed in other ways, they also serve as models for the general game algebra mentioned above. This is at a level of mere actions and strategies. In addition, however, it has gradually become clear that the notions of rationality underlying standard game-theoretic solution concepts such as Nash equilibrium are partly *epistemic* in nature. If I know that you are rational in the sense of always playing that action which is best for you as far as you know, then I will choose my moves accordingly. And as you can be sure that I react that way, you can anticipate on that, and so on. In this way, common knowledge of rationality can lock players into Nash equilibria. Such epistemic considerations become even more pressing when we allow imperfect information in the structure of the game. There is an incipient logical literature on epistemic analysis of games (Stalnaker 1999 is a good example), which also includes other classical themes. E.g., games involve preferences and expectations, and hence they naturally support conditional and default reasoning about what happens in all most expected alternative courses of events. Thus, a join of logic and game theory provides much more delicate models for analyzing what rational players will, or should, do.

First-order logic again We already saw how predicate-logical evaluation games mirror some very general game-theoretic structure. This feeling can even be substantiated further. Van Benthem 2002A proves that every non-valid principle of abstract general game algebra can be refuted in terms of predicate-logical evaluation games. Thus, simple games for predicate logic can have full game-theoretic expressiveness. And we also saw the conceptual lab idea at work in Hintikka's evaluation games with imperfect information.

The longer term Although we presented games as a mid-term activity, *infinite* logic games can occur, too, in argumentation or other tasks. These model activities where finite termination is not necessarily a virtue, such as the running of your operating system, or the pragmatic conventions of a language. This moves us from simple dynamic logics to a branching temporal setting with potentially infinite runs. The most elegant logical paradigm of this sort so far are the linear logic games for communication or co-operation of Abramsky 1996. These were inspired by the realities of distributed computation, which are an essential feature of modern computing. But cf. also Halpern 2001 on the importance of protocols in understanding communication and strategic behaviour generally.

Conclusion The next level of dynamic ambition is the study of short-term or long-term strategic interaction. This is an area where logic meets with both computer science and game theory, with effects which are still unfolding. There is no accepted logical framework yet for all this, but some ingredients are again dynamic and epistemic logic, with additional resources for describing strategies and conditional reasoning about them.

9. Theory Change

The Dynamic Turn also applies to higher logical aggregation levels, such as theories. Indeed, one of the first influential modern dynamic logics was belief revision theory (cf. Gärdenfors and Rott 1995) which describe three main operations on theories: *update* with consistent new information, *revision* when information has to be accommodated which conflicts with the theory so far, and *contraction* removing certain undesired statements from the current theory. All three are ubiquitous in life, as we confront our expectations with surprising observations, and have to rearrange them. They also occur on a grander scale in science, when we change theories that contradict the facts, or even *themselves*. From a logical point of view, belief revision theory filled a gap. The classical foundations of mathematics picture reasoning as accumulation of truths, and sometimes proven consistency. In this light, every inconsistency which comes to light is a catastrophe, which, as Frege put it, can bring the whole house of cards tumbling down. But in reality, we are remarkably good at coping with inconsistent information, and creating new theories, and the dynamic challenge is precisely to find more logic to that. Belief revision theory has interesting links with conditional logic, but these lie beyond our scope here.

A unified picture? The instances of logical dynamics mentioned so far are related, but their connections are far from being well-understood. E.g., merging update logics with belief revision theory is a job still to be done, so that update processes become more realistic as an account of reasoning in games, where one makes observations as moves are played that lead to update in the earlier epistemic-dynamic sense, while also revising one's expectations about the remainder of the game. Such a merge would also make belief revision more realistic in a multi-agent social setting. Complex theories are often produced by communities, and hence one might expect that update and revision on large chunks of data structure also involve issues of social organization. (Cf. Carroll, Hannan and Pólos 2000 on logical theories of the evolution of social organizations.) Many further desiderata for combining logical perspectives can be derived from our presentation. One other striking example is this. Even though classical accounts of theory structure rightly emphasize the importance of the *language* in which the current theory is phrased, belief revision theory so far has had little to say about linguistic strategies for modifying a theory to accommodate prima facie conflicting evidence (Rott 2001). The latter seems much closer to how we often accommodate contradictions in conversation, by shifting meanings and making distinctions so that everyone's statements come out true.

Long-term phenomena once again Long-term phenomena to be explained need not just be patterns of belief revision. Equally worth explaining may be stable features, such as the continued functioning of linguistic or logical conventions like Gricean maxims in conversation. There is an incipient literature on scenarios in *evolutionary game theory* which would explain such forms of stability (cf. van Rooy 2002). Also relevant in this more positive mode are explicit *learning mechanisms* (Kelly 1996), merging ideas from the philosophy of science, mathematical topology, and computer science. Predicate logic is again a prime example for learning theories, witness Osherson,

Stob and Weinstein 1986. Hendricks 2002 makes an extensive plea for the epistemological relevance of this move.

10. Conclusion

It is time to wrap up. For a start, predicate logic has come rather well out of this story – even though we have not even listed all of its lively curriculum vitae. The 1980s and 1990s saw various further twists, such as the study of expressive and computational fine-structure in description logic (Areces 2000), modal reinterpretation of Tarski semantics leading to decidable predicate logics (Németi 1985, van Benthem 1996), and perhaps most intriguingly, the emergence of new interfaces with probability. These started with the famous Zero-One Law (Fagin 1976), up until the discovery of surprising complexity thresholds in the statistical behavior of logic systems (Kirkpatrick and Selman 1994). The other main theme of this paper has been the agenda shifting through the dynamic turn at various logical aggregation levels, making us rethink the various topics that logical analysis is or can be about. Both surveys are by no means complete, and we do not mean to suggest (circumscriptively) that topics not mentioned are topics that do not exist. For instance, our story would be much richer if it also added proof-theoretic perspectives.

We conclude with some general comments. First, there is an issue of *natural boundaries* to logic: but we have not planted any boundary markers so far, as our aim was just to show the broader landscape. Indeed, some of the natural boundaries might come from method rather than territory, witness the constant role of first-order logic in our exploration. There are some presuppositions about logic in our panorama, however, no matter how tentatively construed. One is that logic is both about the mathematics of formal systems and about describing and analyzing real phenomena of reasoning, communication, and interaction. The other is that, on the spectrum from normative a priori to descriptive and empirical, we are moving a bit toward the latter. Logic may not really be empirical cognitive science of reasoning and interaction (facts are strong medicine...): but its inspiration for new theory building certainly derives from observations about natural logic and actual human behaviour. Still, we have not taken a strong stance on this either. In fact, one of the ways in which logic might also interface with reality more is the design stance of computer science: creating *virtual reasoning systems* that can be used in the real world.

As a side purpose of this paper we mention an emphasis on some history of ideas. In addition to basic skills training, it might be a good idea to teach more of the development of logical themes over time - even when we are not sure where we are heading in the long run. The few lines drawn here, no matter how sketchily, across the past 50 or 75 years, give every good reason to celebrate the past, and to expect more good things to come.

References

[1] S. Abramsky, 1996, Semantics of Interaction: an Introduction to Game Semantics, in P. Dybjer and A. Pitts, eds., Proceedings 1996 CLiCS Summer School, Isaac Newton Institute, Cambridge University Press, Cambridge, 1-31.

[2] C. Areces, 2000, *Logic Engineering*, dissertation DS-2000-05, Institute for Logic, Language and Computation, University of Amsterdam.

[3] A. Baltag, L. Moss and S. Solecki, 1998, The Logic of Public Announcements, Common Knowledge and Private Suspicions, *Proceedings TARK 1998*, 43-56, Morgan Kaufmann Publishers, Los Altos. Updated version, 2003, department of computer science, Indiana University, Bloomington, and department of computing, Oxford University.

[4] J. Barwise, ed., 1977, *Handbook of Mathematical Logic*, North-Holland, Amsterdam.

[5] J. Barwise and J. van Benthem, 1999, Interpolation, Preservation, and Pebble Games, *Journal of Symbolic Logic* 64, 881-903.

[6] J. van Benthem, 1982, The Logical Study of Science, *Synthese* 51, 431-472.

[7] J. van Benthem, 1989, Semantic Parallels in Natural Language and Computation, in H-D Ebbinghaus et al., eds., *Logic Colloquium. Granada 1987*, North-Holland, Amsterdam, 331-375.

[8] J. van Benthem, 1996, *Exploring Logical Dynamics*, CSLI Publications, Stanford.

[9] J. van Benthem, 1999-2003, *Logic in Games*, lecture notes, Institute for Logic, Language and Computation, University of Amsterdam.

[10] J. van Benthem, 2002A, Logic Games are Complete for Game Logics, to appear in M. Pauly, ed., special issue of *Studia Logica* on logic and games.

[11] J. van Benthem, 2002B, One is a Lonely Number, invited lecture, Colloquium Logicum, Muenster. Report PP-2003-07, ILLC, University of Amsterdam.

[12] J. van Benthem, 2002C, Structural Properties of Dynamic Reasoning, to appear in J. Peregrin, ed., *Meaning: the Dynamic Turn*, Elsevier, Colchester.

[13] J. van Benthem, 2003A, Is there still Logic in Bolzano's Key?, Tech Report, Institute for Logic, Language and Computation, University of Amsterdam. To appear in Proceedings Bolzano Workshop 2002, Austrian Academy of Sciences.

[14] J. van Benthem, 2003B, Logic and the Dynamics of Information, to appear in L. Floridi, ed., special issue of *Minds and Machines*.

[15] J. van Benthem and A. ter Meulen, eds., 1997, *Handbook of Logic and Language*, Elsevier Science Publishers, Amsterdam.

[16] P. Blackburn, M. de Rijke and Y. Venema, 2001, *Modal Logic*, Cambridge University Press, Cambridge.

[17] G. Carroll, M. Hannan, and L. Plos, 2000, Foundations of a Theory of Social Forms, ERIM Report 2000-29, School of Management, Erasmus University, Rotterdam.

[18] B. ten Cate and Ch. Shan, 2002, Question Answering: from Partitions to Prolog, in U. Egli and Chr. Fermüller, eds., Proceedings of TABLEAUX 2002: *Automated Reasoning with Analytic Tableaux and Related Methods*, Lecture Notes in Artificial Intelligence 2381, Springer Verlag.

[19] H. van Ditmarsch, 2000, *Knowledge Games*, dissertation DS-2000-06, ILLC Amsterdam and University of Groningen.

[20] R. Fagin, 1976, Probabilities on Finite Models, *Journal of Symbolic Logic* 41, 50-58.

[21] R. Fagin, J. Halpern, Y. Moses and M. Vardi,1995, *Reasoning about Knowledge*, The MIT Press, Cambridge (Mass.).

[22] D. Gabbay, 1996, An Overview of Fibred Semantics and the Combination of Logics, *Frontiers of Combining Systems* 1996, 1-55.

[23] P. Gärdenfors and H. Rott, 1995, Belief Revision, in D. M. Gabbay, C. J. Hogger and J. A. Robinson, eds., *Handbook of Logic in Artificial Intelligence and Logic Programming* 4, Oxford University Press, Oxford 1995.

[24] V. Goranko, 2000, The Basic Algebra of Game Equivalences, Report PP-2000-12, Institute of Logic, Language, and Computation, University of Amsterdam. To appear in *Studia Logica*.

[25] J. Groenendijk and M. Stokhof, 1991, Dynamic Predicate Logic, *Linguistics and Philosophy* 14, 39-100.

[26] J. Halpern, 2001, A Computer Scientist Looks at Game Theory, department of computer science, Cornell University. Presented at Siena Summer School on Cognitive Processes and Rationality in Economics.

[27] V. Hendricks, 2002, Active Agents, ΦNEWS 2(2002), 5-41. To appear in J. van Benthem and R. van Rooy, eds., special issue on Information Theories of the *Journal of Logic, Language and Information.*

[28] J. Hintikka, 1973, *Logic, Language Games, and Information*, The Clarendon Press, Oxford.

[29] J. Hintikka and G. Sandu, 1997, Game-Theoretical Semantics, in J. van Benthem and A. ter Meulen, eds., 361-410.

[30] E. Keenan and D. Westerstahl, 1997, Generalized Quantifiers, in J. van Benthem and A. ter Meulen, eds., 1997, 839-893.

[31] K. Kelly, 1996, *The Logic of Reliable Inquiry*, Oxford University Press, Oxford.

[32] S. Kirkpatrick and B. Selman, 1994, Critical Behavior in the Satisfiability of Random Boolean Expressions, *Science* 264, 1297-1301.

[33] Th. Kuipers, 2001, *Structures in Science*, Kluwer Academic Publishers, Dordrecht.

[34] Ch. Lding and Ph. Rohde, 2003, Model Checking and Satisfiability for Sabotage Modal Logic, Lehrstuhl fr Informatik VII, RWTH Aachen.

[35] T. Maibaum and P. Veloso, 1995, On the Modularisation Theorem for Logical Specifications, *Information Processing Letters* 53, 287-293.

[36] I. Mason, 1985, The Metatheory of the Classical Propositional Calculus is not Axiomatizable, *Journal of Symbolic Logic* 50, 451-457.

[37] J. McCarthy, 1980, Circumscription- A Form of Nonmonotonic Reasoning, *Artificial Intelligence* 13, 27-39.

[38] I. Németi, 1985, The Equational theory of Cylindric Relativized Set Algebras is Decidable, Preprint 63/85, Mathematical institute, Hungarian Academy of Sciences, Budapest.

[39] M. Osborne and A. Rubinstein, 1994, *A Course in Game Theory*, The MIT Press, Cambridge (Mass.).

[40] D. Osherson, M. Stob and S. Weinstein 1986, *Systems that Learn*, The MIT Press, Cambridge (Mass.).

[41] R. Parikh, 1985, The Logic of Games and its Applications, *Annals of Discrete Mathematics* 24, 111-140.

[42] M. Przełęcki, 1969, *The Logic of Empirical Theories*, Routledge and Kegan Paul, London.

[43] R. van Rooy, 2002, Quality and Quantity of Information Exchange, to appear in J. van Benthem and R. van Rooy, eds., special issue on Information Theories of the *Journal of Logic, Language and Information.*

[44] H. Rott, 2001, Theoretical Concepts in Flux: Conceptual Knowledge and Theory Change, in R. Eckardt, K. von Heusinger and Ch. Schwarze, eds., *Words in Time - Diachronic Semantics from Different Points of View*, FB Sprachwissenschaft der Universität Konstanz, 279-301.

[45] M. Ryan, 1992, *Ordered Presentations of Theories: Default Reasoning and Belief Revision*, PhD thesis, Department of Computing, Imperial College, London.

[46] V. Sanchez Valencia, 1991, *Studies on Natural Logic and Categorial Grammar*, dissertation, Institute for Logic, Language and Computation, University of Amsterdam. Expanded version to appear with MIT Press, Cambridge (Mass.).

[47] J. Sneed, 1971, *The Logical Structure of Mathematical Physics*, Reidel, Dordrecht.

[48] R. Stalnaker, 1999, Extensive and Strategic Form: Games and Models for Games, *Research in Economics* 53:2, 93-291.

[49] C. Stirling, 1999, Bisimulation, Modal Logic, and Model Checking Games, *Logic Journal of the IGPL* 7, 103-124.

[50] F. Suppes, ed., 1974, *The Structure of Scientific Theories*, University of Illinois Press, Urbana.

[51] F. Veltman, 1996, Defaults in Update Semantics, *Journal of Philosophical Logic* 25, 221-261.

Johan van Benthem,
Amsterdam and Stanford,
June 2003

Josep Maria Font

Generalized Matrices in Abstract Algebraic Logic

Abstract. The aim of this paper is to survey some work done recently or still in progress that applies *generalized matrices* (also called *abstract logics* by some) to the study of sentential logics. My main concern will be to emphasize the links between this line of research and other existing frameworks in Algebraic Logic, either well-established ones (such as the old *theory of logical matrices* and the younger theories of *protoalgebraic logics*, *algebraizable logics* and the associated hierarchy) or really new ones (such as the theory of *algebraizability of Gentzen systems* or the *model theory of equality-free logic*). I would like to convey the idea that the interaction between these neighbouring fields may be specially fruitful, as it seems to be one of the leading forces in the shaping of this emerging field called *Abstract Algebraic Logic.*

Abstract Algebraic Logic (AAL) may be described as today's view of Algebraic Logic. The term began to be used around the mid-nineties, promoted by Wim Blok and Don Pigozzi (the expression "universal algebraic logic" was also used for a short time). In 1997 a *Workshop on Abstract Algebraic Logic* [35] was organized as the closing event of a *Semester in Algebraic Logic and Model Theory* of the *Centre de Recerca Matemàtica* of the *Institut d'Estudis Catalans*, located in Bellaterra, near Barcelona. Selected from its more than thirty talks, twelve papers have been published in two special issues of *Studia Logica* (number 1 of volume 65, 2000, and numbers 1/2 of volume 74, 2003). One of them [38] is a long survey of the field. In the meantime, the encyclopedic monograph [22] by Janusz Czelakowski has appeared. Also the long paper [26] by Czelakowski and Pigozzi contains a readable, substantial introduction to the field; finally, a forthcoming long paper surveying an important part of the subject is Blok and Pigozzi's [6]. With this material now available, I feel safe in limiting my aim to the topic indicated by the title and in not giving too many details; interested readers now have several sources where to find whatever they want to know about the undefined notions I will deal with and about some theorems I am going to mention.

Let me begin with some general reflections.

V. F. Hendricks and *J. Malinowski* (eds.),
Trends in Logic: 50 Years of Studia Logica

Trends in Logic **21**: 57–86, 2003.

1. On Abstract Algebraic Logic

In my view, most of the work done in this area can be organized around two main questions:

Q1. What do we mean when we say that a logic $\mathcal{L}$ can be *algebraized* and that the class K is *its algebraic counterpart*?

Q2. What can we learn about the logic $\mathcal{L}$ or about the class K when they are related as in question (Q1)?

Question (Q1) concerns *the process of algebraization* while question (Q2) concerns the *consequences* of this process. Works on question (Q1) are easily recognized, as they address the problem rather explicitly. Also, many works on question (Q2) have an explicit format; for instance, the Budapest group (ANDRÉKA, NÉMETI, SAIN, etc.) has coined the expression *bridge theorems* for those that take the typical form

$$\mathcal{L} \text{ has property } \boldsymbol{P} \iff \mathsf{K} \text{ has property } \boldsymbol{P'}, \tag{1}$$

where $\boldsymbol{P}$ is a typically logical property, what one usually calls a *metalogical property* (e.g., interpolation), and $\boldsymbol{P'}$ is a typically algebraic property (e.g., amalgamation). There are also papers whose contents is (almost) entirely algebraic, but which can be considered as dealing with question (Q2) if we put them in the proper context.

One of the most interesting facts one recognizes in the historical development of Algebraic Logic, specially in the last ten or twenty years, is that answers to these and related questions have been organized and *have to be organized* at several *levels*.

Through the works of BLOK, PIGOZZI, CZELAKOWSKI and HERRMANN a *hierarchy of logics*, also called *the Leibniz hierarchy*, has emerged. This hierarchy is organized by taking into account the strength of the relationship Logic↔Algebras, or more generally Logic↔Models, that is, by their degree of algebraizability; the term *algebraizable* in the technical sense is applied only to those more restricted classes of logics where these links are very strong, and still with several qualifications: *weak, finite, strong, regular.* The most common answer to question (Q1) is that each algebra of K has a distinguished element 1 such that the following completeness theorem holds.

$$\Gamma \vdash_{\mathcal{L}} \varphi \iff \{\gamma \approx 1 : \gamma \in \Gamma\} \models_{\mathsf{K}} \varphi \approx 1 \tag{2}$$

This is a particular case of the technical notion of the class K being an *algebraic semantics* for $\mathcal{L}$ [3], where a finite set of equations in one variable replaces the single equation $x \approx 1$ exemplified here. Having an algebraic semantics has been shown to be a rather weak kind of algebraizability, and actually not playing a particular rôle in the hierarchy unless it is complemented by stronger conditions; roughly speaking, representability of $\models_{\mathsf{K}}$ inside $\mathcal{L}$ as a kind of converse of (2); the following is the most common particular case:

$$\{\delta_i \approx \varepsilon_i : i \in I\} \models_{\mathsf{K}} \delta \approx \varepsilon \iff \{\delta_i \leftrightarrow \varepsilon_i : i \in I\} \vdash_{\mathcal{L}} \delta \leftrightarrow \varepsilon \tag{3}$$

(in general, a finite set of formulas in two variables replaces the formula $x \leftrightarrow y$ used here). Only if the two representations are inverse to one another in a certain precise, natural sense, we get *algebraizability*, a situation where interesting and strong bridge theorems can be obtained. All this is explained in [38].

Analysing the hierarchy one recognizes that the different levels of algebraizability are reflected in different *levels of complexity of the models*: In the best-behaved cases (the different modalities of algebraizability) we can work with just *algebras* $\boldsymbol{A}$, without further structure. Since algebras carry in themselves their equations, we can work with the equational consequence $\models_{\mathsf{K}}$ *relativized* to a class K of algebras; and it is this consequence that makes the connections and helps in finding the bridge theorems. But in wider classes of logics (equivalential, protoalgebraic) we have to work with *matrices*, i.e., structures $\langle \boldsymbol{A}, F\rangle$ where $\boldsymbol{A}$ is an algebra and $F \subseteq A$. Then the bridge theorems will concern classes of matrices rather than classes of algebras. There is now a reasonable evidence that the widest class of logics for which this scheme seems to work smoothly enough in order to obtain significant bridge theorems concerning matrices is the class of *protoalgebraic logics*. By way of illustration, in [4, 21] it is shown that for protoalgebraic logics the LDT (Local Deduction Theorem) is equivalent to the FEP (filter extension property) for the class of reduced matrix models of the logic; if the logic is finitely algebraizable then the FEP reduces to the more algebraic RCEP (relative congruence extension property) for the equivalent quasivariety K.

When trying to enlarge the domain of applicability of the methods of Abstract Algebraic Logic outside the class of protoalgebraic logics, one can put more structure in the model, and consider *generalized matrices*, that is, systems $\langle \boldsymbol{A}, \mathcal{C}\rangle$ where $\boldsymbol{A}$ is an algebra and $\mathcal{C} \subseteq \mathcal{P}(A)$ is a family of subsets of A. This will be the proper subject of the paper.

So far I have recalled two hierarchies: One among logics, and one among their models. I would like to consider another process of enlarging the scope of Algebraic Logic that has also created a kind of hierarchy: the *notion of sentential logic* itself can also be considered at several levels.

The first level, which I call the *old style*, is to define a logic as a set of *formulas* (satisfying some conditions, of course; for instance invariance under substitutions). At this level, one identifies a logic $\mathcal{L}$ with a set of *theorems.*

The second level, which I prefer to call the *Tarski style* (and resist to call "Polish style"), is to define a logic $\mathcal{L}$ as a *consequence relation* between sets of formulas and formulas, denoted by $\vdash_{\mathcal{L}}$, also satisfying some minimal requirements, for instance invariance under substitutions, or (optionally) finitarity. The consequences of the empty set of assumptions are called *theorems*, and constitute a "logic" in the old style. In this paper the word "logic" will refer to a logic in the second level sense, unless otherwise specified. Considering logics at this level has become absolutely necessary in some situations; for instance, if we want to study the algebraizability and several bridge properties of modal logics: It is well-known that in this domain there are different consequence relations with the same set of theorems, each showing a different behaviour both from the metalogical and from the algebraic points of view; this is for instance the case of the two consequences one can associate with each class of Kripke-style or relational frames, a "global" or strong consequence and a "local" or weak one.

In the finitary case a logic at this level is determined by a set of rules of the form $\varphi_0, \ldots, \varphi_{n-1} \vdash \varphi_n$ through the well-known syntactical notion of *proof.* In particular such a proof system generates all the theorems of the logic. Indeed, in the old style days this was the only purpose of a proof system, and the analysis of the rules that preserve the set of "theorems" led in a further step to the creation of the second level conception. TARSKI in [47] mentions "the axiomatic method" along with "the matrix method" as the two forms of definition of a logic in his sense, and in the first case he assumes substitution is one of the rules of the proof system. In [57] his investigation takes the most general form and he does not assume substitution-invariance; this condition was explicitly incorporated by ŁOŚ and SUSZKO [46] in 1958. RASIOWA does not include invariance under substitutions in her general definition of a logic in the second level sense, but she explicitly assumes such a logic as given by an axiomatic system where axioms and rules are closed under substitution instances; [50, § VIII.5].

There is a third level that has been a natural step when trying to answer question (Q1) for non-protoalgebraic logics. I will call it the *Gentzen style.* Its main object of study are the so-called *Gentzen style rules*, that is, rules of the form

$$\frac{\{\Gamma_i \vdash \varphi_i : i < n\}}{\Gamma_n \vdash \varphi_n} \tag{4}$$

where the Γ_i can be either finite sets or finite sequences of formulas. This style appeared by purely proof-theoretic motivations: the investigation of the "rules" that preserved the set of *sequents* of the logic in the second level sense, that is, the pairs $\langle \Gamma, \varphi \rangle$ such that $\Gamma \vdash_{\mathcal{L}} \varphi$. By using these rules, the old notion of proof, now acting on sequents (commonly written as $\Gamma \vdash \varphi$) as basic linguistic objects, defines the so-called *Gentzen systems*, that is, consequence relations between sets of sequents and sequents, again satisfying some minimal requirements such as invariance under substitutions; additionally, they can satisfy some of the so-called structural rules (idempotency, exchange, weakening, cut, etc.). These are the "logics" in the third level sense.

The relations between the second and the third levels closely parallel those between the first and the second. Moreover, if all structural rules are (at least) admissible, then the "theorems" of a Gentzen system $\mathfrak{G}$, i.e., its *derivable sequents*, define a logic in the sense of the second level by means of

$$\varphi_0, \ldots, \varphi_{n-1} \vdash_{\mathcal{L}} \varphi_n \iff \begin{array}{l} \text{the sequent } \varphi_0, \ldots, \varphi_{n-1} \vdash \varphi_n \\ \text{is derivable in } \mathfrak{G} \end{array} \tag{5}$$

or a similar stipulation; and two different Gentzen systems can have the same derivable sequents but different metalogical properties (for instance, one can be cut-free while the other is not).

In recent years the increased interest on the so-called *substructural logics* has put Gentzen *calculi* in a more central position. Again, we can take a broader view and consider them as just one of the ways of presenting a Gentzen system, the abstract notion whose semantical and algebraic sides can also be studied. This trend began in [40]; among the most recent contributions let me mention [1, 41, 42]. Its study, very interesting by itself, has been also useful for the algebraic study of sentential logics (in the second level sense) that do not successfully fit into the hierarchy of protoalgebraic logics. Moreover, the abstract theory can be generalized and applied not only to sequents of the form $\Gamma \vdash \varphi$ but to arbitrary kinds of sequents $\Gamma \vdash \Delta$, and even to more complicated syntactical objects that have been recently considered, such as *m-sided sequents* and *hypersequents.* The study of the relationship between the abstract algebraic study of such systems and their proof-theoretic properties has just begun, as shows [41], where the connection between protoalgebraicity (applied to m-sided Gentzen systems) and the presence of the cut rule is studied.

Here I will consider only the work done with the restricted kind of sequents first considered, and assume all logics are finitary. In this case, generalized matrices provide one of the most natural notions of model of a logic in the third level sense, and at the same time they are also models of logics in the second level sense in an extremely natural way. Actually, I consider that one of the most interesting facets of the research with generalized matrices is the interaction between these two rôles they can play. I will try to show some results that emphasize this interaction.

2. Prehistory

To be precise, let $\boldsymbol{Fm}$ denote the formula algebra of a certain arbitrary, fixed, algebraic type. A ***sentential logic*** $\boldsymbol{\mathcal{L}}$ is given by a finitary and substitution-invariant consequence relation $\vdash_{\mathcal{L}}$ between sets of formulas and formulas (i.e., $\vdash_{\mathcal{L}} \subseteq P(Fm) \times Fm$) which will be written in infix notation as usual. The set of all ***theories*** of the logic is denoted by $Th\mathcal{L}$, and the associated closure operator by $\mathrm{Cn}_{\mathcal{L}}$; that is, for each set Γ of formulas, $\mathrm{Cn}_{\mathcal{L}}(\Gamma) = \{\varphi \in Fm : \Gamma \vdash_{\mathcal{L}} \varphi\}$ is the theory generated by Γ.

The traditional algebraic model theory of sentential logics is matrix-based. A ***matrix*** is a pair $\langle \boldsymbol{A}, F\rangle$ where $\boldsymbol{A}$ is an algebra of the appropriate type, and $F \subseteq A$, and it is a ***model of*** $\boldsymbol{\mathcal{L}}$ when F is an ***$\mathcal{L}$-filter***, that is, when for any $\varphi_0, \ldots, \varphi_n \in Fm$, if $\varphi_0, \ldots, \varphi_{n-1} \vdash_{\mathcal{L}} \varphi_n$ and $h(\varphi_i) \in F$ for all $i < n$ then $h(\varphi_n) \in F$, for any interpretation h in $\boldsymbol{A}$ (i.e., for any $h \in \mathrm{Hom}(\boldsymbol{Fm}, \boldsymbol{A})$). The set of all $\boldsymbol{\mathcal{L}}$-filters on $\boldsymbol{A}$ is denoted by $\mathcal{F}i_{\mathcal{L}}\boldsymbol{A}$.

The most general notion of ***generalized matrix*** is a pair $\boldsymbol{\mathcal{A}} = \langle \boldsymbol{A}, \mathcal{C}\rangle$ where $\boldsymbol{A}$ is an algebra (of the required type) and $\mathcal{C} \subseteq \mathcal{P}(A)$ is an arbitrary collection of subsets of A. In this sense, they were conceived around 1969 by Wójcicki (in [59, 60]; see also [61, § IV.4]) as generalizations of the idea of matrix, more precisely, as an alternative presentation of the closely related notion of a ***bundle of matrices***, which refers to the family $\{\langle \boldsymbol{A}, F\rangle : F \in \mathcal{C}\}$. Most model-theoretic or universal-algebraic ideas and constructions on matrices can be reproduced or adjusted for generalized matrices. Also, set-theoretically, as a set with a family of subsets, they can be considered under topological intuitions. An ordinary matrix can be identified with the generalized matrix of the form $\langle \boldsymbol{A}, \{F, A\}\rangle$; thus the new theory can contain the classical one to a certain extent. Generalized matrices have reappeared recently in [28] under the name of ***atlases***, where its use is extended to deal with "multiple-conclusion logics" or "Scott consequence relations".

Generalized matrices have a well-known *dual presentation* as pairs $\langle \boldsymbol{A}, \mathrm{C}\rangle$ where $\mathrm{C} : \mathcal{P}(A) \to \mathcal{P}(A)$ is a ***closure operator*** on the power set of A. This was SMILEY's presentation in [56] and MAGARI's in [48] (he used the term *calcoli generali*). The first in-depth study of generalized matrices under this presentation was BROWN's 1969 dissertation [13], where SUSZKO was the principal advisor, and then their joint papers with BLOOM [11, 10, 14]; the last two were published together in 1973 with an interesting preface by SUSZKO. In these they coined the term *abstract logics*. While [11, 14] present the general theory with short examples, [10] studies the abstract logics consisting of a Boolean algebra and the closure operator of filter generation and shows these are characterized, roughly speaking, by the same metalogical properties that determine classical logic, viz. finitarity, the Deduction Theorem[1], and having the classical tautologies as theorems (that these properties characterize classical sentential logic is due to TARSKI). Similar characterizations were obtained in [8, 9] for several natural fragments of intuitionistic logic containing conjunction. This approach was extensively followed by the Barcelona group, see [33] and the many references therein. However, except for [56], all these works prior to 1990 do not try to provide a precise explanation of the relationship between the abstract logics they were considering and the sentential logics they had in mind. They explored the *analogies* between a logic $\mathcal{L}$ presented as the abstract logic $\langle \boldsymbol{Fm}, \mathrm{Cn}_{\mathcal{L}}\rangle$ and the abstract logics of the form $\langle \boldsymbol{A}, \mathrm{Fi}^{\boldsymbol{A}}\rangle$ where $\boldsymbol{A}$ is an algebra of the class naturally associated with the logic (for instance, Boolean algebras or Heyting algebras for classical or intuitionistic logic, respectively) and $\mathrm{Fi}^{\boldsymbol{A}}$ is the corresponding closure operator of filter generation; but again the analogy did not rest on a technical notion of an abstract logic being a model of a sentential logic such as that of WÓJCICKI's.

SMILEY's 1962 discussion [56, pp. 433–435] of the insufficiency of ordinary matrices to model some logics contains the first proposal to use algebras with a closure operator in order to *model the deducibility relation* rather than theoremhood. Unfortunately, SMILEY's proposal was only followed briefly by HARROP in [44, 45] but did not attract any attention from the algebraic logic community (even SMILEY's well-known monograph with SHOESMITH [55] uses only ordinary matrices) and WÓJCICKI did not further develop the theory of generalized matrices beyond his first completeness results in [59, 60]. The stream reappears only in 1991 in the paper [40] published in *Studia Logica*, in a *Special Issue on Algebraic Logic* edited by BLOK and PIGOZZI; indeed, the personal contact between these editors and the authors

[1]See the Appendix for formal statements of some properties.

of [40] was a key point in the change of perspective.

Note that in order to obtain a one-to-one correspondence between the two dual presentations one has to restrict attention to the generalized matrices where the family of subsets $\mathcal{C}$ is a ***closure system*** (it is closed under intersections of arbitrary subfamilies). Moreover, given the context of this discussion in the model theory of sentential logics as defined above, I will also assume that $\mathcal{C}$ is ***inductive*** (closed under unions of upwards-directed subfamilies); dually, this means that C is ***finitary***. These assumptions will hold throughout the paper.

The duality is effected by the following constructions: Given an inductive closure system $\mathcal{C}$ on a set A one obtains a finitary closure operator by defining, for each $X \subseteq A$, $\mathrm{C}(X) = \bigcap\{T \in \mathcal{C} : X \subseteq T\}$. Conversely, for each such C, the family $\mathcal{C} = \{T \subseteq A : \mathrm{C}(T) = T\}$ is an inductive closure system. BULL's 1976 paper [15] exploits this duality as the basis of an abstract analysis of classical logic from the perspective of algebra and topology.

This duality is very convenient. In the development of the general theory, each notion and each situation can be expressed or understood in the presentation that best suits it. We are going to see this in the next sections.

3. Generalized matrices as models of sentential logics

As an easy example of the duality of presentations of generalized matrices mentioned in the previous section, consider the notion of model of a logic: WÓJCICKI's original definition, which uses the closure system, is certainly compact:

DEFINITION 1. *A generalized matrix $\langle \boldsymbol{A}, \mathcal{C}\rangle$ is a* ***model of*** *a sentential logic $\mathcal{L}$ when $\mathcal{C} \subseteq \mathcal{F}i_{\mathcal{L}}\boldsymbol{A}$; that is, when the closure system is just a collection of $\mathcal{L}$-filters in the traditional sense.*

However concise this is, an equivalent definition using the closure operator gives a different and more intuitive view[2]:

PROPOSITION 2. *A generalized matrix $\langle \boldsymbol{A}, \mathrm{C}\rangle$ is a model of $\mathcal{L}$ if and only if it is a* ***model of all Hilbert-style rules of*** $\mathcal{L}$*, in the sense that*

$$\textit{if}\quad \varphi_0,\ldots,\varphi_{n-1} \vdash_{\mathcal{L}} \varphi_n \quad \textit{then}\quad h(\varphi_n) \in \mathrm{C}\big(h(\varphi_0),\ldots,h(\varphi_{n-1})\big) \tag{6}$$

for any interpretation h in $\boldsymbol{A}$ and for any $\varphi_0,\ldots,\varphi_n \in Fm$.

---

[2] I write $\mathrm{C}(a,\ldots,z)$ for $\mathrm{C}(\{a,\ldots,z\})$ and $\mathrm{C}(X,a)$ for $\mathrm{C}(X \cup \{a\})$ when $a,\ldots,z \in A$ and $X \subseteq A$, and use other similar customary abbreviations.

Thus, behind this notion of model as a collection of $\mathcal{L}$-filters, which might look as an innocent generalization, there is a very natural notion of an interpretation satisfying a sequent: We can say that h ***satisfies*** a sequent $\varphi_0, \ldots, \varphi_{n-1} \vdash \varphi_n$ when the right-hand half of (6) holds. As the reader is already guessing, this will be used later on to define the notion of a model of a Gentzen-style rule or of a Gentzen system.

It may be convenient to consider another example. A mapping h is a ***strict homomorphism*** between two generalized matrices $\mathcal{A} = \langle \boldsymbol{A}, \mathcal{C} \rangle$ and $\mathcal{B} = \langle \boldsymbol{B}, \mathcal{D} \rangle$ when it is an algebraic homomorphism $h : \boldsymbol{A} \to \boldsymbol{B}$ such that $\mathcal{C} = \{h^{-1}[G] : G \in \mathcal{D}\}$. The dual expression, however, is logically speaking more significant: A homomorphism $h : \boldsymbol{A} \to \boldsymbol{B}$ is strict between $\mathcal{A}$ and $\mathcal{B}$ if and only if for every $X \cup \{a\} \subseteq A$, $a \in \mathrm{C}(X) \iff h(a) \in \mathrm{D}\big(h[X]\big)$. Of particular interest are the strict *surjective* homomorphisms, which are called ***bilogical morphisms*** in [10, 33]: among other properties, they make the closure systems $\mathcal{C}$ and $\mathcal{D}$ to be isomorphic as complete lattices.

A common way of describing Abstract Algebraic Logic [6, 38] is to say that this theory starts from an abstraction of "the traditional Lindenbaum-Tarski process". As is well-known, this process consists in factoring the formula algebra by congruences associated with theories of the logic, in order to generate a class of algebras that can deserve the title of *the algebraic counterpart of the logic*, i.e., the class of *Lindenbaum-Tarski algebras*. The abstraction can be extended to make it a particular case of a ***process of reduction*** that can be applied to an arbitrary generalized matrix. To implement it we first consider the following

DEFINITION 3. *The* ***Frege relation*** $\Lambda(\mathcal{A})$ *of a generalized matrix* $\mathcal{A}$ *is defined by:* $\langle a, b \rangle \in \Lambda(\mathcal{A}) \iff \mathrm{C}(a) = \mathrm{C}(b)$, $\forall a, b \in A$. *The* ***Tarski congruence*** *of* $\mathcal{A}$, $\widetilde{\boldsymbol{\Omega}}(\mathcal{A})$, *is the largest congruence of* $\boldsymbol{A}$ *that is below* $\Lambda(\mathcal{A})$. *A generalized matrix* $\mathcal{A}$ *is* ***reduced*** *when* $\widetilde{\boldsymbol{\Omega}}(\mathcal{A}) = Id_A$. *The* ***reduction*** *of* $\mathcal{A}$ *is* $\mathcal{A}^* = \big\langle \boldsymbol{A}/\widetilde{\boldsymbol{\Omega}}(\mathcal{A}), \mathcal{C}/\widetilde{\boldsymbol{\Omega}}(\mathcal{A})\big\rangle$, *where* $\mathcal{C}/\widetilde{\boldsymbol{\Omega}}(\mathcal{A}) = \{F/\widetilde{\boldsymbol{\Omega}}(\mathcal{A}) : F \in \mathcal{C}\}$.

The Frege relation is an abstract analogue of the interderivability relation $\dashv\vdash_{\mathcal{L}}$ of sentential logics, and is always an equivalence relation, but may not be a congruence. This is why in general it is strictly necessary to introduce the Tarski congruence, which is the analogue for generalized matrices of the *Leibniz congruence* of an ordinary matrix (and reduces to it when the generalized matrix is an ordinary one). In some respects, generalized matrices and the Tarski operator $\widetilde{\boldsymbol{\Omega}}$ behave on models of arbitrary logics as ordinary matrices and the Leibniz operator $\boldsymbol{\Omega}$ behave on (matrix) models of protoalgebraic logics; for instance, the Tarski operator is always monotonic with respect to the natural ordering between generalized matrices in their closure operator form, while the Leibniz operator is monotonic only on the matrix models of a protoalgebraic logic.

In the literature one can find (at least) the following three general processes for associating a class of algebras with a sentential logic $\mathcal{L}$ using this reduction process:

- If we identify $\mathcal{L}$ with the generalized matrix $\langle \boldsymbol{Fm}, Th\mathcal{L}\rangle$, where $Th\mathcal{L}$ is the family of its *theories*, we can consider its Tarski congruence $\widetilde{\boldsymbol{\Omega}}(\mathcal{L})$ and the so-called ***Lindenbaum-Tarski algebra of*** $\mathcal{L}$, i.e., the quotient algebra $\boldsymbol{Fm}/\widetilde{\boldsymbol{\Omega}}(\mathcal{L})$. This algebra generates the variety $\mathsf{V}(\mathcal{L}) = \mathbb{V}(\boldsymbol{Fm}/\widetilde{\boldsymbol{\Omega}}(\mathcal{L}))$, considered e.g., in [51, 52], where it is called *the variety associated with* $\mathcal{L}$.
- In the classical theory of matrices we find the class $\mathsf{Alg}^*\mathcal{L}$ of *algebraic reducts of reduced matrix models* for $\mathcal{L}$.
- And finally there is the class $\mathsf{Alg}\mathcal{L}$ of *algebraic reducts of reduced (generalized matrices) models of* $\mathcal{L}$. They are called $\mathcal{L}$-***algebras***.

The class $\mathsf{Alg}^*\mathcal{L}$ has been traditionally regarded as the algebraic counterpart of $\mathcal{L}$; however, as I will remark in a moment, there are reasons for considering that it is $\mathsf{Alg}\mathcal{L}$ that better does the job in general. But first let me use the reduction process again to define the notion of *full model*, a special kind of model that shows a remarkably nice behaviour.

DEFINITION 4. *A generalized matrix* $\mathcal{A} = \langle \boldsymbol{A}, C\rangle$ *is a* ***full model of*** $\mathcal{L}$ *when the closure system of its reduction consists of all the* $\mathcal{L}$*-filters of the quotient algebra:* $C/\widetilde{\boldsymbol{\Omega}}(\mathcal{A}) = \mathcal{F}i_{\mathcal{L}}(\boldsymbol{A}/\widetilde{\boldsymbol{\Omega}}(\mathcal{A}))$. *The class of all full models of* $\mathcal{L}$ *is denoted by* $\mathsf{FGMod}\mathcal{L}$. *The set of all full models of* $\mathcal{L}$ *over a fixed algebra* $\boldsymbol{A}$, *which is denoted by* $\mathcal{FGM}od_{\mathcal{L}}\boldsymbol{A}$, *is ordered under the pointwise ordering of the corresponding closure operators (which is dual to the set-inclusion ordering of the corresponding closure systems).*

Among general PROPERTIES of this notion, let me highlight:

1. Full models are *full* in the sense that $\mathcal{A} = \langle \boldsymbol{A}, C\rangle$ is a full model of $\mathcal{L}$ if and only if $C = \{F \in \mathcal{F}i_{\mathcal{L}}\boldsymbol{A} : \widetilde{\boldsymbol{\Omega}}(\mathcal{A}) \text{ is compatible with } F\}$.

   Recall that a congruence θ is *compatible* with a subset F when $a\theta b$ implies $a \in F \iff b \in F$; or, equivalently, when $\theta \subseteq \boldsymbol{\Omega}_{\boldsymbol{A}}(F)$.
2. Full models are *full* in the sense that the class $\mathsf{FGMod}\mathcal{L}$ is the smallest class of models of $\mathcal{L}$ closed under the construction of images and inverse images by strict surjective homomorphisms that contains all models of the form $\langle \boldsymbol{A}, \mathcal{F}i_{\mathcal{L}}\boldsymbol{A}\rangle$ (the so-called *basic full models*).

3. The so-called **Isomorphism Theorem:**

   For any logic $\mathcal{L}$ *and any algebra* $\boldsymbol{A}$, *the Tarski operator* $\widetilde{\boldsymbol{\Omega}}$ *establishes an order isomorphism between the complete lattices* $\mathcal{FGM}od_{\mathcal{L}}\boldsymbol{A}$ *and* $\mathrm{Co}_{\mathsf{Alg}\mathcal{L}}\boldsymbol{A} = \{\theta \in \mathrm{Co}\boldsymbol{A} : \boldsymbol{A}/\theta \in \mathsf{Alg}\mathcal{L}\}$.

What is most remarkable in this theorem is its extreme generality: it holds for *any logic* and *any algebra* (of the relevant similarity type). It can be considered an extension of the isomorphism between filters and relative congruences for finitely algebraizable logics; this last isomorphism, established in 1989 by BLOK and PIGOZZI [3, Thm. 5.1], generalizes the well-known correspondence between filters and congruences in Boolean algebras or in Heyting algebras, and between open filters and congruences in topological or monadic Boolean algebras. Other particular instances of the isomorphism were known before long to algebraic logicians: in RASIOWA's [50] the isomorphism is not explicitly stated, but follows easily from the correspondences between filters and congruences there established, and in 1981 CZELAKOWSKI says, just before proving his Theorem II.2.10 of [17], that it "generalizes some observations made independently by several people". This isomorphism has been an inspiring source of ideas about the relationships between logical and algebraic properties, as well as a fundamental tool for showing non-algebraizability of certain logics.

4. The class $\mathsf{Alg}\mathcal{L}$ is also the class of algebraic reducts of the reduced full models of $\mathcal{L}$.
5. The algebras in $\mathsf{Alg}\mathcal{L}$ can be obtained by factoring a formula algebra of suitable cardinality by the Tarski congruence of some full model; more precisely,

   $$\mathsf{Alg}\mathcal{L} = \mathbb{I}\big\{\boldsymbol{Fm}_\kappa/\widetilde{\boldsymbol{\Omega}}(\boldsymbol{\mathcal{A}}) : \boldsymbol{\mathcal{A}} \text{ full model of } \mathcal{L} \text{ over } \boldsymbol{Fm}_\kappa\ ,\ \kappa \in \mathbb{C}\mathrm{ARD}\big\}$$

   where $\boldsymbol{Fm}_\kappa$ is the formula algebra over κ variables.

Concerning the relationships between the three classes of algebras considered so far there are several general facts and some particular ones:

6. In general $\mathsf{Alg}^*\mathcal{L} \subseteq \mathsf{Alg}\mathcal{L} \subseteq \mathsf{V}(\mathcal{L})$.
7. $\mathsf{Alg}\mathcal{L}$ is the class of subdirect products of algebras in $\mathsf{Alg}^*\mathcal{L}$. Hence the two classes generate the same quasivariety and the same variety.
8. The variety generated by $\mathsf{Alg}^*\mathcal{L}$ and by $\mathsf{Alg}\mathcal{L}$ is $\mathsf{V}(\mathcal{L})$.
9. If $\mathcal{L}$ is protoalgebraic then $\mathsf{Alg}^*\mathcal{L} = \mathsf{Alg}\mathcal{L}$.

Therefore, $\mathsf{Alg}\mathcal{L}$ coincides with the class of "$\mathcal{L}$-algebras" introduced by RASIOWA for the logics she studied in [50] (now commonly called *implicative logics*), and with the *equivalent quasivariety* of the finitely algebraizable logics introduced by BLOK and PIGOZZI in [3].

Two big classes of logics, independent of the just mentioned ones, for which the definition of $\mathcal{L}$-algebras gives a sound result are the following.

10. Let $\mathcal{L}$ be *selfextensional* (see Section 6.1) and have a Conjunction or satisfy the uniterm Deduction Theorem. Then $\mathbf{Alg}\mathcal{L} = \mathbf{V}(\mathcal{L})$.

For non-protoalgebraic logics it is $\mathbf{Alg}\mathcal{L}$ that gives the "right" or expected result, while $\mathbf{Alg}^*\mathcal{L}$ can be a strictly smaller class with no particular metalogical significance. This has been shown to be the case, for instance, in three of the most natural examples of non-protoalgebraic logics, represented in Table 1: The conjunction-disjunction fragment of classical logic $\mathcal{CPC}_{\wedge\vee}$ [40], Belnap's four valued logic (with truth and falsity constants) $\mathcal{B}$ [32],

$\mathcal{L}$	$\mathbf{Alg}\mathcal{L}$	$\mathcal{F}i_{\mathcal{L}}\boldsymbol{A}$ for $\boldsymbol{A} \in \mathbf{Alg}\mathcal{L}$
protoalgebraic		
$\mathcal{IPC}_{\to}$	Hilbert algebras	{implicative filters}
$\mathcal{CPC}$	Boolean algebras	{filters}
global $\mathcal{S}5$	monadic Boolean algebras	{open filters }
local $\mathcal{S}5$	monadic Boolean algebras	{filters}
$\L_{\infty}$	MV-algebras	{implicative filters}
$\mathcal{R}$	R-algebras	$\{F : F$ is a lattice filter and $a \to a \in F \;\; \forall a\}$
$\mathcal{BCK}$	BCK-algebras	{implicative filters}
non-protoalgebraic		
$\mathcal{CPC}_{\wedge\vee}$	distributive lattices	{lattice filters} $\cup \{\emptyset\}$
$\mathcal{B}$	De Morgan algebras	{lattice filters}
$\mathcal{IPC}^*$	pseudo-complemented distributive lattices	{lattice filters}
$\mathcal{WR}$	R-algebras	{lattice filters} $\cup \{\emptyset\}$

Table 1. Some logics, their $\mathcal{L}$-algebras and their reduced full models. See the text for notations and references.

or the implication-less fragment of intuitionistic logic $\mathcal{IPC}^*$ [3, 53]. However, property 9 is not a characterization of protoalgebraicity: there are non-protoalgebraic logics where the two classes coincide, like the inferential version $\mathcal{WR}$ of the well-known system $\mathcal{R}$ of relevance logic, suggested in [62, §2.10] and studied in [39]; this coincides with the semantic consequence one can associate with the ternary relational semantics. While in most of the non-protoalgebraic cases the class $\mathbf{Alg}\mathcal{L}$ has been found to be a variety and thus to coincide with $\mathbf{V}(\mathcal{L})$, for some subintuitionistic logics studied in [12] the situation is quite different; more precisely, the three classes of algebras associated with the global consequence of the class of transitive "intuitionistic-like" Kripke models are all different (this solved one of the open problems in [33]) and both $\mathbf{Alg}^*\mathcal{L}$ and $\mathbf{Alg}\mathcal{L}$ are not even quasivarieties.

It is easy to see that the ***reduced full models*** have a very precise form, namely they are exactly the generalized matrices $\langle \boldsymbol{A}, \mathcal{C}\rangle$ such that $\boldsymbol{A} \in \mathbf{Alg}\mathcal{L}$ and $\mathcal{C} = \mathcal{F}i_{\mathcal{L}}\boldsymbol{A}$. They are also characterized as the *reduced basic full models*, i.e., the generalized matrices of the form $\langle \boldsymbol{A}, \mathcal{F}i_{\mathcal{L}}\boldsymbol{A}\rangle$ that are reduced. Hence, $\boldsymbol{A} \in \mathbf{Alg}\mathcal{L}$ iff $\langle \boldsymbol{A}, \mathcal{F}i_{\mathcal{L}}\boldsymbol{A}\rangle$ is reduced. In examples, one sees that these generalized matrices have both a logical and an algebraic natural interpretation; the reader will not find any surprise in Table 1[3]. The empirical evidence it offers together with some of the previous results, seem to be good enough reasons for considering that reduced full models and $\mathbf{Alg}\mathcal{L}$ give a satisfactory account of the algebraization of (almost ?) any logic, regardless of whether it is protoalgebraic or not, and for calling $\mathbf{Alg}\mathcal{L}$ ***the class of Lindenbaum-Tarski algebras of*** $\mathcal{L}$.

Quite different and a more interesting problem is the characterization of *arbitrary full models*; they exist on any algebra, for instance $\langle \boldsymbol{A}, \mathcal{F}i_{\mathcal{L}}\boldsymbol{A}\rangle$ is always the finest one, but in general they are not so neatly characterized as the reduced ones. One way will be dealt with in the next section.

One kind of the intended characterizations is to describe the full models as certain families of $\mathcal{L}$-filters of some particular form (in contrast to general models, that can be obtained from arbitrary families of $\mathcal{L}$-filters closed under intersection). Experience shows that for most of the best-behaved logics full models are principal filters in the lattice of all $\mathcal{L}$-filters on each algebra. In the next two results from [33] we see that this form of full models can even characterize two main classes in the hierarchy:

---

[3]The two modal logics are the global and the local consequences associated with the class of Kripke models for the logic, $\mathcal{S}5$ in this case; they are sometimes called the normal and the quasi-normal logics. A similar situation arises for every normal "system" (in the old style sense) of modal logic

THEOREM 5. *A logic $\mathcal{L}$ is protoalgebraic if and only if every full model of $\mathcal{L}$ has the form $\langle \boldsymbol{A}, (\mathcal{F}i_{\mathcal{L}}\boldsymbol{A})^F \rangle$ for some $F \in \mathcal{F}i_{\mathcal{L}}\boldsymbol{A}$, where $(\mathcal{F}i_{\mathcal{L}}\boldsymbol{A})^F = \{G \in \mathcal{F}i_{\mathcal{L}}\boldsymbol{A} : F \subseteq G\}$ is the principal filter of $\mathcal{F}i_{\mathcal{L}}\boldsymbol{A}$ generated by F.*

However, not every $\mathcal{L}$-filter generates a full model in the above way; those that do have been studied in [34], where they are called *Leibniz filters* because they can be characterized as the smallest ones among all those having the same Leibniz congruence.

THEOREM 6. *A logic $\mathcal{L}$ is weakly algebraizable if and only if the full models of $\mathcal{L}$ are exactly all models of the form mentioned in the previous theorem.*

The class of weakly algebraizable logics has been studied in depth in [23]; it includes all algebraizable logics, but also others which are not algebraizable, related to certain subtractive varieties. For these logics the construction of Theorem 5 yields a lattice isomorphism between $\mathcal{FGM}od_{\mathcal{L}}\boldsymbol{A}$ and $\mathcal{F}i_{\mathcal{L}}\boldsymbol{A}$, for every $\boldsymbol{A}$. Thus to some extent working with full models amounts to working with filters. Moreover, from this isomorphism and the Isomorphism Theorem of item 3 above an alternative proof of BLOK and PIGOZZI's isomorphism is obtained.

4. Generalized matrices as models of Gentzen systems

This is a specially fruitful line of research concerning the problem of characterization of full models. Notice that generalized matrices, in their disguise as closure operators, satisfy all the so-called structural rules; therefore we will restrict our attention to *Gentzen systems having all structural rules.*

DEFINITION 7. *A generalized matrix $\langle \boldsymbol{A}, \mathrm{C} \rangle$ is a* ***model of a Gentzen-style rule***

$$\frac{\{\Gamma_i \vdash \varphi_i : i < n\}}{\Gamma_n \vdash \varphi_n} \tag{7}$$

when every interpretation h in $\boldsymbol{A}$ that satisfies all sequents in the antecedent (in the sense given after Proposition 2), also satisfies the sequent in the consequent.

$\mathcal{A}$ is a ***model of a Gentzen system*** *$\mathfrak{G}$ when it is a model of all its derivable rules. The class of all such models is denoted by* **Mod**$\mathfrak{G}$.

A rule like (7) is a ***rule of*** $\mathcal{L}$ *when the generalized matrix $\langle \boldsymbol{Fm}, \mathrm{Cn}_{\mathcal{L}} \rangle$ is a model of (7) in the above sense; that is, when for every substitution σ such that $\sigma\Gamma_i \vdash_{\mathcal{L}} \sigma\varphi_i$ for all $i < n$, also $\sigma\Gamma_n \vdash_{\mathcal{L}} \sigma\varphi_n$.*

The previous experience of the Barcelona group with particular logics and particular classes of algebras has shown that some logics characterize their full models in terms of properties that can be expressed by Gentzen-style rules (which are, of course, rules of the logic). To mention just the easiest example: The full models of $\mathcal{IPC}_{\rightarrow}$ are the generalized matrices that satisfy the Deduction Theorem. This has led naturally to the question: *When is the class of full models of a logic completely described by some set of Gentzen-style rules?* The associated technical notion (called "strong adequacy" in [33]) is the following.

DEFINITION 8. *A Gentzen system $\mathfrak{G}$ is **fully adequate** for a sentential logic $\mathcal{L}$ when* $\mathbf{FGMod}\mathcal{L} = \mathbf{Mod}\mathfrak{G}$*; in case $\mathcal{L}$ has no theorems, one has to assume that all generalized matrices considered here have no theorems, either.*

Among the consequences of $\mathfrak{G}$ being fully adequate for $\mathcal{L}$ let me mention:

1. The Gentzen system $\mathfrak{G}$ is ***adequate*** for the sentential logic $\mathcal{L}$, that is, the derivable sequents of $\mathfrak{G}$ define $\mathcal{L}$ as in (5).
2. $\mathcal{L}$ is the weakest logic satisfying the rules of $\mathfrak{G}$.
3. $\mathbf{Alg}\mathcal{L} = \mathbf{Alg}\mathfrak{G}$, the class of algebraic reducts of reduced models of $\mathfrak{G}$.
4. $\boldsymbol{A} \in \mathbf{Alg}\mathcal{L}$ if and only if $\langle \boldsymbol{A}, \mathcal{F}i_{\mathcal{L}}\boldsymbol{A}\rangle$ is a reduced model of $\mathfrak{G}$ (and it is the only one on $\boldsymbol{A}$).

Therefore we have a description of $\mathbf{Alg}\mathcal{L}$ in terms of Gentzen-style rules: Algebras in $\mathbf{Alg}\mathcal{L}$ are exactly those supporting a generalized matrix that is a model of the Gentzen system and is reduced. In case the Gentzen system has a nice presentation, these characterizations can be useful and significant. Examples can be found in Chapter 5 of [33].

Not every logic has a fully adequate Gentzen system, but if one exists then it is unique, so it seems to have some kind of distinctness among all those that are adequate for the logic; however, the deep significance of the notion of full adequacy has not been thoroughly investigated. The question of existence of a fully adequate Gentzen system for a sentential logic has generated other questions, and has opened several lines of research; in particular, it turns out that this problem is related to other, more classical problems of Abstract Algebraic Logic, as I am going to show.

One line of research is to study the full adequacy problem in relation with several *closure properties* of the class $\mathbf{FGMod}\mathcal{L}$ of all full models of $\mathcal{L}$ or of the lattice $\mathcal{FGM}od_{\mathcal{L}}\boldsymbol{A}$ of all full models of $\mathcal{L}$ on a fixed, arbitrary algebra $\boldsymbol{A}$. Some relevant results obtained so far are:

THEOREM 9 ([36]). *For any logic $\mathcal{L}$ the following properties are equivalent:*

(i) *$\mathcal{L}$ has a fully adequate Gentzen system.*

(ii) $\mathbf{FGMod}\mathcal{L}$ *is closed under sub-generalized matrices and under reduced products of generalized matrices.*

(iii) *For any algebra $\boldsymbol{A}$, the lattice $\mathcal{FGMod}_{\mathcal{L}}\boldsymbol{A}$ is closed under arbitrary intersections (hence it is a complete sublattice of the lattice of all generalized matrices over $\boldsymbol{A}$).*

This belongs to the kind of results called bridge theorems at the beginning, as the existence of a fully adequate Gentzen system can be regarded as a metalogical property of the logic while the other items are properties of models.

Another line of research is to relate full models with Deduction Theorems of several kinds.

THEOREM 10 ([36] and [37]). *Let $\mathcal{L}$ be a finitely algebraizable logic. Then:*

1. **FGMod**$\mathcal{L}$ *is closed under sub-generalized matrices if and only if $\mathcal{L}$ has the Local Deduction Theorem (LDT).*
2. *$\mathcal{L}$ has a fully adequate Gentzen system if and only if $\mathcal{L}$ has the Deduction Theorem.*

Besides their interest for the full adequacy problem, these results add to the extensive literature on the several forms of the Deduction Theorem. One of the key tools in this analysis has been the study of Leibniz filters and their relation with full models as described in Theorems 5 and 6. The other key approach is to exploit the presentation of generalized matrices as first-order structures that are the models of a strict universal Horn theory without equality, further extending BLOOM's idea [7], already generalized by BLOK and PIGOZZI in [5]. Briefly: given a generalized matrix $\langle \boldsymbol{A}, \mathrm{C}\rangle$ one considers the structure $\mathfrak{A} = \big\langle \boldsymbol{A}, \{R_m^{\mathfrak{A}} : m \geq 1\}\big\rangle$ with algebraic part $\boldsymbol{A}$ and where R_m is an m-ary relation symbol interpreted as

$$\langle a_1, \ldots, a_m\rangle \in R_m^{\mathfrak{A}} \iff a_m \in \mathrm{C}\big(a_1, \ldots, a_{m-1}\big)$$

for $m > 1$, and $R_1^{\mathfrak{A}} = \mathrm{C}(\emptyset)$. This allows to take advantage of general results on the *model theory of equality-free logic*, recently developed mainly by ELGUETA, DELLUNDE, CZELAKOWSKI and JANSANA. The papers [16, 27, 29, 30] contain a sample of such results and witness the mutual influence between Algebraic Logic and Model Theory.

The preceding results belong to the research line devoted to the theoretical analysis of full adequacy. One can also investigate it for particular logics or groups or logics. One key problem here is, in case we know that certain logic does have a fully adequate Gentzen system, to find a *calculus* or presentation for it with Gentzen-style rules that have a relevant metalogical significance. In Subsection 6.3 I mention two big classes of logics for which the full adequacy problem has a positive solution.

5. Transfer properties

Let's return to the beginning of the paper. There I mention *bridge theorems* (1) in connection with question (Q2). Now I am interested in a more specific class of typical answers to this question, which have been called *transfer theorems.* They are similar in structure to bridge theorems, but in transfer theorems the property $\boldsymbol{P'}$ of (1) is the same as property $\boldsymbol{P}$, and the right-hand side of the equivalence has a local character; thus, their typical form is

$$\boldsymbol{\mathcal{L}} \text{ has property } \boldsymbol{P} \quad \Longleftrightarrow \quad \text{Every object in } \mathsf{K} \text{ has property } \boldsymbol{P}. \tag{8}$$

Obviously, this can only make sense for properties $\boldsymbol{P}$ that can be equally predicated of the logic and of its models; accordingly, one has to shape models in an appropriate way or consider one of their presentations or aspects that can be successfully unified with those of the logic itself.

I think it is no exaggeration to say that *one can view an important number of recent works in Abstract Algebraic Logic as transfer theorems*; to be more precise, as works concerning transfer of properties of the lattice $\langle \mathcal{T}h\boldsymbol{\mathcal{L}}, \cap, \vee \rangle$ of the theories of $\boldsymbol{\mathcal{L}}$ to the lattices $\langle \mathcal{F}i_{\boldsymbol{\mathcal{L}}}\boldsymbol{A}, \cap, \vee \rangle$ of the $\boldsymbol{\mathcal{L}}$-filters on arbitrary algebras; or, equivalently, to the lattices $\langle \mathcal{C}, \cap, \vee \rangle$ of the closure system of arbitrary full models. Some transfer properties hold in general, while some hold only for restricted classes of logics; in some cases equivalent conditions have been found for certain transfers to hold.

To mention only one example: That the lattice of theories is distributive does not imply in general that all filter lattices, or all full models, are distributive; however, this is true inside the class of protoalgebraic logics, as proved in 1984 by CZELAKOWSKI [18][4]. Other celebrated results by CZELAKOWSKI, BLOK, PIGOZZI, HERRMANN and JANSANA are actually transfer theorems concerning *properties of the Leibniz operator*, such as monotonicity, definability, continuity, injectivity, commutativity with inverse endomorphisms, etc. Some of these properties transfer from $\boldsymbol{\Omega}_{\boldsymbol{Fm}}$ on $\mathcal{T}h\boldsymbol{\mathcal{L}}$ to $\boldsymbol{\Omega}_{\boldsymbol{A}}$ on $\mathcal{F}i_{\boldsymbol{\mathcal{L}}}\boldsymbol{A}$ in general, while some do only for protoalgebraic logics; moreover the logics that satisfy some of them constitute well-known classes in the hierarchy: protoalgebraic, equivalential, finitely equivalential, weakly algebraizable, algebraizable, etc. See [38] for more details and references.

Transfer theorems also arise in the context of generalized matrices: if one considers the logic itself as the generalized matrix $\langle \boldsymbol{Fm}, \mathcal{T}h\boldsymbol{\mathcal{L}} \rangle$ then it

[4] He stated and proved it for equivalential logics, but his proof actually works for all protoalgebraic logics.

makes sense to ask whether properties of $\mathcal{L}$ expressed by a Gentzen-style rule do transfer from $\langle \boldsymbol{Fm}, \mathcal{T}h\mathcal{L}\rangle$ to models like $\langle \boldsymbol{A}, \mathcal{F}i_{\mathcal{L}}\boldsymbol{A}\rangle$, or, equivalently, to full models $\langle \boldsymbol{A}, \mathcal{C}\rangle$. This has the supplementary interest that properties expressed by Gentzen-style rules usually have an intuitive and direct interpretation at a metalogical level. Well-known examples are having the Deduction Theorem, having a Disjunction, or the Introduction of Modality. CZELAKOWSKI explicitly proved in [19, 20] that Disjunction and the Deduction Theorem always transfer, that is, that a logic satisfies any of them *if and only if* all its full models satisfy it[5]. In the best-behaved cases (such as the finitely algebraizable logics), the whole full model structure is encoded in the algebraic structure, so that after being transferred to the models the property may become equivalent to a typically algebraic property (and then we have a typical bridge theorem). This is the case, for instance, of the Deduction Theorem, which transfers to the equational consequence relative to K, and there it becomes equivalent to the familiar EDPRC property, [6]. Some of the results establishing bridge theorems between several interpolation and amalgamation properties can be formulated as transfer theorems for carefully formulated forms of amalgamation for matrices; the general theory of these correspondences has been developed in the context of equivalential and algebraizable logics in [24].

I am not aware of much work on the transfer problem *in general*; let me mention a few results. Trivially, all properties expressible as a Hilbert-style rule transfer. Moreover, CZELAKOWSKI's proofs of the transference of the Deduction Theorem or Disjunction can be generalized to obtain the following.

THEOREM 11. *Let $\gamma(x, \vec{y}) \in Fm$ be a formula, and let $\{x_n : n \in \omega\}$ be a denumerable set of variables disjoint from those in $\vec{y}$. Call $\boldsymbol{P}(\gamma)$ the property represented by the following set of Gentzen-style rules.*

$$\left\{ \frac{x_0, \ldots, x_{n-1} \vdash x_n}{\gamma(x_0, \vec{y}), \ldots, \gamma(x_{n-1}, \vec{y}) \vdash \gamma(x_n, \vec{y})} \; : \; n \in \omega \right\} \tag{9}$$

Then the property $\boldsymbol{P}(\gamma)$ transfers (from the logic to all full models).

The case $\gamma(x, y) = y \to x$ essentially accounts for the transfer of the Deduction Theorem, and the case $\gamma(x, y) = x \vee y$ covers the case of Disjunction.

Another very general, specially remarkable fact is the following result by CZELAKOWSKI and PIGOZZI:

---

[5]He did not use the notion of full model, but his results amount to this one.

THEOREM 12 (see [22], § 1.7). *If $\mathcal{L}$ is a protoalgebraic logic, then any property expressed by a universal sentence of the first-order language of lattices transfers (that is, it holds in the lattice of theories if and only if it holds in the lattice of the closure system of every full model).*

We thus see that the transfer of distributivity mentioned before is not an isolated phenomenon, but a particular instance of a more general fact.

A few more transfer results and some problems concerning the congruence property will be mentioned in the next section.

6. Some open problems

This is a relatively new research area, so it is surely full of open problems and ideas for further work. For instance, in Section 4 I have already said that the notion of full adequacy has not been deeply investigated. Here, rather than formulating very specific *problems*, I am going to comment on a number of *topics* where there are several questions still to be scrutinized and deserving, in my opinion, special attention. Some of them are intimately connected, at least judging from the already known facts and the methods of proof hitherto used; but perhaps new ideas will make them differentiate significatively.

6.1. The transfer of congruence properties

This is a particular case of the general problem of transfer of any metalogical property that can be formulated in terms of generalized matrices.

Extending the notation introduced in Theorem 5, for a generalized matrix $\mathcal{A} = \langle \boldsymbol{A}, \mathcal{C} \rangle$ and an $F \in \mathcal{C}$ we put $\mathcal{C}^F = \{G \in \mathcal{C} : F \subseteq G\}$ and $\mathcal{A}^F = \langle \boldsymbol{A}, \mathcal{C}^F \rangle$. If $\mathcal{L}$ is a sentential logic and $T \in Th\mathcal{L}$ is a theory closed under substitutions then $\mathcal{L}^T$ is what is commonly called an axiomatic extension of $\mathcal{L}$; but here closure under substitutions is not required.

DEFINITION 13. *A generalized matrix $\mathcal{A}$ has* **the congruence property** *when $\boldsymbol{\Lambda}(\mathcal{A}) \in \mathrm{Co}\boldsymbol{A}$, that is, when $\widetilde{\boldsymbol{\Omega}}(\mathcal{A}) = \boldsymbol{\Lambda}(\mathcal{A})$; equivalently, when $\mathcal{A}$ is a model of the* **congruence rules***:*

$$\frac{\{x_i \vdash y_i\ ,\ y_i \vdash x_i : i < n\}}{\varpi x_0 \dots x_{n-1} \vdash \varpi y_0 \dots y_{n-1}} \tag{10}$$

for each basic operation symbol ϖ in the language (where n is its arity).

A sentential logic $\mathcal{L}$ is **selfextensional** *when as a generalized matrix it has the congruence property in the just defined sense. That is, when the* *interderivability relation* $\Lambda(\mathcal{L})$ *of the logic has the* **replacement property**.

A generalized matrix $\langle \boldsymbol{A}, C\rangle$ ***has the Fregean property*** *when for each* $F \in C$ *the generalized matrix* $\boldsymbol{A}^F$ *has the congruence property.*

A logic is ***Fregean*** *when as a generalized matrix it has the Fregean property, that is, when the interderivability relation modulo any theory of the logic has the replacement property.*

Selfextensional logics were first considered by WÓJCICKI (see [62, §5.6.5] for references); he has shown that they correspond to the local consequence defined by any class of *referential matrices*, a special kind of generalized matrices that abstract from several kinds of relational semantics. Fregean logics were first considered, in the present sense, in [31]; protoalgebraic Fregean logics were being studied independently at the same time by CZELAKOWSKI and PIGOZZI but their results have been published only recently in [22, 26]. Every Fregean logic is selfextensional; among Fregean logics we find classical and intuitionistic logic, their implicative or equivalential fragments, their axiomatic extensions, and all two-valued logics. Among non-Fregean but selfextensional logics there are the local consequences of normal modal logics, Belnap's logic, and the inferential system $\mathcal{WR}$ of relevance logic already mentioned in Section 3.

One can consider the transfer problem of these two properties: being selfextensional and being Fregean.

The interest in the transfer of the congruence property is not merely abstract. Having it makes life much easier; a generalized matrix is reduced and has the congruence property if and only if $\mathrm{C}(a) = \mathrm{C}(b)$ implies $a = b$; topologically this means being T_0 ($a \neq b \implies \exists F \in C$ such that $a \in F$ and $b \notin F$, or $a \notin F$ and $b \in F$). In generalized matrices with the congruence property the equations valid in its reduction are obtained by computing "modulo C" in their algebra reduct. If $\mathcal{L}$ is a selfextensional logic then an equation $\varphi \approx \psi$ holds in $\mathbf{V}(\mathcal{L})$ if and only if φ and ψ are interderivable modulo $\mathcal{L}$. Fregean logics seem to have rather strong properties and exhibit a nice algebraic behaviour; see next subsection.

The general transfer problems of the congruence property and of the Fregean property, stated in [33], have been open for some time, but were recently solved in the negative independently by BABYONISHEV [2] and by BOU [12]. Since the counterexamples they found are non-protoalgebraic logics, it is still interesting to investigate these transfer problems restricted to certain classes of logics, either to identify classes of logics where the transfer problem has a positive solution, or to find more counterexamples inside these classes. Some results have already been obtained:

THEOREM 14 ([26]). *If $\mathcal{L}$ is a protoalgebraic logic then the Fregean property transfers to all its full models.*

This is actually a consequence of a more general result of the same paper, which roughly speaking says that for protoalgebraic logics any property that can be formulated in terms of Gentzen-style rules that admit an arbitrary number of "side formulas" in the antecedent does transfer from the logic to all its full models; being Fregean is a property of this kind.

THEOREM 15 ([33]). *Let $\mathcal{L}$ be a logic satisfying the uniterm Deduction Theorem, or having a Conjunction. Then the congruence property transfers to all its full models.*

Selfextensional logics are much weaker than Fregean ones, and perhaps adding protoalgebraicity is not enough for the transfer. Actually, every logic with the Deduction Theorem is protoalgebraic, but the proof of the above result works only if the Deduction Theorem holds for a single term connective and cannot be extended to arbitrary protoalgebraic logics. But at present no counterexample is known even inside the much restricted class of finitely algebraizable logics. Hence the transfer problem of the congruence property is still open for the classes in the Leibniz hierarchy.

6.2. The two hierarchies

The main hierarchy that organizes the universe of logics as studied by Abstract Algebraic Logic, the so-called *Leibniz hierarchy*, can be described under several points of view, all concerning notions of the classical theory of matrices: the lattice-theoretic behaviour of the Leibniz operator on $\mathcal{L}$-filters, the definability of the $\mathcal{L}$-filters of the reduced matrices, the closure properties of the class of all reduced matrices of the logic, etc. In view of some already obtained results it seems that another, less complicated hierarchy of logics deserves some attention. It is based on the congruence property. Besides the classes of *selfextensional* and of *Fregean* logics it comprises the following two:

DEFINITION 16. *A logic is **fully selfextensional** when all its full models have the congruence property, and is **fully Fregean** when all its full models have the Fregean property.*

Thus the transfer problems of the congruence property can be paraphrased as the questions of whether every selfextensional logic is fully selfextensional, and of whether every Fregean logic is fully Fregean.

Here the (partly) open problems concern the clarification of the structure of this hierarchy and of its relationships with the Leibniz hierarchy. Every fully Fregean logic is both Fregean and fully selfextensional, but it is not known whether it equals the intersection of these two classes. The class of selfextensional logics includes the other classes, and most probably it does not equal their union (i.e., most probably there are selfextensional logics which are neither Fregean nor fully selfextensional). From the negative solutions to the transfer problems mentioned in Subsection 6.1 it follows that the four classes are different.

In principle it seems that selfextensionality is a property independent of the Leibniz hierarchy: There are selfextensional and non-selfextensional examples in all its subclasses, except for a few cases. The situation is very different for Fregean logics: due to Theorem 14 inside protoalgebraic logics (that is, inside the whole Leibniz hierarchy) the classes of Fregean and of fully Fregean logics coincide. And a few more things are known:

THEOREM 17.

1. *Every protoalgebraic Fregean logic with theorems is regularly algebraizable, that is, it is algebraizable and the designated set of its reduced matrices consists of a single element* [26, 31].
2. *A weakly algebraizable logic is Fregean if and only if it is fully selfextensional* [33].
3. *A fully selfextensional logic is algebraizable if and only if it is weakly algebraizable* [33].

Thus inside weakly algebraizable logics (hence, inside algebraizable logics) the only interesting remaining question is whether every selfextensional such logic is Fregean. This problem was already formulated in [31], and as far as I know, no counterexamples have been presented. In contrast with the strength that being Fregean seems to acquire when this property is coupled with algebraizability, as is clarified in [26, 49], selfextensionality seems much weaker; this makes the absence of counterexamples particularly noteworthy.

A related remaining task, which is interesting by itself but which can also shed some light over the mentioned open problems, is the classification in the two hierarchies of as many logics as possible. This forcefully includes the determination of the full models of these logics. The fact that the majority of logics (partially) classified up to now are either selfextensional, or algebraizable, or both, may just indicate that these logics are among the best known (and easier to study?) ones. The validity and usefulness of this kind of general theories and general classification schemes depends on their ability to deal not only with the well-behaved, central examples, but with the more marginal or even pathological ones[6]. I am convinced that the ex-

[6] For instance, those that are neither protoalgebraic nor selfextensional.

amination of a large number of logics coming from very different contexts will strengthen our understanding of these hierarchies and of the real significance of their subclasses; and as a bonus some of the problems might be solved with natural rather than *ad hoc* counterexamples. Of course, it can also turn out that some solution is affirmative, which would require a general proof; but even in this case the analysis of examples and of why they have certain property or why they do not have it, might help to shape a better view of the situation.

6.3. The variety problem

This problem arises in the theory of algebraizable logics, but has ramifications concerning generalized matrices. If $\mathcal{L}$ is a finitary, finitely algebraizable logic, then the general theory guarantees that its algebraic counterpart $\mathbf{Alg}\mathcal{L}$ (its largest equivalent algebraic semantics) is a quasivariety. However, experience shows that for a very large number of such algebraizable logics this class is indeed a variety (when this happens the logics are called ***strongly algebraizable***). The interesting questions is: *why is it so?* What do these logics have in common? Is there a characterization of such logics? Can we find necessary or sufficient conditions that explain this behaviour? More generally, one can also wonder why is it that $\mathbf{Alg}\mathcal{L}$ is a variety for many other logics, not necessarily algebraizable, not even protoalgebraic.

Some sufficient conditions have already been found in connection with the proof of Theorem 15:

THEOREM 18 ([33]). *Let $\mathcal{L}$ be a selfextensional logic satisfying the uniterm Deduction Theorem, or having a Conjunction. Then $\mathbf{Alg}\mathcal{L}$ is a variety, that is, $\mathbf{Alg}\mathcal{L} = \mathbf{V}(\mathcal{L})$.*

COROLLARY 19. *If $\mathcal{L}$ is a selfextensional and algebraizable logic satisfying the uniterm Deduction Theorem, or having a Conjunction, then $\mathcal{L}$ is strongly algebraizable.*

We have already seen (Theorem 17.1) that if the logic is not merely selfextensional but protoalgebraic and Fregean then it is also regularly algebraizable. Some deep results concerning this case are contained in [26]; all the logics analysed in this paper are shown to be definitionally equivalent to certain axiomatic expansions of the $(\to)$-fragment and of the $(\to, \wedge)$-fragment, respectively, of the intuitionistic propositional logic; see [38, § 4.4] for an overview. In [25] it is shown that the condition that the Deduction Theorem is effected by a single term is essential in the preceding results, by constructing an *ad hoc* example of a Fregean and algebraizable logic with a two-term Deduction Theorem that is not strongly algebraizable.

These are partial results that apply only to selfextensional or Fregean logics, but the phenomenon has been equally observed in non-selfextensional logics. Some of these cases might also be covered by the theory of the *strong version* of a protoalgebraic logic started in [34]: There it is shown that certain protoalgebraic logics that are not algebraizable engender a "strong version" which is at least weakly algebraizable, and has the same class of associated algebras. If in addition the weaker logic falls under one of the previous results, then this class is a variety and the strong version happens to be strongly algebraizable, without falling itself under the previous results. But again this theory is of limited application (it applies only to protoalgebraic logics), and moreover it only gives a sufficient condition. Thus there is plenty of room here for investigation.

Another interesting research would be to obtain more direct proofs of the preceding results, particularly of Theorem 18. The proof in [33] is obtained in a broader context, where it is also shown that the full adequacy problem has a positive solution in these cases:

THEOREM 20. *Let $\mathcal{L}$ be a selfextensional logic satisfying the uniterm Deduction Theorem, or having a Conjunction. Then $\mathcal{L}$ has a fully adequate Gentzen system.*

Actually the finding of such fully adequate Gentzen system is the key point in the proof that the associated class of algebras (which is also associated to the Gentzen system) is a variety. Remark that the statements of Theorems 15 and 18 concern generalized matrices understood as models of sentential logics, but not necessarily as models of Gentzen systems. While this fact emphasizes the usefulness of this double rôle of generalized matrices, it also calls for new proofs that can be developed entirely within the "sentential" level of the theory.

6.4. The notion of G-algebraizability

The just mentioned proof in [33] of Theorem 18 through the existence of a fully adequate Gentzen systems goes farther away from the classical theory of matrices and of generalized matrices. Actually, what one does to infer Theorem 18 from Theorem 20 is to consider the fully adequate Gentzen system under the light of a "third level" version of the notion of algebraizability. Such extension allows one to deal with substructural Gentzen systems, and was introduced by GIL, REBAGLIATO, TORRENS and VERDÚ [42, 43, 53, 54, 58]; a summary appears in Section 4.2 of [38]. In [33] it is shown that the obtained Gentzen system is finitely algebraizable in the sense of REBAGLIATO and VERDÚ, and that its largest equivalent algebraic semantics is precisely the class $\mathsf{V}(\mathcal{L})$; but on the other hand the general theory says this largest equivalent algebraic semantics has to be $\mathsf{Alg}\mathfrak{G}$, which by full adequacy equals $\mathsf{Alg}\mathcal{L}$.

While this situation is a nice example of how to combine work at several levels of the hierarchy of notions of sentential logic described in Section 1, it also suggests the consideration of a new notion of algebraizability for sentential logics:

DEFINITION 21. *A sentential logic is **G-algebraizable** when it has a fully adequate Gentzen system that is algebraizable in the sense of* [53, 54].

Note that this new concept is not an extension of the notion of algebraizable logic; there are logics which are algebraizable in the usual sense but do not have a fully adequate Gentzen system (this can be shown by Theorem 10) and hence are not G-algebraizable; and conversely, some of the known G-algebraizable logics are not algebraizable in the usual sense (actually they are not even protoalgebraic). Examples of the first kind are $\mathcal{BCK}$, $\mathcal{R}$ or $\mathcal{L}_\infty$, and examples of the second kind are $\mathcal{CPC}_{\wedge\vee}$ and $\mathcal{B}$; see Table 1 for notations and references.

In G-algebraizable logics the link with algebraic properties of their associated class of algebras is effected by properties expressible as Gentzen-style rules, while in classically algebraizable logics the link is effected by the equational consequence relative to the class of algebras and two substitution-invariant translations. The in-depth study of the new notion and its interaction with the two hierarchies of Subsection 6.2 will surely be a rich area of research.

6.5. Characterizing the class $\mathsf{Alg}\mathcal{L}$

The most common perspective in Algebraic Logic is to view it from the logical side; i.e., one has a logic and is interested in knowing more things about it through the study of the algebraic properties of a certain class of algebras that bears to the logic some specific relation. Up to now the closest kind of relation is that of being the equivalent algebraic semantics for the logic, in case this one is algebraizable.

But there is a dual perspective, which is to start from a class of algebras and to look for a certain logic that bears to it the said relationship; this could be called *the problem of the logification* of a class of algebras. The theory of algebraizability has found one solution to it: For instance, one can specify necessary and sufficient conditions for a quasivariety to be the equivalent algebraic semantics of a finitary, finitely algebraizable logic, see [22, § 4.6]. An even finer work can be done if the focus is restricted to protoalgebraic Fregean logics: the corresponding algebras form the so-called congruence-orderable relatively point-regular quasivarieties; see [38, § 4.4] for more details and references. Sometimes, also the isomorphism theorem for algebraizable logics can be used to show that a certain class of algebras can never be the equivalent algebraic semantics of any algebraizable logic, as is done in [32] for the variety of De Morgan algebras and in [40] for the variety of distributive lattices.

I have already said that the notion of G-algebraizability just discussed can account for the algebraic character of logics not falling under the algebraizable paradigm. Conversely, it can also give rise to a broader notion of "algebra of logic". What has not been investigated yet, is the dual perspective on this new concept: Is there a characterization of the classes of algebras that can be shown to be the algebraic counterpart of a G-algebraizable logic? My conjecture is that this should be relatively easy if one restricts the search to fully selfextensional logics.

Finally, one can also consider the very general problem: Is there a characterization of those classes of algebras that are of the form $\mathbf{Alg}\mathcal{L}$ for some sentential logic $\mathcal{L}$ of the most arbitrary kind? It is probably too difficult, given its extremely generality, so one can also restrict the search to classes of logics with a good behaviour in one sense or another (the results mentioned two paragraphs above correspond to the restrictions to algebraizable and to protoalgebraic and Fregean logics).

Appendix

Let $\mathcal{L}$ be a sentential logic. It satisfies the ***Deduction Theorem*** when there is a finite set $\Delta(p,q) \subseteq Fm$ such that for all $\Gamma \cup \{\varphi, \psi\} \subseteq Fm$, $\Gamma, \varphi \vdash_{\mathcal{L}} \psi \iff \big(\Gamma \vdash_{\mathcal{L}} \delta(\varphi,\psi)$ for all $\delta(p,q) \in \Delta(p,q)\big)$. If the set Δ contains just one formula, we speak of a *uniterm Deduction Theorem*, in contrast to the *multiterm* case, where the set Δ may contain more than one formula. In the literature this property is also called the *Deduction-Detachment Theorem (DDT)*.

$\mathcal{L}$ has a ***Conjunction*** when there is a binary term (i.e., a formula in two variables) $p \wedge q$ such that $\{\varphi, \psi\} \dashv\vdash_{\mathcal{L}} \varphi \wedge \psi$ for all $\phi, \psi \in Fm$.

$\mathcal{L}$ has a ***Disjunction*** when there is a binary term (i.e., a formula in two variables) $p \vee q$ such that $\Gamma, \varphi \vee \psi \vdash_{\mathcal{L}} \xi \iff (\Gamma, \varphi \vdash_{\mathcal{L}} \xi$ and $\Gamma, \psi \vdash_{\mathcal{L}} \xi)$ for all $\Gamma \cup \{\varphi, \psi\} \subseteq Fm$.

$\mathcal{L}$ satisfies the ***Introduction of Modality*** for a unary connective or term $\Box p$ when $\Gamma \vdash_{\mathcal{L}} \varphi$ implies $\Box\Gamma \vdash_{\mathcal{L}} \Box\varphi$, for all $\Gamma \cup \{\varphi\} \subseteq Fm$.

Acknowledgements

The author has received partial support from the Spanish grant BFM2001-3329 and from the Catalonian grant 2001SGR-00017.

References

[1] ADILLON, R., AND VERDÚ, V. On a contraction-less intuitionistic propositional logic with conjunction and fusion. *Studia Logica, Special Issue on Abstract Algebraic Logic 65*, 1 (2000), 11–30.

[2] BABYONISHEV, S. Strongly Fregean logics. *Reports on Mathematical Logic*. To appear.

[3] BLOK, W., AND PIGOZZI, D. *Algebraizable logics*, vol. 396 of *Mem. Amer. Math. Soc.* A.M.S., Providence, January 1989.

[4] BLOK, W., AND PIGOZZI, D. Local deduction theorems in algebraic logic. In *Algebraic Logic*, H. Andréka, J. D. Monk, and I. Németi, Eds., vol. 54 of *Colloq. Math. Soc. János Bolyai.* North-Holland, Amsterdam, 1991, pp. 75–109.

[5] BLOK, W., AND PIGOZZI, D. Algebraic semantics for universal Horn logic without equality. In *Universal Algebra and Quasigroup Theory*, A. Romanowska and J. D. H. Smith, Eds. Heldermann, Berlin, 1992, pp. 1–56.

[6] BLOK, W., AND PIGOZZI, D. Abstract algebraic logic and the deduction theorem. *Bulletin of Symbolic Logic*. To appear.

[7] BLOOM, S. L. Some theorems on structural consequence operations. *Studia Logica 34* (1975), 1–9.

[8] BLOOM, S. L. Projective and inductive generation of abstract logics. *Studia Logica 35* (1976), 249–255.

[9] BLOOM, S. L. A note on Ψ-consequences. *Reports on Mathematical Logic 8* (1977), 3–9.

[10] BLOOM, S. L., AND BROWN, D. J. Classical abstract logics. *Dissertationes Math. (Rozprawy Mat.) 102* (1973), 43–51.

[11] BLOOM, S. L., BROWN, D. J., AND SUSZKO, R. Some theorems on abstract logics. *Algebra i Logika 9*, 3 (1970), 274–280.

[12] BOU, F. *Implicación estricta y lógicas subintuicionistas.* Master thesis, University of Barcelona, 2001.

[13] BROWN, D. *Abstract logics.* Ph. D. Dissertation, Stevens Institute of Technology, 1969.

[14] BROWN, D. J., AND SUSZKO, R. Abstract logics. *Dissertationes Math. (Rozprawy Mat.) 102* (1973), 9–42.

[15] BULL, R. A. The algebraic foundations of logic. *Reports on Mathematical Logic 6* (1976), 7–28.

[16] CASANOVAS, E., DELLUNDE, P., AND JANSANA, R. On elementary equivalence for equality-free logic. *Notre Dame Journal of Formal Logic 37*, 3 (1996), 506–522.

[17] CZELAKOWSKI, J. Equivalential logics, I, II. *Studia Logica 40* (1981), 227–236 and 355–372.

[18] CZELAKOWSKI, J. Filter distributive logics. *Studia Logica 43* (1984), 353–377.

[19] CZELAKOWSKI, J. Remarks on finitely based logics. In *Models and sets*, G. H. Müller and M. M. Richter, Eds., vol. 1103 of *Lecture Notes in Mathematics*. Springer Verlag, Berlin, 1984, pp. 147–168.

[20] CZELAKOWSKI, J. Algebraic aspects of deduction theorems. *Studia Logica 44* (1985), 369–387.

[21] CZELAKOWSKI, J. Local deductions theorems. *Studia Logica 45* (1986), 377–391.

[22] CZELAKOWSKI, J. *Protoalgebraic Logics*, vol. 10 of *Trends in Logic, Studia Logica Library*. Kluwer Academic Publishers, Dordrecht, 2001.

[23] CZELAKOWSKI, J., AND JANSANA, R. Weakly algebraizable logics. *The Journal of Symbolic Logic 65*, 2 (2000), 641–668.

[24] CZELAKOWSKI, J., AND PIGOZZI, D. Amalgamation and interpolation in abstract algebraic logic. In *Models, algebras and proofs*, X. Caicedo and C. H. Montenegro, Eds., no. 203 in Lecture Notes in Pure and Applied Mathematics Series. Marcel Dekker, New York and Basel, 1998, pp. 187–265.

[25] CZELAKOWSKI, J., AND PIGOZZI, D. Fregean logics with the multiterm deduction theorem and their algebraization. Preprint 433, Centre de Recerca Matemàtica, Bellaterra (Barcelona), February 2000.

[26] CZELAKOWSKI, J., AND PIGOZZI, D. Fregean logics. *Annals of Pure and Applied Logic*. To appear.

[27] DELLUNDE, P., AND JANSANA, R. Some characterization theorems for infinitary universal Horn logic without equality. *The Journal of Symbolic Logic 61*, 4 (1996), 1242–1260.

[28] DUNN, J. M., AND HARDEGREE, G. M. *Algebraic methods in philosophical logic*, vol. 41 of *Oxford Logic Guides*. Oxford Science Publications, Oxford, 2001.

[29] ELGUETA, R. *Algebraic model theory for languages without equality*. Ph. D. Dissertation, University of Barcelona, 1994.

[30] ELGUETA, R. Characterizing classes defined without equality. *Studia Logica 58*, 3 (1997), 357–394.

[31] FONT, J. M. On the Leibniz congruences. In *Algebraic Methods in Logic and in Computer Science*, C. Rauszer, Ed., vol. 28 of *Banach Center Publications*. Polish Academy of Sciences, Warszawa, 1993, pp. 17–36.

[32] FONT, J. M. Belnap's four-valued logic and De Morgan lattices. *Logic Journal of the I.G.P.L. 5*, 3 (1997), 413–440.

[33] FONT, J. M., AND JANSANA, R. *A general algebraic semantics for sentential logics*, vol. 7 of *Lecture Notes in Logic*. Springer-Verlag, 1996. 135 pp. Presently distributed by the Association for Symbolic Logic.

[34] FONT, J. M., AND JANSANA, R. Leibniz filters and the strong version of a protoalgebraic logic. *Archive for Mathematical Logic 40* (2001), 437–465.

[35] FONT, J. M., JANSANA, R., AND PIGOZZI, D., Eds. *Workshop on abstract algebraic logic*, vol. 10 of *Quaderns*. Centre de Recerca Matemàtica, Bellaterra (Spain), 1998. 199 pp.

[36] FONT, J. M., JANSANA, R., AND PIGOZZI, D. Fully adequate Gentzen systems and closure properties of the class of full models. Manuscript, 1999.

[37] FONT, J. M., JANSANA, R., AND PIGOZZI, D. Fully adequate Gentzen systems and the deduction theorem. *Reports on Mathematical Logic 35* (2001), 115–165.

[38] FONT, J. M., JANSANA, R., AND PIGOZZI, D. A survey of abstract algebraic logic. *Studia Logica, Special Issue on Abstract Algebraic Logic, Part II 74*, 1/2 (2003). 13–97.

[39] FONT, J. M., AND RODRÍGUEZ, G. Algebraic study of two deductive systems of relevance logic. *Notre Dame Journal of Formal Logic 35*, 3 (1994), 369–397.

[40] FONT, J. M., AND VERDÚ, V. Algebraic logic for classical conjunction and disjunction. *Studia Logica, Special Issue on Algebraic Logic 50* (1991), 391–419.

[41] GIL, A. J., AND REBAGLIATO, J. Protoalgebraic Gentzen systems and the cut rule. *Studia Logica, Special Issue on Abstract Algebraic Logic 65*, 1 (2000), 53–89.

[42] GIL, A. J., REBAGLIATO, J., AND VERDÚ, V. A strong completeness theorem for the Gentzen systems associated with finite algebras. *Journal of Applied Non-Classical Logics 9*, 1 (1999), 9–36.

[43] GIL, A. J., TORRENS, A., AND VERDÚ, V. On Gentzen systems associated with the finite linear MV-algebras. *Journal of Logic and Computation 7*, 4 (1997), 473–500.

[44] HARROP, R. Some structure results for propositional calculi. *The Journal of Symbolic Logic 30* (1965), 271–292.

[45] HARROP, R. Some forms of models of propositional calculi. In *Contributions to Mathematical Logic*, H. Schmidt, K. Schütte, and H. Thiele, Eds. North-Holland, 1968, pp. 163–174.

[46] ŁOŚ, J., AND SUSZKO, R. Remarks on sentential logics. *Indagationes Mathematicae 20* (1958), 177–183.

[47] ŁUKASIEWICZ, J., AND TARSKI, A. Untersuchungen über den Aussagenkalkül. *Comptes Rendus des Séances de la Société des Sciences et des Lettres de Varsovie, Cl. III 23* (1930), 30–50.

[48] MAGARI, R. Calcoli generali (I,II,III). *Le Matematiche (Catania) 21* (1966), 83–108, 135–149, 237–281.

[49] PIGOZZI, D. Fregean algebraic logic. In *Algebraic Logic*, H. Andréka, J. D. Monk, and I. Németi, Eds., vol. 54 of *Colloq. Math. Soc. János Bolyai*. North-Holland, Amsterdam, 1991, pp. 473–502.

[50] RASIOWA, H. *An algebraic approach to non-classical logics*, vol. 78 of *Studies in Logic and the Foundations of Mathematics*. North-Holland, Amsterdam, 1974.

[51] RAUTENBERG, W. Axiomatizing logics closely related to varieties. *Studia Logica, Special Issue on Algebraic Logic 50* (1991), 607–622.

[52] RAUTENBERG, W. On reduced matrices. *Studia Logica 52* (1993), 63–72.

[53] REBAGLIATO, J., AND VERDÚ, V. On the algebraization of some Gentzen systems. *Fundamenta Informaticae, Special Issue on Algebraic Logic and its Applications 18* (1993), 319–338.

[54] REBAGLIATO, J., AND VERDÚ, V. Algebraizable Gentzen systems and the deduction theorem for Gentzen systems. Mathematics Preprint Series 175, University of Barcelona, June 1995.

[55] SHOESMITH, D., AND SMILEY, T. *Multiple-conclusion logic*. Cambridge University Press, Cambridge, 1978.

[56] SMILEY, T. J. The independence of connectives. *The Journal of Symbolic Logic 27* (1962), 426–436.

[57] TARSKI, A. Über einige fundamentale Begriffe der Metamathematik. *Comptes Rendus des Séances de la Société des Sciences et des Lettres de Varsovie, Cl. III 23* (1930), 22–29.

[58] TORRENS, A. Model theory for sequential deductive systems. Manuscript, 1991.

[59] WÓJCICKI, R. Logical matrices strongly adequate for structural sentential calculi. *Bulletin de l'Académie Polonaise des Sciences, Classe III XVII* (1969), 333–335.

[60] WÓJCICKI, R. Some remarks on the consequence operation in sentential logics. *Fundamenta Mathematicae 68* (1970), 269–279.

[61] WÓJCICKI, R. Matrix approach in the methodology of sentential calculi. *Studia Logica 32* (1973), 7–37.

[62] WÓJCICKI, R. *Theory of logical calculi. Basic theory of consequence operations*, vol. 199 of *Synthese Library*. Reidel, Dordrecht, 1988.

JOSEP MARIA FONT
Faculty of Mathematics, University of Barcelona
Gran Via 585
E-08007 Barcelona, Spain

MELVIN FITTING

Intensional Logic — Beyond First Order

Abstract. Classical first-order logic can be extended in two different ways to serve as a foundation for mathematics: introduce higher orders, type theory, or introduce sets. As it happens, both approaches have natural analogs for quantified modal logics, both approaches date from the 1960's, one is not very well-known, and the other is well-known as something else. I will present the basic semantic ideas of both higher order intensional logic, and intensional set theory. Before doing so, I'll quickly sketch some necessary background material from quantified modal logic. Except for standard material concerning propositional modal logics, the paper is essentially self-contained.

Keywords: intensional logic, modal logic, set theory, type theory, higher order logic, forcing

1. Introduction

A little way into the twentieth century, both Bertrand Russell and Ernst Zermelo discovered the now well-known paradox about the set of all sets not having themselves as a member (or of the predicate that can be predicated of those predicates not predicatable of themselves). Each came up with a solution, and while there are similarities, the differences are striking. Russell, [Rus08], created type theory, which separates predicates into distinct categories—types—each predicate falling into exactly one category. The types themselves have a rather labyrinthine structure, but at least that structure is a well-founded one. That is, if one starts with a predicate of arbitrary type, moves to a predicate of the type of one of its arguments, then to a predicate of the type of an argument of that, and so on, one must eventually 'bottom out.' Zermelo, taking a different route, introduced the cumulative hierarchy of sets, with the universe of sets divided into levels indexed by ordinals, [Zer08, Zer35]. It is more complicated than type theory, in a way, because the levels extend through the entire class of ordinals. On the other hand, it is simpler in that it is cumulative: a set present at any level is also present at all higher ones. This means that, unlike in Russell's theory, we do not find a distinct copy of the natural numbers at each type level—one copy serves for all. Still, as in type theory, a set in Zermelo's hier-

V. F. Hendricks and *J. Malinowski* (eds.),
Trends in Logic: 50 Years of Studia Logica

Trends in Logic **21**: 87–108, 2003.

archy can only have as members things of lower levels, and this (apparently) serves to avoid paradox.

Both type theory and the cumulative hierarchy of sets were introduced to provide a safe foundation for mathematics, and both apparently do that. But since their introduction, the two have gone their separate ways. The cumulative hierarchy of sets allows mathematics to be developed in a simpler and, most feel, more natural way, so type theory has largely been ignored by mathematicians. On the other hand, the constructs of language often seem to have a type structure to them. We say of a person that he is good, or tall (or not). We say of goodness that it is a virtue while tallness is not. We do not say a person is a virtue or that virtue is tall. It is not simply that saying these would be to utter falsehoods—we do not say them because the subject/predicate combinations are inappropriate—saying them results in an ill-typed sentence. Since type theory seems to have reasonable relationships to natural language, it should not come as a surprise to find that it has found important applications in computer science, in the design of programming languages. Imposing a type structure has been found to lead to languages in which certain kinds of programming errors are much harder to make—they become grammatical errors which are easily found by a compiler. Even a language like Lisp, which is untyped, is often used with an implicit typing in mind by the programmer.

The history of modal logic has often paralleled that of classical logic, but with a time delay of a generation or two. Propositional modal logics are now standard tools. First-order modal logics have a remarkably rich structure that is currently being explored by researchers. But what about the analogs of the work of Russell and Zermelo? As it happens, both have modal—intensional—versions, and both are in the literature, though their appearance followed the classical versions by perhaps half a century. Even so, they are not well-known. Indeed, the intensional version of Zermelo's cumulative set hierarchy is in the peculiar position of being well-known, but as something else. It is my aim, in this paper, to sketch the basic ideas. What we will be looking at should be standard tools, and I hope to move them a little in that direction.

Since both intensional type theory and intensional set theory are complex organisms, I think it best to lead up to them gradually, so that not all the essential ideas turn up in a bunch. Consequently I begin with brief sketches of simpler structures involving quantification. These are certainly more familiar to readers. If they are too familiar, I apologize in advance.

2. First-Order Background

I assume everyone reading this paper is thoroughly familiar with propositional modal logics, but there are some points concerning first-order versions that may be less well-known, so a little background discussion of these issues is appropriate. I am only interested in semantics here—axiom systems or other proof procedures are ignored in this discussion. Also, there are two broadly differing approaches to quantified modal semantics, one using counterpart relations, the other not. I follow the tradition, dating from Kripke, that does not use counterpart relations—when one talks of Niagara Falls under other possible circumstances, one does not mean a counterpart of Niagara Falls, one means Niagara Falls. However, for those wishing to read further on the counterpart version, I recommend [Lew68, Lew71] for the early work, and [Kra93, KK01, KK03, Kut02], and the references given in these papers, for recent work. Also, some discussion of relationships between the two approaches can be found in [Fit01].

To keep things relatively simple, throughout this paper I'll assume languages have no constant or function symbols. Formulas are built up from atomic formulas involving relation symbols and variables, using propositional connectives, quantifiers, and modal operators. I'll take some of these as primitive and the others as defined, as convenient.

2.1. Objects

Perhaps the simplest first-order modal assumption in the literature is that quantifiers range over objects that are the same from world to world, and that such objects exist at all worlds. This is the so-called *constant domain* semantics. Let's begin here.

DEFINITION 2.1. The structure $\mathcal{M} = \langle \mathcal{G}, \mathcal{R}, \mathcal{D}, \mathcal{I} \rangle$ is a *constant domain* first-order model if $\mathcal{G}$ is a non-empty set of possible worlds, $\mathcal{R}$ is an accessibility relation on $\mathcal{G}$, $\mathcal{D}$ is a non-empty domain, and $\mathcal{I}$ is an interpretation, assigning to each n-place relation symbol a mapping from possible worlds to n-place relations on $\mathcal{D}$. A *valuation* in $\mathcal{M}$ is a mapping from variables to $\mathcal{D}$. Valuations are not world-dependent, unlike interpretations.

$\mathcal{M}, \Gamma \Vdash_v X$ is intended to symbolize that formula X is true at world Γ of model $\mathcal{M}$ with respect to valuation v. It is characterized as follows.

1. For atomic formulas, $\mathcal{M}, \Gamma \Vdash_v P(x_1, \ldots, x_n)$ if $\langle v(x_1), \ldots, v(x_n) \rangle \in \mathcal{I}(P)(\Gamma)$
2. $\mathcal{M}, \Gamma \Vdash_v X \wedge Y$ iff $\mathcal{M}, \Gamma \Vdash_v X$ and $\mathcal{M}, \Gamma \Vdash_v Y$

3. $\mathcal{M}, \Gamma \Vdash_v \neg X$ iff not-$\mathcal{M}, \Gamma \Vdash_v X$

4. $\mathcal{M}, \Gamma \Vdash_v (\forall x)\Phi$ iff $\mathcal{M}, \Gamma \Vdash_w \Phi$ for every valuation w that differs from v on at most x

5. $\mathcal{M}, \Gamma \Vdash_v \Box X$ iff $\mathcal{M}, \Delta \Vdash_v X$ for every $\Delta \in \mathcal{G}$ such that $\Gamma \mathcal{R} \Delta$

This is a well-known modal semantics—perhaps the best known of those for quantified modal logic—so I need say little more about it here. The next most complicated after the constant domain semantics is the varying domain version, which still assumes quantification is over objects. Since the semantics is similar to that of Definition 2.1, I will only say wherein the differences lie.

DEFINITION 2.2. $\langle \mathcal{G}, \mathcal{R}, \mathcal{D}, \mathcal{I} \rangle$ is a *varying domain* first order model if it meets the conditions of Definition 2.1, except that $\mathcal{D}$ is a *mapping* that assigns to each member of $\mathcal{G}$ some non-empty set (the domain at that possible world), and $\mathcal{I}$ assigns to each n-place relation symbol a mapping from possible worlds to n-place relations on $\cup_{\Gamma \in \mathcal{G}} \mathcal{D}(\Gamma)$.

The definition of truth at a world is also as in Definition 2.1, except for the following.

4. $\mathcal{M}, \Gamma \Vdash_v (\forall x)\Phi$ iff $\mathcal{M}, \Gamma \Vdash_w \Phi$ for every valuation w that differs from v on at most x, *and is such that* $w(x) \in \mathcal{D}(\Gamma)$

With varying domain models, quantification at a world is over the domain associated with that world, and not over a common world-independent domain. One can think of the domain associated with a world as the set of things *actually* existing at that world, and so the semantics is sometimes called *actualist*. In a similar way, constant domain semantics is sometimes called *possibilist*—one thinks of the common domain as what does, or what might exist, where a possible existent is an actual existent in some possible world. Understanding the differences between constant and varying domain models, and their connections with the Barcan formula and its converse, was an early achievement of first-order modal logic.

As it happens, the possibilist semantics, which is the simpler one technically, can serve very well for both possibilist and actualist purposes, a point discussed in both [HC96] and in [FM98]. Suppose we have a varying domain model, $\mathcal{M} = \langle \mathcal{G}, \mathcal{R}, \mathcal{D}, \mathcal{I} \rangle$. Using it, define a new constant domain $\mathcal{D}' = \cup_{\Gamma \in \mathcal{G}} \mathcal{D}(\Gamma)$. Also, introduce a new one-place relation symbol E to play the role of an *existence* predicate, and let $\mathcal{I}'$ be like $\mathcal{I}$ except that it is extended to E by setting $\mathcal{I}'(\mathsf{E})$ to be the mapping sending each possible world

Γ to $\mathcal{D}(\Gamma)$. Now we have a constant domain model, $\mathcal{M}' = \langle \mathcal{G}, \mathcal{R}, \mathcal{D}', \mathcal{I}' \rangle$. Suppose we translate each formula X not containing E to a formula X^{E} by relativizing each quantifier: replace $(\forall x)\Phi$ by $(\forall x)[\mathsf{E}(x) \supset \Phi]$ and $(\exists x)\Phi$ by $(\exists x)[\mathsf{E}(x) \wedge \Phi]$. It is not hard to show that $\mathcal{M}, \Gamma \Vdash_v X$ iff $\mathcal{M}', \Gamma \Vdash_v X^{\mathsf{E}}$. In other words, varying domain semantics can be embedded into constant domain semantics provided we give up the notion that existence must be captured by the existential quantifier.

From now on in this paper, I will only consider constant domain models. As just noted, this is no real restriction, though it certainly simplifies things. But this still leaves the appropriateness of the assumption that quantification is over objects, and this point requires a section to itself.

2.2. Intensions

The objects that quantifiers range over in the previous section are like the Eiffel tower, or George Washington, the same thing from possible world to possible world. But *intensions* are also quite reasonable to work with—the gross domestic product of Denmark, or the Secretary-General of the United Nations, say—varying in designation from circumstance to circumstance. It is clear how to model such things semantically: use functions from possible worlds to objects. But complexities force themselves on us. For instance, if we say the gross domestic product of Denmark currently is $\$171.6 \times 10^9$ (true at the time of writing), we are relating something that is non-rigid to something that is rigid. The gross domestic product of Denmark varies with time—it is non-rigid—while the number 171.6×10^9 is time-independent—it is rigid. The need to talk about rigid and non-rigid items together means we need a more complex notion of atomic formula.

For the rest of this section I'll assume that each relation symbol has a *type* associated with it, where a type is an n-tuple whose entries are either I or O. Informally, O is for *object*, the kind of thing we considered in the previous section, while I stands for *intension*, which we generally intend to be something non-rigid. For example, a type of $\langle I, O \rangle$ would be appropriate for saying the gross domestic product of Denmark is $\$171.6 \times 10^9$ We also need two kinds of variables, call them intensional, or type I, and extensional, or type O. If R is a relation symbol of type $\langle T_1, \ldots, T_n \rangle$, where each T_i is either I or O, then for $R(x_1, \ldots, x_n)$ to be an atomic formula, x_i must be an intensional variable if T_i is I, and extensional if T_i is O. All this is rather straightforward.

If f is an intension, sometimes we want to talk about f itself, and sometimes about what f designates in a particular context. For instance, if we say "the gross domestic product of Denmark is a useful economic measure," we refer to the intension, gross-domestic-product-of-Denmark. But if we say "the gross domestic product of Denmark is currently greater than the gross domestic product of Finland," we refer to the value that gross-domestic-product-of-Denmark designates at the present moment, $\$171.6 \times 10^9$. It is useful to have notation to distinguish these sorts of things, and we use an *extension of* operator, $\downarrow$, for the purpose. If x is an intensional variable, $\downarrow x$ is extensional, while $\downarrow$ is not applicable to extensional variables. The idea is, the value associated with $\downarrow x$ in a context is to be whatever extensional object is designated by the intension associated with x in that context. It is allowed that $\downarrow x$ can fill an O position in an atomic formula, if x can fill an I one. I'll call $\downarrow x$ *relativized*, since its semantical behavior will depend on context. This semantic behavior will be specified below, and it will be clear that it corresponds to the informal ideas just discussed.

As it happens, on combining the machinery discussed so far, we find ambiguities. If we say the Secretary-General of the United Nations cannot be younger than 65, in a sense this is true, because at the time of writing this paper the Secretary-General is Kofi Annan, and he is already 65. But we can also say the Secretary-General of the United Nations might be younger than 65, and this too is plausible because the Secretary-Generalship probably will be held by a younger person someday. Suppose we informally think of P as a relation symbol of type $\langle O \rangle$ intended to say, of its argument, that its age is less than 65. Now, our problem is that if x is an intensional variable whose intended value is the intension Secretary-General-of-the-United-Nations, our first informal assertion above would be formalized as $\neg\Diamond P(\downarrow x)$ while the second becomes $\Diamond P(\downarrow x)$. Since both are plausible, we clearly have a problem. Of course, the problem arises because the action of letting x designate, that is, of evaluating $\downarrow x$, and the action of passing to an alternative possible world, that is, of interpreting $\Diamond$, are not actions that commute.

To disambiguate this, one more piece of machinery is needed. It will play a much bigger role later on when the full type hierarchy is introduced, but as we have just seen, it is needed now too.

DEFINITION 2.3. If $x_1, \ldots, x_n$ is a sequence of variables, of respective types $T_1, \ldots, T_n$, and Φ is a formula, then $\langle \lambda x_1, \ldots, x_n.\Phi \rangle$ is a *predicate abstract*, of type $\langle T_1, \ldots, T_n \rangle$. It can be used like a relation symbol in forming atomic formulas, that is, if $t_1, \ldots, t_n$ are of types $T_1, \ldots, T_n$ respectively, then $\langle \lambda x_1, \ldots, x_n.\Phi \rangle(t_1, \ldots, t_n)$ is an atomic formula.

With this broadening of syntax, the problematic formula $\Diamond P(\downarrow x)$ splits into $\langle \lambda y.\Diamond P(y) \rangle(\downarrow x)$ and $\Diamond \langle \lambda y.P(y) \rangle(\downarrow x)$, where x is an intensional variable whose value is intended to be the Secretary-General-of-the-United-Nations intension. Of course we need a semantics that gives these formulas different behaviors. The following is taken from [Fit01], where all this was introduced under the name **FOIL**, for *first-order intensional logic.* A very full discussion of the essential ideas involved can be found in [FM98]. Note that, following up on earlier discussion, domains of quantification are constant, not varying.

DEFINITION 2.4. A **FOIL** model is a structure $\mathcal{M} = \langle \mathcal{G}, \mathcal{R}, \mathcal{D}_O, \mathcal{D}_I, \mathcal{I} \rangle$ where: $\mathcal{G}$ and $\mathcal{R}$ are as usual, $\mathcal{D}_O$ is a non-empty set, the *object domain*, $\mathcal{D}_I$ is a non-empty set of functions from $\mathcal{G}$ to $\mathcal{D}_O$, the *intension domain*, and $\mathcal{I}$ is an *interpretation* such that, if P is a relation symbol of type $\langle T_1, \ldots, T_n \rangle$ then $\mathcal{I}(P)$ is a mapping from $\mathcal{G}$ to subsets of $\mathcal{D}_{T_1} \times \cdots \times \mathcal{D}_{T_n}$.

A *valuation* in $\mathcal{M}$ is a map that assigns to each type O variable a member of $\mathcal{D}_O$ and to each type I variable a member of $\mathcal{D}_I$. If v is a valuation, let $\hat{v}$ be a mapping on variables *and relativized variables* to functions on worlds, given by: if x is unrelativized, $\hat{v}(x)$ is the constant map assigning to each world Γ the entity $v(x)$, and if $\downarrow x$ is relativized, $\hat{v}(\downarrow x)$ is the map assigning to world Γ the object $v(x)(\Gamma)$.

Truth at a world is defined more or less as expected. There are two kinds of quantifiers, since there are two kinds of variables—if an object variable is involved, quantification is over $\mathcal{D}_O$, and if an intension variable is involved, quantification is over $\mathcal{D}_I$. The main point is that condition 1 of Definition 2.1 is replaced with the following.

1. For atomic formulas,

   (a) if $P(x_1, \ldots, x_n)$ is an atomic formula (so $x_1, \ldots, x_n$ are extensional variables, intensional variables, and relativized intensional variables), $\mathcal{M}, \Gamma \Vdash_v P(x_1, \ldots, x_n)$ iff $\langle \hat{v}(x_1)(\Gamma), \ldots, \hat{v}(x_n)(\Gamma) \rangle \in \mathcal{I}(P)(\Gamma)$

   (b) $\mathcal{M}, \Gamma \Vdash_v \langle \lambda x_1, \ldots, x_n.\Phi \rangle (t_1, \ldots, t_n)$ iff $\mathcal{M}, \Gamma \Vdash_w \Phi$, where w is like v except that $w(x_i) = \hat{v}(t_i)(\Gamma)$, for $i = 1, \ldots, n$.

In **FOIL**, the use of predicate abstraction is essentially as a scope indicator. In $\langle \lambda y.\Diamond P(y) \rangle (\downarrow x)$ the x is being used with broad scope while in $\Diamond \langle \lambda y.P(y) \rangle (\downarrow x)$ it is used with narrow scope. These notions essentially trace back to Russell's work on definite descriptions, [Rus05], but are also significant for intensional issues. The logic **FOIL** is a flexible and natural logic, by design with both extensional and intensional features, but here it is simply presented as a lead-in to what follows, essentially to introduce some notation and basic ideas.

3. Classical type theory

Russell intended to base the foundations of mathematics purely on classical logic. The number 3, for instance, was to be the property that applied to just those properties that applied to three items. But this required that properties be applicable to other properties, and this circumstance eventually lead to paradox. After a variety of other attempts, Russell's eventual way of dealing with paradox was to introduce type theory. The first version, [Rus08], and in *Principia Mathematica*, [WR27], involved *ramified* types, but eventually this evolved into what has become known as the *simple theory of types.* This has been given elegant formulations in more than one way—here I'll quickly sketch one version, as further lead-in to the modal system presented in the next section. The basic idea is to divide the universe of discourse into separate categories, types, with a predicate of one type applicable only to items of certain other designated types. But unlike the baby notion of types from the previous section, now we need predicates, predicates of predicates, predicates of these, and so on. The type structure is intended to avoid any hazardous circularity of reference, thus avoiding paradoxes, while still providing a rich enough structure within which to carry out the development of mathematics.

DEFINITION 3.1. 0 is a type. If $T_1, \ldots, T_n$ are types, $\langle T_1, \ldots, T_n \rangle$ is a type.

The idea is, 0 is intended to be the type of ground-level objects, and $\langle T_1, \ldots, T_n \rangle$ is intended to be the type of a predicate that takes n arguments, of types T_1, ..., T_n respectively.

For each type T I'll assume there are infinitely many variables of that type. The type of a variable will be clear from context, or else I'll simply say in words what the type is. Formulas come next, but their definition is more complex then in a first-order setting.

DEFINITION 3.2. Formulas, terms, and predicate abstracts are characterized (simultaneously) as follows.

1. Let Φ be a formula and x_1, ..., x_n be a sequence of distinct variables of types T_1, ..., T_n respectively. $\langle \lambda x_1, \ldots, x_n.\Phi \rangle$ is a *predicate abstract.* Its type is $\langle T_1, \ldots, T_n \rangle$.

2. *Terms* are either predicate abstracts or variables. (Recall, I'm not considering constant or function symbols in this paper.)

3. If t is a term of type $\langle T_1, \ldots, T_n \rangle$, and $t_1, \ldots, t_n$ is a sequence of terms of types $T_1, \ldots, T_n$ respectively, then $t(t_1, \ldots, t_n)$ is an atomic formula, hence also a formula.

4. If Φ is a formula, so is $\neg\Phi$.

5. If Φ and Ψ are formulas so is $(\Phi \wedge \Psi)$.

6. If Φ is a formula and x is a variable then $(\forall x)\Phi$ is a formula.

Now that we have formulas, the next thing is to have a standard version of a model. In first-order logic there are relation symbols, and so an interpretation function is needed to give them semantic meanings. In the present setup there are no relation symbols as such—there are only variables, and predicate abstracts of non-ground types—so an interpretation function is not needed.

DEFINITION 3.3. Let $\mathcal{D}$ be a non-empty set. For each type T a collection $[\![T, \mathcal{D}]\!]$ is defined as follows.

1. $[\![0, \mathcal{D}]\!] = \mathcal{D}$.

2. $[\![\langle T_1, \ldots, T_n \rangle, \mathcal{D}]\!]$ is the collection of all subsets of $[\![T_1, \mathcal{D}]\!] \times \cdots \times [\![T_n, \mathcal{D}]\!]$.

A member of $[\![T, \mathcal{D}]\!]$ is called an *object of type T over $\mathcal{D}$*. A *valuation* in $\mathcal{D}$ is a mapping v that assigns to each variable x of type T some object of type T over $\mathcal{D}$.

Now we can simultaneously define the notion of truth of a formula over a domain $\mathcal{D}$, and the denotation of a term, all with respect to a valuation. Just as in the previous section, I'll use $\hat{v}$ for the extension of v to more complex things than just variables—in this case, to predicate abstracts. Think of $\hat{v}$ as a mapping from terms to denotations.

DEFINITION 3.4. Let $\mathcal{D}$ be a non-empty set, and let v be a valuation in it.

1. If x is a variable of type T, $\hat{v}(x) = v(x)$, and so $\hat{v}(x) \in [\![T, \mathcal{D}]\!]$.

2. If $\langle \lambda x_1, \ldots, x_n.\Phi \rangle$ is a predicate abstract of type T then $\hat{v}(\langle \lambda x_1, \ldots, x_n.\Phi \rangle)$ is the following member of $[\![T, \mathcal{D}]\!]$:

$$\{\langle w(x_1), \ldots, w(x_n) \rangle \mid \text{valuation } w \text{ agrees with } v \text{ on all variables except possibly } x_1, \ldots, x_n \text{ and } \Phi \text{ is true over } \mathcal{D} \text{ with respect to } w\}$$

3. An atomic formula $t(t_1, \ldots, t_n)$ is true over $\mathcal{D}$ with respect to v, if $\langle \hat{v}(t_1), \ldots, \hat{v}(t_n) \rangle \in \hat{v}(t)$.

4. $\neg\Phi$ is true over $\mathcal{D}$ with respect to v if Φ is not true over $\mathcal{D}$ with respect to v.

5. $\Phi \wedge \Psi$ is true over $\mathcal{D}$ with respect to v if both Φ and Ψ are true over $\mathcal{D}$ with respect to v.

6. $(\forall x)\Phi$ is true over $\mathcal{D}$ with respect to v if Φ is true over $\mathcal{D}$ with respect to every valuation w that agrees with v on all variables except possibly for x.

This is the natural and obvious semantics for the classical typed language that was introduced above. It is well known that validity with respect to this semantics does not correspond to any formal proof procedure. One needs a more general semantics—Henkin models—but these details are more than I want to get into here. Suffice it to say that the result is, in fact, suitable for Russell's purpose. Unfortunately, the result is not appealing to mathematicians. For instance, if T is a type, we can characterize the predicates of type $\langle T \rangle$ that apply to exactly three things of type T, say we call them holding-of-three-things predicates. Then we can introduce a predicate of type $\langle\langle T \rangle\rangle$ that applies to exactly the holding-of-three-things predicates of type $\langle T \rangle$, and we can identify this type $\langle\langle T \rangle\rangle$ predicate with the number 3, but note that it is the number 3 appropriate for type T objects. We actually get multiple copies of the number 3, one for each type T. This kind of thing is unfortunate, and doomed type theory as a mathematical tool, though as I noted earlier, it has found applications in linguistics and computer science.

This is the last of the sections intended to lead up to two specific intensional systems, about to be presented. I will say no more about the type system of this section, since it is subsumed in the system of the next section.

4. Higher-Order Modal Logic

While types present difficulties for mathematical applications, they are natural in other areas, in particular, in the formal treatment of natural language. Since modal constructs come up often in natural language expressions—temporal constructs especially—a combination of typed logic and Kripke structures is a natural thing to consider. In fact, precisely this combination has a venerable history—[Mon60, Mon68, Mon70, Gal75]. Here I present a variation on that work, amounting to a combination of type machinery from Section 3 with intensional machinery from Section 2.2. After the formal machinery has been introduced, I'll look at a few examples.

4.1. The Formal Machinery

The starting point is to extend the definition of type—think of the types of Definition 3.1 as extensional; we now add intensional ones.

DEFINITION 4.1. Extensional and intensional types are characterized as follows.

1. 0 is an *extensional type.*

2. If $T_1, \ldots, T_n$ are types, extensional or intensional, $\langle T_1, \ldots, T_n \rangle$ is an *extensional type.*

3. If T is an extensional type, $\uparrow T$ is an *intensional type.*

Suppose, for example, that people are objects of type T. Then *Russian-population* would designate a property of type $\langle T \rangle$, with the intended value being the set of those people who are Russian citizens, and *Russian* would designate a property of type $\uparrow\langle T \rangle$, the intensional object associating different sets of people with different time points.

A higher-type modal language must be defined. I'll assume there is a distinct set of variables corresponding to each type, and variables are classified as intensional or extensional, according to their types. An extension-of operator on terms, $\downarrow$, is needed, just as in **FOIL**. A modal operator $\Box$ is added (with $\Diamond$ defined). Definition 3.2, for higher-type classical terms, formulas, and abstracts, must be modified to take modal operators and intensional objects into account. I only give the additions and changes to the earlier definition, and do not repeat what is not changed.

DEFINITION 4.2. For formula, term, and predicate abstract, modifying Definition 3.2.

1. If Φ is a formula and $x_1, \ldots, x_n$ is a sequence of distinct variables of types $T_1, \ldots, T_n$ respectively, then $\langle \lambda x_1, \ldots, x_n.\Phi \rangle$ is a term of the intensional type $\uparrow\langle T_1, \ldots, T_n \rangle$.

2. Predicate abstracts and variables are terms. If t is a term of intensional type $\uparrow T$ then $\downarrow t$ is a term of type T.

3. If t is a term of either type $\langle T_1, \ldots, T_n \rangle$ or type $\uparrow\langle T_1, \ldots, T_n \rangle$, and t_1, $\ldots$, t_n is a sequence of terms of types T_1, $\ldots$, T_n respectively, then $t(t_1, \ldots, t_n)$ is an atomic formula.

4. Formulas are built up using propositional connectives and quantifiers exactly as in the classical case.

5. If Φ is a formula, so is $\Box\Phi$.

Case 2 uses the extension-of operator, converting a term of intensional type to a term of extensional type. For instance, if t is the intensional *Russian* example, $\downarrow t$ at a particular time instant should be the set of people who are Russian citizens at that moment. The symbols $\uparrow$ and $\downarrow$ were chosen to suggest that one 'cancels' the other. But note, $\downarrow$ is a symbol of the formal language, while $\uparrow$ is in the metalanguage—part of the typing mechanism.

Next, semantics. We need a modal analog of higher-type domains—Definition 3.3.

DEFINITION 4.3. Let $\mathcal{G}$ be a non-empty set (of possible worlds) and let $\mathcal{D}$ be a non-empty set (the ground-level domain). For each type T, a collection $[\![T, \mathcal{D}, \mathcal{G}]\!]$ is defined as follows.

1. $[\![0, \mathcal{D}, \mathcal{G}]\!] = \mathcal{D}$.

2. $[\![\langle T_1, \ldots, T_n \rangle, \mathcal{D}, \mathcal{G}]\!]$ is the collection of all subsets of $[\![T_1, \mathcal{D}, \mathcal{G}]\!] \times \cdots \times [\![T_n, \mathcal{D}, \mathcal{G}]\!]$.

3. $[\![\uparrow T, \mathcal{D}, \mathcal{G}]\!]$ is the set of all functions from $\mathcal{G}$ to $[\![T, \mathcal{D}, \mathcal{G}]\!]$.

A member of $[\![T, \mathcal{D}, \mathcal{G}]\!]$ is an *object of type T over D and G*. An object is intensional or extensional or according to whether its type is intensional or extensional.

A *valuation* in $\mathcal{D}$ and $\mathcal{G}$ is a mapping v that assigns to each variable of type T some object of type T over $\mathcal{D}$ and $\mathcal{G}$.

Now, finally, simultaneous characterizations of truth for formulas, and designation for terms.

DEFINITION 4.4. Let $\mathcal{M} = \langle \mathcal{G}, \mathcal{R}, \mathcal{D} \rangle$ be a structure, with $\mathcal{G}$ non-empty (the set of possible worlds), $\mathcal{D}$ non-empty (the ground-level domain), and $\mathcal{R}$ a binary relation on $\mathcal{G}$ (accessibility). For each valuation v in $\mathcal{D}$ and $\mathcal{G}$, an associated mapping $\hat{v}$ on terms is defined, and simultaneously the truth-status of a formula with respect to worlds and valuations is defined.

1. If x is a variable then $\hat{v}(x)$ is the constant function mapping each world to $v(x)$.

2. If t is a term of type $\uparrow T$ then $\hat{v}(\downarrow t)$ is the function mapping each world Γ to $v(t)(\Gamma)$.

3. Suppose $\langle \lambda x_1, \ldots, x_n.\Phi\rangle$ is a predicate abstract of type $\uparrow\langle T_1, \ldots, T_n\rangle$. Let f be the function that assigns to each world Γ the set

$$\{\langle w(x_1), \ldots, w(x_n)\rangle \mid w \text{ agrees with } v \text{ on all variables except possibly } x_1, \ldots, x_n, \text{ and } \mathcal{M}, \Gamma \Vdash_w \Phi\}.$$

Then, $\hat{v}(\langle \lambda x_1, \ldots, x_n.\Phi\rangle)$ is the constant function that maps each world to f.

4. For an atomic formula $t(t_1, \ldots, t_n)$,

   (a) If t is of extensional type, $\mathcal{M}, \Gamma \Vdash_v t(t, \ldots, t_n)$ provided $\langle \hat{v}(t_1)(\Gamma), \ldots, \hat{v}(t_n)(\Gamma)\rangle \in \hat{v}(t)(\Gamma)$.

   (b) If t is of intensional type, $\mathcal{M}, \Gamma \Vdash_v t(t_1, \ldots, t_n)$ provided $\mathcal{M}, \Gamma \Vdash_v (\downarrow t)(t_1, \ldots, t_n)$ (Reducing things to the previous case).

5. $\mathcal{M}, \Gamma \Vdash_v \neg\Phi$ if it is not the case that $\mathcal{M}, \Gamma \Vdash_v \Phi$.

6. $\mathcal{M}, \Gamma \Vdash_v \Phi \wedge \Psi$ if $\mathcal{M}, \Gamma \Vdash_v \Phi$ and $\mathcal{M}, \Gamma \Vdash_v \Psi$.

7. $\mathcal{M}, \Gamma \Vdash_v (\forall x)\Phi$ if $\mathcal{M}, \Gamma \Vdash_w \Phi$ for every valuation w that agrees with v on all variables except possibly x.

8. $\mathcal{M}, \Gamma \Vdash_v \Box\Phi$ if $\mathcal{M}, \Delta \Vdash_v \Phi$ for all $\Delta \in \mathcal{G}$ such that $\Gamma\mathcal{R}\Delta$.

This completes specification of the formal syntactic and semantic machinery. Unfortunately it is not enough. Just as with the classical version presented in Section 3, the logic defined by these semantics cannot have a complete proof system. And just as classically, the solution is to introduce 'non-standard' models, analogs of the classical Henkin models. This is not the place to get into details, which are somewhat complex. I refer you to [Fit02], where a full development can be found. The system given here, in turn, is a mild elaboration of an intensional higher-order logic of Montague and Gallin, [Mon60, Mon68, Mon70, Gal75]. The elaboration consists of introducing separate intensional and extensional types. In the Montague/Gallin system all types are intensional, though extensional objects can be introduced as intensions that do not depend on circumstances—rigid intensions, in other words.

4.2. Examples

With respect to a structure $\mathcal{M} = \langle \mathcal{G}, \mathcal{R}, \mathcal{D} \rangle$, there is an obvious distinction between *non-rigidity* and *rigidity*. A member, f, of an intensional type may take on different values at different worlds, in which case we say it is non-rigid. On the other hand, f might well be a constant function, in which case we say it is rigid. In general, the distinction between non-rigidity and rigidity is more than can be captured within our formal language, but a kind of localized version of it can be. Let us say f is *rigid at* Γ if the value that f has at Γ is the same as the value it has at any world accessible from Γ. This is something that can be said within our modal language, and I'll give two ways of doing so.

Let us suppose that, in addition to the machinery outlined in the previous section, we also have an equality predicate for each type, that is, there is a predicate $=$ of type $\langle T, T \rangle$, for each type T. (I know this violates what was said earlier about there being no constant symbols in the language, but please allow this extension, just this once.) I'll assume that $=$ is always interpreted as the equality relation, of appropriate type. The details are straightforward. Now, suppose $f \in [\![\uparrow T, \mathcal{D}, \mathcal{G}]\!]$, and v is a valuation mapping x to f. The following obviously says that f is locally rigid at Γ (in it, y is of type T).

$$\mathcal{M}, \Gamma \Vdash_v \langle \lambda y.\Box(y = \downarrow x) \rangle (\downarrow x)$$

There is another notion, essentially introduced by Gödel in the course of his ontological argument, which I have been calling *stability*. This time suppose g is not just intensional, but is in $[\![\uparrow \langle T \rangle, \mathcal{D}, \mathcal{R}]\!]$. If v maps z to g, the following is what it means to say g is *stable* at Γ (in it, w is of type T).

$$\mathcal{M}, \Gamma \Vdash_v (\forall w)[z(w) \supset \Box z(w)] \wedge (\forall w)[\Diamond z(w) \supset z(w)]$$

One can show that, for g in $[\![\uparrow \langle T \rangle, \mathcal{D}, \mathcal{R}]\!]$, stability at Γ and rigidity at Γ are equivalent. And this can be carried further still, to relate these notions to the vanishing of a *de re*, *de dicto* distinction.

In [Fit02], the higher-order intensional system sketched here was applied to analyze ontological arguments, Gödel's in particular. Subtle distinctions, such as whether Gödel intended intensional or extensional terms at various points in his argument, lead to seriously different results. If you are interested in seeing further examples of the present logic in use, I suggest looking at this material.

5. The Cumulative Set Hierarchy

At about the same time that Russell was developing type theory, Zermelo was introducing axiomatic set theory [Zer08, Zer35]. Behind the scenes, and serving as motivation for Zermelo, was a structuring of sets that has since been made explicit and has come to be known as the *cumulative hierarchy of sets.* This has long been a standard part of set theory courses. I'll just sketch the ideas here. Assume the underlying set theory is Zermelo-Fraenkel (Z-F).

One associates with each ordinal α a set, sometimes denoted V_α, sometimes R_α, as follows.

$$\begin{aligned} V_0 &= \emptyset \\ V_{\alpha+1} &= \text{power set of } V_\alpha \\ V_\lambda &= \cup_{\alpha<\lambda} V_\alpha \text{ for limit ordinals } \lambda \end{aligned}$$

This is not a separation into disjoint categories in the way that type theory is. It is cumulative, in the sense that $\alpha \leq \beta$ implies $V_\alpha \subseteq V_\beta$. Still, if x is in some V_α, there is a smallest ordinal β such that $x \in V_{\beta+1}$. In this case, x is said to be of *rank* β, and dividing the universe of sets up into groupings by rank amounts to sorting it out into things very much like types. If x is of rank β, the ranks of all members of x must be smaller ordinals. Rather nicely, the V_α themselves can be explicitly 'talked about,' since for each α, V_α is a member of $V_{\alpha+1}$; in fact, V_α itself has rank α. This allows the 'internalization' of things.

One can also define a limit—a proper class that is often denoted V_Ω—by setting $x \in V_\Omega$ if $x \in V_\alpha$ for some ordinal α. The members of V_Ω are called the *regular* or *well-founded* sets. It can be shown that if the Z-F axioms are consistent, they remain so when an axiom is added asserting that every set is regular. This additional axiom is quite commonly assumed, in fact. That is, it is standard to assume the universe of sets has a type-like structure, but one that is cumulative, which simplifies the construction of mathematical structures.

Whether mathematicians are explicit about their set-theoretic assumptions or not, it is a fact that essentially all of mathematics can be formalized within the framework sketched above. Indeed, standard mathematical objects tend to come along at 'low' levels in the cumulative hierarchy: the set of natural numbers (thought of as finite von Neumann ordinals) is in $V_{\omega+1}$, for instance. The cumulative hierarchy, and not type theory, has become the standard foundational structure for mathematics.

6. An Intensional Cumulative Set Hierarchy

A modal analog of the cumulative set hierarchy has been around for a long time, but has not generally been recognized to be that. Set-theoretic *forcing* was invented by Paul Cohen in the early 1960's, [Coh63, Coh64, Coh66], and gradually modified to its present form by a number of people, including Dana Scott and Robert Solovay, [SS71], and Joseph Shoenfield, [Sho71]. Although forcing was designed to be a powerful tool for set theorists, it can also be seen as providing us semantically with an intensional set theory. The presentation of forcing in [SF96] is in this direction, but the emphasis there is on the set-theoretic results, and not on the connections with intensional logic. Here I will say a bit more about these connections.

6.1. The General Case

The higher-order intensional system of types presented in Section 4 has both extensional and intensional types. This is convenient for the application for which it was designed, [Fit02]. But as noted earlier, it is also possible to formulate it with only intensional types—identifying extensional objects with rigid intensions. While one can take either route with sets, it is standard to adopt an intension-only approach, and that is what will be done here.

DEFINITION 6.1. Let $\mathcal{G}$ be a non-empty set, intended to be possible worlds. To each ordinal is associated a set, as follows.

$$\begin{aligned} V_0^{\mathcal{G}} &= \emptyset \\ V_{\alpha+1}^{\mathcal{G}} &= \text{ power set of } \mathcal{G} \times V_\alpha^{\mathcal{G}} \\ V_\lambda^{\mathcal{G}} &= \cup_{\alpha<\lambda} V_\alpha^{\mathcal{G}} \text{ for limit ordinals } \lambda \\ V_\Omega^{\mathcal{G}} &= \cup_\alpha V_\alpha^{\mathcal{G}} \text{ where the union is over all ordinals} \end{aligned}$$

Just as in the classical setting of Section 5, this gives us a cumulative hierarchy: $\alpha \leq \beta$ implies $V_\alpha^{\mathcal{G}} \subseteq V_\beta^{\mathcal{G}}$, with a proper class, $V_\Omega^{\mathcal{G}}$ as limit. As before, we can define the rank of a member of $V_\Omega^{\mathcal{G}}$ to be the smallest ordinal α such that $f \in V_{\alpha+1}^{\mathcal{G}}$.

The idea behind the construction above is this: think of $f \in V_\Omega$ as encoding non-rigid membership information—if $\langle \Gamma, g \rangle \in f$, we should think of this as telling us that f has g as a member at world Γ. More formally, to $f \in V_{\alpha+1}^{\mathcal{G}}$ we associate a mapping, denoted f'', that assigns a subset of $V_\alpha^{\mathcal{G}}$ to each member of $\mathcal{G}$, by setting $f''(\Gamma) = \{g \in V_\alpha^{\mathcal{G}} \mid \langle \Gamma, g \rangle \in f\}$. Thus for each 'set' f, the associated mapping f'' is an *extension of f* function.

Assume we have a modal language with only ϵ as a binary relation symbol, written in infix position, and no constant or function symbols. (I use ϵ to distinguish intensional set membership from $\in$, the 'real' set membership relation.) There is only a single kind of quantifier. I'll set up modal models whose domain is the proper class $V_\Omega^{\mathcal{G}}$.

DEFINITION 6.2. Let $\mathcal{G}$ be a non-empty set, and $\mathcal{R}$ be a binary relation on $\mathcal{G}$, giving us a modal frame $\mathcal{M} = \langle \mathcal{G}, \mathcal{R} \rangle$. A *(set) valuation* in $\mathcal{M}$ is a mapping v from variables to members of $V_\Omega^{\mathcal{G}}$. Truth over $\mathcal{M}$ with respect to valuation v is defined as follows.

1. For atomic formulas, $\mathcal{M}, \Gamma \Vdash_v (x \epsilon y)$ iff $v(x) \in [v(y)]''(\Gamma)$.
2. The non-atomic cases are treated as usual in constant domain modal models, Definition 2.1, with quantifiers ranging over $V_\Omega^{\mathcal{G}}$.

One can obviously have $(x \epsilon y)$ true at one world but false at another, with respect to a valuation. That is, these are *intensional* sets. Among these intensional sets there will be some that are rigid. Indeed, we can identify those that are hereditarily rigid with the classical sets of V_Ω. Suppose we define a mapping from V_Ω to $V_\Omega^{\mathcal{G}}$ (making use of the well-foundedness of V_Ω) as follows. For each $d \in V_\Omega$:

$$\hat{d} = \mathcal{G} \times \{\hat{e} \mid e \in d\}$$

It is not hard to see that for each ordinal α, $d \in V_\alpha$ implies $\hat{d} \in V_\alpha^{\mathcal{G}}$. Also, $\hat{d}$ will be rigid, as will its members, members of members, and so on. And obviously, for $d, e \in V_\Omega$, we have $e \in d$ if and only if $\mathcal{M}, \Gamma \Vdash_v (x \epsilon y)$ when $v(x) = \hat{e}$ and $v(x) = \hat{d}$, independently of Γ.

Still, things can be rather badly behaved. One of the basic principles of classical set theory is that sets are completely determined by their membership—the so-called *axiom of extensionality.* It should not be a great surprise to find that this fails badly for intensional sets. Here is an example. It is standard in Z-F to identify numbers with certain sets—von Neumann ordinals—so that $0 = \emptyset$, $1 = \{0\}$, $2 = \{0, 1\}$, and so on. Then there are rigid members $\hat{0}$, $\hat{1}$, and $\hat{2}$ in $V_\Omega^{\mathcal{G}}$, for every choice of $\mathcal{G}$. (In fact, these particular ones are all in $V_3^{\mathcal{G}}$.) Now, say $\mathcal{G}$ consists of just two worlds, Γ and Δ, and $\mathcal{M} = \langle \mathcal{G}, \mathcal{R} \rangle$, where the specification of $\mathcal{R}$ is not relevant. Consider the non-rigid set $f \in V_3^{\mathcal{G}}$ such that $f = \{\langle \Gamma, \hat{0} \rangle, \langle \Gamma, \hat{1} \rangle, \langle \Delta, \hat{1} \rangle\}$. Suppose v is a valuation such that $v(x) = \hat{2}$, $v(y) = f$, and $v(z) = \hat{3}$. Then we have $\mathcal{M}, \Gamma \Vdash_v (\forall w)[(w \epsilon x) \equiv (w \epsilon y)]$, that is, at Γ, $\hat{2}$ and f have the same members. But $\mathcal{M}, \Gamma \Vdash_v (x \epsilon z)$ while $\mathcal{M}, \Gamma \not\Vdash_v (y \epsilon z)$, so even though Γ

'thinks' $\hat{2}$ and f have the same members, it does not 'think' they are equal. Indeed, we should expect something like this because members of $\hat{3}$ must be rigid, but f is not.

The intensional set models constructed above have not been investigated in general. I do think a wider exploration would be of considerable interest. But it is time to turn to their most profound application.

6.2. The Important Special Case

Suppose we think of possible worlds as incomplete states of information, approximations. And suppose we think of a move from a possible world to an accessible possible world as a move that increases information, that improves the approximation. Naturally, as we move from world to accessible world to accessible world, we want 'facts' to accumulate. This forces two conditions on us. The first condition is syntactic: we should confine our attention to 'facts' of the form $\Box X$, because only these say something about arbitrary accessible worlds. The second condition is semantic: we should work with models in which accessibility is transitive (so 'facts' accumulate) and reflexive (so the 'facts' of any world from which we start a sequence of approximations remain with us). That is, we should work with $S4$ models.

Under $S4$ circumstances, positive information, represented by formulas of the form $\Box X$, does accumulate as we move from world to accessible world. But what about negative information—formulas of the form $\neg\Box X$? We cannot expect accumulation for them, of course, or the modal operator would become trivial. But suppose we weaken the requirement by asking only that, for negative information, there should at least be the possibility of it becoming permanent. This can be accomplished by further restricting formulas to the form $\Box\Diamond X$. Since these begin with a necessity symbol, we retain the accumulation of positive information. But also, since $\neg\Box\Diamond X \equiv \Diamond\Box\neg X$, even for negative information we do have the possibility of it becoming permanent—we have the possibility of $\Box\neg X$. All this is quite loose, but it suggests we consider the following mapping from a classical language (without modal operators) into a modal language.

DEFINITION 6.3. For each formula X without modal operators, define a modal formula $\langle\langle X\rangle\rangle$ as follows.

1. $\langle\langle A\rangle\rangle = \Box\Diamond A$ for A atomic
2. $\langle\langle \neg X\rangle\rangle = \Box\Diamond\neg\langle\langle X\rangle\rangle$
3. $\langle\langle X \wedge Y\rangle\rangle = \Box\Diamond(\langle\langle X\rangle\rangle \wedge \langle\langle Y\rangle\rangle)$

4. $\langle\!\langle(\forall x)\Phi\rangle\!\rangle = \Box\Diamond(\forall x)\langle\!\langle\Phi\rangle\!\rangle$

This translation has a number of nice features. First of all, if other connectives and quantifiers are defined in the usual ways, they behave decently. For example, $\langle\!\langle X \supset Y\rangle\!\rangle \equiv \Box\Diamond(\langle\!\langle X\rangle\!\rangle \supset \langle\!\langle Y\rangle\!\rangle)$ is $S4$ valid, as is $\langle\!\langle(\exists x)\Phi\rangle\!\rangle \equiv \Box\Diamond(\exists x)\langle\!\langle\Phi\rangle\!\rangle$. Next, formulas are easily seen to be preserved under passage to alternative worlds, since $\langle\!\langle X\rangle\!\rangle \equiv \Box\langle\!\langle X\rangle\!\rangle$ is $S4$ valid. And negations behave as we wanted, since $\neg\langle\!\langle X\rangle\!\rangle \supset \Diamond\langle\!\langle\neg X\rangle\!\rangle$ is $S4$ valid—negative information has the possibility of becoming permanent. What is more, it can be shown that this is an exact embedding of classical logic into $S4$—X is classically valid iff $\langle\!\langle X\rangle\!\rangle$ is $S4$ valid, [Fit70, SF96].

So, from here on let us require accessibility relations to be reflexive and transitive—frames are for $S4$—and let us consider the behavior of formulas of the form $\langle\!\langle X\rangle\!\rangle$, where X is a formula in the language of classical Z-F set theory. Unfortunately, there is still the problem of the failure of the extensionality axiom—it is a problem that does not get solved this easily. Of course for many purposes one might not want an intensional version of set theory to obey an extensionality principle, but still it is pertinent to ask, what must be done to have it if we want it?

The ingenious solution to restoring the extensionality axiom for rigid and non-rigid sets alike is due to Paul Cohen, [Coh63, Coh64, Coh66], though his work was in a different context. The idea begins with the introduction of a carefully crafted candidate for an equality relation, leading in turn to a better candidate for the membership relation. Let $\approx$ be a new binary relation symbol—we want it to have the following behavior: for every $\Gamma \in \mathcal{G}$,

$$\begin{aligned}\mathcal{M},\Gamma \Vdash_v (f \approx g) \equiv \langle\!\langle &(\forall x)[(x \mathrel{\epsilon} f) \supset (\exists y)(y \mathrel{\epsilon} g \wedge x \approx y)]\wedge \\ &(\forall x)[(x \mathrel{\epsilon} g) \supset (\exists y)(y \mathrel{\epsilon} f \wedge x \approx y)]\rangle\!\rangle\end{aligned}$$

It is not at all obvious that this can be achieved, but in fact it can—one introduces a series of approximations to it, $\approx_\alpha$ for each ordinal α, where $\approx_\alpha$ is defined in terms of $\approx_\beta$ with $\beta < \alpha$, and then one defines $\approx$ as a limit of these $\approx_\alpha$. Once one has this, one introduces a new version of membership, denoted here by $\hat{\in}$, by setting

$$\mathcal{M},\Gamma \Vdash_v (f \mathrel{\hat{\in}} g) \text{ if } \mathcal{M},\Gamma \Vdash_v (\exists x)\langle\!\langle x \approx f \wedge x \mathrel{\epsilon} g\rangle\!\rangle$$

From now on, apply Definition 6.3 with the understanding that the classical atomic formula $(x \in y)$ is to be translated as $\Box\Diamond(x \mathrel{\hat{\in}} y)$. Introducing this more elaborate version of membership does, in fact, give us the validity of $\langle\!\langle X\rangle\!\rangle$, where X is the standard axiom of extensionality, and it does so even though we still have non-rigid intensional sets around.

What is truly remarkable, and in fact is the real significance of this construction, is that we get much more than just the axiom of extensionality. If $\mathcal{M} = \langle \mathcal{G}, \mathcal{R} \rangle$ is any $S4$ frame, and if X is any theorem of classical Z-F set theory, then $\langle\!\langle X \rangle\!\rangle$ is valid in $\mathcal{M}$. Still further, there are particular choices of $S4$ frames that invalidate $\langle\!\langle Y \rangle\!\rangle$, where Y is the continuum hypothesis. This means we have the classical unprovability of the continuum hypothesis! One way of looking at set-theoretic forcing is that it is the exploration of intensional models of set theory.

7. Conclusion

I have presented intensional versions of both higher-order classical logic, and set theory. Both ultimately date from the 1960's and, while intensional higher-order logic has been explored and utilized by a few logicians, not even that can be said for intensional set theory. It has been entirely the provence of classical set-theoreticians who think of it as something else altogether. It is time for these constructs to become better known, and in particular, for intensional set theoretic models to be explored for their own sakes.

References

[Coh63] Paul J. Cohen. The independence of the continuum hypothesis, I. *Proceedings of the National Academy of Sciences, U.S.A.*, 50:1143–1148, 1963.

[Coh64] Paul J. Cohen. The independence of the continuum hypothesis, II. *Proceedings of the National Academy of Sciences, U.S.A.*, 51:105–110, 1964.

[Coh66] Paul J. Cohen. *Set Theory and the Continuum Hypothesis.* W. A. Benjamin, New York, 1966.

[Fit70] Melvin C. Fitting. An embedding of classical logic in S4. *Journal of Symbolic Logic*, 35:529–534, 1970.

[Fit01] Melvin C. Fitting. First-order intensional logic. Forthcoming in *Proceedings of the Tarski Conference*, 2001.

[Fit02] Melvin C. Fitting. *Types, Tableaus, and Gödel's God.* Kluwer, 2002.

[FM98] Melvin C. Fitting and Richard Mendelsohn. *First-Order Modal Logic.* Kluwer, 1998. Paperback, 1999.

[Gal75] D. Gallin. *Intensional and Higher-Order Modal Logic.* North-Holland, 1975.

[HC96] G. E. Hughes and M. J. Cresswell. *A New Introduction to Modal Logic.* Routledge, London, 1996.

[KK01] Marcus Kracht and Oliver Kutz. The semantics of modal predicate logic I, counterpart frames. In Frank Wolter, Heinrich Wansing, Maarten de Rijke, and Michael Zakharyaschev, editors, *Advances in Modal Logic*, volume 3, Stanford, 2001. CSLI Publications.

[KK03] Marcus Kracht and Oliver Kutz. The semantics of modal predicate logic II, modal individuals revisited. In Reinhard Kahle, editor, *Intensionality. Proceedings of a conference held at Munich, 27th–29th October 2000*, Lecture Notes in Logic. A. K. Peters, 2003. To appear.

[Kra93] Marcus Kracht. How completeness and correspondence theory got married. In M. de Rijke, editor, *Diamonds and Defaults*, Dordrecht, 1993. Kluwer.

[Kut02] Oliver Kutz. New semantics for modal predicate logics. In *Foundations of the Formal Sciences II*, Trends in Logic. Kluwer, 2002.

[Lew68] David Lewis. Counterpart theory and quantified modal logic. *Journal of Philosophy*, 65:113–126, 1968.

[Lew71] David Lewis. Counterparts of persons and their bodies. *Journal of Philosophy*, 68:203–211, 1971.

[Mon60] Richard Montague. On the nature of certain philosophical entities. *The Monist*, 53:159–194, 1960. Reprinted in [Tho74], 148–187.

[Mon68] Richard Montague. Pragmatics. pages 102–122. 1968. In *Contemporary Philosophy: A Survey*, R. Klibansky editor, Florence, La Nuova Italia Editrice, 1968. Reprinted in [Tho74], 95–118.

[Mon70] Richard Montague. Pragmatics and intensional logic. *Synthèse*, 22:68–94, 1970. Reprinted in [Tho74], 119–147.

[Rus05] Bertrand Russell. On denoting. *Mind*, 14 (new series):479–493, 1905.

[Rus08] Bertrand Russell. Mathematical logic as based on the theory of types. *American journal of mathematics*, 30:222–262, 1908. Reprinted in [vH67], pp 150-182.

[Sco71] Dana S. Scott, editor. *Axiomatic Set Theory I*, volume 13 of *Symposium in Pure Mathematics*, Providence, R.I., 1971. American Mathematical Society. (Held in Los Angeles, 1967).

[SF96] Raymond M. Smullyan and Melvin C. Fitting. *Set Theory and the Continuum Problem.* Oxford University Press, 1996.

[Sho71] Joseph R. Shoenfield. Unramified forcing. pages 357–381, 1971. In [Sco71].

[SS71] Dana S. Scott and Robert M. Solovay. Boolean-valued models for set theory. 1971. In [Sco71].

[Tho74] Richmond H. Thomason, editor. *Formal Philosophy, Selected Papers of Richard Montague.* Yale University Press, New Haven and London, 1974.

[vH67] Jean van Heijenoort, editor. *From Frege to Gödel.* Harvard University Press, Cambridge, MA, 1967.

[WR27] Alfred North Whitehead and Bertrand Russell. *Principia Mathematica.* Cambridge University Press, Cambridge, 2nd edition, 1927. First edition, 1910.

[Zer08] Ernst Zermelo. Untersuchungen über die Grundlagen der Mengenlehre I. *Mathematische Annalen*, 65:261–281, 1908. English translation in [vH67], pp 199-215.

[Zer35] Ernst Zermelo. Grundlagen einer allgemeinen Theorie der mathematischen Satzsysteme. *Fundamenta Mathematicae*, 25:136–146, 1935.

Robert Goldblatt

Questions of Canonicity

Abstract. We review progress on the long standing question of whether every *canonical* modal logic must be characterized by an *elementary* class of Kripke frames, or equivalently, whether every canonical variety of Boolean algebras with operators must be generating by the complex algebras of an elementary class of relational structures. We verify that this does hold for certain families $\mathbf{RA}_n$, $\mathsf{SNr}_\beta\mathbf{CA}_\alpha$ and $\mathsf{S}\mathfrak{Ra}\mathbf{CA}_\alpha$ of varieties related to relation algebras and cylindric algebras. Canonical extensions of structures are shown to be free objects in certain categories of structures with topology, and to be associated with a monad on the category of sets that generalizes Manes' Theorem to relational structures.

Keywords: Boolean algebra with operators, canonical extension, elementary class, variety, relation algebra, free object.

Overview

This paper is about the concepts of the *canonical structure* of a Boolean algebra with operators, the *canonical extension* of such an algebra and of a relational structure, and the closely related notion of the *canonical frame* of a propositional modal logic. These ideas originated in work of Alfred Tarski and Bjarni Jónsson from the late 1940's.[1]

They have played a significant role in modal and algebraic logic in the subsequent decades, and continue to be of importance. Our aim is to review their place in the study both of modal logics and of varieties of algebras, to present some new results, and to suggest questions for further research.

In the first Section we discuss the concepts of "canonicity" and "elementarity" for logics and varieties, and survey the author's generalizations and strengthenings of the 1973 result of Kit Fine that elementary modal logics are canonical. We also review progress on the fundamental converse question, still unanswered, of whether canonicity implies elementarity. This concerns the extent to which notions of canonical object account for the correspondence manifest in numerous cases between *equationally* definable classes (varieties) of algebras that are closed under canonical extensions, and logics characterized by classes of frames that are *elementary*, i.e. first-order definable.

[1] See [29] for the historical background.

V. F. Hendricks and *J. Malinowski* (eds.),
Trends in Logic: 50 Years of Studia Logica

Trends in Logic **21**: 109–144, 2003.

Section 2 studies three infinite families of canonically-closed varieties of Boolean algebras with operators—called $\mathbf{RA}_n$, $\mathsf{S}\,\mathsf{Nr}_\beta\mathbf{CA}_\alpha$ and $\mathsf{S}\mathfrak{Ra}\mathbf{CA}_\alpha$—which are related to relation algebras and cylindric algebras, and which have been the focus of recent attention [24]. We show that each of these varieties is generated by the class of complex algebras of some elementary class of structures.

Section 3 uses topology and category theory to obtain a characterization of the operation of forming the canonical extension of a relational structure. These extensions are free objects in certain categories of topological structures, and are associated with a monad on the category **Set** of sets and functions. This analysis lifts to the relational structural setting the famous theorem of Manes that the category of compact Hausdorff spaces is isomorphic to the category of algebras for the ultrafilter monad on **Set**.

Section 4 poses questions for further research.

1. Canonicity

By an *operator* on a Boolean algebra we will mean any finitary function that, in each argument, is additive and preserves the least element 0. "Additive" means that it preserves the Boolean sum (join) operation. A *Boolean algebra with operators* (BAO) is, as the name suggests, an algebra comprising a Boolean algebra with a specified collection of operators.

In their seminal work [32] that founded the study of BAO's, Jónsson and Tarski showed that a relational structure $\mathfrak{S}$ of any kind has a *complex algebra* $\mathsf{Cm}\,\mathfrak{S}$, which is a BAO based on the Boolean algebra of all subsets of $\mathfrak{S}$. Each $n+1$-ary relation R of $\mathfrak{S}$ determines an n-ary operator f_R of $\mathsf{Cm}\,\mathfrak{S}$. $f_R(X_1, \dots, X_n)$ is the subset

$$\{x : R(x_1, \dots, x_n, x) \text{ for some } x_1 \in X_1, \dots, x_n \in X_n\}.^2$$

In the converse direction, for each BAO $\mathfrak{A}$ there is a certain structure, which we will denote $\mathbb{C}\mathrm{st}\,\mathfrak{A}$ and call the *canonical structure of* $\mathfrak{A}$, such that there is an isomorphic embedding $\mathfrak{A} \rightarrowtail \mathsf{Cm}\,\mathbb{C}\mathrm{st}\,\mathfrak{A}$ representing $\mathfrak{A}$ as an algebra of subsets of $\mathbb{C}\mathrm{st}\,\mathfrak{A}$. In this way Jónsson and Tarski lifted the Stone representation of Boolean algebras to BAO's, showing that each n-ary operator of $\mathfrak{A}$ is represented by the operator defined from an $n+1$-ary relation of $\mathbb{C}\mathrm{st}\,\mathfrak{A}$.

---

[2] We are following the *algebraist's convention* of assigning a special role to the last coordinate of R. The *logician's convention* is to use the first coordinate and write $R(x, x_1, \dots, x_n)$ in this definition.

We will write $\mathsf{Em}\,\mathfrak{A}$ for the complex algebra $\mathsf{Cm}\,\mathbb{C}\mathsf{st}\,\mathfrak{A}$, which is commonly known as the *canonical embedding algebra*, or *canonical extension* of $\mathfrak{A}$. Jónsson and Tarski called it the *perfect extension* of $\mathfrak{A}$, although to be precise they gave an axiomatic algebraic definition of *a* perfect extension, proving that any two such extensions are isomorphic over $\mathfrak{A}$. Here we will adopt what has become the standard approach of defining $\mathbb{C}\mathsf{st}\,\mathfrak{A}$ as the structure whose members are the ultrafilters of $\mathfrak{A}$, with each operator f of $\mathfrak{A}$ determining the relation R_f of $\mathbb{C}\mathsf{st}\,\mathfrak{A}$ such that

$$R_f(p_1, \ldots, p_{n+1}) \quad \text{iff} \quad \{f_R(x_1, \ldots, x_n) : x_1 \in p_1, \ldots, x_n \in p_n\} \subseteq p_{n+1}.$$

We include here the case $n = 0$ of f being a nullary operation, identifiable with an element d of $\mathfrak{A}$. Then $R_f(p)$ iff $d \in p$, so that R_f becomes the subset $\{p : d \in p\}$ of $\mathbb{C}\mathsf{st}\,\mathfrak{A}$.

It was shown in [32, Theorem 2.18] that any equation satisfied by $\mathfrak{A}$ that is *positive*, in the sense of not involving Boolean complementation, must also be satisfied by $\mathsf{Em}\,\mathfrak{A}$. This was the beginning of many investigations of properties preserved under the passage from $\mathfrak{A}$ to $\mathsf{Em}\,\mathfrak{A}$, investigations that are still ongoing after half a century. We can formulate such questions by asking whether a class $\mathcal{W}$ of algebras is closed under Em, i.e. whether $\mathfrak{A} \in \mathcal{W}$ implies $\mathsf{Em}\,\mathfrak{A} \in \mathcal{W}$. $\mathcal{W}$ is called *canonical* if it has this closure property. Canonicity has been of particular interest for varieties (equational classes) of modal algebras, relation algebras, and cylindric algebras [2, 12, 13, 15, 18, 21, 24, 29, 30, 31].

In the mid-1960's John Lemmon and Dana Scott defined, for each normal propositional modal logic Λ, a certain structure $\mathfrak{S}^\Lambda$ whose points are the maximally Λ-consistent sets of formulas, and showed that $\mathfrak{S}^\Lambda$ can be used as the basis for completeness theorems for many logics [34]. $\mathfrak{S}^\Lambda$ is known as the *canonical frame* for Λ and is commonly invoked for logics whose language has denumerably many variables. But more generally, for each infinite cardinal κ we can define a frame $\mathfrak{S}^\Lambda_\kappa$ whose points are the maximally Λ-consistent sets of formulas from a language with κ-many variables. These frames are intimately connected with the notion of a canonical structure $\mathbb{C}\mathsf{st}\,\mathfrak{A}$, as follows from the fact that the underlying set of $\mathbb{C}\mathsf{st}\,\mathfrak{A}$ is the set of ultrafilters of $\mathfrak{A}$. Maximally consistent sets of formulas correspond to ultrafilters of the Lindenbaum algebra of a logic, and Lindenbaum algebras are the ones that are free. Thus the canonical frame $\mathfrak{S}^\Lambda_\kappa$ is isomorphic to the canonical structure $\mathbb{C}\mathsf{st}\,\mathfrak{A}^\Lambda_\kappa$, where $\mathfrak{A}^\Lambda_\kappa$ is the free algebra on κ many generators in the variety of all algebras that validate the logic Λ.[3]

[3] See [9] for more information on the origin of canonical frames.

$\mathfrak{S}_\kappa^\Lambda$ supports a model that falsifies all non-theorems of Λ, so if it validates Λ, then it characterizes the logic. Thus Λ is called a *canonical logic* if it is valid in $\mathfrak{S}_\kappa^\Lambda$ for all κ. Most completeness theorems using canonical frames proceed by showing that $\mathfrak{S}_\kappa^\Lambda$ satisfies some first-order definable condition that guarantees the validity of Λ (thereby showing Λ is canonical). In 1973 Kit Fine provided a theoretical basis for this procedure by proving the following fundamental theorem for a monomodal logic Λ:

> *if Λ is characterized by some elementary class of frames, then it is canonical, i.e. is valid in $\mathfrak{S}_\kappa^\Lambda$ for all κ.*

[3, Theorem 3]. We will say that Λ is *elementary* if it is characterized by some elementary class of frames. An elementary class is one that is the class of all models of some set of first-order sentences, and can be characterized as a class that is closed under ultraproducts while its *complement* is closed under ultrapowers. Only the closure of the class under ultraproducts is needed for Fine's proof that an elementary logic is canonical.[4]

Fine's theorem has been generalized by the present author in two main directions, firstly broadening the domain of application, and secondly strengthening the conclusion. The broader domain comes from reformulating the result as one about varieties of algebras, and proving it for BOA's of any type. To describe this we will say that a variety $\mathcal{V}$ is *generated by* a class $\mathcal{K}$ of relational structures (of appropriate type) if $\mathcal{V}$ is the smallest variety that includes the class

$$\mathsf{Cm}\,\mathcal{K} = \{\mathfrak{A} : \mathfrak{A} \cong \mathsf{Cm}\,\mathfrak{S} \text{ for some } \mathfrak{S} \in \mathcal{K}\}$$

of all (isomorphic copies of) complex algebras of structures in $\mathcal{K}$. This means that $\mathcal{V} = \mathsf{HSPCm}\,\mathcal{K}$, where H, S, and P denote the operations of closure of a class of algebras under (isomorphic copies of) homomorphic images, subalgebras, and direct products, respectively. We write $\mathsf{Var}\,\mathcal{K}$ for the variety $\mathsf{HSPCm}\,\mathcal{K}$ generated by $\mathcal{K}$. In general a variety may be generated by many different classes of structures: if $\mathcal{K}$ generates $\mathcal{V}$, then so does any class $\mathcal{K}'$ with $\mathcal{K} \subseteq \mathcal{K}' \subseteq \mathbb{S}\mathsf{tr}\,\mathcal{V}$, where

$$\mathbb{S}\mathsf{tr}\,\mathcal{V} = \{\mathfrak{S} : \mathsf{Cm}\,\mathfrak{S} \in \mathcal{V}\}$$

is the class of all *structures for* $\mathcal{V}$, or $\mathcal{V}$*-structures*. A variety is *elementarily generated* if it is equal to $\mathsf{Var}\,\mathcal{K}$ for some elementary class $\mathcal{K}$.

---

[4] Closure under ω-saturated elementary extensions was also used in [3], but that follows from closure under ultraproducts (indeed under ultrapowers) [1, Section 6.1].

THEOREM 1.1. *If $\mathcal{K}$ is closed under ultraproducts, then the variety* $\mathsf{Var}\mathcal{K}$ *generated by $\mathcal{K}$ is canonical, i.e. $\mathfrak{A} \in \mathsf{Var}\mathcal{K}$ implies $\mathbb{C}\mathsf{st}\mathfrak{A} \in \mathcal{K}$.* ■

The replacement of an elementary class in the hypothesis of this theorem by the weaker notion of one closed under ultraproducts is not a true strengthening, because when $\mathcal{K}$ is closed under ultraproducts, the closure of $\mathcal{K}$ under elementary equivalence gives an elementary class that also generates $\mathsf{Var}\mathcal{K}$ [15, p. 581]. The proof of this uses the Keisler-Shelah ultrapower theorem, in a similar manner to the proof of Theorem 1.4 below.

Underlying the proof of Theorem 1.1 is a categorical *duality* between BAO's and relational structures [11, 12, 15] that lifts the correspondences $\mathfrak{A} \mapsto \mathbb{C}\mathsf{st}\mathfrak{A}$ and $\mathfrak{S} \mapsto \mathsf{Cm}\,\mathfrak{S}$ to functors that take any BAO-homomorphism $\mathfrak{A}_1 \to \mathfrak{A}_2$ to a *bounded morphism*[5] $\mathbb{C}\mathsf{st}\mathfrak{A}_2 \to \mathbb{C}\mathsf{st}\mathfrak{A}_1$ of structures, and any bounded morphism $\mathfrak{S}_1 \to \mathfrak{S}_2$ to a BAO-homomorphism $\mathsf{Cm}\,\mathfrak{S}_2 \to \mathsf{Cm}\,\mathfrak{S}_1$. A homomorphism $f : \mathfrak{A}_1 \to \mathfrak{A}_2$ induces the bounded morphism $\mathbb{C}\mathsf{st}\mathfrak{A}_2 \to \mathbb{C}\mathsf{st}\mathfrak{A}_1$ that maps each ultrafilter p of $\mathfrak{A}_2$ to the ultrafilter $f^{-1}p$ of $\mathfrak{A}_1$. A bounded morphism $\theta : \mathfrak{S}_1 \to \mathfrak{S}_2$ induces the homomorphism $\mathsf{Cm}\,\mathfrak{S}_2 \to \mathsf{Cm}\,\mathfrak{S}_1$ that takes each set X in $\mathsf{Cm}\,\mathfrak{S}_2$ to its inverse image $\theta^{-1}X$ in $\mathsf{Cm}\,\mathfrak{S}_1$.

These functors map injections to surjections and vice versa. Thus an injective $\mathfrak{A}_1 \rightarrowtail \mathfrak{A}_2$ induces a surjective $\mathbb{C}\mathsf{st}\mathfrak{A}_2 \twoheadrightarrow \mathbb{C}\mathsf{st}\mathfrak{A}_1$, and hence an injective $\mathsf{Cm}\,\mathbb{C}\mathsf{st}\mathfrak{A}_1 \rightarrowtail \mathsf{Cm}\,\mathbb{C}\mathsf{st}\mathfrak{A}_2$, i.e. $\mathsf{Em}\,\mathfrak{A}_1 \rightarrowtail \mathsf{Em}\,\mathfrak{A}_2$. Similarly a surjective homomorphism $\mathfrak{A}_1 \twoheadrightarrow \mathfrak{A}_2$ lifts to a surjective homomorphism $\mathsf{Em}\,\mathfrak{A}_1 \twoheadrightarrow \mathsf{Em}\,\mathfrak{A}_2$.

Dual to the algebraic class operations H, S, and P are operations $\mathbb{S}$, $\mathbb{H}$, and $\mathbb{U}\mathsf{d}$ of closure of a class of relational structures under "inner" substructures,[6] bounded epimorphic images, and disjoint unions of structures, respectively. We also write $\mathbb{P}\mathsf{u}$ and $\mathbb{P}\mathsf{w}$ for closure of classes of structures under ultraproducts and ultrapowers, and Pu and Pw for closure of classes of *algebras* under these operations.[7] Closure under *ultraroots* is denoted by $\mathbb{R}\mathsf{u}$. Thus $\mathbb{R}\mathsf{u}\mathcal{K}$ is the class

$$\{\mathfrak{S} : \Pi_U \mathfrak{S} \in \mathcal{K} \text{ for some ultrafilter } U\}$$

of all structures $\mathfrak{S}$ that have some ultrapower $\Pi_U\mathfrak{S}$ in $\mathcal{K}$. A class is elementary iff it is closed under ultraproducts and ultraroots. When $\mathcal{K}$ is closed

[5] Also known as a "p-morphism" in the case of modal frames. The technical definition of bounded morphism is given in Section 3.2 below.

[6] A substructure $\mathfrak{S}_1$ of $\mathfrak{S}_2$ is *inner* if the inclusion $\mathfrak{S}_1 \hookrightarrow \mathfrak{S}_2$ is a bounded morphism. The notion is sometimes called "generated subframe" in modal logic.

[7] In general, "sans serif" capitals E, H, P, S... will be used as the first letter in symbolic names for operations on *algebras*, while "blackboard bold" letters $\mathbb{C}$, $\mathbb{E}$, $\mathbb{H}$, $\mathbb{S}$, $\mathbb{U}$... occur likewise in names of operations on *structures*.

under ultraproducts, $\mathbb{R}\mathsf{u}\,\mathcal{K}$ is the closure of K under elementary equivalence, and hence is the smallest elementary class including $\mathcal{K}$ [15, 4.11].

Another important operation is $\mathbb{U}\mathsf{b}$, the closure under *bounded unions*. A structure $\mathfrak{S}$ is the bounded union of structures $\mathfrak{S}_i$ if $\mathfrak{S}$ is the union of these $\mathfrak{S}_i$'s and each $\mathfrak{S}_i$ is an inner substructure of $\mathfrak{S}$. $\mathbb{U}\mathsf{b}$ is dual to the operation of forming subdirect products of algebras [17, p. 415]. Any structure is the bounded union of the inner substructures generated by each of its points. This dualizes the representation of an algebra as a subdirect product of subdirectly irreducible algebras, since the complex algebra of a structure generated by a point is always subdirectly irreducible ([12, Theorem 3.3.1], [39]).

Dual to the notion of the canonical extension $\mathsf{Em}\,\mathfrak{A} = \mathsf{Cm}\,\mathbb{C}\mathsf{st}\,\mathfrak{A}$ of a BAO is the canonical extension $\mathbb{E}\mathsf{x}\,\mathfrak{S} = \mathbb{C}\mathsf{st}\,\mathsf{Cm}\,\mathfrak{S}$ of a structure $\mathfrak{S}$. The points of $\mathbb{E}\mathsf{x}\,\mathfrak{S}$ are the ultrafilters on the underlying set of $\mathfrak{S}$, and by identifying each point of $\mathfrak{S}$ with the principal ultrafilter it generates, we may view $\mathbb{E}\mathsf{x}\,\mathfrak{S}$ as an extension of $\mathfrak{S}$ itself. Note that $\mathsf{Cm}\,\mathbb{E}\mathsf{x}\,\mathfrak{S} = \mathsf{Em}\,\mathsf{Cm}\,\mathfrak{S}$. Any surjective or injective bounded morphism $\mathfrak{S}_1 \to \mathfrak{S}_2$ lifts to a bounded morphism $\mathbb{E}\mathsf{x}\,\mathfrak{S}_1 \to \mathbb{E}\mathsf{x}\,\mathfrak{S}_2$ of the same kind.

Johan van Benthem [43, 3.6] reworked the proof of Fine's theorem to show, for any monomodal logic Λ, that

> if Λ is characterized by some elementary class of frames, then whenever a frame $\mathfrak{S}$ validates Λ, so does $\mathbb{E}\mathsf{x}\,\mathfrak{S}$.

To describe properties of class operations we use the notation $\mathbb{X} \leq \mathbb{Y}$, where $\mathbb{X}$ and $\mathbb{Y}$ are combinations of operations, to mean that $\mathbb{X}\mathcal{C} \subseteq \mathbb{Y}\mathcal{C}$ for all appropriate classes $\mathcal{C}$. Then we put $\mathbb{X} = \mathbb{Y}$ when $\mathbb{X} \leq \mathbb{Y}$ and $\mathbb{Y} \leq \mathbb{X}$.

There are numerous relationships holding between class operations that can be expressed in this notation, such as

$$\mathbb{S}\mathbb{H} \leq \mathbb{H}\mathbb{S}, \quad \mathsf{P}\mathsf{Cm} = \mathsf{Cm}\,\mathbb{U}\mathsf{d}, \quad \mathbb{U}\mathsf{d} \leq \mathbb{U}\mathsf{b} = \mathbb{H}\mathbb{U}\mathsf{d}, \quad \mathbb{P}\mathsf{u}\mathbb{S} \leq \mathbb{S}\mathbb{P}\mathsf{u},$$

and the facts that for any variety of BOA's $\mathcal{V}$, $\mathbb{S}\mathsf{tr}\,\mathcal{V}$ is closed under $\mathbb{S}$, $\mathbb{H}$, $\mathbb{U}\mathsf{d}$, $\mathbb{U}\mathsf{b}$ and $\mathbb{R}\mathsf{u}$, while the complement of $\mathbb{S}\mathsf{tr}\,\mathcal{V}$ is closed under $\mathbb{E}\mathsf{x}$ and $\mathbb{P}\mathsf{w}$. One straightforward application of the duality is that $\mathbb{C}\mathsf{st}\,\mathsf{H}\mathsf{S} \leq \mathbb{S}\mathbb{H}\mathbb{C}\mathsf{st}$. An extensive catalogue of such results is given in [15, 17].

A crucial property connecting canonical extensions with ultrapowers is

$$\mathbb{E}\mathsf{x} \leq \mathbb{H}\mathbb{P}\mathsf{w}, \tag{1.1}$$

which follows from the fact that any structure $\mathfrak{S}$ has an ultrapower $\Pi_U\mathfrak{S}$ for which there is a surjective bounded morphism $\Pi_U\mathfrak{S} \twoheadrightarrow \mathbb{E}\mathsf{x}\,\mathfrak{S}$ (see [12,

Theorem 3.6.1] or [13, Lemma 3.1]). For this to hold it suffices that the ultrafilter U be chosen so that the ultrapower is ω-saturated. This is essentially a structural version of the key construction in the proof of Fine's theorem.

Our first proof of Theorem 1.1, given in [12, Theorem 3.6.7], used a number of ingredients, including a diagonal construction of ultraproducts of subdirectly irreducible quotients of complex algebras and an important result of Jónsson [28] about the representation of subdirectly irreducible algebras in congruence-distributive varieties. It also required (1.i), the relationship $\mathsf{Pu}\,\mathsf{Cm} \leq \mathsf{S}\,\mathsf{Cm}\,\mathbb{P}\mathsf{u}$, and the fact that $\mathsf{Var}\mathcal{K}$ is canonical if, and only if, its class $\mathbb{S}\mathsf{tr}\,\mathsf{Var}\,\mathcal{K}$ of structures is closed under $\mathbb{E}\mathsf{x}$.

A different and more structurally revealing proof of Theorem 1.1 was developed in [13] and analysed further in [15] and [17]. It uses the relationship

$$\mathbb{P}\mathsf{w}\,\mathbb{U}\mathsf{d} \leq \mathbb{H}\,\mathbb{U}\mathsf{d}\,\mathbb{P}\mathsf{u}, \tag{1.2}$$

expressing that an ultrapower of a disjoint union of structures from $\mathcal{K}$ can be obtained as a bounded epimorphic image of a disjoint union of ultraproducts of families of members of $\mathcal{K}$. This in turn is a consequence of the more general result $\mathbb{P}\mathsf{u}\,\mathbb{U}\mathsf{b} \leq \mathbb{U}\mathsf{b}\,\mathbb{P}\mathsf{u}$ [15, Theorem 2.4], which will be used below in Theorem 1.6. By combining (1.i), (1.ii) and the result that $\mathbb{C}\mathsf{st}\,\mathsf{H}\,\mathsf{S} \leq \mathbb{S}\,\mathbb{H}\,\mathbb{C}\mathsf{st}$, it can be proven that any class $\mathcal{K}$ of structures has

$$\mathbb{C}\mathsf{st}\,\mathsf{H}\,\mathsf{S}\,\mathsf{P}\,\mathsf{Cm}\,\mathcal{K} \subseteq \mathbb{S}\,\mathbb{H}\,\mathbb{U}\mathsf{d}\,\mathbb{P}\mathsf{u}\,\mathcal{K},$$

which shows how the canonical structures of BAO's from the variety generated by $\mathcal{K}$ can be built from members of $\mathcal{K}$ (see [15, p. 580] or [17, p. 419]). Then if $\mathcal{K}$ is closed under ultraproducts, so that $\mathbb{P}\mathsf{u}\,\mathcal{K} = \mathcal{K}$, it follows that

$$\mathfrak{A} \in \mathsf{H}\,\mathsf{S}\,\mathsf{P}\,\mathsf{Cm}\,\mathcal{K} \quad \text{implies} \quad \mathbb{C}\mathsf{st}\,\mathfrak{A} \in \mathbb{S}\,\mathbb{H}\,\mathbb{U}\mathsf{d}\,\mathcal{K} \subseteq \mathbb{S}\mathsf{tr}\,\mathsf{Var}\,\mathcal{K},$$

showing how canonical structures mediate between the dual operations on algebras and structures. But then

$$\mathsf{Em}\,\mathsf{Var}\,\mathcal{K} = \mathsf{Cm}\,\mathbb{C}\mathsf{st}\,\mathsf{Var}\,\mathcal{K} \subseteq \mathsf{Cm}\,\mathbb{S}\mathsf{tr}\,\mathsf{Var}\,\mathcal{K} \subseteq \mathsf{Var}\,\mathcal{K},$$

which proves Theorem 1.1. A refinement of this analysis then establishes [15, 4.12] that:

> if $\mathcal{K}$ is closed under ultraproducts, and L is any class satisfying
>
> $$\mathbb{C}\mathsf{st}\,\mathsf{Var}\,K \subseteq L = \mathbb{P}\mathsf{u}\,L \subseteq \mathbb{S}\mathsf{tr}\,\mathsf{Var}\,K,$$
>
> then $M = \mathbb{R}\mathsf{u}\,L$ is an elementary class satisfying $\mathsf{Var}\,K = \mathsf{S}\,\mathsf{Cm}\,M$. In particular, this holds when $L = \mathbb{S}\,\mathbb{H}\,\mathbb{U}\mathsf{d}\,K$ or $L = \mathbb{H}\,\mathbb{S}\,\mathbb{U}\mathsf{d}\,K$.

In fact something much stronger than results like this one and Theorem 1.1 can be proven by these methods. Instead of just showing that some given variety $\mathsf{Var}\mathcal{K}$ is canonical, we can use the assumption that $\mathbb{P}\mathsf{u}\mathcal{K} = \mathcal{K}$ to prove that a certain class of algebras built from $\mathcal{K}$ *is a variety* as well as being canonical. Moreover we can weaken the hypothesis on $\mathcal{K}$ for this purpose, and obtain the following result [15, Theorem 4.5]:

THEOREM 1.2. *If* $\mathbb{P}\mathsf{u}\mathcal{K} \subseteq \mathbb{HSU}\mathsf{d}\mathcal{K}$, *then* $\mathbb{S}\mathsf{Cm}\mathbb{SU}\mathsf{d}\mathcal{K}$ *is a canonical variety (and hence is the variety generated by* $\mathcal{K}$*).* ■

(Since $\mathbb{SU}\mathsf{d} = \mathbb{U}\mathsf{d}\mathbb{S}$ and $\mathsf{Cm}\mathbb{U}\mathsf{d} = \mathbb{P}\mathsf{Cm}$ in general, $\mathbb{S}\mathsf{Cm}\mathbb{SU}\mathsf{d}\mathcal{K}$ is equal to $\mathbb{SP}\mathsf{Cm}\mathbb{S}\mathcal{K}$.) Theorem 1.2 has been be applied to give structural proofs that the classes $\mathbf{RCA}_\alpha$ of representable cylindric algebras and $\mathbf{ICrs}_\alpha$ of cylindric-relativised set algebras of any dimension α form canonical varieties [15, 4.6].

As a consequence of 1.2, if $\mathcal{K}$ is any class of structures closed under $\mathbb{P}\mathsf{u}$, $\mathbb{S}$, and $\mathbb{U}\mathsf{d}$ (e.g. any elementary class closed under $\mathbb{S}$ and $\mathbb{U}\mathsf{d}$), then $\mathbb{S}\mathsf{Cm}\mathcal{K}$ is a canonical variety [15, 4.3]. This has an elegant application to the class **RRA** of representable relation algebras. A *proper relation algebra* is a collection of binary relations on some set X (i.e. a subset of $\mathcal{P}(X \times X)$) that forms a field of sets (hence a Boolean algebra) that is closed under the binary operator $R_1|R_2$ of composition of relations and the unary converse operator R^{-1}, and contains the identity relation on X as a distinguished element. Any algebra isomorphic to a proper relation algebra is called a *representable* relation algebra. Thus members of **RRA** are of the type $\mathfrak{A} = (\mathfrak{A}_0, ;, ^\smile, 1')$ with $\mathfrak{A}_0$ a Boolean algebra having a binary operator ; called *composition*, a unary operator $^\smile$ called *conversion*, and a member 1' of $\mathfrak{A}_0$ called the *identity element.* **RRA** is a subclass of the variety **RA** of *relation algebras*, the latter being defined by a (finite) set of equations [33, 4.1]. **RA** is a canonical variety [33] and $\mathbb{S}\mathsf{tr}\mathbf{RA}$ is an elementary class [36]. Tarski used metamathematical arguments in [40, 2.4] to show that **RRA** is an equational class. Some years later J. D. Monk proved that **RRA** is canonical by exploiting its connections with cylindric algebra theory (see [24, p. 123] for more information about this).

The structures whose complex algebras are of **RA**-type have the form $\mathfrak{S} = (S, C, R, I)$ with $C \subseteq X^3$, $R \subseteq X^2$ and $I \subseteq X$. In the algebra $\mathsf{Cm}\,\mathfrak{S}$, I is the identity element, composition is the operator defined by C, i.e.

$$X;Y = \{z \in S : \exists x \in X\, \exists y \in Y\, C(x, y, z)\},$$

while conversion is defined from R by

$$X^\smile = \{z \in S : \exists x \in X\, (xRz)\}.$$

Jónsson and Tarski showed in [33, 5.8] that **RRA** $= \mathbb{S}\mathsf{Cm}\,\mathcal{GB}$, where $\mathcal{GB}$ is a class of such structures, called *generalized Brandt groupoids*, that are defined by a set of conditions that are expressible in first-order logic. Thus $\mathcal{GB}$ is an elementary class. They also showed that $\mathfrak{S}$ is a generalized Brandt groupoid iff $\mathsf{Cm}\,\mathfrak{S}$ is a relation algebra and the ternary relation C of $\mathfrak{S}$ is functional, in the sense that if $C(x, y, z_1)$ and $C(x, y, z_2)$ then $z_1 = z_2$ (see Theorems 5.5 and 5.7 and comment on page 159 of [33]).

THEOREM 1.3. **RRA** *is a variety and is canonical.*

PROOF. Since $\mathcal{GB}$ is an elementary class it is $\mathbb{P}\mathsf{u}$-closed, so to prove $\mathbb{S}\mathsf{Cm}\,\mathcal{GB}$ is a canonical variety it is enough by Theorem 1.2 to prove that $\mathcal{GB}$ is closed under $\mathbb{U}\mathsf{d}$ and $\mathbb{S}$.

We use the fact just mentioned that $\mathfrak{S} \in \mathcal{GB}$ iff $\mathsf{Cm}\,\mathfrak{S} \in$ **RA** and C is functional in $\mathfrak{S}$. But $\mathsf{Cm}\,\mathfrak{S} \in$ **RA** iff $\mathfrak{S} \in \mathbb{S}\mathsf{tr}$**RA**, and $\mathbb{S}\mathsf{tr}$**RA** is closed under $\mathbb{U}\mathsf{d}$ and $\mathbb{S}$ (indeed this holds of $\mathbb{S}\mathsf{tr}\,\mathcal{V}$ for any variety $\mathcal{V}$). The condition of functionality of C is readily seen to be preserved by disjoint unions, and is preserved by arbitrary substructures, not just inner ones. These observations suffice to show that $\mathbb{U}\mathsf{d}\,\mathcal{GB} = \mathbb{S}\,\mathcal{GB} = \mathcal{GB}$. ∎

Thus **RRA** can be axiomatized by a set $\mathcal{E}$ of equations that are *collectively* canonical in the sense that if an algebra $\mathfrak{A}$ is a model of all members of $\mathcal{E}$, then so is $\mathsf{Em}\,\mathfrak{A}$. Recently Ian Hodkinson and Yde Venema [26] have shown that no set of equational axioms for **RRA** can consist of equations that are *individually* canonical (an equation ϵ being individually canonical if $\mathfrak{A} \models \epsilon$ implies $\mathsf{Em}\,\mathfrak{A} \models \epsilon$).

The closure properties of a class of the form $\mathbb{S}\mathsf{Cm}\,\mathcal{K}$ can be given a finer analysis. Firstly, the condition $\mathbb{P}\mathsf{w}\,\mathcal{K} = \mathcal{K}$ is sufficent on its own to ensure that $\mathbb{S}\mathsf{Cm}\,\mathcal{K}$ is a canonical class of algebras [12, 3.6.3]. Then if $\mathbb{S}\,\mathcal{K} = \mathcal{K}$ as well, $\mathbb{S}\mathsf{Cm}\,\mathcal{K}$ is closed under homomorphic images. Adding $\mathbb{U}\mathsf{d}\,\mathcal{K} = \mathcal{K}$ to these makes $\mathbb{S}\mathsf{Cm}\,\mathcal{K}$ a canonical variety [15, 4.3].

This analysis has application to the class **GRA** of *group relation algebras*. A group $(S, \circ, {}^{-1}, e)$ may be viewed as an **RA**-type structure $\mathfrak{S}$ with ternary relation C being the graph $\{(x, y, x \circ y) : x, y \in S\}$ of the group operation; $R = \{(x, x^{-1}) : x \in S\}$ being the graph of the group inverse; and $I = \{e\}$ where e is the group identity. If $\mathcal{G}\mathsf{p}$ is the class of groups construed as such structures, then **GRA** $= \mathbb{S}\mathsf{Cm}\,\mathcal{G}\mathsf{p}$ by definition. Each member of $\mathsf{Cm}\,\mathcal{G}\mathsf{p}$ is a representable relation algebra, so **GRA** $\subseteq$ **RRA**. But $\mathcal{G}\mathsf{p}$ is an elementary class, hence in particular closed under ultrapowers, so from what was said in the previous paragraph, it follows directly that **GRA** is a canonical class by itself.

$\mathcal{G}p$ is also closed under $\mathbb{S}$. To see this we need to know some detail of what "inner substructure" means in this context. Given structures $\mathfrak{S}_i = (S_i, C_i, R_i, I_i)$ for $i = 1, 2$, if $\mathfrak{S}_1$ is an inner substructure of $\mathfrak{S}_2$ then in particular it is a substructure of $\mathfrak{S}_2$ and the following holds for all $x, y, z \in S_2$:

$$\text{if } C_2(x, y, z) \text{ and } z \in S_1, \text{ then } x, y \in S_1.$$

But if $\mathfrak{S}_2 \in \mathcal{G}p$, with $I_2 = \{e\}$, then for any $z \in S_1$ we have $C_2(z, e, z)$ from the group law $z \circ e = z$, so $e \in S_1$ by the inner substructure condition. Then for any $x \in S_2$, $C_2(x, x^{-1}, e)$ and so $x \in S_1$. Hence $\mathfrak{S}_1 = \mathfrak{S}_2$. This shows that a group structure has no proper inner substructures, and therefore trivially $\mathbb{S}\mathcal{G}p = \mathcal{G}p$. Since $\mathcal{G}p$ is closed under ultraproducts, Theorem 1.2 then yields that the class $\mathsf{S}\mathsf{Cm}\mathbb{U}\mathsf{d}\mathcal{G}p$, i.e. $\mathsf{S}\mathsf{P}\mathsf{Cm}\mathcal{G}p$, is a canonical variety. Hence it is the variety generated by **GRA**.

Tarski [40] proved that **GRA** is a universal class and consists of the simple algebras in $\mathsf{S}\mathsf{P}\mathsf{Cm}\mathcal{G}p$. Steven Givant [8] has shown that this is a manifestation of the following general phenomenon:

> if $\mathcal{W}$ is a class of simple algebras in a discriminator variety $\mathcal{V}$, and there exists a class $\mathcal{K}$ of structures that is closed under ultraproducts and has $\mathsf{Cm}\mathcal{K} \subseteq \mathcal{W} \subseteq \mathsf{S}\mathsf{Cm}\mathcal{K}$, then $\mathsf{S}\mathsf{P}\mathcal{W}$ is a canonical variety and $\mathsf{S}\mathcal{W}$ is the universal class of simple algebras in $\mathsf{S}\mathsf{P}\mathcal{W}$.

■

The second way in which we have generalized the theorem of Fine is to strengthen its conclusion, showing that if a logic Λ is elementary, then it is validated not only by its canonical frames but also by other structures that have some first-order connection with these frames. This relates to the vital question of whether the converse of Fine's theorem is true:

> *must every canonical logic Λ be elementary, i.e. be characterized by some elementary class of frames?*

A natural candidate for an elementary characterizing class here is the class $\mathcal{K}_\omega^\Lambda$ of all frames that are first-order equivalent to the canonical Λ-frame $\mathfrak{S}_\omega^\Lambda$ based on a language with denumerably many variables. $\mathcal{K}_\omega^\Lambda$ is the class of all models of the *first-order theory* of $\mathfrak{S}_\omega^\Lambda$, i.e. the set of all first-order sentences true in $\mathfrak{S}_\omega^\Lambda$. The relevance of this class is shown by the following result:

> if Λ is characterized by some class of frames that is closed under ultraproducts, then it is valid in all members of the elementary class $\mathcal{K}_\omega^\Lambda$ [14, Theorem 11.3.1].

Of course strengthening the conclusion of a statement makes the converse weaker, and the converse of this last theorem is immediate: if Λ is valid in all members of $\mathcal{K}_\omega^\Lambda$ then it is characterized by an elementary class, namely $\mathcal{K}_\omega^\Lambda$, since $\mathfrak{S}_\omega^\Lambda \in \mathcal{K}_\omega^\Lambda$ and all non-theorems of Λ are falsified in $\mathfrak{S}_\omega^\Lambda$.

The algebraic version of this result is formulated for a variety $\mathcal{V}$ of BAO's, and invokes the free $\mathcal{V}$-algebra $\mathfrak{A}_\omega^\mathcal{V}$ on denumerably many generators:

> $\mathcal{V}$ is generated by some class of structures that is closed under ultraproducts if, and only if, it is generated by the elementary class of all models of $Th\,\mathbb{C}\mathsf{st}\mathfrak{A}_\omega^\mathcal{V}$, the first-order theory of the canonical structure $\mathbb{C}\mathsf{st}\mathfrak{A}_\omega^\mathcal{V}$ [15, Theorem 4.15].

Now the models of $Th\,\mathbb{C}\mathsf{st}\mathfrak{A}_\omega^\mathcal{V}$ are just the structures elementarily equivalent to $\mathbb{C}\mathsf{st}\mathfrak{A}_\omega^\mathcal{V}$, and elementary equivalence has a structural characterization, due to Keisler and Shelah: elementarily equivalent structures are those that have isomorphic U-ultrapowers for some ultrafilter U. From this we get another characterization of elementary generation of varieties:

THEOREM 1.4. *A variety $\mathcal{V}$ is elementarily generated if, and only if, every ultrapower of $\mathbb{C}\mathsf{st}\mathfrak{A}_\omega^\mathcal{V}$ is in $\mathbb{S}\mathsf{tr}\,\mathcal{V}$.*

PROOF. Let $\mathcal{K}$ be the elementary class of all models of $Th\,\mathbb{C}\mathsf{st}\mathfrak{A}_\omega^\mathcal{V}$.

If $\mathcal{V}$ is elementarily generated, then as explained above it is generated by $\mathcal{K}$, so $\mathbb{C}\mathsf{st}\mathfrak{A}_\omega^\mathcal{V} \in \mathcal{K} \subseteq \mathbb{S}\mathsf{tr}\,\mathcal{V}$. Since $\mathbb{P}\mathsf{w}\mathcal{K} \subseteq \mathcal{K}$, this puts all ultrapowers of $\mathbb{C}\mathsf{st}\mathfrak{A}_\omega^\mathcal{V}$ into $\mathbb{S}\mathsf{tr}\,\mathcal{V}$.

Conversely, suppose $\mathbb{S}\mathsf{tr}\,\mathcal{V}$ contains the ultrapower $\Pi_U\mathbb{C}\mathsf{st}\mathfrak{A}_\omega^\mathcal{V}$ for any ultrafilter U. If $\mathfrak{S} \in \mathcal{K}$ then by the Keisler-Shelah theorem there is an ultrafilter U such that $\Pi_U\mathfrak{S}$ and $\Pi_U\mathbb{C}\mathsf{st}\mathfrak{A}_\omega^\mathcal{V}$ are isomorphic. By hypothesis $\Pi_U\mathbb{C}\mathsf{st}\mathfrak{A}_\omega^\mathcal{V} \in \mathbb{S}\mathsf{tr}\,\mathcal{V}$, and so $\Pi_U\mathfrak{S} \in \mathbb{S}\mathsf{tr}\,\mathcal{V}$. But $\mathbb{S}\mathsf{tr}\,\mathcal{V}$ is always closed under ultraroots, so this forces $\mathfrak{S} \in \mathbb{S}\mathsf{tr}\,\mathcal{V}$. Thus we have $\mathcal{K} \subseteq \mathbb{S}\mathsf{tr}\,\mathcal{V}$, and so

$$\mathsf{Var}\,\mathcal{K} \subseteq \mathsf{Var}\,\mathbb{S}\mathsf{tr}\,\mathcal{V} \subseteq \mathcal{V}.$$

Hence $\mathsf{Cm}\,\mathbb{C}\mathsf{st}\mathfrak{A}_\omega^\mathcal{V} \in \mathsf{Var}\,\mathcal{K} \subseteq \mathcal{V}$. Since $\mathfrak{A}_\omega^\mathcal{V}$ generates $\mathcal{V}$ (as the free algebra), and is a subalgebra of $\mathsf{Cm}\,\mathbb{C}\mathsf{st}\mathfrak{A}_\omega^\mathcal{V}$ $(= \mathsf{Em}\,\mathfrak{A}_\omega^\mathcal{V})$, it follows that $\mathsf{Cm}\,\mathbb{C}\mathsf{st}\mathfrak{A}_\omega^\mathcal{V}$ generates $\mathcal{V}$. Hence $\mathcal{V} = \mathsf{Var}\,\mathcal{K}$, and $\mathcal{V}$ is elementarily generated by $\mathcal{K}$. ∎

An even stronger generalization of Fine's theorem can be obtained by restricting attention to *quasi-modal* sentences. These are first-order sentences of the syntactic form $\forall v\varphi$, with φ being constructed from amongst atomic formulas and the constants $\bot$ (False) and $\top$ (True) using at most $\wedge$ (conjunction), $\vee$ (disjunction), and *bounded* universal and existential quantifiers

$$\forall v_0 \cdots \forall v_{n-1}(R(v_0, \ldots, v_{n-1}, v) \rightarrow \psi)$$
$$\exists v_0 \cdots \exists v_{n-1}(R(v_0, \ldots, v_{n-1}, v) \wedge \psi)$$

with v distinct from $v_0, \ldots, v_{n-1}$. The relevance of quasi-modal sentences, and the reason for the name, is that they are precisely those first-order sentences whose satisfaction is preserved by the basic modal-validity preserving operations of $\mathbb{S}$, $\mathbb{H}$, and $\mathbb{U}\mathsf{d}$ [44, 12]. As a consequence of Theorem 1.2 we have that

> if $\mathcal{K}$ is the class of all models of a set of quasi-modal first-order sentences, then $\mathsf{SCm}\,\mathcal{K}$ is a canonical variety.

By the *quasi-modal theory* of a structure $\mathfrak{S}$ we mean the set of all quasi-modal first-order sentences that are true in $\mathfrak{S}$.

It transpires that for any logic Λ, there is no quasi-modally-expressible property that can differentiate canonical Λ-frames: the structures $\mathfrak{S}^{\Lambda}_{\kappa}$ have exactly the same quasi-modal first-order theory for all κ. We will denote this unique quasi-modal theory of canonical Λ-frames by Ψ^{Λ}. Moreover, if Λ is not canonical, then it always has a largest canonical proper sublogic Λ^c and a largest elementary sublogic Λ^e (with $\Lambda^e \subseteq \Lambda^c$), and the quasi-modal theories Ψ^{Λ^c} and Ψ^{Λ^c} of these other logics are identical to Ψ^{Λ}. These results are all proven in [19].

Our main strengthening of Fine's theorem is

THEOREM 1.5. *[14, 11.4.2] If a modal logic Λ is characterized by some class of frames that is closed under ultraproducts, then it is valid in all models of the first-order theory Ψ^{Λ} (including all the canonical frames $\mathfrak{S}^{\Lambda}_{\kappa}$).* ■

The algebraic version of this theorem states that if a variety $\mathcal{V}$ of BAO's is generated by some $\mathbb{P}\mathsf{u}$-closed class of structures, then $\mathbb{S}\mathsf{tr}\,\mathcal{V}$ contains all models of the quasi-modal first-order theory $\Psi^{\mathcal{V}}$ of the canonical structures $\mathbb{C}\mathsf{st}\mathfrak{A}^{\mathcal{V}}_{\kappa}$. The proof of these results depends on a careful analysis of the preservation theorem for quasi-modal sentences given in [15, Section 7].[8] This shows that if $\Psi^{\mathcal{K}}$ is the set of all quasi-modal sentences true of a class $\mathcal{K}$ of structures that is closed under ultraproducts, and $Mod\Psi^{\mathcal{K}}$ is the class of all models of $\Psi^{\mathcal{K}}$, then

$$\mathbb{C}\mathsf{st}\,\mathsf{Var}\,\mathcal{K} \subseteq Mod\Psi^{\mathcal{K}} = \mathbb{R}\mathsf{u}\,\mathbb{U}\mathsf{b}\,\mathbb{R}\mathsf{u}\,\mathbb{U}\mathsf{b}\,\mathbb{R}\mathsf{u}\,\mathbb{H}\mathbb{S}\mathcal{K} \subseteq \mathbb{S}\mathsf{tr}\,\mathsf{Var}\,\mathcal{K}.$$

[8] Quasi-modal sentences were called "pseudo-equational" in [14, 15].

It seems reasonable to ask whether an elementary generating class for $\mathcal{V}$ might be $\mathbb{S}\mathsf{tr}\mathcal{V}$, the class of all structures $\mathfrak{S}$ with $\mathsf{Cm}\,\mathfrak{S} \in \mathcal{V}$. The version of this for a modal logic Λ is the class $\mathbb{S}\mathsf{tr}\,\Lambda$ of all structures (frames) that validate Λ. But is it known that there are *canonical* logics Λ for which $\mathbb{S}\mathsf{tr}\,\Lambda$ is not an elementary class. The first such example, due to Fine [3], was an elementary Λ, characterized by a subclass of $\mathbb{S}\mathsf{tr}\,\Lambda$ defined by a single first-order sentence, for which $\mathbb{S}\mathsf{tr}\,\Lambda$ itself is not even closed under elementary equivalence. Another example with a long history is the canonical variety **RRA** of representable relation algebras discussed in Theorem 1.3. Robin Hirsch and Ian Hodkinson have shown [24, 14.2.3] that $\mathbb{S}\mathsf{tr}\,$**RRA** is not closed under ultraproducts.

The full story about elementarity of $\mathbb{S}\mathsf{tr}\,\Lambda$ was revealed by the discovery of van Benthem [42] that for $\mathbb{S}\mathsf{tr}\,\Lambda$ to be an elementary class it suffices that it be closed under elementary equivalence. It follows that another sufficient condition is closure under ultrapowers. His proof involved a model-theoretic compactness argument (discussed in [16]), but a simple structural explanation was then found in the present author's observation that there is an injective bounded morphism

$$\textstyle\prod_U \mathfrak{S}_i \rightarrowtail \prod_U(\coprod_I \mathfrak{S}_i)$$

of any U-ultraproduct of a family $\{\mathfrak{S}_i : i \in I\}$ of structures into the associated U-ultrapower of their disjoint union $\coprod_I \mathfrak{S}_i$. This makes the ultraproduct isomorphic to an inner substructure of the ultrapower, and establishes the relationship

$$\mathbb{P}\mathsf{u} \leq \mathbb{S}\mathbb{P}\mathsf{w}\mathbb{U}\mathsf{d} \tag{1.3}$$

(see [12, 3.8.3] for details). Now if a class of the form $\mathbb{S}\mathsf{tr}\,\Lambda$ or $\mathbb{S}\mathsf{tr}\,\mathcal{V}$ is closed under ultrapowers, then since it is always closed under $\mathbb{S}$ and $\mathbb{U}\mathsf{d}$, it follows from (1.iii) that it is also closed under ultraproducts. But $\mathbb{S}\mathsf{tr}\,\Lambda$ (or $\mathbb{S}\mathsf{tr}\,\mathcal{V}$) is always closed under ultraroots $\mathbb{R}\mathsf{u}$, so this now shows that it is an elementary class.

Note that if $\mathbb{S}\mathsf{tr}\,\mathcal{V}$ is closed under ultrapowers, and hence is an elementary class, it does not necessarily follow that $\mathcal{V}$ is canonical. The point is that $\mathbb{S}\mathsf{tr}\,\mathcal{V}$ might not generate $\mathcal{V}$, so Theorem 1.1 does not apply. If this is the case, then no class of structures can generate $\mathcal{V}$, and $\mathcal{V}$ is what is known as an *incomplete* variety. This notion corresponds to that of a logic that is not characterized by its class $\mathbb{S}\mathsf{tr}\,\Lambda$ of validating frames, and so has no completeness theorem with respect to the relational semantics. An extreme example, and the first one discovered, is the tense logic of S. K. Thomason

[41] that is consistent but not validated by any frame. This defines a non-trivial variety for which $\mathbb{S}\mathsf{tr}\,\mathcal{V}$ is the empty class.

In view of (1.iii) it is natural to consider whether $\mathbb{P}\mathsf{u}$-closure can be replaced by $\mathbb{P}\mathsf{w}$-closure in variations on Fine's theorem:

THEOREM 1.6. *A variety of BAO's is generated by some $\mathbb{P}\mathsf{u}$-closed class of structures if, and only if, it is generated by a class that is closed under $\mathbb{P}\mathsf{w}$ and $\mathbb{U}\mathsf{d}$.*

PROOF. Let $\mathcal{V}$ be generated by $\mathcal{K}$, and suppose that $\mathcal{K}$ is closed under $\mathbb{P}\mathsf{w}$ and $\mathbb{U}\mathsf{d}$. Then as $\mathcal{K} \subseteq \mathbb{S}\mathcal{K} \subseteq \mathsf{Var}\mathcal{K}$, $\mathcal{V}$ is also generated by $\mathbb{S}\mathcal{K}$, so it suffices to show that $\mathbb{S}\mathcal{K}$ is $\mathbb{P}\mathsf{u}$-closed. But in general $\mathbb{P}\mathsf{u}\mathbb{S} \leq \mathbb{S}\mathbb{P}\mathsf{u}$, so using (1.iii) we get that

$$\mathbb{P}\mathsf{u}\mathbb{S}\mathcal{K} \subseteq \mathbb{S}\mathbb{P}\mathsf{u}\mathcal{K} \subseteq \mathbb{S}\mathbb{S}\mathbb{P}\mathsf{w}\mathbb{U}\mathsf{d}\mathcal{K} \subseteq \mathbb{S}\mathcal{K}$$

as desired, because $\mathbb{S}\mathbb{S} = \mathbb{S}$ and $\mathcal{K}$ is closed under $\mathbb{P}\mathsf{w}$ and $\mathbb{U}\mathsf{d}$.

Conversely, let $\mathcal{V}$ be generated by $\mathcal{K}$ with $\mathcal{K}$ closed under $\mathbb{P}\mathsf{u}$. Then $\mathcal{V}$ is generated by $\mathbb{H}\mathbb{U}\mathsf{d}\mathcal{K}$, which is the same as $\mathbb{U}\mathsf{b}\mathcal{K}$, so it suffices to prove that the latter is closed under $\mathbb{P}\mathsf{w}$ and $\mathbb{U}\mathsf{d}$. Now as an instance of $\mathbb{P}\mathsf{u}\mathbb{U}\mathsf{b} \leq \mathbb{U}\mathsf{b}\mathbb{P}\mathsf{u}$ (see after (1.ii)) we have that $\mathbb{P}\mathsf{w}\mathbb{U}\mathsf{b} \leq \mathbb{U}\mathsf{b}\mathbb{P}\mathsf{u}$, so as $\mathbb{P}\mathsf{u}\mathcal{K} \subseteq \mathcal{K}$ in this case we immediately get $\mathbb{P}\mathsf{w}\mathbb{U}\mathsf{b}\mathcal{K} \subseteq \mathbb{U}\mathsf{b}\mathcal{K}$.

But in general $\mathbb{U}\mathsf{d}\mathbb{H} \leq \mathbb{H}\mathbb{U}\mathsf{d}$ and $\mathbb{U}\mathsf{d}\mathbb{U}\mathsf{d} = \mathbb{U}\mathsf{d}$, so $\mathbb{U}\mathsf{d}\mathbb{U}\mathsf{b} = \mathbb{U}\mathsf{d}\mathbb{H}\mathbb{U}\mathsf{d} \leq \mathbb{H}\mathbb{U}\mathsf{d}\mathbb{U}\mathsf{d} = \mathbb{H}\mathbb{U}\mathsf{d} = \mathbb{U}\mathsf{b}$, showing that $\mathbb{U}\mathsf{b}\mathcal{K}$ is closed under $\mathbb{U}\mathsf{d}$ for any class $\mathcal{K}$. ■

The converse of Fine's theorem remains an open question. Is every canonical logic elementary, and more generally is every canonical variety of BAO's generated by some $\mathbb{P}\mathsf{u}$-closed class of structures? In view of the analysis given here, we see that to answer this converse question negatively it would be enough to exhibit a canonical logic Λ that was not valid in some frame elementarily equivalent to $\mathfrak{S}^{\Lambda}_{\omega}$, or even more weakly was not valid in some model of the theory Ψ^{Λ} (see 1.5). Another perspective is provided by the following parallel to Theorem 1.6:

THEOREM 1.7. *A variety $\mathcal{V}$ of BAO's is canonical if, and only if, it is generated by a class that is closed under $\mathbb{E}\mathsf{x}$ and $\mathbb{U}\mathsf{d}$.*

PROOF. Let $\mathcal{V}$ be canonical. Then $\mathcal{V}$ is generated by its class $\mathbb{S}\mathsf{tr}\,\mathcal{V}$ of structures. For if $\mathfrak{A} \in \mathcal{V}$ then $\mathfrak{A}$ is a subalgebra of $\mathsf{Cm}\mathbb{C}\mathsf{st}\mathfrak{A} = \mathsf{Em}\mathfrak{A} \in \mathcal{V}$, which shows that $\mathbb{C}\mathsf{st}\mathfrak{A} \in \mathbb{S}\mathsf{tr}\,\mathcal{V}$. Hence $\mathcal{V} = \mathbb{S}\mathsf{Cm}\mathbb{S}\mathsf{tr}\,\mathcal{V}$. But $\mathbb{S}\mathsf{tr}\,\mathcal{V}$ is always closed under $\mathbb{U}\mathsf{d}$, and in this case is closed under $\mathbb{E}\mathsf{x}$, for if $\mathfrak{S} \in \mathbb{S}\mathsf{tr}\,\mathcal{V}$, then $\mathsf{Cm}\mathbb{E}\mathsf{x}\mathfrak{S} = \mathsf{Em}\,\mathsf{Cm}\,\mathfrak{S} \in \mathcal{V}$, showing that $\mathbb{E}\mathsf{x}\mathfrak{S} \in \mathbb{S}\mathsf{tr}\,\mathcal{V}$.

Conversely, let $\mathcal{V}$ be generated by $\mathcal{K}$ with $\mathcal{K}$ closed under $\mathbb{E}\mathsf{x}$ and $\mathbb{U}\mathsf{d}$. Then we show that $\mathcal{V}$ is closed under Em. If $\mathfrak{A} \in \mathcal{V}$, then

$$\mathfrak{A} \in \mathsf{HSPCm}\mathcal{K} = \mathsf{HSCm}\mathbb{U}\mathsf{d}\mathcal{K} \subseteq \mathsf{HSCm}\mathcal{K},$$

as $\mathcal{K}$ is $\mathbb{U}\mathsf{d}$-closed. Thus $\mathfrak{A}$ is a homomorphic image of some BAO $\mathfrak{B}$ that is a subalgebra of $\mathsf{Cm}\,\mathfrak{S}$ for some $\mathfrak{S} \in \mathcal{K}$. This gives a diagram of homomorphisms

$$\mathfrak{A} \twoheadleftarrow \mathfrak{B} \rightarrowtail \mathsf{Cm}\,\mathfrak{S},$$

which lifts by duality to

$$\mathsf{Em}\,\mathfrak{A} \twoheadleftarrow \mathsf{Em}\,\mathfrak{B} \rightarrowtail \mathsf{Em}\,\mathsf{Cm}\,\mathfrak{S}.$$

But $\mathbb{E}\mathsf{x}\mathfrak{S} \in \mathcal{K}$ by $\mathbb{E}\mathsf{x}$-closure of $\mathcal{K}$, so $\mathsf{Em}\,\mathsf{Cm}\,\mathfrak{S} = \mathsf{Cm}\,\mathbb{E}\mathsf{x}\mathfrak{S} \in \mathsf{Cm}\mathcal{K} \subseteq \mathcal{V}$. Closure of $\mathcal{V}$ under S and H then guarantees that $\mathsf{Em}\,\mathfrak{A} \in \mathcal{V}$ as desired. ∎

Comparing Theorems 1.6 and 1.7 shows that a canonical variety that is not elementarily generated, if there is such a thing, would be one that has a $\mathbb{E}\mathsf{x}$-$\mathbb{U}\mathsf{d}$-closed generating class but no $\mathbb{P}\mathsf{w}$-$\mathbb{U}\mathsf{d}$-closed generating class.

Yet another perspective on the converse question is provided by considering *finite* validating frames for a logic Λ. We proved in [19, 9.2] that if Λ is elementary then it is valid in any ultraproduct of finite Λ-frames. The algebraic version is that

> if a variety $\mathcal{V}$ is elementarily generated, then $\mathbb{S}\mathsf{tr}\,\mathcal{V}$ is closed under ultraproducts of its finite members.

The reason for this is that if $\mathfrak{S}$ is a finite member of $\mathbb{S}\mathsf{tr}\,\mathcal{V}$, then the finite algebra $\mathsf{Cm}\,\mathfrak{A}$ is a homomorphic image of the free $\mathcal{V}$-algebra $\mathfrak{A}^{\mathcal{V}}_{\omega}$, so $\mathbb{E}\mathsf{x}\mathfrak{S} = \mathbb{C}\mathsf{st}\,\mathsf{Cm}\,\mathfrak{S}$ is isomorphic to an inner substructure of $\mathbb{C}\mathsf{st}\,\mathfrak{A}^{\mathcal{V}}_{\omega}$ and hence is a model of the quasi-modal first-order theory $\Psi^{\mathcal{V}}$ of $\mathbb{C}\mathsf{st}\,\mathfrak{A}^{\mathcal{V}}_{\omega}$. But in the finite case, $\mathbb{E}\mathsf{x}\mathfrak{S}$ is isomorphic to $\mathfrak{S}$. Thus all finite $\mathcal{V}$-structures are models of $\Psi^{\mathcal{V}}$, so any ultraproduct of finite $\mathcal{V}$-structures is a model of $\Psi^{\mathcal{V}}$, and therefore belongs to $\mathbb{S}\mathsf{tr}\,\mathcal{V}$ (see remark after Theorem 1.5).

The situation is well illustrated by the variety **RRA**, which is canonical and elementarily generated. Hence $\mathbb{S}\mathsf{tr}\,$**RRA** is closed under ultraproducts of *finite* structures, but not closed under arbitrary ultraproducts because it is not elementary.

Thus to show that a canonical variety $\mathcal{V}$ is not elementarily generated it would be enough to show that it is not closed under ultraproducts of its finite members. Equivalently, to show that a canonical logic Λ is not elementary it would be enough to show that it is falsifiable in some ultraproduct of finite Λ-frames.

2. Three Cases of Elementary Generation

All canonical varieties $\mathcal{V}$ of BAO's that have been studied to date are known to be elementarily generated. Indeed most have been defined in the first place by equations that are preserved by the $\mathsf{Em}\,\mathfrak{A}$ construction, with the proof of this consisting of a demonstration that the canonical structure $\mathbb{C}\mathsf{st}\,\mathfrak{A}$ satisfies some elementary conditions that force $\mathsf{Em}\,\mathfrak{A}$ to validate these defining equations for $\mathcal{V}$.

However there are some classes of algebras that have turned out to be canonical varieties, but which were defined in some structural way that is not obviously equational or expressible by first-order conditions on generating structures. Such examples arise particularly in the area of relation algebras and cylindric algebras, involving varieties that are defined by some representability conditions, as in the case of **RRA** we discussed above.

In this section we consider three such types of canonical variety whose elementary generation has not previously been shown in the literature. To show this we call on the lessons of Section 1, from which we learned that to prove that variety $\mathcal{V}$ is elementarily generated it is enough to prove that it is generated by a $\mathbb{P}\mathsf{u}$-closed class, or that $\mathbb{S}\mathsf{tr}\,\mathcal{V}$ contains all ultrapowers of $\mathbb{C}\mathsf{st}\,\mathfrak{A}^{\mathcal{V}}_{\omega}$.

2.1. $\mathbf{RA}_n$

The classes $\mathbf{RA}_n$ for $3 \leq n \leq \omega$ were introduced by Roger Maddux in [37].[9] $\mathbf{RA}_\omega$ is identical to **RRA**, so we will focus on the case of finite n. Algebras for these classes are of the same similarity type $\mathfrak{A} = (\mathfrak{A}_0, ;, \breve{}, 1')$ as for **RRA**, with $\mathfrak{A}_0$ a Boolean algebra having operators ; (binary) and $\breve{}$ (unary) and a distinguished element 1'. The structures dual to these algebras have the form $\mathfrak{S} = (S, C, R, I)$ with $C \subseteq S^3$, $R \subseteq S^2$ and $I \subseteq S$, as discussed prior to Theorem 1.3.

There is a canonical variety **NA** of *nonassociative relation algebras* defined by equations that are weaker than those for **RA** (so $\mathbf{RA} \subseteq \mathbf{NA}$) but strong enough to imply that if x is an atom of $\mathfrak{A}$ then so is $x^{\smile}$, and $x^{\smile\smile} = x$ [36, 3.4]. This implies that if $\mathsf{Cm}\,\mathfrak{S} \in \mathbf{NA}$, then the relation R of $\mathfrak{S}$ is total and functional and defines a function of period two. We can replace R by this function $f : S \to S$ that has $ffx = x$ [35, p. 56]. Then the conversion operation in $\mathsf{Cm}\,\mathfrak{S}$ is just given by f-images:

$$X^{\smile} = \{fx : x \in X\}.$$

[9] They were originally called MA_n.

$\mathbb{S}\mathrm{tr}\,\mathbf{NA}$ is an elementary class [35, 36].

Maddux introduced the notion of an *n-dimensional basis* for an atomic **NA**-algebra as a collection of $n \times n$ matrices of atoms of $\mathfrak{A}$ with certain properties (see below). $\mathbf{RA}_n$ is defined as the class of subalgebras of those complete and atomic **NA**-algebras that have an n-dimensional basis.

Now if $\mathfrak{A}$ is complete and atomic, then $\mathfrak{A}$ is isomorphic to $\mathsf{Cm}\,\mathfrak{S}$ where $\mathfrak{S}$ is a structure based on the set of atoms of $\mathfrak{A}$.[10] Then matrices of atoms of $\mathfrak{A}$ become matrices of points of $\mathfrak{S}$, and conditions on atoms of $\mathfrak{A}$ become conditions expressible in the language of $\mathfrak{S}$. We will define bases in these terms here. An $n \times n$ matrix will be described as $a = (a_{ij})$, where a_{ij} is the ij-th entry of a.

An *n-dimensional basis* for a structure of type $\mathfrak{S} = (S, C, f, I)$ is a set M of $n \times n$ matrices of elements of S satisfying the following.

(a) If $a \in M$, then $a_{ii} \in I$, $fa_{ij} = a_{ji}$ and $C(a_{ik}, a_{kj}, a_{ij})$ for all $i, j, k < n$.

(b) If $a \in M$, $x, y \in S$, $i, j < n$ and $C(x, y, a_{ij})$, then for all $k < n$ with $k \notin \{i, j\}$ there exists some $b \in M$ with $b_{ik} = x$, $b_{kj} = y$, and $b_{lm} = a_{lm}$ whenever $k \notin l, m$ (so b agrees with a outside of row k and column k).

(c) For all $x \in S$ there is some $a \in M$ with $a_{01} = x$.

Let $\mathcal{B}_n$ be the class of all structures $\mathfrak{S}$ such that $\mathfrak{S} \in \mathbb{S}\mathrm{tr}\,\mathbf{NA}$ (i.e. $\mathsf{Cm}\,\mathfrak{S} \in \mathbf{NA}$) and $\mathfrak{S}$ has an n-dimensional basis. Then $\mathbf{RA}_n = \mathsf{S}\,\mathsf{Cm}\,\mathcal{B}_n$ by definition.

We are going to show that $\mathcal{B}_n$ is closed under $\mathbb{P}\mathrm{u}$, $\mathbb{S}$ and $\mathbb{U}\mathrm{d}$. Hence by Theorem 1.2, $\mathsf{S}\,\mathsf{Cm}\,\mathcal{B}_n$ is a canonical variety, and so is the variety generated by $\mathcal{B}_n$. This achieves our goal of showing that $\mathbf{RA}_n$ has a $\mathbb{P}\mathrm{u}$-closed generating class, and therefore an elementary one. But it also provides a new proof of the result [37] that $\mathbf{RA}_n$ is a variety and is canonical.

$\mathbb{S}\mathrm{tr}\,\mathbf{NA}$ is closed under $\mathbb{S}$ and $\mathbb{U}\mathrm{d}$, since **NA** is a variety, and is $\mathbb{P}\mathrm{u}$-closed because it is elementary. Thus the burden of the proof is to show that the property of a structure having an n-dimensional basis is preserved by these three constructions.

For the case of $\mathbb{P}\mathrm{u}$, an $n \times n$ matrix will be viewed as an n^2-tuple

$$(a_{00}, \ldots, a_{ij}, \ldots, a_{n-1\,n-1})$$

indexed by the pairs ij with $i, j < n$ (ordered lexicographically, say). Then a set M of $n \times n$ matrices over S becomes an n^2-ary relation $M \subseteq S^{n^2}$. For

[10] The definition of this structure appears in Section 4 below.

a fixed n, the three conditions (a)–(c) can be expressed in the first-order language of the structure $(\mathfrak{S}, M)$. For instance (c) is expressible by

$$\forall x \exists a_{00} \ldots \exists a_{n-1\,n-1} \, (M(a_{00}, \ldots, a_{n-1\,n-1}) \wedge x = a_{01})$$

(or equivalently $\forall x \exists a_{00} \ldots \exists a_{n-1\,n-1} M(a_{00}, x, a_{03}, \ldots, a_{n-1\,n-1})$). Conjoining the three sentences expressing (a)–(c) gives a single first-order sentence φ_n that is true in $(\mathfrak{S}, M)$ iff M is an n-dimensional basis for $\mathfrak{S}$. Then regarding M as a relation variable we form the *second-order* sentence $\exists M \varphi_n$ which it true in $\mathfrak{S}$ iff $\mathfrak{S}$ has an n-dimensional basis.

But syntactically $\exists M \varphi_n$ is a Σ^1_1 second-order sentence, and it is a standard fact that the truth of such sentences is preserved by ultraproducts [1, 4.1.14]. Therefore $\mathcal{B}_n$ is closed under ultraproducts.

For the case of closure under $\mathbb{S}$ we use the fact that, in this context, if $\mathfrak{S}_1$ is an inner substructure of $\mathfrak{S}_2$ then it is a substructure of $\mathfrak{S}_2$ such that $f_2^{-1}(S_1) \subseteq S_1$, i.e. $f_2 x \in S_1$ implies $x \in S_1$ for all $x \in S_2$, and the following holds for all $x, y, z \in S_2$:

$$\text{if } C_2(x, y, z) \text{ and } z \in S_1, \text{ then } x, y \in S_1$$

(as noted above in the discussion of **GRA**). The condition on f_2 ensures that S_1 is closed under f_2, for if $x \in S_1$, then $f_2 f_2 x \in S_1$ as f_2 is of period two, hence $f_2 x \in S_1$ by the inner condition.

If $\mathfrak{S}_1$ is an inner substructure of $\mathfrak{S}_2$ and $\mathfrak{S}_2$ has an n-dimensional basis M_2, let M_1 be the set of matrices from M_2 whose entries all belong to S_1. Then M_1 is an n-dimensional basis for $\mathfrak{S}_1$. The key to proving this is the observation that as soon as a matrix $a \in M_2$ has *at least one* entry in S_1 then all its entries are in S_1, so $a \in M_1$. For, if $a_{ij} \in S_1$, then for any $k < n$ we have $C_2(a_{ik}, a_{kj}, a_{ij})$ by (a), and hence $a_{ik}, a_{kj} \in S_1$ by the inner substructure condition. Thus all members of the row and column of a_{ij} are in S_1. Now we can repeat that argument for each member of the row (or column) of a_{ij} to deduce that all entries of a are in S_1. Using this observation it is now straightforward to show that the conditions (a)–(c) are preserved in passing from M_2 to M_1. Thus $\mathcal{B}_n$ is closed under $\mathbb{S}$.

For the case of $\mathbb{U}\mathrm{d}$, let $\mathfrak{S}$ be the disjoint union of a family $\{\mathfrak{S}_\lambda : \lambda \in I\}$ with each $\mathfrak{S}_\lambda$ having an n-dimensional basis M_λ. $\mathfrak{S}$ is the union of a family of pairwise disjoint copies of the $\mathfrak{S}_\lambda$'s, so we can assume that these are already pairwise disjoint. Each is then an inner substructure of $\mathfrak{S}$. The union of all the M_λ's is easily checked to be an n-dimensional basis for $\mathfrak{S}$.

This completes the proof that $\mathcal{B}_n$ is closed under $\mathbb{P}\mathrm{u}$, $\mathbb{S}$ and $\mathbb{U}\mathrm{d}$, and hence that $\mathbf{RA}_n$ is an elementarily generated canonical variety. Note that we have not shown that $\mathbb{S}\mathrm{tr}\,\mathbf{RA}_n$ is an elementary class, and indeed this is not in general true. The intersection of all the $\mathbf{RA}_n$'s is the variety $\mathbf{RRA}$ [37], so the intersection of all the $\mathbb{S}\mathrm{tr}\,\mathbf{RA}_n$'s is $\mathbb{S}\mathrm{tr}\,\mathbf{RRA}$ which is not an elementary class. Since the intersection of a family of elementary classes is elementary, some $\mathbb{S}\mathrm{tr}\,\mathbf{RA}_n$ must fail to be elementary. In fact there must be some m such that $\mathbb{S}\mathrm{tr}\,\mathbf{RA}_n$ fails to be elementary for all $n > m$. This is because the $\mathbb{S}\mathrm{tr}\,\mathbf{RA}_n$'s form a nested sequence with $\mathbb{S}\mathrm{tr}\,\mathbf{RA}_n \supseteq \mathbb{S}\mathrm{tr}\,\mathbf{RA}_{n+1}$, so if there were arbitrarily large n with $\mathbb{S}\mathrm{tr}\,\mathbf{RA}_n$ elementary, then $\mathbb{S}\mathrm{tr}\,\mathbf{RRA}$ would be an intersection of elementary classes.

2.2. $\mathsf{S}\,\mathsf{Nr}_\beta\mathbf{CA}_\alpha$

Let α and β be ordinals with $\beta < \alpha$. $\mathbf{CA}_\alpha$ is the variety of *cylindric algebras of dimension* α, and consists of BAO's of the type $\mathfrak{A} = (\mathfrak{A}_0, c_{ij}, d_{ij})_{i,j<\alpha}$, with each c_{ij} a unary operator called a *cylindrification* and each d_{ij} a member of of $\mathfrak{A}_0$ called a *diagonal element.* $\mathbf{CA}_\alpha$ can be defined by positive equations, and is a canonical variety for which $\mathbb{S}\mathrm{tr}\,\mathbf{CA}_\alpha$ is an elementary class [21, 2.7].

The members of $\mathbb{S}\mathrm{tr}\,\mathbf{CA}_\alpha$ are structures of the form $\mathfrak{S} = (S, R_i, E_{ij})_{i,j<\alpha}$ with R_i an equivalence relation on S, and E_{ij} a subset of S that is the diagonal element d_{ij} in $\mathsf{Cm}\,\mathfrak{S}$. Cylindrifications in $\mathsf{Cm}\,\mathfrak{S}$ are defined by

$$c_i X = \{y \in S : \exists x \in X(xR_i y)\}.$$

If $\mathfrak{A}$ is as above, then the algebra

$$\mathsf{Rd}_\beta\mathfrak{A} = (\mathfrak{A}_0, c_{ij}, d_{ij})_{i,j<\beta}$$

will be called the *β-reduct* of $\mathfrak{A}$. It is a member of $\mathbf{CA}_\beta$ and any of its subalgebras is a *subreduct* of $\mathfrak{A}$. The set

$$Cl_{\alpha-\beta}\mathfrak{A} = \{x \text{ in } \mathfrak{A}_0 : c_i x = x \text{ for all } i \in \alpha - \beta\}$$

is the underlying set of a subalgebra of $\mathsf{Rd}_\beta\mathfrak{A}$, comprising those elements that are left fixed by all cylindrifications c_i with $i \geq \beta$. This subalgebra is called the *neat β-reduct* of $\mathfrak{A}$, and denoted $\mathsf{Nr}_\beta\mathfrak{A}$. $\mathsf{Nr}_\beta\mathbf{CA}_\alpha$ is the class of all neat β-reducts of members of $\mathbf{CA}_\alpha$, and its closure $\mathsf{S}\,\mathsf{Nr}_\beta\mathbf{CA}_\alpha$ under subalgebras is the class of all *neat β-subreducts* of $\mathbf{CA}_\alpha$-alebras.

$\mathsf{S}\,\mathsf{Nr}_\beta\mathbf{CA}_\alpha$ was shown by Monk to be a canonical variety by direct algebraic constructions proving that it is closed under H, S, P and Em (see [21, 2.6.32 and 2.7.24]). These varieties play a significant role in cylindric algebra theory. In particular, $\mathsf{S}\,\mathsf{Nr}_\beta\mathbf{CA}_{\beta+\omega}$ coincides with the class $\mathbf{RCA}_\beta$ of all representable cylindric algebras of dimension β.

Now fix a member $\mathfrak{B} = (\mathfrak{B}_0, c_{ij}, d_{ij})_{i,j<\beta}$ of $\mathsf{S}\,\mathsf{Nr}_\beta\mathbf{CA}_\alpha$. We will show that any ultrapower of $\mathbb{C}\mathsf{st}\,\mathfrak{B}$ belongs to $\mathbb{S}\mathsf{tr}\,\mathsf{S}\,\mathsf{Nr}_\beta\mathbf{CA}_\alpha$. Symbolically, this says that

$$\mathbb{P}\mathsf{w}\,\mathbb{C}\mathsf{st}\,\mathsf{S}\,\mathsf{Nr}_\beta\mathbf{CA}_\alpha \subseteq \mathbb{S}\mathsf{tr}\,\mathsf{S}\,\mathsf{Nr}_\beta\mathbf{CA}_\alpha.$$

By our Theorem 1.4 that is (more than) enough to show that $\mathsf{S}\,\mathsf{Nr}_\beta\mathbf{CA}_\alpha$ is elementarily generated.

Take $\mathfrak{A} \in \mathbf{CA}_\alpha$ such that $\mathfrak{B}$ is a subalgebra of the neat β-reduct $\mathsf{Nr}_\beta\mathfrak{A}$. Then $\mathfrak{B}$ is a subalgebra of the β-reduct $\mathsf{Rd}_\beta\mathfrak{A}$ of $\mathfrak{A}$, and $\mathfrak{B}_0$ is a Boolean subalgebra of $\mathfrak{A}_0$ included in $Cl_{\alpha-\beta}\mathfrak{A}$. To establish some notation, put

$$\mathbb{C}\mathsf{st}\,\mathfrak{B} = (S^{\mathfrak{B}}, R_i^{\mathfrak{B}}, E_{ij}^{\mathfrak{B}})_{i,j<\beta}$$
$$\mathbb{C}\mathsf{st}\,\mathfrak{A} = (S^{\mathfrak{A}}, R_i^{\mathfrak{A}}, E_{ij}^{\mathfrak{A}})_{i,j<\alpha}$$

and then

$$\mathbb{R}\mathsf{d}_\beta\mathbb{C}\mathsf{st}\,\mathfrak{A} = (S^{\mathfrak{A}}, R_i^{\mathfrak{A}}, E_{ij}^{\mathfrak{A}})_{i,j<\beta}.$$

Here $S^{\mathfrak{B}}$ is the set of all ultrafilters of $\mathfrak{B}_0$, and $S^{\mathfrak{A}}$ the set of ultrafilters of $\mathfrak{A}_0$. The structure $\mathbb{R}\mathsf{d}_\beta\mathbb{C}\mathsf{st}\,\mathfrak{A}$ is by definition the *β-reduct* of the canonical structure $\mathbb{C}\mathsf{st}\,\mathfrak{A}$. The relation $R_i^{\mathfrak{A}}$ is defined by

$$pR_i^{\mathfrak{A}}q \text{ iff } \{c_i x : x \in p\} \subseteq q.$$

By duality, the inclusion of $\mathfrak{B}$ into $\mathsf{Rd}_\beta\mathfrak{A}$ induces a surjection $\theta : S^{\mathfrak{A}} \twoheadrightarrow S^{\mathfrak{B}}$, given by $\theta(x) = x \cap B_0$, where B_0 is the underlying set of $\mathfrak{B}_0$. θ is a bounded morphism from $\mathbb{C}\mathsf{st}\,\mathsf{Rd}_\beta\mathfrak{A}$ onto $\mathbb{C}\mathsf{st}\,\mathfrak{B}$.

Now let U be an ultrafilter on a set I. For a structure $\mathfrak{S}$, the elements of the U-ultrapower $\Pi_U\mathfrak{S}$ will be denoted f^U, where $f \in S^I$. Here $f^U = \{g \in S^I : \{k \in I : f(k) = g(k)\} \in U\}$. The ultrapower construction lifts the bounded epimorphism $\theta : \mathbb{C}\mathsf{st}\,\mathsf{Rd}_\beta\mathfrak{A} \to \mathbb{C}\mathsf{st}\,\mathfrak{B}$ just defined to the map

$$\theta_U : \Pi_U\mathbb{C}\mathsf{st}\,\mathsf{Rd}_\beta\mathfrak{A} \to \Pi_U\mathbb{C}\mathsf{st}\,\mathfrak{B}$$

having $\theta_U(f^U) = (\theta \circ f)^U$. θ_U is a bounded epimorphism, and so by duality it induces an injective $\mathbf{CA}_\beta$-homomorphism

$$\theta_U^+ : \mathsf{Cm}\,\Pi_U\mathbb{C}\mathsf{st}\,\mathfrak{B} \rightarrowtail \mathsf{Cm}\,\Pi_U\mathbb{C}\mathsf{st}\,\mathsf{Rd}_\beta\mathfrak{A},$$

which acts by $X \mapsto \theta_U^{-1}(X)$.

Now because reduct formations just involve deleting operations and relations without changing the underlying set of the algebra or structure, it is evident that they commute with many operations. In particular we can show such facts as $\mathbb{C}\mathrm{st}\,\mathsf{Rd}_\beta\mathfrak{A} = \mathbb{R}\mathrm{d}_\beta\mathbb{C}\mathrm{st}\,\mathfrak{A}$, $\Pi_U\mathbb{R}\mathrm{d}_\beta\mathfrak{S} = \mathbb{R}\mathrm{d}_\beta\Pi_U\mathfrak{S}$, and $\mathsf{Cm}\,\mathbb{R}\mathrm{d}_\beta\mathfrak{S} = \mathsf{Rd}_\beta\mathsf{Cm}\,\mathfrak{S}$. Using these to re-write the codomain of θ_U^+ we get

$$\theta_U^+ : \mathsf{Cm}\,\Pi_U\mathbb{C}\mathrm{st}\,\mathfrak{B} \rightarrowtail \mathsf{Rd}_\beta\mathsf{Cm}\,\Pi_U\mathbb{C}\mathrm{st}\,\mathfrak{A}.$$

Since $\mathbf{CA}_\alpha$ is canonical, $\mathbb{C}\mathrm{st}\,\mathfrak{A}$ belongs to the elementary class $\mathbb{S}\mathrm{tr}\,\mathbf{CA}_\alpha$, and therefore so does the ultrapower $\Pi_U\mathbb{C}\mathrm{st}\,\mathfrak{A}$. Hence $\mathsf{Cm}\,\Pi_U\mathbb{C}\mathrm{st}\,\mathfrak{A}$ belongs to $\mathbf{CA}_\alpha$. So, if we can show that

(†) θ_U^+ maps $\mathsf{Cm}\,\Pi_U\mathbb{C}\mathrm{st}\,\mathfrak{B}$ into $Cl_{\alpha-\beta}\mathsf{Cm}\,\Pi_U\mathbb{C}\mathrm{st}\,\mathfrak{A}$,

this will make $\mathsf{Cm}\,\Pi_U\mathbb{C}\mathrm{st}\,\mathfrak{B}$ isomorphic to a subalgebra of the neat β-reduct $\mathsf{Nr}_\beta\mathsf{Cm}\,\Pi_U\mathbb{C}\mathrm{st}\,\mathfrak{A}$, and hence a member of $\mathsf{S}\,\mathsf{Nr}_\beta\mathbf{CA}_\alpha$. That puts $\Pi_U\mathbb{C}\mathrm{st}\,\mathfrak{B}$ into $\mathbb{S}\mathrm{tr}\,\mathsf{S}\,\mathsf{Nr}_\beta\mathbf{CA}_\alpha$, showing that the latter contains all ultrapowers of $\mathbb{C}\mathrm{st}\,\mathfrak{B}$, as desired.

For each element X of $\mathsf{Cm}\,\Pi_U\mathbb{C}\mathrm{st}\,\mathfrak{B}$, put $X^+ = \theta_U^+(X) = \theta_U^{-1}(X)$. To prove (†) we have to show that $c_iX^+ = X^+$ in $\mathsf{Cm}\,\Pi_U\mathbb{C}\mathrm{st}\,\mathfrak{A}$ for any $i \geq \beta$. Let R_i^U be the relation in the ultrapower $\Pi_U\mathbb{C}\mathrm{st}\,\mathfrak{A}$ that determines c_i. By definition,

$$f^U R_i^U g^U \quad \text{iff} \quad \{k \in I : f(k) R_i^{\mathfrak{A}} g(k) \text{ in } \mathbb{C}\mathrm{st}\,\mathfrak{A}\} \in U.$$

Since $\mathsf{Cm}\,\Pi_U\mathbb{C}\mathrm{st}\,\mathfrak{A}$ is a cylindric algebra it is immediate that $X^+ \subseteq c_iX^+$. For the converse inclusion, let $g^U \in c_iX^+$, so that $f^U R_i^U g^U$ for some $f^U \in X^+$. Hence $\theta_U(f^U) \in X$ and the set

$$J = \{k \in I : f(k) R_i^{\mathfrak{A}} g(k)\}$$

belongs to U. Now for any $k \in J$, if $x \in \theta(f(k)) = f(k) \cap B_0$, then $c_ix \in g(k)$ as $x \in f(k)$ and $f(k)R_i^{\mathfrak{A}}g(k)$; while $x \in B_0 \subseteq Cl_{\alpha-\beta}\mathfrak{A}$ so $c_ix = x$; hence $x \in g(k) \cap B_0 = \theta(g(k))$. This proves that $\theta(f(k)) \subseteq \theta(g(k))$ and therefore that $\theta(f(k)) = \theta(g(k))$ as both are ultrafilters of $\mathfrak{B}_0$. Thus the functions $\theta \circ f$ and $\theta \circ g$ agree on the set $J \in U$, which is enough to force $(\theta \circ f)^U = (\theta \circ g)^U$, i.e. $\theta_U(f^U) = \theta_U(g^U)$. But $\theta_U(f^U) \in X$, and so $g^U \in \theta_U^{-1}(X) = X^+$.

This establishes that $c_iX^+ = X^+$, completing the proof of (†), hence the proof that $\Pi_U\mathbb{C}\mathrm{st}\,\mathfrak{B}$ is a $\mathsf{S}\,\mathsf{Nr}_\beta\mathbf{CA}_\alpha$-structure, and thereby completing the proof that $\mathsf{S}\,\mathsf{Nr}_\beta\mathbf{CA}_\alpha$ is elementarily generated.

2.3. $\mathsf{S}\mathfrak{Ra}\mathbf{CA}_\alpha$

From any cylindric algebra $\mathfrak{A} = (\mathfrak{A}_0, c_{ij}, d_{ij})_{i,j<\alpha}$ with $\alpha \geq 3$ an $\mathbf{RA}$-type algebra $\mathfrak{Ra}\mathfrak{A}$ can be defined on the underlying set of the neat 2-reduct $\mathsf{Nr}_2\mathfrak{A}$

of $\mathfrak{A}$. This uses the *substitution operators* of $\mathfrak{A}$: for any distinct $i, j < \alpha$, let $s^i_j b = c_i(d_{ij} \cdot b)$. Then s^i_j is an operator on $\mathfrak{A}_0$ that is *completely additive* in the sense that if B is any subset that has a join ΣB in $\mathfrak{A}_0$, then $s^i_j \Sigma B = \Sigma\{s^i_j b : b \in B\}$. This follows from the complete additivity of the cylindrifications c_i of $\mathfrak{A}$ and the Boolean principle that $a \cdot \Sigma B = \Sigma\{a \cdot b : b \in B\}$ (i.e. complete additivity of the Boolean product (meet) operation).

The set $Cl_{\alpha-2}\mathfrak{A}$ underlying $\mathsf{Nr}_2\mathfrak{A}$ is closed under the operations ; , $\breve{}$, 1' defined by

$$a;b = c_2(s^1_2 a \cdot s^0_2 b), \qquad a^{\smile} = s^2_0 s^0_1 s^1_2 a, \qquad 1' = d_{01}. \tag{2.4}$$

From the complete additivity of c_i and s^i_j it follows readily that $\breve{}$ is completely additive and ; is completely additive in each of its arguments. Define

$$\mathfrak{Ra}\mathfrak{A} = (Cl_{\alpha-2}\mathfrak{A}, \ ; \ , \ \breve{} \ , 1').$$

Henkin and Tarski, who developed the $\mathfrak{Ra}$ construction, proved that $\mathfrak{Ra}\mathfrak{A}$ is a relation algebra whenever $\alpha \geq 4$ [22, 5.3.8]. For $\alpha = 3$ this can fail (see [24, p. 191] for discussion). For $\alpha \geq 3$, the class $\mathsf{S}\mathfrak{Ra}\mathbf{CA}_\alpha$ is a canonical variety, as may be shown by a similar proof to that for $\mathsf{S}\mathsf{Nr}_\beta\mathbf{CA}_a$ [24, 5.48].

In order to give a structural proof that $\mathsf{S}\mathfrak{Ra}\mathbf{CA}_\alpha$ is elementarily generated, we introduce a new construction $\mathbb{M}\mathfrak{S}$ of $\mathbf{CA}_\alpha$-structures that dualizes the operations (2.iv) on cylindric algebras, and which plays an analogous role for $\mathsf{S}\mathfrak{Ra}\mathbf{CA}_\alpha$ to that played by $\mathbb{R}\mathsf{d}_\beta$ for $\mathsf{S}\mathsf{Nr}_\beta\mathbf{CA}_\alpha$. First note that these operations (2.iv) are defined on the whole of the Boolean algebra $\mathfrak{A}_0$, so we can form the algebra

$$\mathsf{M}\mathfrak{A} = (\mathfrak{A}_0, \ ; \ , \ \breve{} \ , 1').$$

Then $\mathfrak{Ra}\mathfrak{A}$ is the subalgebra of $\mathsf{M}\mathfrak{A}$ based on $Cl_{\alpha-2}\mathfrak{A}$. (The letter M could stand for "monoid", since 1' is an identity element for the operation ;, which is associative when $\alpha \geq 4$.)

Now for a $\mathbf{CA}_\alpha$-structure $\mathfrak{S} = (S, R_i, E_{ij})_{i,j<\alpha}$, put

$$\mathbb{M}\mathfrak{S} = (S, C, R, E_{01}),$$

where $C \subseteq S^3$ and $R \subseteq S^2$ are defined by

$$C(x, y, z) \quad \text{iff} \quad z \in \{x\}; \{y\} \text{ in } \mathsf{M}\,\mathsf{Cm}\,\mathfrak{S},$$
$$xRy \quad \text{iff} \quad y \in \{x\}^{\smile} \text{ in } \mathsf{M}\,\mathsf{Cm}\,\mathfrak{S}.$$

LEMMA 2.1. $\mathsf{Cm}\,\mathbb{M}\mathfrak{S} = \mathsf{M}\,\mathsf{Cm}\,\mathfrak{S}$.

PROOF. $\mathsf{Cm}\,\mathbb{M}\mathfrak{S}$ and $\mathsf{M}\,\mathsf{Cm}\,\mathfrak{S}$ are both based on the Boolean powerset algebra of S, and both have E_{01} as their distinguished element 1'. Let f be the binary operator on $\mathsf{Cm}\,\mathbb{M}\mathfrak{S}$ determined by C, i.e.

$$f(X,Y) = \{z \in S : \exists x \in X\, \exists y \in Y\, C(x,y,z)\}.$$

But since the operation $X;Y$ given by the definition of $\mathsf{M}\,\mathsf{Cm}\,\mathfrak{S}$ is completely additive in X and in Y we get

$$X;Y = \bigcup_{x\in X, y\in Y} \{x\};\{y\} = \{z \in S : \exists x \in X\, \exists y \in Y\, (z \in \{x\};\{y\})\,\}$$

which is equal to $f(X,Y)$ by definition of C.

Similarly, if $gX = \{y \in S : \exists x \in X\,(xRy)\}$ is the unary operator on $\mathsf{Cm}\,\mathbb{M}\mathfrak{S}$ determined by R, then by complete additivity of $\breve{}$ in $\mathsf{M}\,\mathsf{Cm}\,\mathfrak{S}$ we get $X^{\smile} = \bigcup_{x\in X}\{x\}^{\smile} = gX$. Thus the operators of $\mathsf{Cm}\,\mathbb{M}\mathsf{S}$ and $\mathsf{M}\,\mathsf{Cm}\,\mathfrak{S}$ are identical. ∎

LEMMA 2.2. *$\Pi_U\mathbb{M}\mathfrak{S} = \mathbb{M}\Pi_U\mathfrak{S}$, for any ultrafilter U.*

PROOF. Let U be an ultrafilter on set I. The ultrapowers $\Pi_U\mathfrak{S}$ and $\Pi_U\mathbb{M}\mathfrak{S}$, and the structure $\mathbb{M}\Pi_U\mathfrak{S}$, are all based on the ultrapower $\Pi_U S$ of set S. We have $\Pi_U\mathbb{M}\mathfrak{S} = (\Pi_U S, C^U, R^U, E^U_{01})$, where

$$\begin{array}{ll} C^U(f^U, g^U, h^U) & \text{iff } \{k \in I : C(f(k), g(k), h(k))\} \in U, \\ f^U R g^U & \text{iff } \{k \in I : f(k) R g(k)\} \in U, \\ f^U \in E^U_{01} & \text{iff } \{k \in I : f(k) \in E_{01}\} \in U. \end{array}$$

$\Pi_U\mathfrak{S} = (\Pi_U S, R^U_i, E^U_{ij})_{i,j<\alpha}$ is defined likewise. Then

$$\mathbb{M}\Pi_U\mathfrak{S} = (\Pi_U S, C', R', E^U_{01}),$$

where C' and R' are defined from $\mathsf{M}\,\mathsf{Cm}\,\Pi_U\mathfrak{S}$ by the $\mathbb{M}$ construction. So all we have to show is that $C' = C^U$ and $R' = R^U$.

Now the relation $C(x,y,z)$ in $\mathbb{M}\mathfrak{S}$ is definable by a formula $\varphi_C(x,y,z)$ in the first-order language of any $\mathbf{CA}_\alpha$-structure $\mathfrak{S}$. From the way that the cylindrification c_i is defined from the relation R_i, we have that $u \in s^i_j\{v\}$ in $\mathsf{Cm}\,\mathfrak{S}$ iff $\exists w(wR_iu$ and $w \in E_{ij} \cap \{v\})$, which is equivalent to $(vR_iu \wedge v \in E_{ij})$. So for $\varphi_C(x,y,z)$ we can use the (quasi-modal) formula

$$\exists u(uR_2z \wedge xR_1u \wedge x \in E_{12} \wedge yR_0u \wedge y \in E_{02}).$$

By the fundamental theorem of satisfaction in ultraproducts (Łoś's theorem),

$$\Pi_U\mathfrak{S} \models \varphi_C(f^U, g^U, h^U) \quad \text{iff} \quad \{k : \mathfrak{S} \models \varphi_C(f(k), g(k), h(k))\} \in U,$$

which just says that

$$C'(f^U, g^U, h^U) \quad \text{iff} \quad \{k : C(f(k), g(k), h(k))\} \in U,$$

implying that $C' = C^U$. Similarly there is a formula $\varphi_R(x, y)$ in the language of $\mathbf{CA}_\alpha$-structures that defines the binary relation R of $\mathbb{M}\mathfrak{S}$, and this can be used in the same way to show that $R' = R^U$. ∎

LEMMA 2.3. $\mathbb{M}\mathbb{C}\mathsf{st}\mathfrak{A} = \mathbb{C}\mathsf{st}\,\mathsf{M}\mathfrak{A}$.

PROOF. Let $\mathbb{C}\mathsf{st}\,\mathsf{M}\mathfrak{A} = (S_0, C_0, R_0, E_0)$, where S_0 is the set of ultrafilters of $\mathfrak{A}_0$, C_0 is the ternary relation on S_0 defined by the operator ; of $\mathsf{M}\mathfrak{A}$, R_0 the binary relation defined by $^\smile$, and $E_0 = \{p \in S_0 : 1' \in p\}$ is the set (unary relation) determined by the identity element 1' of $\mathsf{M}\mathfrak{A}$..

C_0 and R_0 in turn define operators ; , $^\smile$ on the power set of S_0. All that we will need to know about those operators is how they act on singleton subsets $\{p\}$ of S_0, which is, by definition,

$$\{p\};\{q\} = \{r \in S_0 : C_0(p, q, r)\}, \tag{2.5}$$

$$\{p\}^\smile = \{q \in S_0 : pR_0q\}. \tag{2.6}$$

Next let $\mathbb{M}\mathbb{C}\mathsf{st}\mathfrak{A} = (S_0, C, R, E)$. Here C, R, E are defined from the operations ; , $^\smile$, 1' of $\mathsf{M}\,\mathsf{Cm}\,\mathbb{C}\mathsf{st}\mathfrak{A} = \mathsf{M}\,\mathsf{Em}\,\mathfrak{A}$, and hence from the cylindric algebra operations of $\mathsf{Em}\,\mathfrak{A}$. In particular, E is the 01-diagonal element of $\mathsf{Em}\,\mathfrak{A}$, which is the set $\{p \in S_0 : d_{01} \in p\}$. Since $d_{01} = 1'$ (2.iv), this shows that $E = E_0$.

The relation C is defined by

$$C(p, q, r) \quad \text{iff} \quad r \in c_2(s_2^1\{p\} \cdot s_2^0\{q\}) \text{ in } \mathsf{Em}\,\mathfrak{A}.$$

To show that $C = C_0$ we introduce the algebra $\mathfrak{A}^+ = (\mathfrak{A}, ;, ^\smile, 1')$, the expansion of $\mathfrak{A}$ itself by the operations of (2.iv). The $\mathbf{CA}_\alpha$-reduct of $\mathsf{Em}\,\mathfrak{A}^+$ is just $\mathsf{Em}\,\mathfrak{A}$, and the $\{;, ^\smile, 1'\}$-reduct is $\mathsf{Em}\,\mathsf{M}\mathfrak{A}$. Since the equations in (2.iv) are all positive (do not involve Boolean complementation), they are preserved by canonical extensions, and so they continue to hold in $\mathsf{Em}\,\mathfrak{A}^+$. Therefore

$$\{p\};\{q\} = c_2(s_2^1\{p\} \cdot s_2^0\{q\}) \text{ in } \mathsf{Em}\,\mathfrak{A}^+.$$

Thus $r \in c_2(s_2^1\{p\} \cdot s_2^0\{q\})$ iff $r \in \{p\};\{q\}$ iff (by 2.v) $C_0(p,q,r)$. This proves $C = C_0$.

Similarly, R is defined by

$$pRq \quad \text{iff} \quad q \in s_0^2 s_1^0 s_2^1\{p\} \text{ in } \mathsf{Em}\mathfrak{A}.$$

But $\{p\}^{\smile} = s_0^2 s_1^0 s_2^1\{p\}$ in $\mathsf{Em}\mathfrak{A}^+$, by preservation of the equations (2.iv) again, so from (2.vi) we get pRq iff pR_0q, and so $R = R_0$. ■

We can now go ahead and apply the three Lemmas 2.1–2.3 to show that if $\mathfrak{B} \in \mathsf{S}\mathfrak{Ra}\mathbf{CA}_\alpha$ then any ultrapower $\Pi_U\mathbb{C}\mathbf{st}\,\mathfrak{B}$ of the canonical structure of $\mathfrak{A}$ is a $\mathsf{S}\mathfrak{Ra}\mathbf{CA}_\alpha$-structure, i.e.

$$\mathbb{P}\mathsf{w}\mathbb{C}\mathsf{st}\,\mathsf{S}\mathfrak{Ra}\mathbf{CA}_\alpha \subseteq \mathbb{S}\mathsf{tr}\,\mathsf{S}\mathfrak{Ra}\mathbf{CA}_\alpha,$$

implying the desired result that $\mathsf{S}\mathfrak{Ra}\mathbf{CA}_\alpha$ is elementarily generated. The reasoning is exactly parallel to the $\mathsf{S}\,\mathsf{Nr}_\beta\mathbf{CA}_\alpha$ case, and makes use of some of the work already done for that case.

Since $\mathfrak{B} \in \mathsf{S}\mathfrak{Ra}\mathbf{CA}_\alpha$, there is some $\mathfrak{A} \in \mathbf{CA}_\alpha$ such that $\mathfrak{B}$ is a subalgebra of $\mathfrak{Ra}\mathfrak{A}$, and hence of $\mathsf{M}\mathfrak{A}$. By duality this induces the bounded epimorphism $\theta : \mathbb{C}\mathsf{st}\,\mathsf{M}\mathfrak{A} \twoheadrightarrow \mathbb{C}\mathsf{st}\,\mathfrak{B}$, given by $\theta(x) = x \cap B_0$, where B_0 is the underlying set of $\mathfrak{B}$. Then for any ultrafilter U, θ lifts to the bounded epimorphism

$$\theta_U : \Pi_U\mathbb{C}\mathsf{st}\,\mathsf{M}\mathfrak{A} \twoheadrightarrow \Pi_U\mathbb{C}\mathsf{st}\,\mathfrak{B}$$

having $\theta_U(f^U) = (\theta \circ f)^U$, which in turn induces the injective homomorphism

$$\theta_U^+ : \mathsf{Cm}\,\Pi_U\mathbb{C}\mathsf{st}\,\mathfrak{B} \rightarrowtail \mathsf{Cm}\,\Pi_U\mathbb{C}\mathsf{st}\,\mathsf{M}\mathfrak{A}$$

acting by $X \mapsto \theta_U^{-1}(X)$. Using 2.3, 2.2 and then 2.1, we rewrite the codomain of θ_U^+, to get

$$\theta_U^+ : \mathsf{Cm}\,\Pi_U\mathbb{C}\mathsf{st}\,\mathfrak{B} \rightarrowtail \mathsf{M}\,\mathsf{Cm}\,\Pi_U\mathbb{C}\mathsf{st}\,\mathfrak{A}.$$

But, as explained in Section 2.2, $\mathsf{Cm}\,\Pi_U\mathbb{C}\mathsf{st}\,\mathfrak{A} \in \mathbf{CA}_\alpha$ because $\mathbf{CA}_\alpha$ is canonical and $\mathbb{S}\mathsf{tr}\,\mathbf{CA}_\alpha$ is elementary. So, if we can show that

(‡) θ_U^+ maps $\mathsf{Cm}\,\Pi_U\mathbb{C}\mathsf{st}\,\mathfrak{B}$ into $Cl_{\alpha-2}\mathsf{Cm}\,\Pi_U\mathbb{C}\mathsf{st}\,\mathfrak{A}$,

this will make $\mathsf{Cm}\,\Pi_U\mathbb{C}\mathsf{st}\,\mathfrak{B}$ isomorphic to a subalgebra of $\mathfrak{Ra}\mathsf{Cm}\,\Pi_U\mathbb{C}\mathsf{st}\,\mathfrak{A}$, and hence a member of $\mathsf{S}\mathfrak{Ra}\mathbf{CA}_\alpha$, putting $\Pi_U\mathbb{C}\mathsf{st}\,\mathfrak{B}$ into $\mathbb{S}\mathsf{tr}\,\mathsf{S}\mathfrak{Ra}\mathbf{CA}_\alpha$ as desired. But since $B_0 \subseteq Cl_{\alpha-2}\mathfrak{A}$, (‡) is just the case $\beta = 2$ of the result (†) proven in Section 2.2.

Finally, we note that $\mathbb{S}\mathsf{tr}\,\mathsf{S}\mathfrak{Ra}\mathbf{CA}_\alpha$ is not itself elementary in general. For infinite α this follows because in that case $\mathsf{S}\mathfrak{Ra}\mathbf{CA}_\alpha = \mathbf{RRA}$ [24, p. 192].

For $3 \leq n < \omega$, the $\mathbf{S}\mathfrak{RaCA}_n$'s form a nested sequence with $\mathbf{S}\mathfrak{RaCA}_n \supseteq \mathbf{S}\mathfrak{RaCA}_{n+1}$, and with the intersection of the sequence being **RRA**. From the fact that $\mathbb{S}\mathrm{tr}\mathbf{RRA}$ is not elementary it then follows, analogously to the $\mathbf{RA}_n$ case, that for cofinitely many n, $\mathbb{S}\mathrm{tr}\mathbf{S}\mathfrak{RaCA}_n$ must fail to be elementary.

3. Canonical Extensions as Free Objects

In this final Section we disuss the sense in which the operation $\mathbb{E}\mathrm{x}\mathfrak{S}$ can be characterized in category-theoretic terms by a universal property of freeness. A natural setting for this involves structures that carry a topology. Before describing that setting we first review the analogous categorical characterization of algebras constructed from formulas.

3.1. Lindenbaum Algebras are Free

Let Λ be a normal propositional modal logic. For each set X, the Lindenbaum algebra construction produces a Λ-algebra $F^\Lambda(X)$, built from the language for Λ that takes X as its set of propositional variables. Two formulas φ, ψ of this language are said to be *Λ-equivalent* if the biconditional $\varphi \leftrightarrow \psi$ is a theorem of Λ. This defines an equivalence relation on the set of formulas that is a congruence for the logical connectives. $F^\Lambda(X)$ is the resulting quotient algebra, based on the set of equivalence classes $|\varphi|$ for all formulas φ. In $F^\Lambda(X)$, $|\varphi| = 1$ precisely when φ is a Λ-theorem.

We will now write $\mathcal{U}\mathfrak{A}$ for the underlying set of an algebra $\mathfrak{A}$. A function of the form $f : X \to \mathcal{U}\mathfrak{A}$ is a *valuation*, assigning a value in $\mathfrak{A}$ to each variable from X. This extends to give a value $f(\varphi) \in \mathcal{U}\mathfrak{A}$ to each formula by interpreting the connectives of φ by the corresponding operations in $\mathfrak{A}$. φ is *valid in* $\mathfrak{A}$ when $f(\varphi) = 1$ for all such valuations. $\mathfrak{A}$ is a *Λ-algebra* when all Λ-theorems are valid in $\mathfrak{A}$.

As is well known, $F^\Lambda(X)$ is a Λ-algebra. It has a special valuation $\eta_X : X \to \mathcal{U}F^\Lambda(X)$ defined by $\eta_X(p) = |p|$ for all $p \in X$. Then $\eta_X(\varphi) = |\varphi|$ for all formulas φ, showing that $F^\Lambda(X)$ invalidates all non-theorems of Λ under η_X. If $\mathfrak{A}$ is any Λ-algebra, then a valuation $f : X \to \mathcal{U}\mathfrak{A}$ will assign the same value to any two Λ-equivalent formulas, so a function $f^+ : \mathcal{U}F^\Lambda(X) \to \mathcal{U}\mathfrak{A}$ is well-defined by putting $f^+(|\varphi|) = f(\varphi)$. Then f^+ is a homomorphism from $F^\Lambda(X)$ to $\mathfrak{A}$ that has $f^+(|p|) = f(p)$ for all variables $p \in X$, and is the only homomorphism with this property. This is because f^+ is determined by its values on the $|p|$'s, since they generate the algebra $F^\Lambda(X)$. To sum up:

(3.i) For any Λ-algebra $\mathfrak{A}$ and any function $f : X \to \mathcal{U}\mathfrak{A}$, there is exactly one homomorphism $f^+ : F^\Lambda(X) \to \mathfrak{A}$ such that $\mathcal{U}f^+ \circ \eta_X = f$:

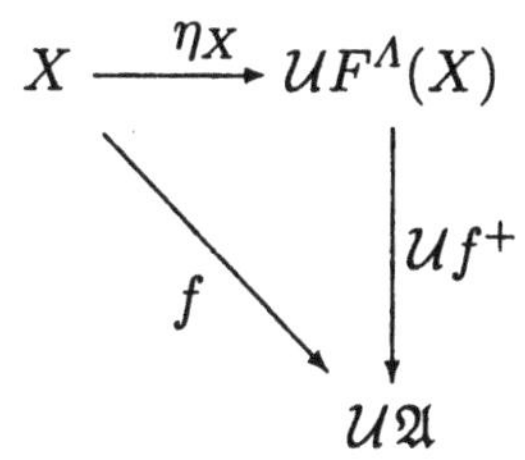

Here we use the notation $\mathcal{U}f^+$ when f^+ is being considered as a set function, forgetting that it is also a homomorphism. In universal algebraic terms, (3.i) states that the algebra $F^\Lambda(X)$ is *free over* X *in the variety* $\mathcal{V}^\Lambda$ *of all* Λ *algebras.* In categorical terms, the maps $\mathfrak{A} \mapsto \mathcal{U}\mathfrak{A}$ and $g \mapsto \mathcal{U}g$ constitute the *forgetful functor* $\mathcal{V}^\Lambda \to$ **Set**, where **Set** is the category of sets and functions, while $\mathcal{V}^\Lambda$ is viewed as a category whose arrows are the homomorphisms between Λ-algebras. In categorical terms, (3.i) states that the pair $(F^\Lambda(X), \eta_X)$ is *free over* X *with respect to the functor* $\mathcal{U}$. This property characterizes the pair uniquely up to a unique isomorphism, by a standard argument (see Theorem 3.2 below).

3.2. Topological Structures

By a *type* we will mean a function of the form $\tau : I \to \{1, 2, 3, \ldots\}$. A *structure of type* τ has the form $\mathfrak{S} = (S, \{R_i : i \in I\})$ with R_i being a $\tau_i + 1$-ary relation on S. Then the complex algebra $\mathsf{Cm}\,\mathfrak{S}$ is a BAO *of type* τ, having the τ_i-ary operator f_{R_i} corresponding to each R_i, as defined in Section 2. The *dual* operation $f^\delta_{R_i}$ to f_{R_i} is defined by

$$f^\delta_{R_i}(X_1, \ldots, X_{\tau_i}) = -f_{R_i}(-X_1, \ldots, -X_{\tau_i}).$$

Given τ-structures $\mathfrak{S}_k = (S_k, \{R^k_i : i \in I\})$ for $k = 1, 2$, a *weak morphism* $\theta : \mathfrak{S}_1 \to \mathfrak{S}_2$ is a function $\theta : S_1 \to S_2$ satisfying, for each $i \in I$,

$$R^1_i(x_1, \ldots, x_{\tau_i+1}) \quad \text{implies} \quad R^2_i(\theta(x_1), \ldots, \theta(x_{\tau_i+1})).$$

A *bounded morphism* is a weak morphism that also satisfies

> $R^2_i(y_1, \ldots, y_{\tau_i}, \theta(x))$ implies $R^1_i(x_1, \ldots, x_{\tau_i}, x)$ for some $x_1, \ldots, x_{\tau_i} \in S_1$ such that $\theta(x_j) = y_j$ for all $j \leq \tau_i$.

By a *topological* τ*-structure* we will mean a τ-structure $\mathfrak{S}$ that carries a topology $\mathcal{T}^\mathfrak{S}$ such that the following hold for each $i \in I$:

- if $X_1, \dots, X_{\tau_i}$ are $\mathcal{T}^{\mathfrak{S}}$-open subsets of S, then so are $f_{R_i}(X_1, \dots, X_{\tau_i})$ and $f^{\delta}_{R_i}(X_1, \dots, X_{\tau_i})$;
- for each $x \in S$, the set $R_i^{-1}(x) = \{(x_1, \dots, x_{\tau_i}) : R_i(x_1, \dots, x_{\tau_i}, x)\}$ is closed in the product topology on S^{τ_i}.

The canonical extension $\mathbb{E}\mathrm{x}\,\mathfrak{S}$ of any τ-structure becomes a topological structure under the *Stone topology*, which has as base the sets $\|X\|$ for all $X \subseteq S$, where $\|X\|$ is the set of ultrafilters on S to which X belongs. The Stone topology makes $\mathbb{E}\mathrm{x}\,\mathfrak{S}$ into a compact, Hausdorff and totally connected space (i.e. a *Stone space* [27]).

There is a natural injection $\eta_{\mathfrak{S}} : \mathfrak{S} \to \mathbb{E}\mathrm{x}\,\mathfrak{S}$ for which $\eta_{\mathfrak{S}}(x) = \{X : x \in X\}$ is the principal ultrafilter on S generated by x. $\eta_{\mathfrak{S}}$ is continuous with respect to the given topology on $\mathfrak{S}$ and the Stone topology on $\mathbb{E}\mathrm{x}\,\mathfrak{S}$, and is always a weak morphism. But $\eta_{\mathfrak{S}}$ is not in general bounded. In fact it is bounded precisely when the inverse-image sets $R_i^{-1}(x)$ are all finite.

We can now state an analogue of (3.i) for topological τ-structures $\mathfrak{S}$:

THEOREM 3.1. *For any compact Hausdorff τ-structure $\mathfrak{T}$ and any continuous function θ from $\mathfrak{S}$ to $\mathfrak{T}$, there is a unique continuous function $\theta^+ : \mathbb{E}\mathrm{x}\,\mathfrak{S} \to \mathfrak{T}$ such that $\theta^+ \circ \eta_{\mathfrak{S}} = \theta$:*

$$\begin{array}{ccc}
\mathfrak{S} & \xrightarrow{\ \eta_{\mathfrak{S}}\ } & \mathbb{E}\mathrm{x}\,\mathfrak{S} \\
 & \underset{\theta}{\searrow} & \big\downarrow \theta^+ \\
 & & \mathfrak{T}.
\end{array}$$

Moreover, if θ is a weak or bounded morphism, then so is θ^+, respectively.

■

The function θ^+ is provided by the standard topological theory of convergence of ultrafilters. An ultrafilter on $\mathfrak{T}$ *converges* to a point y of $\mathfrak{T}$ if it contains all of the open neighbourhoods of y in $\mathfrak{T}$. Since $\mathfrak{T}$ is compact and Hausdorff, every ultrafilter on $\mathfrak{T}$ converges to exactly one point, defining a map $\varepsilon_{\mathfrak{T}} : \mathbb{E}\mathrm{x}\,\mathfrak{T} \to \mathfrak{T}$. But by duality, θ lifts to the map $\mathbb{E}\mathrm{x}\,\theta : \mathbb{E}\mathrm{x}\,\mathfrak{S} \to \mathbb{E}\mathrm{x}\,\mathfrak{T}$ such that for each ultrafilter p on $\mathfrak{S}$, $\mathbb{E}\mathrm{x}\,\theta(p)$ is the ultrafilter $\{Y : \theta^{-1}Y \in p\}$ on $\mathfrak{T}$. Then $\theta^+(p)$ is defined to be the unique point of $\mathfrak{T}$ to which $\mathbb{E}\mathrm{x}\,\theta(p)$ converges. Thus $\theta^+ = \varepsilon_{\mathfrak{T}} \circ \mathbb{E}\mathrm{x}\,\theta$.

Now $\mathbb{E}\mathrm{x}\,\theta$ is continuous, weak, or bounded whenever θ has the corresponding property. But $\varepsilon_{\mathfrak{T}}$ is always a continuous bounded morphism for compact Hausdorff $\mathfrak{T}$, and so Theorem 3.1 follows from these observations. It leads to the following characterization of the canonical structure $\mathbb{E}\mathrm{x}\mathfrak{S}$.

THEOREM 3.2. *Let $\mathfrak{S}^*$ be a compact Hausdorff τ-structure with a continuous weak morphism $\eta : \mathfrak{S} \to \mathfrak{S}^*$ such that for any compact Hausdorff τ-structure $\mathfrak{T}$ and any continuous weak morphism $\theta : \mathfrak{S} \to \mathfrak{T}$ there is a unique continuous weak morphism $\theta^* : \mathfrak{S}^* \to \mathfrak{T}$ with $\theta^* \circ \eta = \theta$. Then there exists a unique homeomorphic isomorphism $\rho : \mathbb{E}\mathrm{x}\mathfrak{S} \to \mathfrak{S}^*$ such that $\rho \circ \eta_{\mathfrak{S}} = \eta$.*

PROOF. Consider the diagram

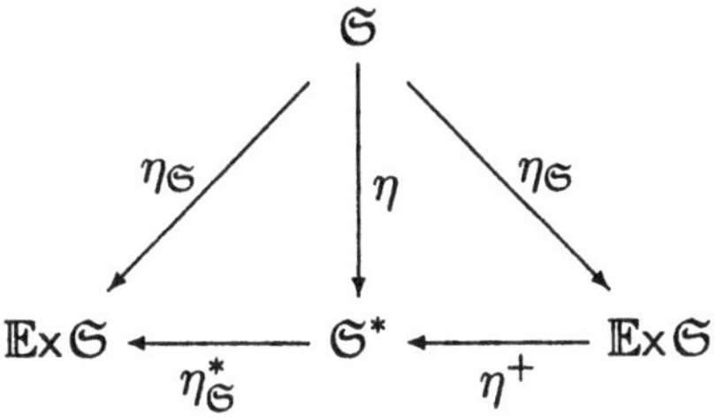

η^+ is the continuous weak morphism resulting from taking η as θ in Theorem 3.1, while $\eta^*_{\mathfrak{S}}$ is the one resulting from taking $\eta_{\mathfrak{S}}$ as θ in the assumed property of $(\mathfrak{S}^*, \eta)$. The composition $\eta^*_{\mathfrak{S}} \circ \eta^+ : \mathbb{E}\mathrm{x}\mathfrak{S} \to \mathbb{E}\mathrm{x}\mathfrak{S}$ is a continuous weak morphism, and hence by the uniqueness expressed in Theorem 3.1 must be the identity function on $\mathbb{E}\mathrm{x}\mathfrak{S}$. Similarly, $\eta^+ \circ \eta^*_{\mathfrak{S}}$ is the identity function on $\mathfrak{S}^*$. Thus we put $\rho = \eta^+$ to obtain a bijective continuous weak morphism whose inverse $\eta^*_{\mathfrak{S}}$ is also a continuous weak morphism. This ensures that ρ is a homeomorphism of topological spaces and an isomorphism of relational structures. ∎

Given the importance of bounded morphisms in our theory, it may seem surprising that only weak morphisms are involved in this characterization of canonical extensions of structures. But the explanation lies in the observation that in order for a bijection $\theta : \mathfrak{S}_1 \to \mathfrak{S}_2$ to be an isomorphism it suffices that both θ and θ^{-1} be *weak* morphisms, for that is enough to ensure that in general

$$R_i^1(x_1, \dots, x_{\tau_i+1}) \quad \text{iff} \quad R_i^2(\theta(x_1), \dots, \theta(x_{\tau_i+1})).$$

Theorem 3.1 also admits a more categorical description. Let $\mathbf{TopSt}_\tau$ be the category of topological τ-structures and continuous weak morphisms, and $\mathbf{CHSt}_\tau$ its full subcategory of compact Hausdorff structures. Then 3.1 states that the pair $(\mathbb{E}\mathrm{x}\mathfrak{S}, \eta_{\mathfrak{S}})$ is *free over $\mathfrak{S}$ with respect to the inclusion functor* $\mathbf{CHSt}_\tau \to \mathbf{TopSt}_\tau$. The assignments $\mathfrak{S} \mapsto \mathbb{E}\mathrm{x}\mathfrak{S}$ and $\theta \mapsto \mathbb{E}\mathrm{x}\theta$

give a functor $\mathbb{E}\mathrm{x} : \mathbf{TopSt}_\tau \to \mathbf{CHSt}_\tau$ that is left adjoint to this inclusion, making $\mathbf{CHSt}_\tau$ into what is known as a *reflective* subcategory of $\mathbf{TopSt}_\tau$ [23, 36.1].

There appears to be no corresponding analysis of the subcategories of $\mathbf{TopSt}_\tau$ and $\mathbf{CHSt}_\tau$ whose arrows are bounded morphisms. The obstacle is that the functions $\eta_\mathfrak{S}$ are in general only weak, and not bounded, morphisms, so do not belong to those subcategories.

A different, non-topological, approach to characterizing $\mathbb{E}\mathrm{x}\,\mathfrak{S}$ can be developed from categorical structure associated with $\mathbb{E}\mathrm{x}$ as a functor on the category of τ-structures and weak morphisms (technically, this categorical structure is known as a *monad*). By a *retraction* onto a structure $\mathfrak{S}$ we will mean any function $\rho : \mathbb{E}\mathrm{x}\,\mathfrak{S} \to \mathfrak{S}$ such that $\rho \circ \eta_\mathfrak{S}$ is the identity function on $\mathfrak{S}$, i.e. $x = \rho(\eta_\mathfrak{S}(x))$ in general. Given retractions ρ_1, ρ_2 onto structures $\mathfrak{S}_1, \mathfrak{S}_2$ respectively, a function $\theta : S_1 \to S_2$ is called *retraction preserving* if $\theta \circ \rho_1 = \rho_2 \circ \mathbb{E}\mathrm{x}\,\theta$:

$$\begin{array}{ccc}
\mathbb{E}\mathrm{x}\,\mathfrak{S}_1 & \xrightarrow{\mathbb{E}\mathrm{x}\,\theta} & \mathbb{E}\mathrm{x}\,\mathfrak{S}_2 \\
\Big\downarrow{\scriptstyle \rho_1} & & \Big\downarrow{\scriptstyle \rho_2} \\
\mathfrak{S}_1 & \xrightarrow{\theta} & \mathfrak{S}_2
\end{array}$$

There is a natural retraction onto any canonical extension $\mathbb{E}\mathrm{x}\,\mathfrak{S}$ given by the function $\mu_\mathfrak{S} : \mathbb{E}\mathrm{x}\mathbb{E}\mathrm{x}\,\mathfrak{S} \to \mathbb{E}\mathrm{x}\,\mathfrak{S}$ having, for any ultrafilter q on $\mathbb{E}\mathrm{x}\,\mathfrak{S}$,

$$\mu_\mathfrak{S}(q) = \{X \subseteq S : \|X\| \in q\},$$

where $\|X\| = \{p \in \mathbb{E}\mathrm{x}\,\mathfrak{S} : X \in p\}$ as above. $\mu_\mathfrak{S}$ is a bounded morphism of τ-structures.

By a *retraction structure* we will mean a τ-structure $\mathfrak{S}$ with a specified retraction $\rho_\mathfrak{S} : \mathbb{E}\mathrm{x}\,\mathfrak{S} \to \mathfrak{S}$ which is itself a bounded morphism that is retraction preserving from $(\mathbb{E}\mathrm{x}\,\mathfrak{S}, \mu_\mathfrak{S})$ to $(\mathfrak{S}, \rho_\mathfrak{S})$:

$$\begin{array}{ccc}
\mathbb{E}\mathrm{x}\mathbb{E}\mathrm{x}\,\mathfrak{S} & \xrightarrow{\mathbb{E}\mathrm{x}\,\rho_\mathfrak{S}} & \mathbb{E}\mathrm{x}\,\mathfrak{S} \\
\Big\downarrow{\scriptstyle \mu_\mathfrak{S}} & & \Big\downarrow{\scriptstyle \rho_\mathfrak{S}} \\
\mathbb{E}\mathrm{x}\,\mathfrak{S} & \xrightarrow{\rho_\mathfrak{S}} & \mathfrak{S}
\end{array}$$

The category $\mathbf{Ret}_\tau$ has retraction structures as objects and retraction preserving weak morphisms as arrows, while $\mathbf{St}_\tau$ is the category of τ-structures and weak morphisms. It turns out that for any τ-structure $\mathfrak{S}$, the canonical extension $\mathbb{E}\mathrm{x}\,\mathfrak{S}$ is a retraction structure under the retraction $\mu_\mathfrak{S}$. Thus $\mathbb{E}\mathrm{x}$ becomes a functor from $\mathbf{St}_\tau$ to $\mathbf{Ret}_\tau$ that is left adjoint to the forgetful functor $\mathcal{U} : \mathbf{Ret}_\tau \to \mathbf{St}_\tau$ which forgets about retractions and retraction preservation. The result which guarantees this is the following analogue of Theorem 3.1.

THEOREM 3.3. *For any retraction τ-structure $\mathfrak{T}$ and any weak morphism θ from $\mathfrak{S}$ to $\mathcal{U}\mathfrak{T}$, there is a unique retraction reserving weak morphism $\theta^+ : \mathbb{E}\mathrm{x}\,\mathfrak{S} \to \mathfrak{T}$ such that $\mathcal{U}\theta^+ \circ \eta_\mathfrak{S} = \theta$:*

$$\begin{array}{ccc} \mathfrak{S} & \xrightarrow{\eta_\mathfrak{S}} & \mathbb{E}\mathrm{x}\,\mathfrak{S} \\ & \searrow^{\theta} & \downarrow{\scriptstyle \mathcal{U}\theta^+} \\ & & \mathcal{U}\mathfrak{T}. \end{array}$$

■

Now if $\mathfrak{S}$ is a compact Hausdorff τ-structure, then there is a function $\varepsilon_\mathfrak{S} : \mathbb{E}\mathrm{x}\,\mathfrak{S} \to \mathfrak{S}$, discussed earlier in relation to Theorem 3.1, with $\varepsilon_\mathfrak{S}(p)$ being the unique point of $\mathfrak{S}$ to which the ultrafilter p converges. $\varepsilon_\mathfrak{S}$ is a retraction and a bounded morphism for which the diagram

$$\begin{array}{ccc} \mathbb{E}\mathrm{x}\,\mathbb{E}\mathrm{x}\,\mathfrak{S} & \xrightarrow{\mathbb{E}\mathrm{x}\,\varepsilon_\mathfrak{S}} & \mathbb{E}\mathrm{x}\,\mathfrak{S} \\ {\scriptstyle \mu_\mathfrak{S}}\downarrow & & \downarrow{\scriptstyle \varepsilon_\mathfrak{S}} \\ \mathbb{E}\mathrm{x}\,\mathfrak{S} & \xrightarrow{\varepsilon_\mathfrak{S}} & \mathfrak{S} \end{array}$$

commutes. Thus $\mathfrak{S}$ is a retraction structure under $\varepsilon_\mathfrak{S}$. The assignment $\mathfrak{S} \mapsto (\mathfrak{S}, \varepsilon_\mathfrak{S})$ gives rise to an isomorphism between the categories $\mathbf{CHSt}_\tau$ and $\mathbf{Ret}_\tau$. This generalizes the celebrated result of Manes (see [38] or [27, Section III.2]) that the category of compact Hausdorff topological spaces and continuous functions is isomorphic to the category of algebras for the ultrafilter monad on $\mathbf{Set}$.

Details of the theory surveyed in Section 3.2, with proofs of Theorems 3.1 and 3.3, will be given in [10].

4. Further Questions

The material reviewed in this paper suggests further problems for research. In the absence of a resolution of the converse of Fine's theorem, we may seek partial solutions, in the form of theorems showing that canonicity implies elementarity for various classes of logics or varieties. There are a number of such theorems in the literature. For example, Fine [4] showed this for the class of monomodal logics that are characterised by some class of transitive frames that is closed under subframes (not just inner ones). The transitivity restriction here was removed by Frank Wolter in his dissertation (see [46, Theorem 3.11]). Wolter also proved [45] that canonicity implies elementarity for all normal extensions of the standard bimodal linear tense logic.

In a different direction, a general result can be obtained by considering *atom structures*. If a BAO $\mathfrak{A}$ is atomic, it determines a structure $\mathbb{A}\mathrm{t}\mathfrak{A}$ based on the set of atoms of $\mathfrak{A}$, with each operator f of $\mathfrak{A}$ determining the relation R_f of $\mathbb{A}\mathrm{t}\mathfrak{A}$ having

$$R_f(x_1, \ldots, x_{n+1}) \quad \text{iff} \quad f(x_1, \ldots, x_n) \geq x_{n+1}.$$

For any variety $\mathcal{V}$, the class $\mathbb{A}\mathrm{t}\mathcal{V}$ of such atom structures of atomic members of $\mathcal{V}$ is closed under ultraproducts and includes $\mathbb{S}\mathrm{tr}\mathcal{V}$. We showed in [15] that if a canonical variety has $\mathbb{A}\mathrm{t}\mathcal{V} = \mathbb{S}\mathrm{tr}\mathcal{V}$, then $\mathbb{A}\mathrm{t}\mathcal{V}$ is an elementary class generating $\mathcal{V}$. However there are canonical varieties, and even elementarily generated ones, for which $\mathbb{A}\mathrm{t}\mathcal{V} \supsetneq \mathbb{S}\mathrm{tr}\mathcal{V}$. Examples include the variety of algebras for the modal logic K4+$(\Box\Diamond\varphi \to \Diamond\Box\varphi)$ [15, p. 592], and the variety **RRA** [25].

Another condition that is sufficient for canonicity of $\mathcal{V}$ to imply elementary generation is *singleton-persistence.* This means that if $\mathcal{V}$ contains the subalgebra of $\mathsf{Cm}\,\mathfrak{S}$ generated by the atoms (i.e. the singleton subsets of $\mathfrak{S}$), then it contains $\mathsf{Cm}\,\mathfrak{S}$ itself. Singleton persistence is strong enough to force $\mathbb{S}\mathrm{tr}\mathcal{V}$ to be an elementary class [18].

The question remains: are there other results on of this kind, providing conditions on some class of varieties, or some family of classes of structures that generate varieties, that ensure that canonicity implies elementary generation? For instance, can we show this for varieties that are generated by their finite members? Another important question concerns *how much* canonicity is enough. Is validity of a logic Λ in the canonical structure $\mathfrak{S}_\omega^\Lambda$ sufficient to make Λ be elementary? If so, then by Fine's theorem validity of Λ in $\mathfrak{S}_\omega^\Lambda$ would imply its validity in $\mathfrak{S}_\kappa^\Lambda$ for all infinite κ. But not even that is known.

We may also raise these questions in contexts other than that of BAO's, such for distributive lattices with operators, where canonical extensions have been studied in [6, 7] and for which a duality can be developed using partially ordered relational structures [12, 20]. Then there are other kinds of intensional logic that have relational semantics and notions of canonical model, such as intuitionistic logic (with or without modalities), and various species of substructural logic (linear logic, relevant logic etc).

Then there is the question of whether the categorical/topological description of $\mathbb{E}\mathrm{x}\mathfrak{S}$ in Section 3 can be adapted to these other contexts. Finally, it is notable that there is now available a notion of canonical extension for any kind of lattice with additional operations [5]. A duality theory for such algebras has yet to be developed, and it remains to be seen whether an analogue of $\mathbb{E}\mathrm{x}$ can be successful formulated at that level of generality.

References

[1] C. C. Chang and H. J. Keisler. *Model Theory.* North-Holland, Amsterdam, 1973.

[2] Maarten de Rijke and Yde Venema. Sahlqvist's theorem for Boolean algebras with operators with an application to cylindric algebras. *Studia Logica*, 54:61–78, 1995.

[3] Kit Fine. Some connections between elementary and modal logic. In Stig Kanger, editor, *Proceedings of the Third Scandinavian Logic Symposium*, pages 15–31. North-Holland, 1975.

[4] Kit Fine. Logics containing $K4$. Part II. *The Journal of Symbolic Logic*, 50(3):619–651, 1985.

[5] Mai Gehrke and John Harding. Bounded lattice expansions. *Journal of Algebra*, 239:345–371, 2001.

[6] Mai Gehrke and Bjarni Jónsson. Bounded distributive lattices with operators. *Mathematica Japonica*, 40(2):207–215, 1994.

[7] Mai Gehrke and Bjarni Jónsson. Monotone bounded distributive lattice expansions. *Mathematica Japonica*, 52(2):197–213, 2000.

[8] Steven Givant. Universal classes of simple relation algebras. *The Journal of Symbolic Logic*, 64(2):575–589, 1999.

[9] Robert Goldblatt. Mathematical modal logic: A view of its evolution. To appear. Manuscript available at `http://www.mcs.vuw.ac.nz/~rob`.

[10] Robert Goldblatt. Topological frames and retraction. In preparation.

[11] Robert Goldblatt. *Metamathematics of Modal Logic*. PhD thesis, Victoria University, Wellington, February 1974. Included in [14].

[12] Robert Goldblatt. Varieties of complex algebras. *Annals of Pure and Applied Logic*, 44:173–242, 1989.

[13] Robert Goldblatt. On closure under canonical embedding algebras. In H. Andréka, J.D. Monk, and I. Németi, editors, *Algebraic Logic*, volume 54 of *Colloquia Mathematica Societatis János Bolyai*, pages 217–229. North-Holland, 1991.

[14] Robert Goldblatt. *Mathematics of Modality*. CSLI Lecture Notes No. 43. CSLI Publications, Stanford, California, 1993. Distributed by Cambridge University Press.

[15] Robert Goldblatt. Elementary generation and canonicity for varieties of Boolean algebras with operators. *Algebra Universalis*, 34:551–607, 1995.

[16] Robert Goldblatt. Reflections on a proof of elementarity. In Jelle Gerbrandy, Maarten Marx, Maarten de Rijke, and Yde Venema, editors, *JFAK. Essays Dedicated to Johan van Benthem on the Occasion of his 50th Birthday*. Amsterdam University Press, 1999. Electronically available at: `http://turing.wins.uva.nl/~j50/cdrom/`.

[17] Robert Goldblatt. Algebraic polymodal logic: A survey. *Logic Journal of the IGPL*, Special Issue on Algebraic Logic edited by István Németi and Ildikó Sain, 8(4):393–450, July 2000. Electronically available at: `http://www3.oup.co.uk/igpl/Volume_08/Issue_04/`.

[18] Robert Goldblatt. Persistence and atomic generation for varieties of Boolean algebras with operators. *Studia Logica*, 68(2):155–171, 2001.

[19] Robert Goldblatt. Quasi-modal equivalence of canonical structures. *The Journal of Symbolic Logic*, 66:497–508, 2001.

[20] G. Hansoul. A duality for Boolean algebras with operators. *Algebra Universalis*, 17:34–49, 1983.

[21] Leon Henkin, J. Donald Monk, and Alfred Tarski. *Cylindric Algebras I*. North-Holland, 1971.

[22] Leon Henkin, J. Donald Monk, and Alfred Tarski. *Cylindric Algebras II*. North-Holland, 1985.

[23] H. Herrlich and G. Strecker. *Category Theory*. Allyn and Bacon, Newton, MA, 1973.

[24] Robin Hirsch and Ian Hodkinson. *Relational Algebras by Games*, volume 147 of *Studies in Logic*. Elsevier, 2002.

[25] Ian Hodkinson. Atom structures of cylindric algebras and relation algebras. *Annals of Pure and Applied Logic*, 89:117–148, 1997.

[26] Ian Hodkinson and Yde Venema. Canonical varieties with no canonical axiomatisation. Electronically available at `http://www.doc.ic.ac.uk/~imh`, 2003.

[27] P. T. Johnstone. *Stone Spaces*. Cambridge University Press, 1982.

[28] Bjarni Jónsson. Algebras whose congruence lattices are distributive. *Math. Scand.*, 21:110–121, 1967.

[29] Bjarni Jónsson. A survey of Boolean algebras with operators. In *Algebras and Orders*, volume 389 of *NATO ASI Series*, pages 239–286. Kluwer Academic Publishers, 1993.

[30] Bjarni Jónsson. On the canonicity of Sahlqvist identities. *Studia Logica*, 53:473–491, 1994.

[31] Bjarni Jónsson. The preservation theorem for canonical extensions of Boolean algebras with operators. In K. A. Baker and R. Wille, editors, *Lattice Theory and its Applications, in Celebration of Garrett Birkhoff's 80th Birthday*, volume 23 of *Research and Expositions in Mathematics*, pages 121–130. Heldermann Verlag, 1995.

[32] Bjarni Jónsson and Alfred Tarski. Boolean algebras with operators, part I. *American Journal of Mathematics*, 73:891–939, 1951.

[33] Bjarni Jónsson and Alfred Tarski. Boolean algebras with operators, part II. *American Journal of Mathematics*, 74:127–162, 1952.

[34] E. J. Lemmon and D. Scott. Intensional logic. Preliminary draft of initial chapters by E. J. Lemmon, Stanford University (later published as *An Introduction to Modal Logic*, American Philosophical Quarterly Monograph Series, no. 11 (ed. by Krister Segerberg), Basil Blackwell, Oxford, 1977), July 1966.

[35] Roger Maddux. *Topic in Relation Algebras*. PhD thesis, University of California, Berkeley, 1978.

[36] Roger Maddux. Some varieties containing relation algebras. *Transaction of the American Mathematical Society*, 272(2):501–526, 1982.

[37] Roger Maddux. A sequent calculus for relation algebras. *Annals of Pure and Applied Logic*, 25:73–101, 1983.

[38] Ernest G. Manes. *Algebraic Theories*. Springer-Verlag, 1976.

[39] Giovanni Sambin. Subdirectly irreducible modal algebras and initial frames. *Studia Logica*, 62:269–282, 1999.

[40] Alfred Tarski. Contributions to the theory of models III. *Koninklijkle Nederlandse Akademie van Wetenschappen, Proceedings, Series A*, 58:56–64, 1955. (Indagationes Mathematicae, vol. 17).

[41] S. K. Thomason. Semantic analysis of tense logic. *The Journal of Symbolic Logic*, 37:150–158, 1972.

[42] J. F. A. K. van Benthem. Modal formulas are either elementary or not $\Sigma\Delta$-elementary. *The Journal of Symbolic Logic*, 41:436–438, 1976.

[43] J. F. A. K. van Benthem. Some kinds of modal completeness. *Studia Logica*, 39:125–141, 1980.

[44] J. F. A. K. van Benthem. *Modal Logic and Classical Logic*. Bibliopolis, Naples, 1983.

[45] Frank Wolter. Properties of tense logics. *Mathematical Logic Quarterly*, 42:481–500, 1996.

[46] Frank Wolter. The structure of lattices of subframe logics. *Annals of Pure and Applied Logic*, 86:47–100, 1997.

ROBERT GOLDBLATT
Centre for Logic, Language and Computation
Victoria University
P.O. Box 600, Wellington, New Zealand.
www.mcs.vuw.ac.nz/~rob
Rob.Goldblatt@vuw.ac.nz

V. Marra, D. Mundici

Łukasiewicz Logic and Chang's MV Algebras in Action

Abstract. We survey recent developments of Łukasiewicz propositional logic and its algebraic counterpart, Chang's MV algebras.

Keywords: Łukasiewicz logic, MV algebras, MV partitions, Multisets, Locally finite MV algebras, Searching with lies, ℓ-groups, Fans

1. Prologue

On page 5 of his paper *"L'oeuvre scientifique de Jan Łukasiewicz dans le domaine de la logique mathématique"*, Fund. Math., volume 44 (1957) pp. 1–11, Andrzej Mostowski wrote:

> ... A l'heure actuelle il n'est pas possible d'émettre une opinion définitive si les logiques multivalentes ne resteront qu'une construction théorique philosophique, ou si elles trouveront des applications en dehors de la logique la plus abstraite.

While this remark agrees with the status of Łukasiewicz logic in the year 1957,[1] in the last 15 years the number of papers devoted to applications of many-valued Łukasiewicz logic has been rapidly increasing. The space devoted to the topic in surveys and monographs has increased proportionally. To quote a few selected examples, the second edition of the Handbook of Philosophical Logic has a long additional chapter, by Hähnle [56], on infinite-valued proof-theory. Urquhart's classical chapter has been updated and expanded [101]. Hájek's monograph [57] devotes ample space to Łukasiewicz logic, and so does Gottwald's book [54]. The monograph [28] is entirely concerned with Łukasiewicz logic and its algebraic counterpart, the variety of Chang's MV algebras. The second volume of the Handbook of Measure Theory [86] includes several chapters on MV-algebraic measure theory. As shown in the monograph [39] and in the pioneering textbook [93],

---

[1]The first published proof of the completeness theorem for the infinite-valued Łukasiewicz calculus appeared only one year later, in [94].

V. F. Hendricks and *J. Malinowski* (eds.),
Trends in Logic: 50 Years of Studia Logica

Trends in Logic **21**: 145–192, 2003.

MV algebras also afford an important specimen of "quantum structures"; for a general survey see [85]. One can also find in the literature several surveys treating more specialized topics. These range from the game-theoretic interpretation of infinite-valued logic, with applications to error-correcting codes and fault-tolerant search [25, 87], to the relationship between MV algebras, lattice-ordered groups, and toric varieties [66, 65]. The literature has expanded so rapidly that in the year 2000 the AMS Classification Index introduced the special item 06D35 for MV algebras.

◇

Rather than recapitulating well known and easily accessible facts featuring in the aforementioned works, in this paper we shall present Łukasiewicz propositional[2] logic and MV algebras *in action*, both as powerful tools to give solutions to specific problems, and as a natural conceptual framework to generalize fundamental mathematical notions to a many-valued setting. Surely enough, during our excursion we shall come across almost all classical results on Łukasiewicz logic. On the other hand, we shall also familiarize with such novel topics as non-Boolean partitions and their joint refinements, finite and infinite multisets, and others.[3] This having been said, we feel that beneath such a variety of mathematical landscapes there runs a single unifying stream of thoughts. As a matter of fact, throughout this paper we shall be concerned with one and the same theory, albeit one that is just being born: the generalization of sets, partitions, and functions to a many-valued setting. As highlights of the programme, we offer an analysis of joint refinability of many-valued partitions, and an MV-algebraic Stone representation theorem for infinite multisets. We trust readers will acknowledge that such generalizations, while technically rather demanding, are conceptually basic. Accordingly, we begin to embroider our paper around one of the single most important ideas in modern mathematics, that of a function between finite sets.

Acknowledgements. The first author is partially supported by the research training network COMBSTRU (Combinatorial Structure of Intractable Problems) of the European Community, contract no. HPRN-CT-2002-00278. The second author is partially supported by the Italian MURST Project on Many-Valued Logic, ex 40%.

---

[2]From now on the adjective "propositional" shall be tacitly understood.

[3]We shall preferably report on topics that are only touched upon, if not completely omitted, in previous surveys. Our paper may also be regarded as a partial update of the last section of the monograph [28].

2. A naive approach to non-Boolean partitions

For X and Y finite sets, let $f\colon X \to Y$ be a function. The action of f can be faithfully described through the familiar device known as "definition by cases". In logical terms, the latter amounts to a conjunction of implications

$$(\text{if } x \in X_1 \text{ then } y \in Y_1) \ \text{ and } \ \ldots \ \text{ and } \ (\text{if } x \in X_n \text{ then } y \in Y_n) \qquad \text{(i)}$$

satisfying certain additional conditions:

- The functional nature of f demands that $\pi = \{X_1, \ldots, X_n\}$ be a partition into nonempty blocks of the domain X.
- The single-valued nature of f demands that each Y_i be a singleton subset of Y. Equivalently, we ask that $\alpha = \{Y_1, \ldots, Y_n, Y_{n+1}, \ldots, Y_m\}$ be an atomic partition of Y — 'atomic' meaning that all blocks are singletons.[4]
- In order for such a definition by cases to be uniquely determined, π should not be replaceable by a coarser partition of X.[5] There is a unique π satisfying this condition. Equivalently, we ask that the function $\hat{f}\colon \pi \to \alpha$ defined by $\hat{f}(X_i) = Y_i$ be injective.

In summary, $f\colon X \to Y$ *is uniquely determined by an injection* $\hat{f}\colon \pi \rightarrowtail \alpha$ *of a partition of* X *into the unique atomic partition of* Y, *and conversely.*

Suppose one wished to replace the above sets X and Y with "many-valued sets" — objects whose characteristic functions range in $[0,1] \subseteq \mathbb{R}$, as opposed to $\{0,1\}$. Since finite sets do not live in isolation, any reasonable many-valued generalization should encompass the logico-algebraic machinery accompanying finite sets. Thus, one may conveniently recall that to every finite set X there is canonically associated a finite Boolean algebra $\mathcal{P}(X)$ — the set of its subsets equipped with union, intersection, complement, and with the $\emptyset$ and X itself as distinguished elements. Conversely, to every finite Boolean algebra B one canonically associates the finite set $\mathcal{A}(B)$ of its atoms. Further, it is well-known that to every function $f\colon X \to Y$ one can associate a corresponding homomorphism of Boolean algebras $\mathcal{P}(f)\colon \mathcal{P}(Y) \to \mathcal{P}(X)$ (in the other direction). Similarly, every homomorphism of Boolean algebras is carried by $\mathcal{A}$ onto an appropriate contravariant function between the

[4]In so doing, we pay the (modest) price of introducing unused blocks $\{Y_{n+1}, \ldots, Y_m\}$. They witness the generally non-surjective action of f.

[5]That is, the blocks X_i are to be precisely the nonempty fibers of f.

corresponding sets. It turns out that $\mathcal{P}$ (equivalently, $\mathcal{A}$) is a duality between finite sets and functions, and finite Boolean algebras and homomorphisms. While this is just folklore, the functor $\mathcal{P}$ allows us to reformulate within the category of finite Boolean algebras the above analysis of $f\colon X \to Y$. We can thus speak of a partition π of the Boolean algebra B induced by an embedding $f\colon A \to B$, of the blocks of π and their properties, *et cetera.* We omit the trivial details of the translation.

To get things going we also note that there is another duality phenomenon connected with Boolean algebras. Every finite Boolean algebra is the Lindenbaum algebra of a theory in classical propositional logic over finitely many variables.[6] Again, this is folklore, and completely straightforward. However, it shows that when at the beginning of this Section we considered the function $f\colon X \to Y$, *in a precise technical sense we postulated classical propositional logic as an underlying framework.*

It turns out that Łukasiewicz logic does provide a natural framework for a theory of many-valued sets and partitions. Both notions depend on a substantial portion of the theory of Łukasiewicz logic and MV algebras. We thus interject a first compact introduction to a minimal body of knowledge leading to the notion of many-valued partition.

3. MV algebras and MV partitions

RUNNING NOTATION. Throughout this paper, we let $\mathbb{Z}$, $\mathbb{Q}$ and $\mathbb{R}$ stand for the set of integer, rational and real numbers, respectively. A natural number is an element of $\mathbb{N} = \{1, 2, 3, \ldots\}$. We let $\mathbb{Z}^+ = \{0, 1, 2, \ldots\}$, and similarly for $\mathbb{R}^+$, $\mathbb{Q}^+$.

3.1. Chang's Completeness Theorem

MV algebras are a generalization of Boolean algebras. They were introduced by Chang to give a syntax-free proof of the completeness theorem for the infinite-valued Łukasiewicz calculus [21, 22, 23]. We refer to [28] for a self-contained account, where proofs of all basic results are given in full detail.

An *MV algebra* $A = (A, 0, \neg, \oplus)$ is an Abelian monoid $(A, 0, \oplus)$ equipped with an operation $\neg$ such that $\neg\neg x = x$, $x \oplus \neg 0 = \neg 0$ and $\neg(y \oplus \neg x) \oplus y = \neg(x \oplus \neg y) \oplus x$.

---

[6]This can also be formulated as a categorical equivalence, but we do not insist on this point.

The defining equations of MV algebras are meant to capture *some* properties of the real unit interval $[0,1]$ equipped with the operations

$$\neg x = 1 - x \quad \text{and} \quad x \oplus y = \min(1, x+y). \tag{ii}$$

Chang's Completeness Theorem shows that these equations in fact capture *all* equational properties of $[0,1]$:

THEOREM 3.1. [22], [28, Theorems 2.5.3 and 4.5.2] An equation is satisfied in the MV algebra $[0,1]$ iff it is satisfied in all MV algebras.

A Boolean algebra is an MV algebra satisfying the additional equation $x \oplus x = x$. By a *homomorphism* $\eta\colon A \to A'$ we mean a function such that $\eta(0) = 0, \eta(\neg x) = \neg\eta(x)$ and $\eta(x \oplus y) = \eta(x) \oplus \eta(y)$ for all $x, y \in A$. An MV algebra A is *semisimple* iff for each nonzero $x \in A$ there is a homomorphism $\eta\colon A \to [0,1]$ with $\eta(x) \neq 0$.

The intuition that A is semisimple iff it has no infinitesimal elements can be made precise [28, Proposition 3.6.4]. Trivially, the MV algebra $[0,1]$ is semisimple. More generally, any MV algebra of $[0,1]$-valued functions over some set X, with the pointwise MV operations of $[0,1]$, is semisimple. The following theorem states that there are no other examples of semisimple MV algebras.

THEOREM 3.2. [22], [28, Corollary 3.6.8] The following conditions are equivalent for any MV algebra A:

(i) A is semisimple.

(ii) Up to isomorphism, A is an MV algebra of $[0,1]$-valued functions over some set X.

(iii) Up to isomorphism, A is a separating[7] MV algebra of continuous $[0,1]$-valued functions over some compact Hausdorff space X.

Intuitively, semisimple MV algebras stand to $[0,1]$-valued functions as Boolean algebras stand to $\{0,1\}$-valued functions.

Since MV algebras are defined by equations, for any $n \in \mathbb{N}$ the free n-generated MV algebra Free_n exists. Throughout, we let $\mathcal{M}_n$ denote the MV algebra of all continuous (finitely) piecewise linear functions $f\colon [0,1]^n \to [0,1]$ such that every linear piece of g has integer coefficients. By Theorem

[7]That is, for any two distinct points $x, y \in X$ there is a function p in the algebra such that $p(x) \neq p(y)$.

3.1, and standard universal algebraic machinery, Free_n is isomorphic to the (possibly proper) subalgebra of $\mathcal{M}_n$ consisting of functions $d\colon [0,1]^n \to [0,1]$ obtainable from the projection functions $x_1, \ldots, x_n$ by pointwise application of the $\neg$ and $\oplus$ operations — in other words, the MV-definable functions. It follows, in particular, that Free_n is semisimple, [28, Proposition 3.1.8, Theorem 3.6.9]. Further information about Free_n is given by *McNaughton's Theorem* [67], to the effect that in fact *any* function $f \in \mathcal{M}_n$ is MV-definable: f is obtainable from projection functions by pointwise application of the $\neg$ and $\oplus$ operations.[8] Accordingly, an element of $\mathcal{M}_n$ is called a *McNaughton function.* Now McNaughton's Theorem and Chang's Completeness Theorem synergically provide complete information about finitely generated free MV algebras: Free_n is canonically isomorphic to $\mathcal{M}_n$, with free generators represented by projection functions. It should be noted here that elements of Free_n are nothing else but n-variable formulæ in the Łukasiewicz infinite-valued calculus, up to logical equivalence. Ideals J correspond to (deductively closed) theories Θ, and the quotient of Free_n by J is the Lindenbaum algebra of Θ. We shall return to these topics in Section 8.

From the continuity of McNaughton functions one immediately deduces the following strengthening of Chang's Completeness Theorem.

COROLLARY 3.3. An equation is satisfied in the rational MV algebra $[0,1]\cap\mathbb{Q}$ iff it is satisfied in all MV algebras.

3.2. Underlying MV algebraic order, relations with ℓ-groups

Let A be an MV algebra, and let us write 1 as an abbreviation of $\neg 0$. For any $x, y \in A$ we write $x \leq y$ iff $\neg x \oplus y = 1$. Then $\leq$ is a partial order relation, called the *natural order* of A. Indeed, as proved by Chang [21], the natural order endows A with a *distributive lattice structure*, where the join $x \vee y$ and the meet $x \wedge y$ are given by[9]

$$x \vee y = \neg(\neg x \oplus y) \oplus y \text{ and } x \wedge y = \neg(\neg x \vee \neg y). \tag{iii}$$

An *ideal* J in an MV algebra A is a subset of A containing the zero element, closed under the $\oplus$ operation, and also closed under minorants ($x \leq y \in J$ implies $x \in J$). The *principal ideal* generated by an element $a \in A$ is the smallest ideal of A containing a as an element. *Wójcicki's Theorem* now reads as follows.

---

[8] See [75] for a constructive proof.

[9] See [28, Proposition 1.5.1] for a proof.

THEOREM 3.4. [28, Theorem 3.4.9, Lemma 3.6.6 and references on page 76] Let J be a principal ideal in Free_n. Then the quotient MV algebra Free_n/J is semisimple.

A fundamental property of every MV algebra A is that A carries a genuine group-theoretical addition. This is so because A lives in (categorical) symbiosis with a unique lattice-ordered Abelian group. An Abelian group G is said to be *lattice-ordered* (for short, G is an *ℓ-group*) iff G is also equipped with a lattice structure $(G, \vee, \wedge)$ obeying the distributivity conditions $t + (x \vee y) = (t + x) \vee (t + y)$ and $t + (x \wedge y) = (t + x) \wedge (t + y)$. Throughout, we let $G^+ = \{g \in G \mid 0 \leq g\}$. An element $u \in G$ is said to be a *strong unit* iff for all $g \in G^+$ there exists $n \in \mathbb{N}$ such that $nu \geq g$. We say (G, u) is a *(unital) ℓ-group*, but omit the adjective and write G for short when u is understood. By a *(unital) morphism* $\varphi\colon (G, u) \to (G', u')$ we mean a group homomorphism $\varphi\colon G \to G'$ that also preserves the lattice structure and satisfies the condition $\varphi(u) = u'$.

THEOREM 3.5. [71], [28, Corollary 7.1.8, Lemma 2.1.3 (iii)] For any unital ℓ-group (G, u), let $\Gamma(G, u)$ be the unit interval $[0, u] = \{h \in G \mid 0 \leq h \leq u\}$ equipped with the operations $\neg g = u - g$ and $g \oplus h = u \wedge (g + h)$. We then have:

(i) The algebra $A = ([0, u], 0, \neg, \oplus) = \Gamma(G, u)$ is an MV algebra.

(ii) If $\Gamma(\varphi)$ denotes the restriction to $[0, u]$ of the morphism $\varphi\colon (G, u) \to (G', u')$, Γ is a categorical equivalence between unital ℓ-groups and MV algebras.

(iii) The natural order of A agrees with the lattice order of G.

Therefore, every MV algebra A can be identified with the unit interval of a *unique* unital ℓ-group G, in symbols,

$$A = \Gamma(G, u) = [0, u], \qquad \text{(iv)}$$

whence A inherits by restriction the genuine addition of G, as a partial operation in A.

3.3. A definition of many-valued partitions

A (Boolean) *partition* of a set X is a collection of subsets

$$\pi = \{X_1, X_2, \ldots, X_n\}$$

satisfying the following conditions.

Incompatibility.	No two distinct blocks overlap.
Exhaustiveness.	The union of the X_i is the set X.
Irredundancy.	No block is empty.

Let $G = \mathbb{Z}^X$ be the additive group of all integer-valued functions over X, addition being defined pointwise. Identifying each block X_j with its characteristic function $\chi_j \colon X \to \{0,1\}$ — an element of G — the *first two* conditions above are equivalent to

Partition of unity.	The blocks sum up to one, i.e. $\chi_1 + \cdots + \chi_n = \mathbf{1}$, where $\mathbf{1} \in G$ is the function constantly equal to 1.

All *three* assumptions of incompatibility, exhaustiveness, and irredundancy are jointly expressed by

Independent partition of unity.	The blocks form a linearly independent partition of unity, in the $\mathbb{Z}$-module G.

In the language of Boolean algebras, a partition of an algebra B is a collection of nonzero elements such that their join is 1 (the top element of B), and the meet of any two distinct elements is zero. In the context of Łukasiewicz logic and MV algebras there does exist a satisfactory many-valued generalization of the notion of partition, where the above properties are preserved:

DEFINITION 3.6. [76, 78] Let A be an MV algebra and (G, u) its associated unital ℓ-group as in (iv). An *MV partition* π of A is a finite subset $\{h_1, \ldots, h_n\} \subseteq A$, together with a function assigning to each h_i a natural number $m_i \in \mathbb{N}$, $i \in \{1, \ldots, n\}$, satisfying the following conditions:

(*) $m_1 h_1 + \cdots + m_n h_n = 1 = u$, where $+$ denotes addition in G.

(**) The set $\{h_1, \ldots, h_n\}$ (equivalently, the set $\{m_1 h_1, \ldots, m_n h_n\}$) is linearly independent over $\mathbb{Z}$. In other words, whenever $0 = z_1 h_1 + \cdots + z_n h_n \in G$ with integer coefficients z_i, then all z_i's are zero.

Thus, an MV partition of an MV algebra A is a *multiset* of "linearly independent elements of A whose sum equals 1". This latter statement only makes sense with reference to the underlying $\mathbb{Z}$-module structure of the unital ℓ-group G to A via (iv). Here is a case when a definition crucially hinges upon a structure theorem.

In keeping with standard notation for multisets, we shall write

$$\pi = \{h_1^{m_1}, \ldots, h_n^{m_n}\}$$

to convey complete information about the MV partition π. Each element h_i is called a *block* of π. Note that $m_i h_i$ is an element of A. The integers m_i are the *multiplicities* of π. By condition (**) they are uniquely determined.

The notion of MV partition is fundamental in at least one respect. In close analogy with finite-dimensional C^*-algebras (cf. the Antefact, at the end of this paper), MV partitions provide elementary building blocks for any MV algebra. We shall develop this theme for the next few pages. The crux of the matter is that MV partitions admit joint refinements.

DEFINITION 3.7. [76, 78] Let A be an MV algebra, and let G be its corresponding ℓ-group with strong unit u as in (iv). For any two MV partitions of A,

$$\pi_1 = \{h_1^{m_1}, \ldots, h_n^{m_n}\} \text{ and } \pi_2 = \{g_1^{e_1}, \ldots, g_p^{e_p}\},$$

we say that π_2 is a *refinement* of π_1, written $\pi_2 \leq \pi_1$, iff each block h_j is obtainable as a linear combination in G with nonnegative integer coefficients of the blocks g_i's. In other words, the submonoid of G generated by $\{g_1, \ldots, g_p\}$ contains the set $\{h_1, \ldots, h_n\}$. A *joint refinement* of π_1 and π_2 is an MV partition π of A such that $\pi \leq \pi_1$ and $\pi \leq \pi_2$.

Again via Theorem 3.5, the joint refinability of any two MV partitions will be shown to depend on a nontrivial property of ℓ-groups.

THEOREM 3.8. [64] Let G be an ℓ-group. Given $P = \{p_1, \ldots, p_n\} \subseteq G^+ - \{0\}$, there exists a finite $C \subseteq G^+ - \{0\}$ satisfying the following two conditions:

1. C is linearly independent (in the $\mathbb{Z}$-module G).
2. Each p_i is (uniquely, by condition 1) expressible as a linear combination of elements of C, with nonnegative integer coefficients. Equivalently, P is in the submonoid generated by C.

The original proof in [64] is completely elementary, but rather involved. Here we sketch another proof that — while still far from straightforward — has a more geometric flavour, for it relies on the representation theory of ℓ-groups. In order to do this, we need to introduce a fair amount of background on geometric representations of ℓ-groups and MV algebras, but much of this material is of fundamental importance for the whole paper, and not only in connection with the joint refinability problem. We shall strive to make our presentation as concise as possible — full details can be found in the references provided below. We shall return to the proof of Theorem 3.8 in Section 5, and to joint refinements in Subsection 5.4.

4. Fans, polyhedral complexes, unimodularity.

Rational polyhedral geometry is a key tool, among others, of the modern theory of both MV algebras and Abelian ℓ-groups.[10] We shall need the rudiments of both the affine and the homogeneous theory — the former applies to MV algebras, the latter to ℓ-groups. To avoid tedious repetitions, we spell out the homogeneous variant only, leaving the easy adaptation for the affine case to the reader. An excellent general introduction to polyhedral geometry is [105]. References to more specific topics are postfixed to the title of each subsection.

4.1. Basic notions [44]

The geometric objects we are about to define are embedded into the real vector space $\mathbb{R}^n$, for a fixed $n \in \mathbb{N}$. Given a subset $S \subseteq \mathbb{R}^n$, we define the *positive hull* of S (over $\mathbb{R}$) as

$$\operatorname{pos} S = \{\mathbf{x} \mid \mathbf{x} = \sum_{i=1}^{k} r_i \mathbf{s}_i \text{ for some } k \in \mathbb{N},\, r_i \in \mathbb{R}^+,\, \mathbf{s}_i \in S\}.$$

A subset of $\mathbb{R}^n$ is *homogeneous convex* iff it is of the form $\operatorname{pos} S$ for some $S \subseteq \mathbb{R}^n$. A *(rational polyhedral) cone* σ in $\mathbb{R}^n$ is the positive hull $\sigma = \operatorname{pos} F$ of a finite subset $F \subseteq \mathbb{Q}^n$. Since we are dealing with a homogeneous notion, the distinction between *rational* and *integral* polyhedral cones is immaterial.[11] The *dimension* of a cone τ, denoted $\dim \tau$, is the dimension of the real vector space spanned by τ. A vector $\mathbf{v} \in \tau$ is a *vertex* of τ iff $\mathbf{v} \in \mathbb{Z}^n$, $\mathbf{v}$ is *primitive* — i. e., its coordinates are relatively prime — and whenever $\mathbf{v} \in \operatorname{pos} S$ for some $S \subseteq \tau$, it follows that $\mathbf{v} \in S$. The set of vertices of τ is finite, and is denoted $\operatorname{vert} \tau$. We say τ is *simplicial* iff $\operatorname{vert} \tau$ is linearly independent (over $\mathbb{R}$), in which case we use the notation $\tau = \langle \operatorname{vert} \tau \rangle$. If τ is simplicial, one has $\dim \tau = |\operatorname{vert} \tau|$. A *face* of a simplicial cone τ is the positive hull of a subset of $\operatorname{vert} \tau$, hence again a simplicial cone. A *facet* of τ is then a face of dimension $\dim \tau - 1$, or *codimension* 1. Faces of dimension 1 are precisely vertices. The faces of a simplicial cone τ form a bounded distributive lattice under inclusion, the *face lattice* of τ. A *(rational polyhedral) fan* in $\mathbb{R}^n$ is a finite set Σ of cones such that

1. Every face of every cone of Σ belongs to Σ.

---

[10] It also provides a fully expressive language for toric varieties, [44].

[11] This is not true in the affine case — integral affine polyhedral geometry is usually harder than its rational counterpart.

2. Any two cones of Σ intersect in a common face.[12]

The *support* of Σ, denoted $|\Sigma|$, is the union of all its cones. A *(rational) homogeneous polyhedral set* is the support of a fan. The k*-skeleton* of a fan Σ, denoted $\Sigma^{(k)}$, is the set of k-dimensional cones in Σ. It is not itself a fan, for it does not contain, together with a cone, all of its faces. A fan Σ is *simplicial* iff all its cones are simplicial. A cone τ carries a natural fan structure Σ_τ which is given by the collection of all its faces. With this notation, $|\Sigma_\tau| = \tau$. We shall not be pedantic about the distinction between Σ_τ and τ. The affine analogue of a cone in $\mathbb{R}^n$ is a *(rational) polyhedron* in $\mathbb{R}^{n-1}$ — notice that dimension drops by a unit. The affine analogue of a simplicial cone is a simplex. The affine analogue of a fan is a *(rational) polyhedral complex* Σ, which is a *simplicial complex* precisely when all polyhedra $\sigma \in \Sigma$ are simplices. A *(rational) polyhedral set* is the support of a polyhedral complex.
Given two fans Σ and Δ in $\mathbb{R}^n$, Δ is a *subdivision* of Σ, or it *refines* Σ, iff $|\Sigma| = |\Delta|$ and every cone of Δ is contained in some cone of Σ. We write $\Delta \leq \Sigma$ to denote the fact that Δ refines Σ. If this is the case, every cone of Σ is a union of cones of Δ. More generally, we say that Σ is *inscribed* in Δ iff every cone of Σ is a union of cones of Δ, whence we have the (possibly strict) inclusion $|\Sigma| \subseteq |\Delta|$[13]. If R is a homogeneous polyhedral set, we say that R is *inscribed* in Δ iff R is a union of cones of Δ. The *fundamental cone* of $\mathbb{R}^n$ is the simplicial cone whose vertices are precisely the vectors in the canonical basis of $\mathbb{R}^n$, written $\sigma = \langle \mathbf{e}_1, \ldots, \mathbf{e}_n \rangle$. The *fan of the fundamental cone* σ is the fan Σ_σ.

The affine counterparts of the notions introduced in the preceding paragraph are *subdivision* or *refinement*, *inscription* or *polyhedral subcomplex*, *fundamental simplex*, and *simplicial complex of the fundamental simplex*.

Polyhedral complexes and fans can be simplicialized without adding new vertices, for instance through a generalization of the elementary plane construction of 'adding noncrossing diagonals' to a polygon.

Fact (Existence of simplicial subdivisions). Given a fan (polyhedral complex) Σ, there exists a simplicial fan (simplicial complex) Δ with the same 1-skeleton (0-skeleton) of Σ such that $\Delta \leq \Sigma$.

---

[12] We have not formally defined 'face' for a generally non-simplicial polyhedral cone. We rely on the reader's geometric intuition. See [105] for details.

[13] Inscription yields the notion of substructure for fans, so that we could have also called Δ a *subfan* of Σ.

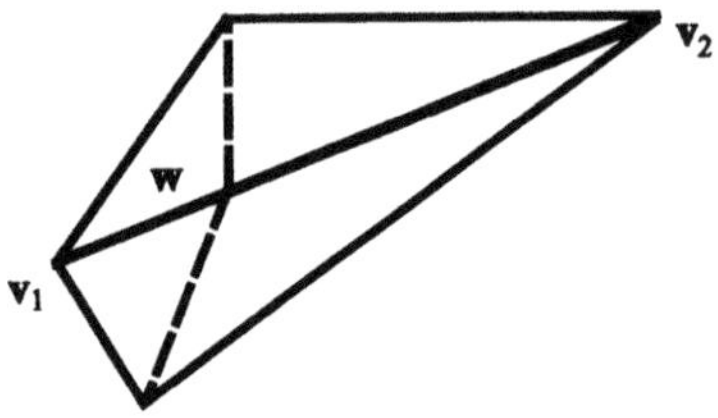

Figure 1. A stellar subdivision of a simplicial complex along $\langle \mathbf{v}_1, \mathbf{v}_2 \rangle$ through $\mathbf{w}$ (affine case). In the homogeneous case, the subdivision is *barycentric* iff $\mathbf{w} = \mathbf{v}_1 + \mathbf{v}_2$.

4.2. Unimodular cones and subdivisions [44]

A simplicial cone $\sigma = \langle \mathbf{v}_1, \ldots, \mathbf{v}_m \rangle \subseteq \mathbb{R}^n$ is *unimodular* (alternative terminology, *regular*) iff $\{\mathbf{v}_1, \ldots, \mathbf{v}_m\}$ can be completed to a $\mathbb{Z}$-module basis of $\mathbb{Z}^n \subseteq \mathbb{R}^n$. A fan Σ is *unimodular* iff all its cones are unimodular. If $m = n$, then σ is unimodular iff $\{\mathbf{v}_1, \ldots, \mathbf{v}_n\}$ group-theoretically generates $\mathbb{Z}^n$, which can be rephrased saying that the $n \times n$ matrix whose columns are given by $\{\mathbf{v}_1, \ldots, \mathbf{v}_n\}$ has determinant of absolute value 1. For the affine version, a simplex $\sigma = \langle \mathbf{v}_1, \ldots, \mathbf{v}_m \rangle \subseteq \mathbb{R}^{n-1}$ is *unimodular* iff the cone $\tau = \langle \phi(\mathbf{v}_1), \ldots, \phi(\mathbf{v}_m) \rangle$ in $\mathbb{R}^n$ determined by the embedding $\phi : \mathbb{R}^{n-1} \rightarrowtail \mathbb{R}^n$ given by

$$(0 \quad \cdots \quad 0 \;\; 1 \;\; 0 \quad \cdots \quad 0) \overset{\phi}{\mapsto} (0 \quad \cdots \quad 0 \;\; 1 \;\; 0 \quad \cdots \quad 0 \;\; 1)$$

is unimodular. Not only can every fan be simpicialized without introducing new vertices — additional subdivisions allow to refine any given fan to a unimodular fan.

Fact (Existence of unimodular subdivisions). Given a simplicial fan (simplicial complex) Σ, there exists a unimodular fan (unimodular simplicial complex) Δ such that $\Delta \leq \Sigma$.

It should be remarked that, unlike the case of simplicial subdivisions, it is in general necessary to introduce additional vertices in order to unimodularize a given fan. In both cases, there seems to be no known canonical way to choose the additional 'diagonals', or to unimodularize a given fan. We now introduce an important notion of refinement originally due to Alexander [4, 5]. Let $\sigma = \langle \mathbf{v}_1, \ldots, \mathbf{v}_m \rangle$ be an m-dimensional simplicial cone of a simplicial fan Σ in $\mathbb{R}^n$. Suppose $\tau = \langle \mathbf{v}_{i_1}, \ldots, \mathbf{v}_{i_d} \rangle$ is a d-dimensional face of σ. Let $\mathbf{w}$ be a vector in the relative interior[14] of τ. The *star* in Σ of a face τ of σ, denoted

[14]That is, the interior of τ as a closed set of the real vector space spanned by τ itself.

star τ, is the set of cones in Σ having τ as a face — it is not a fan. The *closed star* of τ is the fan cstar $\tau = \{\delta \mid \delta$ is a face of a cone $\theta \in$ star $\tau\}$. The *(Alexander) stellar subdivision* (of Σ along τ through $\mathbf{w}$) is the fan obtained from Σ removing star τ and joining $\mathbf{w}$ to the boundary fan (cstar τ − star τ) of the star of τ. (See Figure 4.2.) If Δ is obtained from Σ via a finite number of stellar subdivisions, then $\Sigma \leq \Delta$ holds, but the converse fails — not every refinement can be obtained via stellar subdivisions. Let again τ be a face of σ, as in the above. We call $\mathbf{w} = \sum_{j=1}^{d} \mathbf{v}_{i_j}$ the *barycenter* of τ. The fan obtained from Σ by stellar subdivision along τ through $\mathbf{w}$ is a *barycentric stellar subdivision of* Σ. We write $\Delta \preceq \Sigma$ to denote the fact that Δ is obtained from Σ by a finite number (possibly zero) of stellar barycentric subdivisions. If all barycenters belong to 2-dimensional cones, we write $\Delta \preceq_2 \Sigma$ and speak of *binary starring*, for simplicity. It is easy to show but important to realize that if $\Delta \preceq \Sigma$ and Σ is unimodular, then Δ is unimodular — *barycentric subdivisions preserve unimodularity.* This fails for general subdivisions, even when they are topologically stellar.[15]

4.3. Schauder bases and the De Concini-Procesi Lemma [73, 84, 44]

Let Σ be a unimodular fan in $\mathbb{R}^n$. Let $\rho = \langle \mathbf{r} \rangle \in \Sigma^{(1)}$ (that is, ρ is an integral ray). The *Schauder hat* at ρ over Σ is the unique continuous (finitely) piecewise linear homogeneous function $h_{\mathbf{r}} : |\Sigma| \to \mathbb{R}$ such that

1. $h_{\mathbf{r}}(\mathbf{r}) = 1$,
2. $h_{\mathbf{r}}(\mathbf{v}) = 0$ for every vertex $\mathbf{v} \neq \mathbf{r}$ of any cone in $\Sigma^{(1)}$,
3. $h_{\mathbf{r}}$ is homogeneous linear on each cone of Σ.

As a consequence of the unimodularity of Σ, a Schauder hat is guaranteed to have linear pieces with integral coefficients. We denote by $\mathbf{H}_\Sigma$ the set of all Schauder hats at the vertices of the unimodular fan Σ, and call it the *Schauder basis* over Σ. We say that $|\Sigma|$ is the *support* of $\mathbf{H}_\Sigma$. Conversely, given a Schauder basis $\mathbf{H}$ in $\mathbb{R}^n$, there exists a unique unimodular fan Σ in $\mathbb{R}^n$, denoted $\Sigma_{\mathbf{H}}$, such that $\mathbf{H} = \mathbf{H}_\Sigma$. In other words, Schauder bases and unimodular fans are dual concepts, a circumstance we shall often exploit in

---

[15]The barycenter of a rational simplex is known to geometers of numbers as the *(multidimensional) Farey mediant* of its vertices, a classical notion intimately connected with (multidimensional) continued fraction expansions [19]. We shall not be able to develop this interesting theme on this occasion.

the following.[16] Affine Schauder hats and bases are defined via a straightforward translation of the above homogeneous variant. One technical tool we shall need is a simple result originally proved by De Concini and Procesi in a different context [33].

Fact (The De Concini-Procesi Lemma). Let Σ be a unimodular fan in $\mathbb{R}^n$. Let $H \subseteq \mathbb{R}^n$ be an integral homogeneous hyperplane. There exists a refinement $\Sigma_H \preceq_2 \Sigma$ such that $H \cap |\Sigma|$ is a union of cones of Σ_H, i.e., it is inscribed in Σ_H.

Remark. An elementary self-contained proof of the De Concini-Procesi Lemma was given in [84]. The proof shows that in order to obtain the desired refinement Σ_H, it suffices to apply binary starrings *only* to couples of vertices of the current fan Σ_i that lie on opposite sides of the hyperplane H. Let us call such vertices *in H-opposition.* While one may certainly apply additional binary starrings to couples of vertices that are *not* in H-opposition, and still obtain a fan Σ_H satisfying the lemma, all such subdivisions can always be dispensed with.[17]

As an immediate corollary, we deduce the following statement.

Fact (The De Concini-Procesi Lemma, second version). Let Σ and Δ be fans in $\mathbb{R}^n$ with the same support — $|\Sigma| = |\Delta|$ — and assume Σ is unimodular. There exists a unimodular fan Θ such that

1. $\Theta \preceq_2 \Sigma$;
2. $\Theta \leq \Delta$.

4.4. Geometry of MV algebras and ℓ-groups [28, 32, 48][18]

Throughout, by an *ℓ-function* of n variables we mean an integral (finitely) piecewise linear homogeneous continuous map $f : \mathbb{R}^n \to \mathbb{R}$ such that each of its pieces has integral coefficients. Thus, an ℓ-function is just the homogeneous version of a McNaughton function. We write $\mathcal{A}_n$ to denote the set

[16] Notice that in order to obtain a genuine duality, one would need a definition of Schauder basis independent of the notion of unimodular fan. There is one such definition, but it is far from obvious. Lack of space prevents us from pursuing this interesting matter further in this paper.

[17] We shall have to make use of this observation in the proof of Lemma 5.3 below.

[18] The theory we summarize here is best assembled from the original references given in Table 4.4. At the time of writing, there is no single monograph that gives a full self-contained treatment of the matter, although [28] is essentially complete on the MV-algebraic front.

of all ℓ-functions of n variables over $\mathbb{R}^n$. The projection functions, denoted $\pi_i\colon \mathbb{R}^n \to \mathbb{R}$, are ℓ-functions. When equipped with pointwise addition and order, $\mathcal{A}_n$ is an ℓ-group. Thus, $\mathcal{A}_n$ is the homogeneous analogue of $\mathcal{M}_n$, cf. Section 3.1. Beynon [13] proved the ℓ-analogue of McNaughton's Theorem, namely, *projection functions generate the ℓ-group $\mathcal{A}_n$*. Together with Weinberg's result [102] that $\mathbb{Z}$ *generates the variety of ℓ-groups*, this yields a crucial representation theorem to the effect that $\mathcal{A}_n$ is canonically isomorphic to the free ℓ-group over n variables, free generators being represented by projection functions. Again, we discussed the MV-algebraic (i.e., affine) variant of this statement in Subsection 3.1.

If f is an ℓ-function over $\mathbb{R}^n$, its *zero set* is the closed homogeneous[19] subset $\mathbf{Z}(f) = \{\mathbf{x} \in \mathbb{R}^n \mid f(\mathbf{x}) = 0\}$. An *$\ell$-ideal* of an ℓ-group G is a sublattice subgroup $\mathfrak{o}$ of G that is *convex*, i.e., $x, y \in \mathfrak{o}$, $z \in G$, and $x \leq z \leq y$ imply $z \in \mathfrak{o}$. Kernels of homomorphisms between ℓ-groups are the same thing as ℓ-ideals. We denote the quotient ℓ-group of G modulo its ℓ-ideal $\mathfrak{o}$ by $G/\mathfrak{o}$. Given an ℓ-ideal $\mathfrak{o} \subseteq \mathcal{A}_n$, we associate to it the closed homogeneous subset of $\mathbb{R}^n$

$$\mathbb{V}(\mathfrak{o}) = \{\mathbf{x} \in \mathbb{R}^n \mid \text{ for every } f \in \mathcal{A}_n\,,\ \mathbf{x} \in \mathbf{Z}(f)\}\,.$$

We call $\mathbb{V}(\mathfrak{o})$ the *vanishing locus* or the *localization* or the *zero set* of $\mathfrak{o}$. It is easy to check that $\mathbb{V}(\mathfrak{o})$ is indeed closed and homogeneous. Conversely, given a subset $S \subseteq \mathbb{R}^n$, we can define the *ℓ-ideal over S* or the *vanishing ℓ-ideal of S*, namely

$$\mathbb{I}(S) = \{f \in \mathcal{A}_n \mid \text{ for all } \mathbf{x} \in S\,,\ \mathbf{x} \in \mathbf{Z}(f)\}\,.$$

It is easy to check that $\mathbb{I}(S)$ is indeed an ℓ-ideal, and that $\mathbb{I}(S) = \mathbb{I}(\tilde{S})$ if $\tilde{S}$ is the topological closure of the homogeneous closure of S. As a minimal amount of ring-theoretical experience will suggest, $\mathfrak{o} \subseteq \mathbb{I}(\mathbb{V}(\mathfrak{o}))$ always holds, but the inclusion may well be strict. The characterization of those ℓ-ideals $\mathfrak{o}$ of $\mathcal{A}_n$ such that $\mathfrak{o} = \mathbb{I}(\mathbb{V}(\mathfrak{o}))$ is the exact ℓ-analogue of Hilbert's Nullstellensatz. It is due to Beynon [14], even though the mathematical essence of the result was already with Yosida [104]. An ℓ-group G is *Archimedean* iff for every $x, y \in G$ with $0 < x \leq y$, there exists $m \in \mathbb{N}$ such that mx is incomparable to y. Then $\mathcal{A}_n/\mathfrak{o}$ is Archimedean iff $\mathfrak{o} = \mathbb{I}(\mathbb{V}(\mathfrak{o}))$, [14]. Results of the Nullstellensatz type deal with maximality conditions on congruences. As to principal congruences, Baker proved [9] that if $\mathfrak{o}$ is a principal ℓ-ideal of $\mathcal{A}_n$, the quotient ℓ-group $\mathcal{A}_n/\mathfrak{o}$ is Archimedean.

---

[19] An arbitrary subset of $\mathbb{R}^n$ is called *homogeneous* iff it contains, along with a given point, the unique ray through that point springing from the origin.

It is routine to give a similar treatment for MV algebras: The zero set of a McNaughton function in $\mathcal{M}_n$ is a closed subset of $[0,1]^n$; if J is an ideal of $\mathcal{M}_n$, its localization $\mathbb{V}(J)$ is a closed subset of $[0,1]^n$; if $S \subseteq [0,1]^n$, the set $\mathbb{I}(J)$ of McNaughton functions vanishing over J is an ideal of $\mathcal{M}_n$; Archimedean ℓ-groups are the counterpart of semisimple MV algebras;[20] the MV-Nullstellensatz amounts to Theorem 3.2, and may thus be credited to Chang; and Wójcicki's Theorem 3.4 is the MV-analogue of Baker's result on principal congruences in [9].

An ℓ-group P is said to be *projective* iff for every ℓ-homomorphism $h\colon P \to G_1$ and for every onto ℓ-homomorphism $s\colon G_2 \twoheadrightarrow G_1$, there exists an ℓ-homomorphism $k\colon P \to G_2$ such that $s \circ k = h$. Similarly, upon restriction to the appropriate categories, one can speak of projective MV algebras or, equivalently, projective unital ℓ-groups. On the other hand, an ℓ-group is *finitely presented* iff it is the quotient of a finitely generated free ℓ-group modulo a finitely generated (equivalently,[21] principal) ℓ-ideal. Similarly, one can define finitely presented MV algebras. While projectivity and finite presentability are standard notions in any variety, Beynon proved the peculiarly ℓ-theoretical result that *a finitely generated ℓ-group P is projective iff P is finitely presentable*, [15]. This fails for MV algebras: on general universal algebraic grounds, every finitely generated projective MV algebra is trivially finitely presentable, but the converse does not hold.[22]

Table 4.4 contains fairly complete information about the geometric representation theory of finitely generated ℓ-groups and MV algebras.[23]

5. Simplicial approximations of ℓ-groups and joint refinements of MV partitions

We return to the proof of Theorem 3.8. Except for Subsection 5.4, this material has not previously appeared in print. We closely follow [63].

---

[20] In fact, the latter also are sometimes called Archimedean in the literature.

[21] In both the varieties of ℓ-groups and MV algebras, proving this equivalence is a simple exercise.

[22] For a detailed counterexample based on a purely topological obstruction, see [66]. A very simple counterexample based on a purely arithmetic obstruction is given by the (unique, up to isomorphism) finite MV algebra $Ł_2$ with 3 elements. While $Ł_2$ is obviously finitely presentable, it is not projective, as is easily seen. This phenomenon is a consequence of the arithmetic enrichment of classical logic brought forth by (Łukasiewicz) polyvalence — cf. the remarks following Theorem 6.5 below.

[23] The synopsis is made possible by Theorem 3.5.

MV		ℓ	
Chang's Theorem (1959)	[22]	**Weinberg's Theorem** (1963)	[102]
The variety of MV algebras is generated by $[0,1] \cap \mathbb{Q}$. (Corollary 3.3.)		The variety of ℓ-groups is generated by $\mathbb{Z}$. (Corollary 5.5.)	
McNaughton's Theorem (1951)	[67]	**Beynon's Theorem, I** (1974)	[13]
Every McNaughton function of n variables belongs to $\mathcal{M}_n$. (Theorem 8.1.)		Every ℓ-function of n variables belongs to $\mathcal{A}_n$. (Subsection 4.4, *passim.*)	
Free representation (1951-59)	[22, 67]	**Free representation** (1963-74)	[102, 13]
$\mathcal{M}_n$ is the free MV algebra over n free generators, i.e. projection functions. (Subsection 3.1, *passim.*)		$\mathcal{A}_n$ is the free ℓ-group over n free generators, i.e. projection functions. (Subsection 4.4, *passim.*)	
MV Nullstellensatz (1959)	[104, 22]	**ℓ-Nullstellensatz** (1975)	[14]
TFAE: 1. A is fin. gen. semisimple. 2. $\mathbb{I}(\mathbb{V}(J)) = J$ if $A \cong \mathcal{M}_n/J$. (Theorem 3.2.)		TFAE: 1. G is fin. gen. Archimedean. 2. $\mathbb{I}(\mathbb{V}(\mathfrak{o})) = \mathfrak{o}$ if $G \cong \mathcal{A}_n/\mathfrak{o}$. (Subsection 4.4, *passim.*)	
Wójcicki's Theorem (1973)	[103]	**Baker's Theorem** (1968)	[9]
Every finitely presented MV algebra is semisimple. (Theorem 3.4.)		Every finitely presented ℓ-group is Archimedean. (Subsection 4.4, *passim.*)	
?	?	**Beynon's Theorem, II** (1977)	[15]
? (Cf. footnote 22.)		Every finitely presented ℓ-group is projective. (Subsection 4.4, *passim.*)	

Table 1. A synopsis of the geometric representation theory of MV algebras and ℓ-groups. The references in parentheses should aid the reader in locating statements within the body of this paper.

5.1. First steps

Fix an ℓ-group G. We shall associate to a finite set $P \subseteq G^+ - \{0\}$ an abstract simplicial complex induced by the order structure of G. Recall that an *abstract* (always finite, in this paper) *simplicial complex* $\mathcal{S}_V$ with vertex set V is a collection of subsets of V that includes all singletons and is closed upon taking subsets — if $X \in \mathcal{S}_V$ and $Y \subseteq X$, then $Y \in \mathcal{S}_V$. An $(m+1)$-subset of $\mathcal{S}_V$ is an *m-simplex*, and $\emptyset$ is the unique simplex of any simplicial complex of dimension -1. There is a fundamental invariant attached to $\mathcal{S}_V$. The *f-vector* [97, 105] of $\mathcal{S}_V$, denoted $f_{\mathcal{S}_V}$, is the eventually zero sequence

$$f_{\mathcal{S}_V} = (n_0, n_2, \ldots, n_k, 0, \ldots),$$

where $n_i \in \mathbb{N}$ for $i \in \{0, \ldots, k\}$, and n_i is the number of i-simplices of $\mathcal{S}_V$. The *dimension* of $\mathcal{S}_V$, denoted $\dim \mathcal{S}_V$, is k — in words, the dimension of any one of its max-dimensional simplices. Let $\mathcal{F}$ be any set, finite or infinite, of f-vectors of simplicial complexes. The collection $\mathcal{F}$ can be lexicographically ordered by

$$(n_0, n_1, \ldots, n_k, 0, \ldots) \leq (m_0, m_1, \ldots, m_h, 0, \ldots)$$

iff $h > k$, or $h = k$ and there exists $i \in \{0, \ldots, h\}$ such that

$$n_j = m_j \text{ for all } j \in \{i+1, i+2, \ldots, h\} \quad \text{and} \quad n_i < m_i .$$

Then the chain $\mathcal{F}$ is well-ordered, and one can implement inductions over it. This we shall have occasion to do in a moment. Whenever we write an inequality between two f-vectors, it is understood that we are using lexicographic order.

DEFINITION 5.1. *(The simplicial complex structure of a finite set of elements)* Given a finite subset $P = \{p_1, \ldots, p_n\}$ of $G^+ - \{0\}$, let $\mathcal{S}_P \subseteq \mathcal{P}(P)$ be defined by

$$S \in \mathcal{S}_P \quad \text{iff} \quad S = \emptyset \text{ or } \bigwedge S > 0.$$

It is clear that $\mathcal{S}_P$ is an abstract simplicial complex. We denote its f-vector by f_P in order to avoid the awkward $f_{\mathcal{S}_P}$.

As a final preliminary step, we generalize Alexander's barycentric stellar subdivisions (cf. Subsection 4.2) to the abstract context of ℓ-groups. The operation we shall introduce stands to ℓ-groups as subtraction[24] stands to $\mathbb{Z}$. Accordingly, it accounts for the interaction between algebra and order.

---

[24] Indeed, one could have also reasonably called this basic operation *generalized subtraction.*

DEFINITION 5.2. *(Alexander's starring in ℓ-groups)* Let $P = \{p_1, \dots, p_n\} \subseteq G^+ - \{0\}$. Let $S = \{s_1, \dots, s_k\}$ be a simplex of $\mathcal{S}_P$. Set $S^* = \{s_1^*, \dots, s_k^*, \bigwedge S\} - \{0\}$, where $s_i^* = s_i - \bigwedge S$ for each $i \in \{1, \dots, k\}$. Further, let $P^* = (P - S) \cup S^*$. We say that the abstract simplicial complex $\mathcal{S}_{P^*}$ is obtained from $\mathcal{S}_P$ via *stellar ℓ-subdivision* (in G along S). By abuse of language, we apply the same terminology to P and P^*. If Q is obtained from P by a finite (possibly zero) number of stellar ℓ-subdivisions, we write $Q \preceq P$. We write $Q \preceq_2 P$ to denote that $Q \preceq P$ and all stellar subdivisions are along 1-simplices, in which case we simply speak of *binary ℓ-starring.*

5.2. A lemma of rational polyhedral geometry

We are now ready to tackle a key technical lemma. We shall give a fully detailed proof, for this result embodies the essential reason of why Theorem 3.8 holds.

For the sake of greater clarity, let us discuss some terminological issues related to the proof below. In keeping with previous notation, we denote by $\mathbf{H}_\Lambda$ the set of Schauder hats over a unimodular fan Λ. We shall make crucial use of the f-vector f_Λ. By the latter we mean the f-vector of the unimodular fan Λ, *regarded as a simplicial complex* (k-simplex $=$ $(k+1)$-cone). Thus, "f_Λ" makes perfect sense. According to Definition 5.1, however, "$f_{\mathbf{H}_\Lambda}$" also makes perfect sense, provided it is understood that the underlying ℓ-group G is the projective $\mathcal{A}_n / \mathbb{I}(|\Lambda|)$. A moment's reflection shows that, in fact, $f_{\mathbf{H}_\Lambda} = f_\Lambda$. Below we shall use f_Λ because the latter is independent of the theory of ℓ-groups. Thus, *the lemma we are about to prove is a fact of rational polyhedral geometry.*

LEMMA 5.3. *(Localized f-vector)* Let Σ be a (necessarily unimodular) fan inscribed in the fan of the fundamental cone $\sigma = \langle \mathbf{e}_1, \dots, \mathbf{e}_n \rangle \subseteq \mathbb{R}^n$, where $\{\mathbf{e}_i\}$ is the canonical basis of $\mathbb{R}^n$. Assume $\mathbf{e}_i \in \Sigma$ for all $i \in \{1, \dots, n\}$. Let $\mathbf{H}_\Sigma = \{h_1, \dots, h_n\}$ be the set of Schauder hats over Σ. Let $\mathbf{z} = (z_1, \dots, z_n) \in \mathbb{Z}^n - \{0\}$, and set $\mathbf{g} = z_1\pi_1 + \cdots + z_n\pi_n$, where $\pi_i : \mathbb{R}^n \to \mathbb{R}$ is the ith projection function. Then $\mathbf{g}$ is an integral homogeneous linear function, and its zero set $\mathbf{Z}(\mathbf{g})$ is a homogeneous rational hyperplane in $\mathbb{R}^n$. Let $R = \mathbf{Z}(\mathbf{g}) \cap |\Sigma|$, whence R is a rational homogeneous polyhedral set. There exist unimodular fans Δ and Θ in $\mathbb{R}^n$ satisfying the following conditions:

1. $\Delta \preceq_2 \Sigma$ and Θ is inscribed in Δ.

2. $R = |\Theta|$.

3. $f_\Theta < f_\Sigma$.

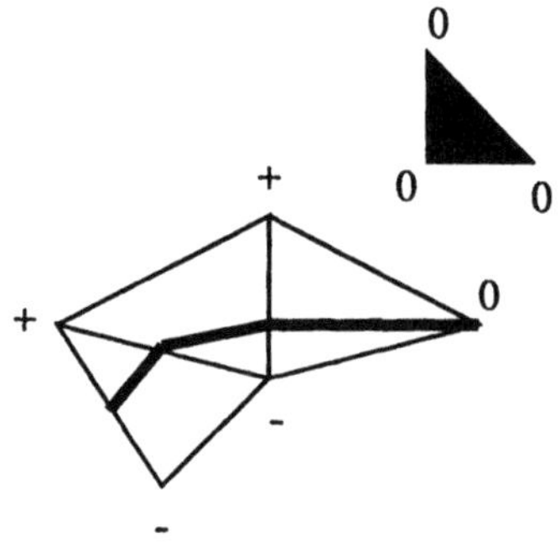

Figure 2. An affine Section of the zero set (shown in thick black) of a linear equation $\sum z_i h_i$ in Schauder hats $\{h_i\}$, cf. Lemma 5.3. The affine vertices corresponding to the h_i's are labelled with the sign of z_i. The lemma shows that the f-vector of an appropriate unimodular triangulation of the zero set is smaller (in the lexicographic order) than the f-vector of the original unimodular complex Σ, and that such a triangulation is obtainable as a subcomplex of a barycentric stellar subdivision of Σ.

PROOF. By the De Concini-Procesi Lemma, there exists a unimodular fan Λ such that $\Lambda \preceq_2 \Sigma$ and R is inscribed in Λ. Let us display the sequence of refinements leading to Λ as

$$\Lambda_0 = \Sigma \succeq_2 \Lambda_1 \succeq_2 \cdots \succeq_2 \Lambda_s = \Lambda \ ,$$

where each Λ_i is obtained from Λ_{i-1} via a single application of binary starring. Given $i \in \{0, \ldots, s\}$, let us say that a pair of Schauder hats $a, b \in \mathbf{H}_{\Lambda_i}$ is *in* $\mathbf{g}$-*opposition* iff $\mathbf{g}(\mathbf{v}_a)\mathbf{g}(\mathbf{v}_b) < 0$, where $\mathbf{v}_a$ and $\mathbf{v}_a$ denote the vertices of a and b, respectively. (Thus, informally, a and b are separated by the integral homogeneous hyperplane $\mathbf{Z}(\mathbf{g})$). Let us now say that Λ is *minimal* iff for every $i \in \{0, \ldots, s-1\}$, the binary starring operation $\Lambda_i \succeq_2 \Lambda_{i+1}$ involves a pair of vertices whose corresponding Schauder hats are in $\mathbf{g}$-opposition. By the De Concini-Procesi Lemma, there exists a minimal Δ such that $\Delta \preceq_2 \Sigma$ and R is inscribed in Δ. The latter condition implies that there exists a unique fan Θ inscribed in Δ such that $R = |\Theta|$. We claim $f_\Theta < f_\Sigma$.[25] To prove this, let us say that a $(t-1)$-simplex of the simplicial complex Σ — that is, a t-cone $\tau = \langle \mathbf{e}_{i_1}, \ldots, \mathbf{e}_{i_t} \rangle \in \Sigma$ — is *cited* iff $z_{i_1}, \ldots, z_{i_t}$ are not all zero. Consequently, we say that τ is *not cited* iff $z_{i_1}, \ldots, z_{i_t}$ are all zero. Let us display the f-vectors of Θ and Σ as $f_\Theta = (m_0, \ldots, m_h, 0, \ldots)$

[25] *Caution.* This condition may well fail if one does not choose Δ minimal in the above argument.

and $f_\Sigma = (n_0, \ldots, n_k, 0, \ldots)$, with $m_h \neq 0$ and $n_k \neq 0$. We first remark that $h \leq k$, because Θ is inscribed in Δ, and the latter refines Σ — hence $\dim|\Theta| = h \leq \dim|\Sigma| = k$. If $h < k$, we are done; hence assume $h = k$. Let $i_0 \in \{0, \ldots, k\}$ be the largest index such that there exists a cited i_0-simplex $C \in \Sigma$. We show that $n_j = m_j$ for all $j \in \{i_0 + 1, \ldots, k\}$. Indeed, by our choice of i_0, every j-simplex of Σ is not cited. Therefore, since Δ is minimal, every j-simplex of Σ is again a j-simplex of Θ. Conversely, any $C \in \Theta$ which is not a simplex of Σ, having arisen from a stellar subdivision, is contained in a cited simplex D of Σ. Since D is at most i_0-dimensional, C is itself at most i_0-dimensional. Thus $n_j = m_j$ holds for all $j \in \{i_0 + 1, \ldots, k\}$. We next show that $n_{i_0} < m_{i_0}$. Let us say that there are $n_{i_0}^{(1)}$ cited i_0-simplices in Σ, and $n_{i_0}^{(0)}$ not cited i_0-simplices in Σ, whence $n_{i_0}^{(0)} + n_{i_0}^{(1)} = n_{i_0}$. We claim $n_{i_0}^{(0)} = m_{i_0}$. First notice that each one of the $n_{i_0}^{(0)}$ not cited i_0-simplices of Σ is again an i_0-simplex of Θ, because Δ is minimal. Thus $n_{i_0}^{(0)} \geq m_{i_0}$. Suppose C were an i_0-simplex of Θ which is not an i_0-simplex of Σ (*absurdum hypothesis*). The intersection of all simplices of Σ containing C is a simplex $D \in \Sigma$ containing C, and $\dim D \geq i_0$. Then D must be a cited simplex of Σ, for $D \neq C$ by our *absurdum hypothesis*, and thus D must have been subdivided. By our choice of i_0, this implies $\dim D \leq i_0$, hence $\dim D = i_0$. Since D is cited, one has $\dim(|D| \cap \mathbf{Z}(\mathbf{g})) < \dim(D)$ — this statement is immediately seen to hold for any cited simplex of Σ. Since $|C| \subseteq (|D| \cap \mathbf{Z}(\mathbf{g}))$, one has $\dim C \leq \dim(|D| \cap \mathbf{Z}(\mathbf{g}))$. Hence $\dim C < \dim D = i_0$, a contradiction. We have thus proved that $n_{i_0}^{(0)} = m_{i_0}$ holds. Since, by our choice of i_0, $n_{i_0}^{(1)} \geq 1$, our claim is settled — $f_\Theta < f_\Sigma$. ■

5.3. Every ℓ-group is approximable by its simplicial subgroups.

Using the previous lemma, we sketch a proof of Theorem 3.8. One can actually derive a stronger statement.

THEOREM 5.4. Let $P = \{p_1, \ldots, p_n\} \subseteq G^+ - \{0\}$ be finite, nonempty, and linearly dependent. We show that there exists a finite $C \subseteq G^+ - \{0\}$ satisfying the following conditions.

1. $C \preceq_2 P$.
2. C is linearly independent.
3. $f_C < f_P$.

Hence Theorem 3.8 follows.

SKETCH OF PROOF. By induction on the set of f-vectors f_P, ordered lexicographically. The basis of the induction is trivial. Assume the result is true for all $Q \subseteq G^+ - \{0\}$ with $f_Q < f_P$. The statement is local, whence we may safely assume that G be generated by P. Let $\phi : \mathcal{A}_n \twoheadrightarrow G$ be the onto ℓ-homomorphism mapping the free generator π_i to p_i, for $i \in \{1, \dots, n\}$, and let $\mathfrak{o} = \operatorname{Ker} \phi$. Let I_0 be the set of those relations of the form $\pi_{i_1} \wedge \cdots \wedge \pi_{i_t}$ which are satisfied by P in G,[26] i.e. such that

$$p_{i_1} \wedge \cdots \wedge p_{i_t} = 0 \ .$$

Then I_0 is finite, and thus generates a principal ℓ-ideal $\mathfrak{s}$ of $\mathcal{A}_n$. One has $\mathbb{V}(\mathfrak{o}) \subseteq \mathbb{V}(\mathfrak{s})$. Notice that $\mathbb{V}(\mathfrak{s})$ is the support of a fan inscribed in the fan of the fundamental cone of $\mathbb{R}^n$. Let us call this fan Σ, and let $\mathbf{H}$ be the Schauder basis over Σ.

If P is linearly dependent, there is an integral linear form $\mathbf{g} = z_1\pi_1 + \cdots + z_n\pi_n$ in $\mathbb{R}^n$ such that $\mathbb{V}(\mathfrak{o}) \subseteq \mathbf{Z}(\mathbf{g})$. By the foregoing, $\mathbb{V}(\mathfrak{o}) \subseteq \mathbf{Z}(\mathbf{g}) \cap |\Sigma|$. Let $R = \mathbf{Z}(\mathbf{g}) \cap |\Sigma|$.

By an application of Lemma 5.3, there exist unimodular fans Δ and Θ, Θ inscribed in Δ, such that $\Delta \preceq_2 \Sigma$ and $R = |\Theta|$. Let $\mathbf{K}$ be the set of Schauder hats over Δ. By a routine application of the Wójcicki-Baker Theorem (cf. Table 4.4) it follows that if $k \in \mathbf{K}$ is such that its vertex $\mathbf{v}_k$ satisfies $\mathbf{v}_k \notin R$, then $\phi(k) = 0$ in G — that is $k \in \operatorname{Ker} \phi$, or $k \in \mathfrak{o}$. Upon setting $Q = \mathbf{K}/\mathfrak{o} - \{0\}$, it now follows that P lies in the submonoid of G generated by Q. Our final claim is that $f_Q < f_P$. Indeed, $f_\Theta < f_\Sigma$ by Lemma 5.3. Since $f_\Theta = f_\mathbf{K}$ and $f_\Sigma = f_\mathbf{H}$ (cf. the remark before the proof of Lemma 5.3), we also have $f_\mathbf{K} < f_\mathbf{H}$. We have $f_\mathbf{K} \geq f_Q$, because f_Q only depends upon (the subsets S with $\bigwedge S = 0$ of) the sublattice of G generated by Q, and the latter is a homomorphic image of the sublattice of $\mathcal{A}_n / \mathbb{I}(|\Theta|)$ generated by $\mathbf{K}$. But we also have $f_\mathbf{H} = f_P$. Indeed, by construction P and $\mathbf{H}$ satisfy the same lattice-theoretical words, because $|\Sigma| = R = \mathbb{V}(\mathfrak{s})$, and therefore $f_\mathbf{H} = f_P$. Hence $f_Q < f_P$, and the proof is complete by induction. ■

The following classical result of Weinberg [102] is an immediate consequence of Theorem 3.8:

COROLLARY 5.5. The variety of ℓ-groups is generated by the additive group of integers.

[26]In other words, these relations describe the non-faces of the abstract simplicial complex $\mathcal{S}_P$ (see Definition 5.1). The ℓ-ideal they generate is an ℓ-theoretical analogue of the defining ideal of the Stanley-Reisner ring of an abstract simplicial complex [97]. We shall not be able to expand on this interesting analogy in this paper.

Indeed, it suffices to show that if an equation in the language of ℓ-groups fails in some ℓ-group G, then it fails in $\mathbb{Z}$. As fully detailed in [63], if an ℓ-group word does not evaluate to zero in G, it does not evaluate to zero in a simplicial subgroup of G (with the simplicial order), hence it does not evaluate to zero in $\mathbb{Z}$. Before returning to MV algebras, let us show that, on the basis of Theorem 3.8, this result of Weinberg can be significantly strengthened to a fully general structural result about ℓ-groups.

The additive group $\mathbb{Z}^n$ of integral n-vectors can be regarded as an ℓ-group upon stipulating $(z_1, \ldots, z_n) \geq 0$ iff $z_i \geq 0$ for every $i \in \{1, \ldots, n\}$. A *simplicial group* is an ℓ-group isomorphic (as an ℓ-group) to $\mathbb{Z}^n$. We shall identify a simplicial group S with $\mathbb{Z}^n$, and call the canonical ($\mathbb{Z}$-module) basis of $\mathbb{Z}^n$ the *simplicial basis* of S. Let now G be an arbitrary ℓ-group, and let $E \subseteq G^+ - \{0\}$ be a finite set. Then E generates a subgroup of G, denoted $[E]$. If E is linearly independent, $[E]$ is free Abelian of rank $|E|$. Furthermore, with the order inherited from G, $[E]$ is a rank $|E|$ simplicial group with simplicial basis E. We call $[E]$ a *simplicial subgroup* of G. Let $\mathcal{B}(G)$ denote the set of all simplicial subgroups of G. Then $\mathcal{B}(G)$ can be partially ordered setting $S_1 \leq S_2$ iff S_1 is a subgroup of S_2. Theorem 3.8 now simply says that the poset $\mathcal{B}(G)$ is upward directed.[27] We thus obtain:

THEOREM 5.6. Let G be an ℓ-group. Then G is the directed union (in the category of partially ordered Abelian groups) of the upward directed poset $\mathcal{B}(G)$ of its simplicial subgroups.

5.4. MV partitions admit joint refinements.

We return to MV algebras. We shall see that MV partitions indeed admit joint refinements. In fact, more is true. Just like simplicial groups are elementary building blocks for ℓ-groups, any MV algebra is approximated by its MV partitions.

COROLLARY 5.7. Let A be an MV algebra with corresponding unital ℓ-group (G, u).

(i) Any two MV partitions π_1 and π_2 of A have a joint refinement.

(ii) For definiteness,[28] assume A is countable. Then there exists a sequence $\pi_1, \pi_2, \ldots$ of MV partitions of A, such that each π_{i+1} is a refinement

---

[27] Notice that Theorem 5.4 gives more precise information that is not used in deriving the statement that follows.

[28] In the general case one replaces sequences by directed systems.

of π_i, and for each $a \in A$ there is an integer j such that a is in the submonoid of G generated by the blocks of π_j.

PROOF. We closely follow the proof in [76]. (i) Let

$$\pi_1 = \{h_1^{m_1}, \ldots, h_q^{m_q}\} \text{ and } \pi_2 = \{g_1^{e_1}, \ldots, g_p^{e_p}\}$$

for integers m_i, $e_j \geq 1$. By Theorem 3.8 there exists an independent subset $\{b_1, ..., b_n\}$ of G such that $0 \leq b_1, ..., b_n$, and each h_i, as well as each g_j, belongs to the submonoid of G generated by $\{b_1, ..., b_n\}$. We can safely assume the set $\{b_1, ..., b_n\}$ be *minimal*, meaning that the submonoid generated by every $(n-1)$-element subset of $\{b_1, ..., b_n\}$ omits some h_i or some g_j. Since u is in the positive span of π, and the latter is in the positive span of $\{b_i\}$, u is in the positive span of $\{b_1, \ldots, b_n\}$. Since the latter set is linearly independent in G, there exist uniquely determined nonnegative integers $p_1, \ldots, p_n \in \mathbb{Z}^+$ such that $\sum_{i=1}^n p_i b_i = u$. Hence $\{b_1^{p_1}, ..., b_n^{p_n}\}$ is an MV partition of A jointly refining π_1 and π_2[29]. (ii) Let $1 = c_1, c_2, c_3, \ldots$ be an enumeration of all elements of A, where 1 is the unit element of A. We shall inductively construct the required sequence of partitions π_n as follows. *Basis:* $\pi_1 = \{1^1\}$. *Induction step:* By hypothesis, we have a partition π_n of A such that each $a \in \{c_1, ..., c_n\}$ belongs to the monoid of G generated by the blocks of π_n. Let Q be the multiset $\{c_{n+1}^1, (\neg c_{n+1})^1\}$. (To avoid trivialities, we assume $c_{n+1} \neq \neg c_{n+1}$.) Since $c_{n+1} \oplus \neg c_{n+1} = 1$ and $c_{n+1} \odot \neg c_{n+1} = 0$, we easily obtain $c_{n+1} + \neg c_{n+1} = 1$. Thus Q satisfies the first condition in Definition 3.6. Since, as we have noted, for the applicability of (i) there is no need to assume that $\{c_{n+1}, \neg c_{n+1}\}$ is an independent set, we obtain from (i) a partition π_{n+1} refining π_n and such that c_{n+1} (as well as $\neg c_{n+1}$) is in the submonoid generated by the elements of π_{n+1}. ■

6. Duality between multisets and locally finite MV algebras

In Section 7 below we shall revisit many-valued partitions from a combinatorial viewpoint. As a tool we shall need to deal with finite multisets. In the present section, following [29], we shall introduce an interesting category of (finite and infinite) multisets, having deep relations with MV algebras.

The notion of multiset is commonplace in mathematics, [3, 98]. In this paper, for instance, multisets have already occurred in the definition of MV partition, and shall surface naturally again in the definition of many-valued

[29] Note that we have made no use of the assumption that π_1 and π_2 satisfy the second condition in Definition 3.6.

logical consequence (cf. Section 8 below), where premises must be understood as a multiset.

By contrast, the subtler notion of "multiset morphism" is not so frequently encountered. The mathematician wishing to introduce a convincing category of infinite multisets is confronted with the nontrivial problem of preserving the familiar constructions and closure properties of sets. Thus, one might reasonably expect that multisets: (1) allow all infinite products and coproducts; (2) arise as inverse limits of finite multisets; (3) and, last but not least, afford a generalized Stone representation theorem relating them to some class of algebraic structures extending Boolean algebras.

Following [29], we shall present a category of multisets satisfying the above conditions. In order to deal with infinite multiplicities, we shall use supernatural numbers, an expedient "ideal" generalization of natural numbers. We shall then equip supernatural numbers with their natural topology, and define a multiset as a continuous map f from a Boolean space X to the supernatural numbers. Morphisms shall be taken care of in Definition 6.3.

6.1. Supernatural Numbers

DEFINITION 6.1. Following [29], by a *supernatural number* we mean a function

$$\nu\colon \{2,3,5,7,11,\ldots\} \to \{0,1,\ldots,\infty\}.$$

For any two supernatural numbers ν and μ, we write $\nu \le \mu$ iff $\nu(p) \le \mu(p)$ for each prime number p. Then the set of supernatural numbers forms a (complete) lattice denoted $\boldsymbol{G}$. Under possibly different names, supernatural numbers are often used as classifiers ([95, 8, 40]; also see [46, Chapter XIII, Section 85]). A supernatural number ν is said to be *finite* iff ∞ does not belong to the range of ν, and $\nu(p)$ is zero for all primes p, up to a finite number of possible exceptions. Regarding ν as a list of exponents for the sequence of prime numbers, we may safely identify each natural number n with its corresponding finite supernatural number ν_n. In particular, $\nu_1(p) = 0$ for all prime numbers p. Under the identification

$$n \leftrightarrow \nu_n, \tag{v}$$

the lattice of finite supernatural numbers coincides with the distributive lattice $\boldsymbol{N_d} = \{1,2,3,\ldots\}$ of natural numbers equipped with divisibility.

As is not hard to see, $\boldsymbol{G}$ provides a concrete realization of the lattice $\mathfrak{I}(\boldsymbol{N_d})$ of ideals of $\boldsymbol{N_d}$.

LEMMA 6.2. The lattices $\mathfrak{I}(\boldsymbol{N_d})$ and $\boldsymbol{G}$ are isomorphic, in such a way that natural numbers (more precisely, principal ideals of $\boldsymbol{N_d}$) correspond to finite supernatural numbers.

Using the identification (v), we further endow $\boldsymbol{G}$ with the topology having as an open basis all sets of the form

$$U_n = \{\nu \in \boldsymbol{G} \mid \nu \geq \nu_n\}, \quad n \in \mathbb{N}.$$

By abuse of notation, $\boldsymbol{G}$ shall also stand for the topological space of supernatural numbers.

6.2. Multisets

The following is a generalization of the notion of finite multiset.

DEFINITION 6.3. The *category of multisets* is defined as follows:

Objects: All pairs (S, σ), where S is a Boolean space (i.e. a totally disconnected compact Hausdorff space) and σ is a continuous map from S into $\boldsymbol{G}$.

Morphisms: Given objects (T, τ) and (U, υ), a morphism $\varphi\colon (T, \tau) \to (U, \upsilon)$ is a continuous function $\varphi\colon T \to U$ such that $\tau \geq \upsilon\varphi$.

As is standard, we let morphisms in the *category of Boolean spaces* be continuous maps. Then Boolean spaces form a full reflexive subcategory of multisets by identifying a Boolean space S with the multiset (S, o), where $o(x) = \nu_1$ for all $x \in S$. The left adjoint is given by the forgetful functor.

By a *finite multiset* we understand a pair (X, σ) where X is a discrete finite set, and $\sigma\colon X \to \boldsymbol{N_d}$ assigns a (finite super)natural number to each $x \in X$. As a particular case of Definition 6.3, under the identification (v) one immediately sees that a morphism $\varphi\colon (T, \tau) \to (U, \upsilon)$ of finite multisets is a multiplicity decreasing function, in the sense that for all $x \in T$, (the natural number corresponding to) $\upsilon(\varphi(x))$ is a divisor of $\tau(x)$. We shall return to finite multisets in Section 7.

6.3. Locally finite MV algebras and multisets

The celebrated Stone duality extends to a duality between multisets and locally finite MV algebras. According to standard universal-algebraic lexicon, an MV algebra is *locally finite* iff all its finitely generated subalgebras are finite.[30] All Boolean algebras are locally finite MV algebras.

A proof of the following characterization can be found in [29].

LEMMA 6.4. For any MV algebra A the following conditions are equivalent:

(i) A is locally finite.

(ii) For each prime ideal J of A, the quotient A/J is isomorphic to a subalgebra of $\mathbb{Q} \cap [0,1]$, where $\mathbb{Q}$ denotes the rational numbers.

(iii) There is a set I such that A is isomorphic to a subalgebra of the MV algebra $(\mathbb{Q} \cap [0,1])^I$ formed by functions of finite range.

The relationship between MV algebras and multisets is fully clarified by the next result.

THEOREM 6.5. [29]

(i) There is a natural equivalence between the opposite of the category of multisets and the category of locally finite MV algebras. Under this equivalence, finite multisets correspond to finite MV algebras.

(ii) The equivalence extends Stone's duality between Boolean spaces and Boolean algebras, in the sense that multisets of the form (S, o) correspond to Boolean algebras.

(iii) The category of multisets has all limits and colimits. Hence, in particular, it has all infinite products and coproducts.

The original Stone theorem established a duality between algebras of two-valued logic and certain spaces of continuous $\{0,1\}$-valued functions. The present generalization shows that locally finite algebras of Łukasiewicz infinite-valued logic are dual to certain spaces of continuous multiplicity-valued functions. This arithmetical enrichment of topological structures arising from logic is, in our opinion, an interesting aspect of the above duality.

In [29] the authors apply Theorem 6.5 together with Lemma 6.4 to deduce:

---

[30] *Warning.* In the MV-algebraic literature, the term "locally finite" is sometimes (misleadingly) used as a synonym of "having no proper homomorphic images". This is confusing, because there exist infinite simple MV algebras with one generator (see, for instance, [28, Corollary 7.2.7]).

COROLLARY 6.6. For every multiset (X, σ) there is an inverse system of finite multisets, indexed by a directed poset, such that (X, σ) is isomorphic to the inverse limit of this system.

Thus, Theorem 6.5 and the latter corollary witness that the category of multisets arising from Definition 6.3 fulfills the list of *desiderata* set forth in the introduction to this section.

7. Matching MV partitions with finite multisets

The notion of MV partition given by Definition 3.6 was shown in the foregoing to be of crucial importance in analyzing the local structure of MV algebras and ℓ-groups. However, that notion shows serious shortcomings when it is specialized to finite multisets via Theorem 6.5. Most of the duality phenomena that make Boolean partitions ubiquitous in mathematics are not exhibited by MV partitions in the sense of Definition 3.6. Specifically, such MV partitions: (1) do not arise as nonempty fibers of morphisms; (2) are not dual to subalgebras; (3) do not allow an arrow-theoretically sound notion of equivalence relation. In this section we shall present a strengthening of the notion of MV partition that satisfies these requirements *in the finite case.*[31] Due to lack of space, we shall have to concentrate on *desideratum* (1), but the interested reader should have little difficulty in verifying (2) and (3).

7.1. One-one and onto morphisms

The material that follows is a short extract of a much more extensive treatment which is currently in progress [31]. Specializing the contents of Definition 6.3, we shall concern ourselves with a category MVSet whose typical object is a finite[32] multiset — a function $\alpha\colon A \to \mathbb{N}$ with finite domain — and whose typical morphism $f\colon \alpha \to \gamma$ is a function on the underlying sets such that for all $a \in A$, $\gamma \circ f(a)$ *divides* $\alpha(a)$. Notice that a morphism from α into itself is the identity on α iff it is the identity on the set A, and $f\colon \alpha \to \gamma$ is an isomorphism iff the underlying function is a multiplicity-preserving bijection. In the light of Theorem 6.5, MVSet can be thought of as a many-valued analogue of the category of finite sets.

[31]While it would be desirable to generalize such considerations to larger classes of MV algebras, at the time of writing it is not clear how one could do so.

[32]Throughout this section, all multisets are finite, whence we drop the adjective from now on.

It turns out that a duality principle underlies the mathematics of MVSet.

DEFINITION 7.1. Let $f: \alpha \to \gamma$ be a morphism of multisets, and let $\mathrm{Dom}\,\alpha = A$, $\mathrm{Dom}\,\gamma = C$.

1. We call f *weakly injective* iff $f: A \to C$ is injective, and *weakly surjective* iff $f: A \to C$ is surjective.

2. We call f *strongly injective* iff it is weakly injective and $\gamma \circ f = \alpha$, and *strongly surjective* iff it is weakly surjective and, for all $c \in C$, $\gamma(c) = \gcd\{\alpha(a) \mid f(a) = c\}$.

The next proposition yields evidence to support the claim that the notions just introduced are the natural ones.[33]

PROPOSITION 7.2. Let $f: \alpha \to \gamma$ be a morphism of multisets. There exists a unique weakly surjective-strongly injective factorization $f = i_s \circ p_w$. There exists a unique strongly surjective-weakly injective factorization $f = i_w \circ p_s$.

It is clear that, in general, f has no strongly surjective-strongly injective factorization. On the other hand, f always admits a weakly surjective-weakly injective factorization, but, in general, this factorization is not unique.

7.2. Many-valued cardinalities are partitions of integers

The many-valued analog of a natural number turns out to be an important tool in the investigation of multisets. For our purposes, it is useful to regard natural numbers as complete isomorphism invariants of finite sets. The corresponding many-valued notion hinges upon a classical mathematical concept [6]. Recall that a *partition of a (nonnegative) integer* $q \in \mathbb{Z}^+$ is, for $q \neq 0$, a weakly decreasing finite sequence λ of integers $\lambda_1 \geq \lambda_2 \geq \cdots \geq \lambda_t \geq 0$, λ_i, $t \in \mathbb{N}$, such that $q = \sum_{i=1}^{t} \lambda_i$. If $q = 0$, by definition q admits the unique partition 0. One calls λ_i a *part* of the partition. The *frequency* of λ_i in λ is the number of occurrences of λ_i in the sequence $\lambda_1 \geq \lambda_2 \geq \cdots \geq \lambda_t$.

A moment's reflection shows that a complete invariant for the isomorphism type of a multiset is the unique partition whose parts are the multiplicities of α. Accordingly, the *cardinality* of $\alpha \neq \emptyset$, denoted $||\alpha||$, is the

[33]In categorical language, weak surjective morphisms are simply epi arrows, while strongly surjective morphisms are extremal, regular or strong arrows — in MVSet all these strengthenings of "epi arrow" coincide. A similar remark applies to weak and strong injections.

partition that describes its isomorphism type, with $||\emptyset||$ being the unique partition of 0.[34] We let $\mathbb{MVN}$ denote the set of all partitions of integers.

7.3. MV partitions, revisited

Just like morphisms (cf. Definition 7.1), partitions of multisets come in two brands.

DEFINITION 7.3. A *weak MV partition of* α is multiset π whose underlying elements $\beta_1, \beta_2, ..., \beta_k$ are multisets, the *blocks* of π, satisfying the following conditions.

1. If $A = \operatorname{Dom}\alpha$ and $B_i = \operatorname{Dom}\beta_i$, $i \in \{1, \ldots, k\}$, then $\{B_1, \ldots, B_k\}$ is a Boolean partition of A into k nonempty blocks.

2. For all $a \in B_i$, $\beta_i(a)\pi(\beta_i) = \alpha(a)$.

A weak MV partition $\pi = \{\beta_1^{t_1}, \ldots, \beta_k^{t_k}\}$ of α is a *strong MV partition* iff $t_i = \gcd\{\alpha(a) \mid a \in B_i\}$, $i \in \{1, \ldots, t\}$.

Remark. It should be noted that every weak MV partition according to the above is an MV partition according to Definition 3.6, while the converse fails.

In the Boolean case, one implements a taxonomy of partitions in terms of a single invariant, the number of blocks of the partition. In MVSet one replaces $\mathbb{N}$ with $\mathbb{MVN}$. Let π be a (weak or strong) MV partition of α, and let $\kappa \in \mathbb{MVN}$. We say that π is a κ*-partition* of α iff $||\pi|| = \kappa$. If $||\alpha|| = \nu$, we write

$$\left\{ \begin{matrix} \nu \\ \kappa \end{matrix} \right\}, \quad \left\{\left\{ \begin{matrix} \nu \\ \kappa \end{matrix} \right\}\right\}$$

to denote the number of strong (respectively, weak) κ-partitions of α.

7.4. MV partitions are fibers of surjections

Closing a circle of ideas originating from the introduction to this section, we can now relate MV partitions to morphisms. To this end, we shall determine the number of surjections $f: \alpha \to \gamma$. If $\nu \in \mathbb{MVN}$ and $||\alpha|| = \nu$, we let $\nu!$ denote the *factorial* of ν, i.e., the number of isomorphisms of α with itself. Clearly $\nu!$ does not depend on the choice of α, and equals the product of the factorials of the frequencies of the parts of ν. One can now prove:

[34] There is no natural notion of strong cardinality for multisets, for 'isomorphism' is a self-dual concept under the weak-strong duality described in Subsection 7.1.

PROPOSITION 7.4. Let α and γ be multisets with $||\alpha|| = \nu$ and $||\gamma|| = \chi$. The number of strong (weak) surjections from α to γ is

$$\left\{ \begin{matrix} \nu \\ \chi \end{matrix} \right\} \chi! \quad \text{and} \quad \left\{\!\!\left\{ \begin{matrix} \nu \\ \chi \end{matrix} \right\}\!\!\right\} \chi! \quad ,$$

respectively.

Thus, strong (weak) MV partitions stand to strong (weak) surjections precisely as Boolean partitions stand to onto functions. More generally, it is shown in [31] that the notions of MV partition given by Definition 7.3 fulfill the further requirements set forth in the introduction to this section.

Once enough evidence is available to validate the basic definitions, one can investigate MVSet as a mathematically interesting domain of discourse in its own right. By way of an *addendum*, we close this section with a cursory excursion along this path. Consider the well known polynomial expansion

$$x^n = \sum_{k \in \mathbb{N}} \left\{ \begin{matrix} n \\ k \end{matrix} \right\} (x)_k \quad ,$$

where $(x)_k = x(x-1)\cdots(x-(k-1))$ is the falling factorial, and $\{ {n \atop k} \}$ is the number of k-partitions of a set with n elements. Consider again multisets α and γ with $||\alpha|| = \nu$ and $||\gamma|| = \chi$. Let χ^ν be the number of morphisms $f \colon \alpha \to \gamma$, $(\chi)_\nu$ the number of such strongly injective morphisms, and $((\chi))_\nu$ the weak analog of $(\chi)_\nu$. As proved in [31], one has

$$\chi^\nu = \sum_{\kappa \in \mathbf{MVN}} \left\{ \begin{matrix} \nu \\ \kappa \end{matrix} \right\} ((\chi))_\kappa \quad .$$

Applying the weak-strong duality principle, one has *gratis*:

$$\chi^\nu = \sum_{\kappa \in \mathbf{MVN}} \left\{\!\!\left\{ \begin{matrix} \nu \\ \kappa \end{matrix} \right\}\!\!\right\} (\chi)_\kappa \quad .$$

Thus, one deduces a duality identity with tautological classical counterpart:

$$\sum_{\kappa} \left\{ \begin{matrix} \alpha \\ \kappa \end{matrix} \right\} ((\gamma))_\kappa = \sum_{\kappa} \left\{\!\!\left\{ \begin{matrix} \alpha \\ \kappa \end{matrix} \right\}\!\!\right\} (\gamma)_\kappa \quad .$$

A handful of experiments should convince the reader that this identity displays nontrivial arithmetical content, and its proof in a classical framework would necessarily be *ad hoc*. In sharp contrast, the proof in MVSet is in fact so natural and straightforward as to border on triviality, [31].

8. Infinite-valued *Calculus Ratiocinator*

If MV algebras are thought of as one face of a coin, the other face naturally is the syntactic-algorithmic machinery of the infinite-valued Łukasiewicz calculus, the Lindenbaum algebra construction being the appropriate tool to flip the coin around at leisure. In fact, the presentation of an MV algebra A as the Lindenbaum algebra of some theory Θ allows one to do "computations over A", upon regarding Θ as a convenient symbolic representation of A.[35]

8.1. Formulæ and their meaning as $[0,1]$-valued functions

Over the MV algebra $[0,1]$ we have already defined the operations $x \oplus y = \min(1, x+y)$ and $\neg x = 1 - x$, as well as the operations $x \vee y = \max(x,y) = \neg(\neg x \oplus y) \oplus y$, and $x \wedge y = \neg(\neg x \vee \neg y)$. We shall also use the operations $x \odot y = \neg(\neg x \oplus \neg y)$, $x \to y = \neg x \oplus y$, $x \leftrightarrow y = (x \to y) \odot (y \to x)$.

Upon fixing a countable set of symbols $\Sigma = \{\neg, \oplus, (,), X_1, X_2, \ldots\}$, called the *alphabet*, we inductively define the set Form of *formulæ* starting from the *propositional variables* $X_1, X_2, \ldots$, and then repeatedly applying the *connectives* of negation $(\neg)$ and disjunction $(\oplus)$, using parentheses to avoid ambiguities. We let Form_n denote the set of formulæ in the variables $X_1, \ldots, X_n$.

To give meaning to formulæ in Form_n, $n \in \mathbb{N}$, we proceed as follows.[36] To begin with, let us agree to call $[0,1]$ the set of *truth-values*. The transformation

$$\psi \in \mathsf{Form}_n \mapsto \mathsf{Int}_\psi \colon [0,1]^n \to [0,1]$$

from formulæ to functions is inductively defined by:

- The ith variable X_i is transformed by Int into the ith projection function x_i.
- $\mathsf{Int}_{\neg\psi} = 1 - \mathsf{Int}_\psi$
- $\mathsf{Int}_{\omega\oplus\psi} = \min(1, \mathsf{Int}_\omega + \mathsf{Int}_\psi)$.

McNaughton's representation theorem, which we informally quoted several times, is a generalization of the well known fact that Boolean formulæ precisely represent Boolean functions.

---

[35] As a set of formulæ, Θ is a (possibly infinite) string of symbols from a fixed alphabet Σ.

[36] Historically, Łukasiewicz's infinite-valued calculus [99] was based on the connectives $\neg$ and $\to$. Following standard practice in the algebraization of many-valued logics [57], we take disjunction $(\oplus)$ and negation $(\neg)$ as basic.

THEOREM 8.1. [67], [28, Proposition 4.5.5] A function $p\colon [0,1]^n \to [0,1]$ arises as Int_ψ for some formula $\psi(X_1, \ldots, X_n)$ iff p is a McNaughton function, i.e., p is continuous, (finitely) piecewise linear and each one of its linear pieces has integer coefficients.

Incidentally, let us note that formulæ in the infinite-valued calculus of Łukasiewicz can express Condition (*) in the definition 3.6 of MV partition:

DEFINITION 8.2. Given variables $Z_1, \ldots, Z_n$, the formula (the MV-term)

$$JUSTONE(Z_1, \ldots, Z_n)$$

is the $\odot$-conjunction of the following formulæ:

$$\left\{\begin{array}{ccccccccc} Z_1 & \oplus & Z_2 & \oplus & \cdots & \oplus & Z_{n-1} & \leftrightarrow & \neg Z_n \\ Z_1 & \oplus & Z_2 & \oplus & \cdots & \oplus & Z_n & \leftrightarrow & \neg Z_{n-1} \\ \cdots & & \cdots & & \cdots & & \cdots & & \cdots \\ Z_1 & \oplus & Z_3 & \oplus & \cdots & \oplus & Z_n & \leftrightarrow & \neg Z_2 \\ Z_2 & \oplus & Z_3 & \oplus & \cdots & \oplus & Z_n & \leftrightarrow & \neg Z_1 \end{array}\right.$$

Formula $JUSTONE(Z_1, \ldots, Z_n)$ is an infinite-valued generalization of the well known Boolean formula $(Z_1 \vee \ldots \vee Z_n) \wedge \left(\bigwedge_{1\leq i<j\leq n} \neg(Z_i \wedge Z_j)\right)$. [37]

8.2. The infinite-valued tautology problem is co-NP-complete

Each formula ψ is a string of symbols over the alphabet Σ, representing a McNaughton function. How complex is the *tautology* problem, i.e., the problem of deciding whether a given formula represents the constant function 1? As a simple consequence of the Completeness Theorem 3.1 one gets:

LEMMA 8.3. There is an effective enumeration of all formulæ that are tautologies.

Using Corollary 3.3 one can routinely prove:

LEMMA 8.4. There is an effective enumeration of all formulæ that are not tautologies.

Thus, *the tautology problem for the infinite-valued Łukasiewicz calculus is Turing-decidable.* A detailed analysis of the fine structure of McNaughton functions leads to a much more precise statement.

[37] See [71, Proposition 3.3] for details

THEOREM 8.5. [72], [28, Theorem 9.3.8] The problem of deciding whether a formula ψ is a tautology in the Łukasiewicz infinite-valued calculus is co-NP-complete.

The result shows that the decision problem of the Łukasiewicz infinite-valued calculus has in fact the same computational complexity as its two-valued counterpart.

Is there a genuine *calculus ratiocinator* for Łukasiewicz infinite-valued propositional logic? Several researchers are striving to give a satisfactory answer to this imprecise question. Hypersequent calculi, and their variants, may provide a key tool [1, 47].

8.3. MV algebras = Lindenbaum algebras of Łukasiewicz calculus

Form its very origins, formal logic has been an attempt to understand the subtle concept of logical consequence; it still is. The next definition reduces the notion of infinite-valued consequence to the much simpler notion of tautology.

DEFINITION 8.6. Let $\Theta \subseteq \mathsf{Form}$ be a set of formulæ. We say that a formula ψ is a *consequence*[38] of Θ,written $\Theta \vdash \psi$, iff for some multiset μ of elements of Θ, the formula $(\bigodot \mu) \to \psi$ is a tautology.[39] We denote by $\Theta^{\vdash}$ the *deductive closure* of Θ, i.e. the set of all its consequences.[40] A deductively closed set of formulæ $\Theta = \Theta^{\vdash}$ is called a *theory*.

Two formulæ ω and ψ of Form_n are said to be *logically equivalent*, in symbols, $\omega \equiv \psi$, iff $\omega \leftrightarrow \psi$ is a tautology. Generalizing the notion of logical equivalence for any set Θ of formulæ, two formulæ ψ, ω are said to be Θ-equivalent, in symbols $\psi \equiv_\Theta \omega$, iff $\Theta \vdash (\psi \leftrightarrow \omega)$. We let $[\omega]_\Theta$ denote the $\equiv_\Theta$-equivalence class of ω.

PROPOSITION 8.7. [28, Corollary 4.4.4]

(i) The set of Θ-equivalence classes of formulæ, equipped with the natural action of the $\oplus$ and $\neg$ connectives, is an MV algebra, the Lindenbaum algebra of Θ, denoted $\mathsf{Form}/\equiv_\Theta$.

(ii) We have $\Theta \vdash \psi \to \omega$ iff $[\psi]_\Theta \leq [\omega]_\Theta$ in the natural MV-algebraic order of $\mathsf{Form}/\equiv_\Theta$.

---

[38] "Syntactic" consequence in [28].

[39] Here $\bigodot \mu$ is the formula obtained by taking the $\odot$-conjunction of as many copies of each formula β in μ, as given by the multiplicity of β in the multiset μ.

[40] We stipulate that $\emptyset^{\vdash}$ coincides with the set of tautologies.

Since $\equiv_\Theta$ is the same as $\equiv_{\Theta^\vdash}$, using the full strength of the notion of free algebra, we obtain the following result:

THEOREM 8.8. [28, Theorem 4.6.9] Up to isomorphism, every countable[41] MV algebra A is the Lindenbaum algebra of some theory Θ in the infinite-valued calculus of Łukasiewicz over the variables $X_0, X_1, \ldots$.

Thus, MV algebras stand to the infinite-valued calculus exactly as Boolean algebras stand to the classical propositional calculus.

8.4. The meaning of Łukasiewicz truth-values

In order to give a natural interpretation of non-Boolean truth-values in Łukasiewicz logic, (a problem that vexed Łukasiewicz himself) following [74] we shall show that two-valued logic stands to many-valued logic as the Twenty Questions game stands to the Twenty Questions game with lies/errors. The latter was introduced by Rényi and, independently, Ulam, as follows. Two players fix a finite set S. The first player — called the Responder — chooses a number $x \in S$, and the second player — the Investigator — must find x by asking a minimum number of questions, to which the Responder can only answer yes or no, *being allowed to lie up to l times.*[42]

Regarding the search of x as a deduction process, let us see what happens after t questions have been asked and answers $A_1, \ldots, A_t$ have been obtained. If no lies are allowed, the elements of S can be partitioned in two classes: those satisfying all answers and those falsifying more than 0 answers — the latter being excluded from the search. In case $l > 0$, an admissible number could also happen to falsify j answers, with $0 < j \leq l$.

To record this state of affairs, the Investigator's state of knowledge s assigns to each $y \in S$ the number j of answers among $A_1, \ldots, A_t$ falsified by y: whenever $j > l$, we will identify j with $l+1$. Suppose we normalize j dividing it by $l+1$, and define $1 - j/(l+1)$ as the *truth-value* of y in the state s. Then the $l+2$ truth-values are given by $0, 1/(l+1), 2/(l+1), \ldots, 1$.

Direct inspection shows that s coincides with the Łukasiewicz conjunction $s_1 \odot s_2 \odot \cdots \odot s_t$ of the states of knowledge individually arising from each answer. Since conjunctions of equal answers (to the same repeated question) are more informative than single answers, conjunction is not idempotent. By writing $s \odot t = 0$ we can express the fact that two states s and t are incompatible.

---

[41] For uncountable MV algebras one just needs a richer supply of variables.

[42] Questions are identified with subsets of S.

We say that a state s is *sharper than* s', (or, s' is *coarser than* s) iff $s(y) \leq s'(y)$ for all y in S. States of knowledge become sharper and sharper during the game. The *initial* state 1, in which no number falsifies any answer, is the coarsest state of all; the zero state is the sharpest; these states do exist in any Rényi-Ulam game.

By definition, Łukasiewicz *negation* $\neg s$ is the coarsest state which is incompatible with s. One immediately has $\neg s = 1 - s$.

We can also define *disjunction* by $r \oplus s = \neg(\neg r \odot \neg s)$. Equipped with these operations the set of states becomes an MV algebra. Inequalities between states of knowledge can now be expressed by equalities, replacing $s \leq t$ by $s \odot \neg t = 0$.

Using Corollary 3.3 one immediately sees that *tautologies in Łukasiewicz logic coincide with those formulæ that are true in every possible game, independently of the number of lies.*[43] Upon adding to the MV equations the idempotence law $x \odot x = x$, we precisely obtain those equations that are valid in all games without lies.

Summing up, in the framework of Rényi-Ulam games with l lies over the search space S we have:

1. The Łukasiewicz truth-values
$$0, \frac{1}{l+1}, \frac{2}{l+1}, \ldots, \frac{l}{l+1}, 1$$
measure (in units of $l+1$) the distance of each possible candidate $x \in S$ from the condition of falsifying too many answers.

2. Answers can be formalized in Łukasiewicz $(l+2)$-valued logic.

3. States of knowledge are Łukasiewicz conjunctions of answers.

4. The natural order between states of knowledge is expressible via Łukasiewicz negation, together with conjunction.

5. In order to accommodate the truth-values needed for all possible games ($l = 0, 1, 2, \ldots$), one must use all rationals in the unit interval.

6. Tautologies of Łukasiewicz logic correspond to those equations between states of knowledge holding in any possible game, independently of the number l of lies.

[43]Each formula ϕ of Łukasiewicz logic is read as the equation $\phi = 1$; conversely, every equation $\phi = \psi$ is read as the formula $\phi \leftrightarrow \psi$. Tautologies correspond to valid equations.

In his book [100] Ulam posed the problem of minimizing the number of questions in a game of Twenty Questions with lies. These games had also been considered earlier by Rényi in [90]. The literature on Rényi-Ulam games is rich and rapidly expanding. The logical interpretation of these games in terms of infinite-valued Łukasiewicz logic was originally given in [74]. Besides providing a natural interpretation of many-valued logic, Rényi-Ulam games are mathematically interesting *in se.* They naturally pertain to Berlekamp's theory of error-correcting communication with feedback, [12]. Optimal searching strategies when all questions are asked non-adaptively correspond to optimal error-correcting codes. We refer to the surveys [87, 26] for further information. Remarkably enough, adaptivity (even a minimum amount of it [24, 27]) results in searching stategies with the smallest possible number of questions — the *a priori* minimum given by the sphere packing lower bound.

9. Other research topics in many-valued logic

Because of space constraints, in this survey we have omitted several important topics of current research on Łukasiewicz logic and MV algebras. Here is a (necessarily incomplete) list:

1. As witnessed by the Handook of Measure Theory [86], MV algebraic measure theory has reached an independent status. Several results in classical Boolean algebraic measure theory á la Carathéodory have been provided an MV-algebraic extension. This is the case, e.g., of the Loomis-Sikorski theorem [37, 79], the Hewitt-Yosida decomposition theorem, the Vitali-Hahn-Saks-Nikodym theorem, and the Nikodym boundedness theorem on extension of measures, [11]. The Kolmogorov construction, the mean value theorem, the laws of large numbers, and many others, have their MV-algebraic counterparts, [93, 92, 38]. The handbook chapter [20] is devoted to extensions of classical constructions and results about σ-fields of sets to the context of σ-complete MV algebras of $[0,1]$-valued functions.

2. As a particularly interesting new development in MV-algebraic probability theory, (see [92] for background) in his paper [91] Riečan defines entropy for *dymanical systems* on an *arbitrary* MV algebra A.[44]

[44]Thus, A need not be σ-complete, nor must it be closed under "product", as is assumed in [88, 89].

An intermediate step is the definition of entropy of a partition of unity in an MV algebra A equipped with a finitely additive functional $\tau\colon A \to [0,1]$. For partitions of this sort, joint refinability is ensured by the Riesz interpolation property, via Theorem 3.5(iii). Generalizing the well known Kolmogorov-Sinaj theorem, Riečan proves that entropy is a complete invariant for equivalence classes of MV dynamical systems. In the particular case when A is a σ-field of sets, the above construction yields the Kolmogorov-Sinaj entropy. If A is the MV algebra of all $[0,1]$-valued measurable functions over a measurable space then the above construction coincides with the Maličký modification of the Kolmogorov-Sinaj entropy (see [62, 93]).

3. Several papers deal with the problem of enriching MV algebras with extra operations, as required by various applications. For instance, in order to deal with independence in MV-algebraic probability theory, [92], one may use a σ-complete MV algebra A (living in symbiosis with its corresponding unital ℓ-group (G,u)) and assume A carries an extra operation $*$ that distributes over the addition of G. By Theorem 3.2, the semisimplicity of A implies that $*$ necessarily boils down to pointwise multiplication of functions. Montagna [68, 69] establishes various categorical equivalences between the resulting varieties of MV algebras with product, and other mathematical (ring-theoretic) structures. In [35] the authors explore the algebraic relationships between classes of MV algebras with product and classes of f-rings. For related work on MV modules see [36].

   Instead of multiplication one may consider tensor products, [79]. Upon tensoring the blocks h_i of an MV partition with elements e_i of a free MV algebra one obtains a useful "if-then-else" approximation $\sum_i h_i \otimes e_i$ of a function defined by "overlapping cases", as discussed in the introduction of this paper. See [80].

   The paper [70] considers several kinds of extensions of MV algebras by extra operations. The authors show that, in many cases, the forgetful functor has a left adjoint, and hence for every MV algebra A there is a "most general" enriched algebra which is generated by A under the added extra operation.

4. As is well known, quantification over the variables does not increase the expressive power of the two-valued propositional calculus. By contrast, existential quantification of a single variable in Łukasiewicz infinite-valued calculus allows the definition of all continuous piecewise linear

functions $g: [0,1]^n \to [0,1]$ with *rational* coefficients. See [2].

5. We shall only give a very partial survey of Jakubík's recent work on MV algebras. In [59, 60] Jakubík proves a Cantor-Bernstein theorem for certain general classes of MV algebras. For related work also see [34]. In [61] he investigates Dedekind completions (i.e., MacNeille completions) of MV algebras. Main result: Suppose A is a strong subdirect product of MV algebras A_i $(i \in I)$. Then its Dedekind completion $D(A)$ is isomorphic to the direct product of MV algebras $D(A_i)$. MV completions are also the main object of the paper [10]. By making a systematic use of Theorem 3.5, the authors transfer the theory of ℓ-convergence and Cauchy structures from ℓ-groups to MV algebras. They consider three particular MV convergences and their corresponding Cauchy completions. The Cauchy completion arising from order convergence coincides with the Dedekind-MacNeille completion of an MV-algebra. The Cauchy completion arising from polar convergence yields a new proof of the existence and uniqueness of the lateral completion of an MV algebra. Finally, the Cauchy completion arising from α-convergence yields the cut completion of an MV algebra.

6. As shown in [30], locally finite MV algebras yield a complete classifier for Boolean algebras with an automorphism having no infinite orbits. Specifically, by a *Boolean algebra with a distinguished automorphism* we understand a pair (B, β) such that B is a Boolean algebra and β is an automorphism of B, considered as a new unary operation. Let A be the category whose objects are Boolean algebras with a distinguished automorphism, and whose morphisms are Boolean homomorphism preserving the automorphism. Let us denote by B the full subcategory of A given by all pairs (B, β) such β is a *locally finite* automorphism (i.e., an automorphism without infinite orbits). [45] In [30] one can find a proof that there are functors $\mathcal{G}$ from the category LFMV of locally finite MV algebras to B, and $\mathcal{F}$ from B to LFMV, such that the composite functor $\mathcal{F}\mathcal{G}$ is naturally equivalent to the identity functor of LFMV. Further, the functor $\mathcal{G}$ is faithful, the functor $\mathcal{F}$ is full, and each object $A \in \mathsf{LFMV}$ is isomorphic to $\mathcal{F}(B, \beta)$ for some object (B, β) in B.

7. In 1961, Higman proved that every finitely generated (not necessarily Abelian) group with a recursively enumerable set of defining relations

[45] One easily sees that β is a locally finite automorphism of a Boolean algebra B iff all finitely generated subalgebras of B are finite, that is, iff B is locally finite in the usual algebraic sense.

can be embedded into a finitely presented group [58]. The result immediately yields the celebrated Novikov-Boone Theorem as a corollary — the word problem for finitely presented groups is insoluble. The ℓ-analogue of Higman's result has been sought for several years, albeit with little success.[46] In [50] it is proved that, under admittedly restrictive hypotheses, such an ℓ-analogue holds. Namely, let A be an Abelian ℓ-group whose underlying group has finite rank. Suppose A is finitely generated and has a recursively enumerable set of defining relations. Then there exists a (provably non-Abelian, if A is not finitely presented) ℓ-group G such that A embeds into G. As a very special case of this result, in [50] one finds an explicit finite presentation of a necessarily non-Abelian ℓ-group G_π that effectively encodes the continued fraction expansion of π. Since finitely generated ℓ-ideals are principal, there is in fact a single word in the language of ℓ-groups that achieves the encoding. Thus, while π is not algebraic, it is "ℓ-algebraic" — the language of ℓ-groups is strictly more expressive than the language of rings of polynomials over $\mathbb{Z}$.

8. It was observed in [71] that no maximally consistent theory in the infinite-valued calculus of Łukasiewicz can be Gödel incomplete. In [82] it is shown that Gödel incomplete *prime*[47] theories Θ do exist. But there is no Gödel incomplete prime theory Θ having a finite number of variables. More generally, a side product of the main result in [50] shows (in algebraic language) that whenever a theory Θ over finitely many variables has the finite rank property — the Abelian group-theoretical rank of the unital ℓ-group arising from Θ via the Lindenbaum and Γ functors is finite — that theory cannot exhibit Gödel incompleteness.[48]

9. As remarked in the introduction, MV algebras constitute interesting examples of "quantum structures". Thus, e.g., an MV algebra is the same thing as a lattice-ordered effect algebra in which disjoint elements are orthogonal [45]. Several important properties of MV algebras can

---

[46] It is known that the ℓ-analogue of the Novikov-Boone Theorem holds, [49].

[47] In the sense that for any two formulæ ϕ, ψ either $\Theta \models \phi \to \psi$ or $\Theta \models \psi \to \phi$. Equivalently, the Lindenbaum algebra of Θ is totally ordered.

[48] Prime theories in a finite number of variables have the finite rank property, but the converse fails. Notice that by Theorem 3.5 the finite rank property is an intrinsic feature of the theory Θ, hence a genuinely logical (many-valued) notion. However, at the time of writing its general significance is admittedly unclear, in sharp contrast with the situation for prime theories.

be generalized to larger classes of quantum structures [39, 38].

10. Antefact

The revival of Łukasiewicz infinite-valued logic has a major but seldom mentioned antefact. In 1976, Elliott [42] published his celebrated classification of Bratteli AF C^*-algebras [16], those C^*-algebras arising as limits of ascending sequences of finite-dimensional C^*-algebras, all with the same unit.[49] While Elliott used a partial semigroup as a classifier, subsequent work by Effros, Handelman, Shen, Goodearl and others [41, 51, 52], and, independently and in a different context, Grillet [55], proved that Elliott's classifier can be equivalently replaced by a countable directed, unperforated, partially ordered Abelian group having the Riesz interpolation property together with a distinguished strong unit.[50] Further, the classifying functor for every AF C^*-algebra D is given by (a suitable order-theoretic enrichment of) Grothendieck's group $K_0(D)$. Thus Elliott's classification cast a bridge between operator algebras in Hilbert space and countable partially ordered Abelian groups.

Ten years after Elliott's classification a categorical equivalence Γ was established between the variety of MV algebras, and the category of lattice-ordered Abelian groups with a distinguished strong unit, [71] (this is Theorem 3.5 in the present paper). It follows that the composite functor $\Gamma \circ K_0$ yields an MV-algebraic classification of those AF C^*-algebras whose Murray-von Neumann order of projections is a lattice. This includes commutative AF C^*-algebras (corresponding to countable Boolean algebras), finite-dimensional C^*-algebras (corresponding to finite MV algebras), Glimm's UHF algebras [40], (corresponding to rational subalgebras of $[0, 1]$), and many other AF C^*-algebras studied in the literature. As we have seen in Section 8, countable MV algebras are the Lindenbaum algebras of Łukasiewicz infinite-valued calculus over countably many variables. Therefore, the natural deductive-algorithmic machinery deriving from Łukasiewicz logic turns out to have a counterpart for AF C^*-algebras. As a first consequence, one has a natural framework for the investigation of decidability and undecidability problems about algebras of operators [83, 18, 17, 81]. Further, one

[49] AF C^*-algebras (where AF stands for "approximately finite-dimensional") allow a mathematically rigorous treatment of infinite spin systems in quantum statistical mechanics. They are also studied for their own sake, as a small yet interesting subclass of C^*-algebras.

[50] See [53] for a paper that puts the original Effros-Handelman-Shen result in algebraic perspective, and provides further references.

may import free MV algebras as universal C^*-algebraic objects [71].

But the connection between AF C^*-algebras and Łukasiewicz logic is a two-way bridge. In the other direction, the approximability of every AF C^*-algebra by its finite-dimensional subalgebras naturally leads to a fundamental problem: Can every MV algebra be recovered from an appropriate class of elementary building blocks by a limit process? The reader who has followed us so far will easily recognize that precisely this problem and its multifarious ramifications have been our *leit-motiv* throughout this paper.

References

[1] S. Aguzzoli, A. Ciabattoni, Finiteness in infinite-valued logic, *J. Logic Language and Information*, 9:5-29, 2000.

[2] S. Aguzzoli, D. Mundici, Weierstrass approximation theorem and Łukasiewicz formulæ with one quantified variable, In: *Beyond Two: Theory and applications of multiple-valued logic,* E. Orłowska, M. Fitting, Eds., Physica-Verlag, Springer, Heidelberg, NY, (2003), pp. 315-335.

[3] M. Aigner, *Combinatorial Theory*, Springer-Verlag, Berlin - Heidelberg - New York, 1979.

[4] J. W. Alexander, Combinatorial analysis situs, *Trans. Amer. Math. Soc.*, 28:301–329, 1926.

[5] J. W. Alexander, The combinatorial theory of complexes, *Ann. of Math. (2)*, 31:292–320, 1930.

[6] G. E. Andrews, *The theory of partitions*, Encyclopedia of Mathematics and its Applications, Addison-Wesley Publishing Company, Reading, 1976.

[7] M. Artin, A. Grothendieck, J. Verdier, *SGA 4 (1963-64), Lecture Notes in Mathematics,* 269, Springer-Verlag, 1972.

[8] R. Baer, Abelian groups without elements of finite order, *Duke Math. J.*, 3: 68–122, 1937.

[9] K. A. Baker, Free vector lattices, *Canad. J. Math.*, 20:58-66, 1968.

[10] R. Ball, G. Georgescu, I. Leuştean, Cauchy completions of MV-algebras, *Algebra Unversalis,* 47:367–407, 2002.

[11] G. Barbieri, H. Weber, Measures on Clans and on MV-algebras, pp. 911-945, in [86].

[12] E. R. Berlekamp, Block coding for the binary symmetric channel with noiseless, delayless feedback. In: *Error-correcting Codes.* (Mann, H. B., Ed.) Wiley, New York, (1968), p. 330-335.

[13] W. M. BEYNON, Combinatorial aspects of piecewise linear maps, *J. London Math. Soc.* (2) 31:719-727, 1974.

[14] W. M. BEYNON, Duality theorems for finitely generated vector lattices, *Proc. London Math. Soc.* (3) 31:114-128, 1975.

[15] W. M. BEYNON, Applications of duality in the theory of finitely generated lattice-ordered abelian groups, *Canad. J. Math.* 29(2):243-254, 1977.

[16] O. BRATTELI, Inductive limits of finite-dimensional C^*-algebras, *Trans. Amer. Math. Soc.*, 171:195-234, 1972.

[17] O. BRATTELI, P. JOERGENSEN, H. H. KIM, F.ROUSH, Non-stationarity of isomorphism between AF algebras defined by stationary Bratteli diagrams, *Ergodic Theory Dyn. Syst.,* 20: 1639-1656, 2000.

[18] O. BRATTELI, P. JOERGENSEN, H. H. KIM, F. ROUSH, Decidability of the isomorphism problem for stationary AF-algebras and the associated ordered simple dimension groups. *Ergodic Theory Dyn. Syst.*, 21:1625-1655, 2002.

[19] A. J. BRENTJES, *Multidimensional continued fraction algorithms*, Mathemathical Centre Tracts, 145, Amsterdam, 1981.

[20] D. BUTNARIU, E. P. KLEMENT, Triangular norm-based measures, pp. 947-1010, in [86].

[21] C. C. CHANG, Algebraic analysis of many-valued logics, *Trans. Amer. Math. Soc.*, 88: 467–490, 1958.

[22] C. C. CHANG, A new proof of the completeness of the Łukasiewicz axioms, *Trans. Amer. Math. Soc.*, 93: 74–80, 1959.

[23] C. C. CHANG, The writing of the MV-algebras, *Studia Logica*, special issue on Many-valued Logics, (D. Mundici, Ed.,) 61: 3-6, 1998.

[24] F. CICALESE, D. MUNDICI, Perfect two fault-tolerant search with minimum adaptiveness, *Advances in Applied Mathematics*, 25:65-101, 2000.

[25] F. CICALESE, D. MUNDICI, Learning and the art of fault-tolerant guesswork, Handbook Chapter, In: *Perspectives on Adaptivity and Learning,* I. Stamatescu et al., Eds., Springer, (2003), pp. 117-143.

[26] F. CICALESE, D. MUNDICI, U. VACCARO Rota-Metropolis cubic logic and Ulam-Rényi games, In: *Algebraic Combinatorics and Computer Science: a tribute to Gian-Carlo Rota,* (H. Crapo, D. Senato, Eds.), Springer-Italia, Milan, (2001), pp. 197–244.

[27] F. CICALESE, D. MUNDICI, U. VACCARO, Least adaptive optimal search with unreliable tests. *Theoretical Computer Science*, 270:877-893, 2002.

[28] R. CIGNOLI, I. M. L. D'OTTAVIANO, D. MUNDICI, *Algebraic Foundations of Many-Valued Reasoning*, Kluwer, Dordrecht, 2000.

[29] R. Cignoli, E. Dubuc, D. Mundici, Extending Stone duality to multisets and locally finite MV-algebras, submitted for publication.

[30] R. Cignoli, E. Dubuc, D. Mundici, An MV-algebraic Invariant for Boolean Algebras with a finite-orbit Automorphism, *Tatra Mountains Math. J.*, Special issue in honor J.Jakubík, to appear.

[31] O. M. D'Antona, V. Marra, The combinatorics of MV sets, in preparation.

[32] M. R. Darnel, *Theory of lattice-ordered groups*, Marcel Dekker Inc., New York, 1995.

[33] C. De Concini, C. Procesi, *Complete symmetric varieties. II. Intersection theory*, In *Algebraic groups and related topics (Kyoto/Nagoya, 1983)*, 481–513, North-Holland, Amsterdam, 1985.

[34] A. De Simone, D. Mundici, M. Navara, A Cantor-Bernstein theorem for σ-complete MV-algebras, *Czech. Math. J.*, to appear in 2003.

[35] A. Di Nola, G. Georgescu, Ideals, ℓ-rings and MV^*-algebras. *Math. Slovaca* 51: 479-506, 2001.

[36] A. Di Nola, P. Flondor, I. Leuştean, MV-modules, *J. Algebra,* to appear.

[37] A. Dvurečenskij, Loomis-Sikorski theorem for σ-complete MV-algebras and ℓ-groups. *J. Australian Math. Soc., A,* 68:261-277, 2000.

[38] A. Dvurečenskij, Measures on Quantum Structures, pp. 827–868, in [86]

[39] A. Dvurečenskij, S. Pulmannová, *New Trends in Quantum Structures,* Kluwer, Dordrecht, 2000.

[40] E. G. Effros, *Dimensions and C^*-algebras*, American Mathematical Society, Providence, RI, 1980.

[41] E. G. Effros, D. E. Handelman, C-L. Shen, Dimension groups and their affine representations. *Am. J. Math.,* 102:385-407, 1980.

[42] G. A. Elliott, On the classification of inductive limits of sequences of semisimple finite-dimensional algebras, *Journal of Algebra*, 38:29-44, 1976.

[43] F. Esteva, L. Godo, F. Montagna, The $L\Pi$ and $L\Pi 1/2$ logics: two complete fuzzy systems joining Łukasiewicz and product logics, *Arch. Math. Logic,* 40: 39-67, 2001.

[44] G. Ewald, *Combinatorial convexity and algebraic geometry,* Springer-Verlag, Berlin, Heidelberg, New York. (Graduate Texts in Mathematics, 1996, vol. 168)

[45] D.J. Foulis, MV and Heyting effect algebras. Invited papers dedicated to Maria Luisa Dalla Chiara, Part II. *Found. Phys.,* 30:1687–1706, 2000.

[46] L. Fuchs, *Infinite Abelian groups*, Vol. 2, Academic Press, New York, 1973.

[47] D. GABBAY, G. METCALFE, N. OLIVETTI, Sequent and hypersequent calculi for abelian and Łukasiewicz logics, Manuscript.

[48] A. M. W. GLASS, *Partially ordered groups.* World Scientific Publishing Co. Inc., River Edge, NJ, 1999.

[49] A. M. W. GLASS, Y. GUREVICH, The word problem for lattice-ordered groups, *Trans. Amer. Math. Soc.* 280:127-138, 1983.

[50] A. M. W. GLASS, V. MARRA, Embedding finitely generated Abelian lattice-ordered groups: Higman's Theorem and a realisation of π, to appear in *J. London Math. Soc.*

[51] K. R. GOODEARL, *Von Neumann regular rings,* Monographs and Studies in Mathematics. 4. London - San Francisco - Melbourne: Pitman Publishing Ltd, 1979.

[52] K. R. GOODEARL, D. E. HANDELMAN, J. W. LAWRENCE, Affine representations of Grothendieck groups and applications to Rickart C^*-algebras and $\aleph_0$-continuous regular rings, *Mem. Am. Math. Soc.* 234, 1980.

[53] K. R. GOODEARL, F. WEHRUNG, Representations of distributive semilattices in ideal lattices of various algebraic structures, *Algebra Universalis,* 45:71-102, 2001.

[54] S.GOTTWALD, *A treatise on many-valued logics.* Studies in Logic and Computation. 9. Baldock: Research Studies Press. 2000.

[55] P. A. GRILLET, Directed colimits of free commutative semigroups, *J. Pure Appl. Algebra,* 9:73-87, 1976.

[56] R. HÄHNLE, Advanced many-valued logics, In: *Handbook of Philosophical Logic,* D.M.Gabbay, F.Guenthner (eds.,) 2nd edition, Vol. 2, 2001, pp.297–395.

[57] P. HÁJEK, *Metamathematics of fuzzy logic,* Kluwer, Dordrecht, 1998.

[58] G. HIGMAN, Subgroups of finitely presented groups, *Proc. Royal Soc. London, Series A*, 262:455-475, 1961.

[59] J. JAKUBÍK, A theorem of Cantor-Bernstein type for orthogonally σ-complete pseudo MV-algebras, *Tatra Mt. Math. Publ.,* 22:91-103, 2001.

[60] J. JAKUBÍK, Convex mappings of archimedean MV-algebras, *Math. Slovaca,* 51:383-391, 2001.

[61] J. JAKUBÍK, Strong subdirect products of MV-algebras, *Math. Slovaca,* 51:507–520, 2001.

[62] P. MALIČKÝ, B. RIEČAN, On the entropy of dynamical systems. In: *Proc. Conf. Ergodic Theory and related Topics II*, (H. Michel, ed.) Teubner, Leipzig (1986), 135 - 138.

[63] V. MARRA, *Non-Boolean partitions. A mathematical investigation through lattice-ordered Abelian groups and MV algebras*, Ph.D. thesis, Department of Computer Science, University of Milan, 2002, pp. 87.

[64] V. MARRA, Every Abelian lattice-ordered group is ultrasimplicial, *Journal of Algebra*, 225: 872-884, 2000.

[65] V. MARRA, D. MUNDICI, Combinatorial fans, lattice-ordered groups, and their neighbours: a short excursion, *Semin. Lothar. de Combinatoire*, 47: Article b47f, 2002 (an electronic journal).

[66] V. MARRA, D.MUNDICI, MV-algebras and Abelian ℓ-groups: a fruitful interaction, In: *Ordered Algebraic Structures*, honoring Paul Conrad for his 80th Birthday, J. Martinez, Ed., Kluwer, 2002, pp. 57–88.

[67] R. MCNAUGHTON, A theorem about infinite-valued sentential logic, *The Journal of Symbolic Logic*, 16: 1-13, 1951.

[68] F. MONTAGNA, An algebraic approach to propositional fuzzy logic, *Journal of Logic, Language and Information*, 9: 91-124, 2000.

[69] F. MONTAGNA, Functorial representation theorems for MV-Delta algebras with additional operators, *Journal of Algebra*, 238:99-125, 2001.

[70] F.MONTAGNA, G. PANTI, Adding structure to MV-algebras. *J. Pure Appl. Algebra*, 164:365-387, 2001.

[71] D. MUNDICI, Interpretation of AF C^*-algebras in Łukasiewicz sentential calculus, *Journal of Functional Analysis*, 65: 15-63, 1986.

[72] D. MUNDICI, Satisfiability in many-valued sentential logic is NP-complete, *Theoretical Computer Science*, 52: 145-153, 1987.

[73] D. MUNDICI, Farey stellar subdivisions, ultrasimplicial groups and K_0 of AF C^*-algebras, *Advances in Mathematics*, 68: 23-39, 1988.

[74] D. MUNDICI, The logic of Ulam's game with lies, In: Knowledge, Belief and Strategic Interaction, *Cambridge Studies in Probability, Induction, and Decision Theory*, (1992), pp. 275-284.

[75] D. MUNDICI, A constructive proof of McNaughton's Theorem in infinite-valued logics. *The Journal of Symbolic Logic*, 59: 596-602, 1994.

[76] D. MUNDICI, Uncertainty measures in MV-algebras, and states of AF C^*-algebras, *Notas de la Sociedad de Matemáticas de Chile*, (Special issue in memoriam Rolando Chuaqui), 15: 42-54, 1996.

[77] D. MUNDICI, Łukasiewicz normal forms and toric desingularizations, In: *Proceedings of Logic Colloquium 1993*, Keele, England, (W. Hodges et al., Eds.,) Oxford University Press, (1996), pp. 401–423.

[78] D. MUNDICI, Nonboolean partitions and their logic, In: *First Springer-Verlag Forum on Soft Computing.* Prague, August 1997, *Soft Computing*, 2: 18-22, 1998.

[79] D. MUNDICI, Tensor Products and the Loomis-Sikorski theorem for MV-algebras, *Advances in Applied Mathematics,* 22: 227-248, 1999.

[80] D. MUNDICI, If-then-else and rule extraction from two sets of rules, In: *From Synapses to Rules,* Proceedings of an International Workshop held at the Center for Physics "E. Majorana", Erice, Italy. B. Apolloni et al., Eds., Kluwer/Plenum, NY, (2002), pp. 87–108.

[81] D. MUNDICI, Simple Bratteli diagrams with a Gödel-incomplete C^*-equivalence problem, *Trans. Amer. Math. Soc.,* to appear.

[82] D. MUNDICI, G. PANTI, Decidable and undecidable prime theories in infinite-valued logic, *Annals of Pure and Applied Logic*, 108:269–278, 2001.

[83] D. MUNDICI, G. PANTI, The equivalence problem for Bratteli diagrams. Technical Report no. 259, University of Siena (Italy), 1993. Unpublished, 7 pages.

[84] G. PANTI, A geometric proof of the completeness of the Łukasiewicz calculus, *J. Symbolic Logic*, 60(2):563–578, 1995.

[85] G. PANTI, Multi-valued Logics. Handbook chapter, In: *Quantified Representation of Uncertainty and Imprecision*, vol. 1, (Smets, P., Ed.), Kluwer, Dordrecht, (1998), pp. 25-74.

[86] E. PAP, Editor, *Handbook of Measure Theory, I,II*, North-Holland, Amsterdam, 2002.

[87] A. PELC, Searching games with errors: fifty years of coping with liars, *Theoretical Computer Science,* 270: 71-109, 2002.

[88] J. PETROVIČOVÁ, On the entropy of partitions in product MV-algebras, *Soft Computing,* 4:41 - 44, 2000.

[89] J. PETROVIČOVÁ, On the entropy of dynamical systems in product MV-algebras. *Fuzzy Sets and Systems,* 121:347 - 351, 2001.

[90] A. RÉNYI, *Napló az információelméletrH ol*, Gondolat, Budapest, 1976. (English translation: *A Diary on Information Theory*, J.Wiley and Sons, New York, 1984).

[91] B. RIEČAN, On the Kolmogorov-Sinaj entropy on MV-algebras, manuscript.

[92] B. RIEČAN, D.MUNDICI, Probability on MV-algebras, Handbook Chapter, In: *Handbook of Measure Theory*, E. Pap, Ed., North-Holland, Amsterdam, (2002), pp. 869–909.

[93] B. RIEČAN, T. NEUBRUNN, *Integral, measure, and ordering*, Kluwer, Dordrecht, 1997.

[94] A. ROSE, J.B. ROSSER, Fragments of many-valued statement calculi, *Trans. Am. Math. Soc.*, 87:1-53, 1958.

[95] J.-P. SERRE, Cohomologie Galoisienne. Cours au Collège de France, 1962-1963. *Lecture Notes in Mathematics*, Vol. 5, 4th edition, Springer-Verlag, Berlin, 1973.

[96] S. S. SHATZ, *Profinite groups, arithmetic, and geometry*, Annals of Mathematics Studies, Vol. 67, Princeton University Press, Princeton, NJ, 1972.

[97] R. P. Stanley, *Combinatorics and commutative algebra*, second ed., Progress in Mathematics, vol. 41, Birkhäuser Boston Inc., Boston, MA, 1996.

[98] R. P. Stanley, *Enumerative Combinatorics, Volume I*, Cambridge Studies in Advanced Mathematics, 49, Cambridge University Press, Cambridge, 1997.

[99] A. Tarski, J. Łukasiewicz, Investigations into the Sentential Calculus, In: *Logic, Semantics, Metamathematics*, Oxford University Press, 1956, pp. 38–59. Reprinted by Hackett Publishing Company, Indianapolis, 1983.

[100] S. Ulam, *Adventures of a Mathematician*, Scribner's, New York, 1976.

[101] A. Urquhart, Basic many-valued logic, In: *Handbook of Philosophical Logic*, D.M.Gabbay, F.Guenthner (eds.) 2nd edition, Vol. 2, (2001), pp. 249–295.

[102] E. C. Weinberg, Free lattice-ordered abelian groups, *Math. Ann.*, 151(1963), 187-199.

[103] R. Wójcicki, On matrix representations of consequence operations of Łukasiewicz's sentential calculi, *Z. Math. Logik Grundl. Math.*, 19:239-247, 1973.

[104] K. Yosida, On vector lattice with a unit, *Proc. Imp. Acad. Tokyo*, 17:121-124, 1941.

[105] G. M. Ziegler, *Lectures on polytopes*, Springer-Verlag, New York, 1995.

Vincenzo Marra
Institut für Informatik und Mathematik
Freie Universität Berlin
Takustr. 9
D-14195 Berlin, Germany
marra@math.fu-berlin.de

Daniele Mundici
Department of Mathematics
University of Florence
Viale Morgagni 67/A
I-50134 Florence, Italy
mundici@math.unifi.it

Hiroakira Ono

Substructural Logics and Residuated Lattices — an Introduction

Abstract. This is an introductory survey of substructural logics and of residuated lattices which are algebraic structures for substructural logics. Our survey starts from sequent systems for basic substructural logics and develops the proof theory of them. Then, residuated lattices are introduced as algebraic structures for substructural logics, and some recent developments of their algebraic study are presented. Based on these facts, we conclude at the end that substructural logics are logics of residuated structures, and in this way explain why sequent systems are suitable for formalizing substructural logics.

Keywords: substructural logics, sequent systems, residuated structures, varieties of residuated lattices

1. Introduction

Substructural logics are logics lacking some or all of the structural rules when they are formalized in sequent systems. They cover many of the well-known nonclassical logics, e.g. Lambek calculus for categorial grammar (with no structural rules), linear logic (with only the exchange rule), BCK-logic and Łukasiewicz's many-valued logics (lacking the contraction rule), and relevant logics (lacking the weakening rule). The purpose of the study of substructural logics is to introduce a uniform framework in which various kinds of nonclassical logics that originated from different motivations can be discussed together, and to find common features among them, taking structural rules for a clue.

This is a brief introductory survey of substructural logics and residuated lattices. Residuated lattices are structures that have been studied by algebraists since the 1930s, but the study has been revived recently as a study of mathematical structures for substructural logics. In the present paper, we will mainly concentrate on examining basic ideas of substructural logics in relation to residuated lattices, and will clarify what substructural logics are and why formalizations in sequent systems are essential. The paper is far from a comprehensive survey and will not touch on technical details in

---

V. F. Hendricks and *J. Malinowski* (eds.),
Trends in Logic: 50 Years of Studia Logica

Trends in Logic **21**: 193–228, 2003.

most cases. An algebraic study of substructural logics in the present paper will be developed in depth and in details in the book [18] in preparation by the author with P. Jipsen and T. Kowalski. For general information on substructural logics see [10], which is the first book on this topic, and also [38, 39].

Since basic substructural logics are formalized as sequent systems, our survey starts from these sequent systems and their proof theory. Then we will discuss the algebraic study of substructural logics based on residuated lattices, which has made rapid progress in recent years. It turns out that universal algebra offers useful and powerful tools in developing the study. These algebraic apsects of substructural logics give us a wider and deeper understanding of them.

For some major reference books about algebraic studies of particular logics in the scope of the present paper, see for instance, [41] for linear logic, [15] for fuzzy logics, [9] for many-valued logics, and [1] and [2] for relevant logics.

My sincere thanks go to Peter Jipsen for his valuable comments on earlier drafts of the paper. Many thanks are also due to my friends and colleagues, in particular Tomasz Kowalski and Tadeusz Litak.

2. Substructural Logics

In this section we introduce "basic" substructural logics as sequent systems. They are obtained from the sequent system **LK** for classical logic or **LJ** for intuitionistic logic by restricting their structural rules. In particular the roles of the structural rules will be examined.

2.1. Gentzen's sequent systems and structural rules

We begin with some explanations of sequent systems **LK** and **LJ** for classical propositional logic and intuitionistic logic, respectively, which were introduced by G. Gentzen [12] in the middle of the 1930s.

Here we consider the language $\mathcal{L}$ of **LK** and **LJ** which consists of logical connectives $\wedge, \vee, \rightarrow$ and $\neg$. A *sequent* of **LK** is an expression of the form $\alpha_1, \ldots, \alpha_m \Rightarrow \beta_1, \ldots, \beta_n$, with $m, n \geq 0$, whose intuitive meaning is that "$\beta_1 \vee \ldots \vee \beta_n$ *follows from assumptions* $\alpha_1, \ldots, \alpha_m$". In this sequent, $\alpha_1, \ldots, \alpha_m$ and $\beta_1, \ldots, \beta_n$ are called the *antecedent* and the *succedent*, respcetively. In the following, Greek capital letters Σ, Λ, Γ etc. denote (finite, possibly empty) sequences of formulas. Initial sequents of **LK** are sequents of the form $\alpha \Rightarrow \alpha$. The rules of inference of **LK** given below can be divided into three categories, i.e. structural rules, cut rule and rules for logical connectives.

Structural rules: These rules determine the meaning of *commas* in sequents. We will discuss this in detail later.

Weakening rules:

$$\frac{\Gamma, \Sigma \Rightarrow \Delta}{\Gamma, \alpha, \Sigma \Rightarrow \Delta} \; (w \Rightarrow) \qquad \frac{\Gamma \Rightarrow \Lambda, \Theta}{\Gamma \Rightarrow \Lambda, \alpha, \Theta} \; (\Rightarrow w)$$

Contraction rules:

$$\frac{\Gamma, \alpha, \alpha, \Sigma \Rightarrow \Delta}{\Gamma, \alpha, \Sigma \Rightarrow \Delta} \; (c \Rightarrow) \qquad \frac{\Gamma \Rightarrow \Lambda, \alpha, \alpha, \Theta}{\Gamma \Rightarrow \Lambda, \alpha, \Theta} \; (\Rightarrow c)$$

Exchange rules:

$$\frac{\Gamma, \alpha, \beta, \Sigma \Rightarrow \Delta}{\Gamma, \beta, \alpha, \Sigma \Rightarrow \Delta} \; (e \Rightarrow) \qquad \frac{\Gamma \Rightarrow \Lambda, \alpha, \beta, \Theta}{\Gamma \Rightarrow \Lambda, \beta, \alpha, \Theta} \; (\Rightarrow e).$$

Cut rule: Usually the cut rule is regarded as one of the structural rules. But, for convenience's sake, we separate the cut rule here from the other structural rules.

$$\frac{\Gamma \Rightarrow \alpha, \Theta \quad \Sigma, \alpha, \Pi \Rightarrow \Delta}{\Sigma, \Gamma, \Pi \Rightarrow \Delta, \Theta}$$

Rules for logical connectives: Similarly to the structural rules, there exist right and left rules for each logical connective. Taken together, they describe the role of each connective.

$$\frac{\Gamma \Rightarrow \alpha, \Theta \quad \Pi, \beta, \Sigma \Rightarrow \Delta}{\Pi, \alpha \to \beta, \Gamma, \Sigma \Rightarrow \Delta, \Theta} \; (\to\Rightarrow) \qquad \frac{\Gamma, \alpha \Rightarrow \beta, \Theta}{\Gamma \Rightarrow \alpha \to \beta, \Theta} \; (\Rightarrow\to)$$

$$\frac{\Gamma, \alpha, \Sigma \Rightarrow \Delta}{\Gamma, \alpha \wedge \beta, \Sigma \Rightarrow \Delta} \; (\wedge 1 \Rightarrow) \qquad \frac{\Gamma, \beta, \Sigma \Rightarrow \Delta}{\Gamma, \alpha \wedge \beta, \Sigma \Rightarrow \Delta} \; (\wedge 2 \Rightarrow)$$

$$\frac{\Gamma \Rightarrow \Lambda, \alpha, \Theta \quad \Gamma \Rightarrow \Lambda, \beta, \Theta}{\Gamma \Rightarrow \Lambda, \alpha \wedge \beta, \Theta} \; (\Rightarrow \wedge)$$

$$\frac{\Gamma, \alpha, \Sigma \Rightarrow \Delta \quad \Gamma, \beta, \Sigma \Rightarrow \Delta}{\Gamma, \alpha \vee \beta, \Sigma \Rightarrow \Delta} \; (\vee \Rightarrow)$$

$$\frac{\Gamma \Rightarrow \Lambda, \alpha, \Theta}{\Gamma \Rightarrow \Lambda, \alpha \vee \beta, \Theta} \; (\Rightarrow \vee 1) \qquad \frac{\Gamma \Rightarrow \Lambda, \beta, \Theta}{\Gamma \Rightarrow \Lambda, \alpha \vee \beta, \Theta} \; (\Rightarrow \vee 2)$$

$$\frac{\Gamma \Rightarrow \alpha, \Theta}{\neg\alpha, \Gamma \Rightarrow \Theta} \; (\neg \Rightarrow) \qquad \frac{\Gamma, \alpha \Rightarrow \Theta}{\Gamma \Rightarrow \neg\alpha, \Theta} \; (\Rightarrow \neg)$$

Proofs and the provability of formulas in **LK** are defined in the usual way. In standard Hilbert-style formulations, "implication" plays usually a special role, different from other logical connectives. This comes from the fact that *modus ponens* is a single rule of inferences in the standard formulation, which is of the form "*from* α *and* $\alpha \rightarrow \beta$, *infer* β". On the other hand, in sequent systems as above, none of the logical connectives have particular roles. We can see indeed that for a given logical connective $\sharp$, rules for $\sharp$ are expressed by using only $\sharp$, without using other connectives. Such a formulation becomes possible only with the help of two metalogical symbols, arrow $\Rightarrow$ and comma. Roles of comma in relation to the arrow are described in the form of structural rules. Since rules for logical connectives are described separately from each other, we can get important syntactic results such as cut elimination theorems, the subformula property and so on.

Sequents of the sequent system **LJ** for intuitionistic logic are expressions of the form $\alpha_1, \ldots, \alpha_m \Rightarrow \beta$, where $m \geq 0$ and β may be empty. Initial sequents and rules of inference of **LJ** are obtained from those of **LK** given above, by deleting first both ($\Rightarrow c$) and ($\Rightarrow e$), and then assuming that both Λ and Θ are empty and that Δ consists of at most one formula.

2.2. Roles of the structural rules

To understand the roles of the structural rules, we will give here an example of a proof of the distributive law in **LJ**, in which both contraction and weakening rules are used in an essential way:

$$
\dfrac{\dfrac{\dfrac{\dfrac{\dfrac{\dfrac{\dfrac{\alpha \Rightarrow \alpha}{\alpha, \beta \Rightarrow \alpha}\,(w \Rightarrow) \quad \dfrac{\beta \Rightarrow \beta}{\alpha, \beta \Rightarrow \beta}\,(w \Rightarrow)}{\alpha, \beta \Rightarrow \alpha \wedge \beta}}{\alpha, \beta \Rightarrow (\alpha \wedge \beta) \vee (\alpha \wedge \gamma)} \quad \dfrac{\dfrac{\dfrac{\alpha \Rightarrow \alpha}{\alpha, \gamma \Rightarrow \alpha}\,(w \Rightarrow) \quad \dfrac{\gamma \Rightarrow \gamma}{\alpha, \gamma \Rightarrow \gamma}\,(w \Rightarrow)}{\alpha, \gamma \Rightarrow \alpha \wedge \gamma}}{\alpha, \gamma \Rightarrow (\alpha \wedge \beta) \vee (\alpha \wedge \gamma)}}{\alpha, \beta \vee \gamma \Rightarrow (\alpha \wedge \beta) \vee (\alpha \wedge \gamma)}}{\alpha \wedge (\beta \vee \gamma), \beta \vee \gamma \Rightarrow (\alpha \wedge \beta) \vee (\alpha \wedge \gamma)}}{\alpha \wedge (\beta \vee \gamma), \alpha \wedge (\beta \vee \gamma) \Rightarrow (\alpha \wedge \beta) \vee (\alpha \wedge \gamma)}}{\alpha \wedge (\beta \vee \gamma) \Rightarrow (\alpha \wedge \beta) \vee (\alpha \wedge \gamma)}\,(c \Rightarrow)}{}
$$

A careful inspection of some parts of the above proof tells us that any sequent of the form $\delta, \varphi \Rightarrow \delta \wedge \varphi$ can be derived by using the weakening rule (see the upper parts), and also that any sequent of the form $\delta \wedge \varphi \Rightarrow \psi$ can be derived from a sequent $\delta, \varphi \Rightarrow \psi$ by using the contraction rule (see the lower four lines, replacing α by δ, $\beta \vee \gamma$ by φ and $(\alpha \wedge \beta) \vee (\alpha \wedge \gamma)$ by ψ). Therefore, in both **LK** and **LJ**, a sequent $\delta, \varphi \Rightarrow \psi$ is provable if and only if $\delta \wedge \varphi \Rightarrow \psi$ is provable. In fact, the if-part is obtained in the following way:

$$
\dfrac{\delta, \varphi \Rightarrow \delta \wedge \varphi \quad \delta \wedge \varphi \Rightarrow \psi}{\delta, \varphi \Rightarrow \psi}\,(cut).
$$

By generalizing this argument (and considering the dual argument between disjunctions and commas in the right-hand side of sequents, in case of **LK**), we have the following proposition. It says that in both **LK** and **LJ**, where we have both weakening and contraction rules, *commas in the left-hand side of a sequent mean conjunctions*, and moreover in **LK** *commas in the right-hand side mean disjunctions.*

PROPOSITION 2.1. *A sequent $\alpha_1, \ldots, \alpha_m \Rightarrow \beta_1, \ldots, \beta_n$ is provable in* **LK** *if and only if the sequent $\alpha_1 \wedge \ldots \wedge \alpha_m \Rightarrow \beta_1 \vee \ldots \vee \beta_n$ is provable in it. This holds also for* **LJ** *if $n \leq 1$.*

Then, the following questions will come naturally to mind:

- What do commas mean in a sequent system lacking either or both of the weakening and contraction rules?
- In general, how and in which respects does the existence of structural rules affect logical properties?

To answer these questions, we need to examine the roles of each of the left structural rules again.

1) Exchange rule ($e \Rightarrow$):

$$\frac{\Gamma, \alpha, \beta, \Sigma \Rightarrow \Delta}{\Gamma, \beta, \alpha, \Sigma \Rightarrow \Delta}$$

The exchange rule ($e \Rightarrow$) allows us to use assumptions, i.e. formulas in the left-hand side, in an arbitrary order.

2) Weakening rule ($w \Rightarrow$):

$$\frac{\Gamma, \Sigma \Rightarrow \Delta}{\Gamma, \alpha, \Sigma \Rightarrow \Delta}$$

The weakening rule ($w \Rightarrow$) allows us to add any redundant assumption. In other words, when a sequent $\Pi \Rightarrow \Theta$ is proved in a system that has no weakening rule, every assumption (i.e. every formula in Π) must be used *at least once* in a proof of $\Pi \Rightarrow \Theta$.

3) Contraction rule ($w \Rightarrow$):

$$\frac{\Gamma, \alpha, \alpha, \Sigma \Rightarrow \Delta}{\Gamma, \alpha, \Sigma \Rightarrow \Delta}$$

The contraction rule allows us to use each assumption more than once. Thus, when a sequent is proved in a system lacking the contraction rule, each of its assumptions is used *at most once* in its proof.

Therefore, in a sequent system with all of these structural rules, if a given sequent $\alpha_1, \ldots, \alpha_m \Rightarrow \beta$ is provable, it means that β can be derived from $\alpha_1, \ldots, \alpha_m$ by using them in arbitrary order and an arbitrary number of times including none.

Roughly speaking, *substructural logics* are logics lacking some or all of these structural rules, when they are formulated as sequent systems. The above explanation suggests that they are logics *sensitive* to the number and order of occurrences of assumptions. By this reason, they are sometimes called *resource-sensitive logics*. In particular, when a sequent system under consideration has the exchange rule, but neither the weakening nor the contraction rule, every assumption must be used once and only once to derive a conclusion. This is a basic idea of *linear logic* (without exponentials), introduced by J.-Y. Girard, which is obtained from **LK** by deleting both weakening and contraction rules. Also, in relevant logics, where weakening rules are not allowed, every assumption must be used at least once in derivations, and thus no redundant assumptions, i.e. no *irrelevant* assumptions, are used.

2.3. Comma, fusion and implication

The above argument suggests that commas on the left-hand side of sequents do not behave like conjunctions when some of structural rules are lacking, while they are identified with conjunctions in **LK** and **LJ** as shown in Proposition 2.1. Then, what does each comma mean in such a situation? To see this in a more explicit way, let us introduce a new logical connective $*$ (or $\otimes$) which *represents* a comma in substructural logics. This connective $*$ is sometimes called the *fusion* or the *multiplicative conjunction*, while the usual conjunction is called the *additive* conjunction, to distinguish it from the multiplicative one.[1] We assume the following rules for $*$.

$$\frac{\Gamma, \alpha, \beta, \Sigma \Rightarrow \Delta}{\Gamma, \alpha * \beta, \Sigma \Rightarrow \Delta}\ (* \Rightarrow) \qquad \frac{\Gamma \Rightarrow \alpha, \Lambda \quad \Sigma \Rightarrow \beta, \Theta}{\Gamma, \Sigma \Rightarrow \alpha * \beta, \Lambda, \Theta}\ (\Rightarrow *)$$

Then, as one might expect, the following holds.

PROPOSITION 2.2. *In the sequent system having only rules for $*$ and cut rule, a sequent $\alpha_1, \ldots, \alpha_m \Rightarrow \beta$ is provable if and only if $\alpha_1 * \ldots * \alpha_m \Rightarrow \beta$ is provable.*

We show moreover an important relationship between fusion and implication.

LEMMA 2.3. *A sequent $\alpha * \beta \Rightarrow \gamma$ is provable if and only if $\alpha \Rightarrow \beta \to \gamma$ is provable.*

Note that to show this, it is not necessary to use any structural rule except the cut rule. Suppose that $\alpha * \beta \Rightarrow \gamma$ is provable. Then $\alpha, \beta \Rightarrow \gamma$ is also provable by Proposition 2.2. Using ($\Rightarrow\to$) we have that $\alpha \Rightarrow \beta \to \gamma$ is provable. Conversely, suppose that $\alpha \Rightarrow \beta \to \gamma$ is provable. Then $\alpha, \beta \Rightarrow \gamma$ is provable as the following proof shows, and hence $\alpha * \beta \Rightarrow \gamma$ is provable.

---

[1]Similarly, commas in the right-hand side of sequents when they are allowed, will be denoted by $+$ and called *multiplicative disjunction*. We will consider in this subsection only sequents with a single formula in the conclusion, and thus consider only $*$. This is partly for the sake of simplicity, but also because we don't know exactly how to attach adequate and comprehensible meaning to $+$.

$$\cfrac{\alpha \Rightarrow \beta \to \gamma \qquad \cfrac{\beta \Rightarrow \beta \quad \gamma \Rightarrow \gamma}{\beta \to \gamma, \beta \Rightarrow \gamma}}{\alpha, \beta \Rightarrow \gamma}\ (cut)$$

Lemma 2.3 with Proposition 2.2 says that deletion or addition of structural rules has a significant effect on the "meaning" of implication. In algebraic terms, what is shown in the above lemma means that implication is the *residual* of fusion, or $*$ and $\to$ form a *residuated pair*. This fact will be the basis of our algebraic study of substructural logics in Sections 4 and 5.

2.4. Propositional constants

Sometimes it is convenient to add propositional constants when introducing formal systems. For instance, we use propositional constants $\top$ and $\bot$ to denote the constantly *true* and *false* propositions, respectively. When we introduce them, we need to add the following initial sequents for $\top$ and $\bot$:

1. $\Gamma \Rightarrow \top$,
2. $\Gamma, \bot, \Sigma \Rightarrow \Delta$.

Here, Γ, Σ and Δ may be empty. When a system under consideration has weakening rules, they can be replaced by weaker initial sequents $\Rightarrow \top$ and $\bot \Rightarrow$, respectively. On the other hand, if a system doesn't have them, constants defined by these weaker initial sequents behave in a different way. Now let us introduce additional new propositional constants, denoted by 1 and 0. We assume the following initial sequents and rules of inference for them:

3. $\Rightarrow 1$,
4. $0 \Rightarrow$,

$$\frac{\Gamma, \Sigma \Rightarrow \Delta}{\Gamma, 1, \Sigma \Rightarrow \Delta}\ (1w) \qquad\qquad \frac{\Gamma \Rightarrow \Lambda, \Theta}{\Gamma \Rightarrow \Lambda, 0, \Theta}\ (0w)$$

Intuitively, constants 1 and 0 denote the "empty sequence of formulas" in the left-hand side of an arrow and in the right-hand side, respectively. Also we can see that 1 (0) is the *weakest* (*strongest*) proposition among provable formulas (contradictory formulas, respectively). Here, by a contradictory formula, we mean a formula α such that $\alpha \Rightarrow$ is provable. When we have the constant 0, we may define the negation $\neg\alpha$ of a formula α by $\alpha \to 0$, and can dispense with rules for $\neg$. Note here that $\top$ (0) is logically equivalent to $\neg \bot$ ($\neg 1$, respectively). These four propositional constants are used in the standard formulation of linear logic and relevant logics.[2] Using weakening rules, we can show that $\top$ ($\bot$) is logically equivalent to 1 (0, respectively). Conversely, if $\top$ is equal to 1, then by using the initial sequent 1, the rule $(1w)$ and the cut rule, we can derive the weakening rule $(w \Rightarrow)$.

2.5. Basic substructural logics

We introduce here sequent systems for several basic substructural logics. They are obtained from either **LK** or **LJ** by deleting some or all of the structural rules.

Let **FL** be the sequent system obtained from **LJ** by deleting all of the structural rules and then adding rules for $*$. The name **FL** comes from *full Lambek calculus*, which is an extension of the sequent system for categorial grammar, introduced by J. Lambek [26]. Here, we need a comment regarding the definition of implication. Since we don't assume exchange rule in **FL**, it is more natural to introduce two kinds of implication / and \, which are called in a more suitable way, left and right *residuals*, respectively. More precisely, rules for left residuals / and right residuals \ are given as follows.

$$\frac{\Gamma \Rightarrow \alpha \quad \Pi, \beta, \Sigma \Rightarrow \delta}{\Pi, \beta/\alpha, \Gamma, \Sigma \Rightarrow \delta}\ (/ \Rightarrow) \qquad \frac{\Gamma, \alpha \Rightarrow \beta}{\Gamma \Rightarrow \beta/\alpha}\ (\Rightarrow /)$$

$$\frac{\Gamma \Rightarrow \alpha \quad \Pi, \beta, \Sigma \Rightarrow \delta}{\Pi, \Gamma, \alpha\backslash\beta, \Sigma \Rightarrow \delta}\ (\backslash \Rightarrow) \qquad \frac{\alpha, \Gamma \Rightarrow \beta}{\Gamma \Rightarrow \alpha\backslash\beta}\ (\Rightarrow /)$$

It is obvious that if a system has exchange rule then we can show that β/α and $\alpha\backslash\beta$ are equivalent. In such a case, we denote it as $\alpha \to \beta$. Corresponding to Lemma 2.3, we have the following.

LEMMA 2.4. *In* **FL**, *the following three conditions are mutually equivalent. For all formulas α, β and γ,*
1. $\alpha * \beta \Rightarrow \gamma$ *is provable,*
2. $\alpha \Rightarrow \gamma/\beta$ *is provable,*
3. $\beta \Rightarrow \alpha\backslash\gamma$ *is provable.*

It is natural to introduce two kinds of negations $0/\alpha$ and $\alpha\backslash 0$ of a formula α in **FL**, whose rules are given as follows:

---

[2]The reader should be cautious of the usage of symbols for propositional constants in the literature, as the same symbols sometimes denote different constants in other papers.

$$\frac{\Gamma \Rightarrow \alpha}{0/\alpha, \Gamma \Rightarrow} (\neg 1 \Rightarrow) \qquad \frac{\Gamma, \alpha \Rightarrow}{\Gamma \Rightarrow 0/\alpha} (\Rightarrow \neg 1)$$

$$\frac{\Gamma \Rightarrow \alpha}{\Gamma, \alpha\backslash 0 \Rightarrow} (\neg 2 \Rightarrow) \qquad \frac{\alpha, \Gamma \Rightarrow}{\Gamma \Rightarrow \alpha\backslash 0} (\Rightarrow \neg 2)$$

We introduce several sequent systems for substructural logics that are extensions of **FL**. Let e, w and c denote the exchange rule $(e \Rightarrow)$, the weakening rule $(w \Rightarrow)$ and the contraction rule $(c \Rightarrow)$, respectively. We denote sequent systems obtained from **FL** by adding some of these *left* structural rules, by attaching corresponding subscripts e, w, c to **FL**. For example, $\mathbf{FL_e}$ is **FL** with the exchange rule, which is equal to intuitionistic (multiplicative, additive) linear logic, and $\mathbf{FL_{ew}}$ is **FL** with both exchange and weakening rules. Extensions of $\mathbf{FL_{ew}}$ are discussed extensively in [37, 22] and [36]. See also [16], in which $\mathbf{FL_{ew}}$ is called the monoidal logic. Important classes of extensions of $\mathbf{FL_{ew}}$ are Hájek's fuzzy logics and Łukasiewicz's many-valued logics as discussed in §5. The following is easily shown.

LEMMA 2.5. *For all formulas α and β, the sequent $\alpha * \beta \Rightarrow \alpha \wedge \beta$ is provable in $\mathbf{FL_w}$, and also the sequent $\alpha \wedge \beta \Rightarrow \alpha * \beta$ is provable in $\mathbf{FL_c}$.*

Therefore, in $\mathbf{FL_{cw}}$ $\alpha * \beta$ and $\alpha \wedge \beta$ are equivalent. Since $\alpha \wedge \beta \Rightarrow \beta \wedge \alpha$ is provable, $\alpha * \beta \Rightarrow \beta * \alpha$ is also provable, from which it follows that the exchange rule can be derived in $\mathbf{FL_{cw}}$. Thus, $\mathbf{FL_{cw}}$ is a sequent system for intuitionistic logic.

In the same way as above, we can introduce some of sequent systems obtained from **LK** by restricting the structural rules. When a system lacks the exchange rules, there are many possibilities of introducing rules for fusions and implications, but we have no reasonable criteria of making a choice of proper ones. Thus, we consider in the following only sequent systems with exchange rules. Let $\mathbf{CFL_e}$ be the sequent system obtained from **LK** by deleting both weakening and contraction rules. It has both left and right exchange rules. (The letter C of $\mathbf{CFL_e}$ comes from "classical type".) It is essentially equivalent to the multiplicative, additive linear logic **MALL** introduced by Girard [13]. The sequent systems $\mathbf{CFL_{ew}}$ and $\mathbf{CFL_{ec}}$ are obtained from $\mathbf{CFL_e}$ by adding (left and right) weakening rules and contraction rules, respectively.

In the following, we often identify a sequent system with the *logic* determined by it, i.e. the set of all formulas provable in it. Here, we say that a formula α is provable in a sequent system **L** if the sequent $\Rightarrow \alpha$ is provable in it. Also, we call those logics determined by various sequent systems introduced in this subsection, including **LK** and **LJ**, *basic substructural logics.*[3] The system $\mathbf{CFL_{ew}}$ and a set theory based on it were already studied by V. Grišin from the mid 1970s (see e.g. [14]). Also the system $\mathbf{CFL_{ec}}$ is essentially the same as the system **LR** studied by [27], which is known as the relevant logic **R** without the distributive law. It is easy to see the following.

LEMMA 2.6. *The system* $\mathbf{CFL_e}$ *is equivalent to the system* $\mathbf{FL_e}$ *with initial sequents of the form* $\neg\neg\alpha \Rightarrow \alpha$. *Precisely speaking, for each formula* β, β *is provable in* $\mathbf{CFL_e}$ *if and only if it is provable in* $\mathbf{FL_e}$ *by using any sequent of the form* $\neg\neg\alpha \Rightarrow \alpha$ *as additional initial sequents. The same relation holds between* $\mathbf{CFL_{ew}}$ *and* $\mathbf{FL_{ew}}$, $\mathbf{CFL_{ec}}$ *and* $\mathbf{FL_{ec}}$, *and* **LK** *and* **LJ**.

3. Proof theory of Substructual Logics

In this section, we will discuss the cut elimination theorem for substructural logics and its logical consequences, including decidability results. It will be clarified how weakening and contraction rules play key roles in them. Topics touched in this section are discussed in [33] in full details (see also [32] for more information on decision problems).

3.1. Cut elimination theorems

A proof containing no applications of the cut rule is called a *cut-free* proof. The cut elimination theorem for a given sequent system **L** says that any sequent which is provable in **L** always has a cut-free proof in **L**. To get a sequent system for which the cut elimination theorem holds is quite important, since many important results follow as its consequences. To be precise, from the cut elimination theorem the *subformula property* follows in most cases, which says that for any sequent $\Gamma \Rightarrow \Delta$ if it is provable then it has such a proof that every formula appearing in it is a subformula of some formula in $\Gamma \Rightarrow \Delta$. Then, from the subformula property important results like decidability and Craig's interpolation theorem follow. In other words, even if we have the cut elimination theorem for a given propositional sequent system **L**, **L** may be undecidable because of the lack of the subformula property.

[3] One of earliest attempts of considering all of these basic substructural logics is made in [30]. The nomenclature of basic substructural logics taken here is proposed in the paper, and then is modified into the present form in [31]. The name "substructural logics" was suggested by K. Došen at the first conference of the topics held at Tübingen in 1990 (see [10]).

Also it may happen that while the cut elimination theorem doesn't hold, the subformula property holds and therefore both decidability and Craig's interpolation theorem follow from it. An example of the latter is a sequent system for modal logic **S5** introduced by Ohnishi and Matsumoto. Though the cut elimination theorem doesn't hold for it, M. Takano proved that every application of cut rule can be restricted in such a way that the cut formula is a subformula of formulas in the lower sequent. Using this we can derive the subformula property. For details, see e.g. [33].

THEOREM 3.1. *The cut elimination theorem holds for all sequent systems of basic propositional logics except* $\mathbf{FL_c}$, *i.e. for* $\mathbf{FL}$, $\mathbf{FL_e}$, $\mathbf{FL_w}$, $\mathbf{FL_{ew}}$, $\mathbf{FL_{ec}}$, $\mathbf{LJ}$, $\mathbf{CFL_e}$, $\mathbf{CFL_{ew}}$, $\mathbf{CFL_{ec}}$, *and* $\mathbf{LK}$.

The proof of the cut elimination theorem for these sequent systems goes essentially in the same way as Gentzen's original proof. But we make some remarks on how the presence or the lack of structural rules has an effect on the proof.

We start from Gentzen's original proof of the cut elimination theorem for **LJ**, which has all structural rules. The cut rule of **LJ** can be formulated in the following form:

$$\frac{\Gamma \Rightarrow \alpha \quad \alpha, \Pi \Rightarrow \delta}{\Gamma, \Pi \Rightarrow \delta}$$

To prove the cut elimination theorem, we replace all occurrences of applications of the cut rule in a given proof by the following *mix rule*, which is a generalized form of the cut rule:

$$\frac{\Gamma \Rightarrow \alpha \quad \Sigma \Rightarrow \delta}{\Gamma, \Sigma_\alpha \Rightarrow \delta}$$

where Σ contains at least one occurrence of the formula α, and Σ_α denotes the sequence of formulas obtained from Σ by deleting *all* occurrences of α. It is easy to see that any application of the cut rule can be replaced by an application of the mix rule, and vice versa, with the help of structural rules. Therefore, the new sequent system, obtained from **LJ** by replacing the cut rule by the mix rule, is also a system for intuitionistic logic. Thus, it is enough to prove the mix elimination theorem (for the new system), instead of showing the cut elimination theorem for **LJ**. This is proved by using double induction on the *grade* of the mix formulas and the *rank* of a given proof.

But, why is it necessary to replace the cut rule by the mix rule? The reason is found in the presence of contraction rules. The basic idea of eliminating cut rules is either to *push up* each application of the cut rule by exchanging the order of applications of rules, or to replace cut formulas by simpler ones. This idea works well for all cases except the case shown below, where we want to push the cut rule up so that the left contraction rule will be applied after the cut rule:

$$\dfrac{\Gamma \Rightarrow \alpha \quad \dfrac{\alpha, \alpha, \Pi \Rightarrow \delta}{\alpha, \Pi \Rightarrow \delta}\ (c \Rightarrow)}{\Gamma, \Pi \Rightarrow \delta}\ (cut)$$

But this is impossible, since the lower application of the cut rule in the following proof is not *simpler* than the one in the above.

$$\dfrac{\dfrac{\Gamma \Rightarrow \alpha \quad \dfrac{\Gamma \Rightarrow \alpha \quad \alpha, \alpha, \Pi \Rightarrow \delta}{\alpha, \Gamma, \Pi \Rightarrow \delta}\ (cut)}{\Gamma, \Gamma, \Pi \Rightarrow \delta}\ (cut)}{\Gamma, \Pi \Rightarrow \delta}\ (e \Rightarrow)(c \Rightarrow)$$

To resolve this difficulty, we need to introduce the mix rule. This observation shows that when a system lacks the contraction rule, e.g. systems like $\mathbf{FL_e}$ and $\mathbf{FL_{ew}}$, our basic idea mentioned above works fully and hence we can prove the cut elimination theorem directly. (As a matter of fact, in such a system the cut rule cannot be replaced by the mix rule, because of the lack of contraction rules.)

On the other hand, to show the cut elimination theorem for a sequent system which has both exchange and contraction rules but doesn't have weakening rules, e.g. systems like $\mathbf{FL_{ec}}$ and $\mathbf{CFL_{ec}}$, we need to take the following generalized form of the mix rule instead of the original mix rule:

$$\frac{\Gamma \Rightarrow \alpha \quad \Sigma \Rightarrow \delta}{\Gamma, \tilde{\Sigma}_\alpha \Rightarrow \delta}$$

where Σ contains at least one occurrence of the formula α, and $\tilde{\Sigma}_\alpha$ denotes a sequence of formulas obtained from Σ by deleting an arbitrary number, but *at least one* of occurrences of α. For more details of the cut elimination theorem for these basic substructural logics, see [33]. An algebraic proof of the cut elimination theorem for them is given in [5]. As a collorary of Theorem 3.1, we have the following. We note that Craig's interpolation theorem can be shown by using Maehara's method. See e.g. [33] for the details.

THEOREM 3.2. *Both the subformula property and Craig's interpolation theorem hold for* **FL**, $\mathbf{FL_e}$, $\mathbf{FL_w}$, $\mathbf{FL_{ew}}$, $\mathbf{FL_{ec}}$, **LJ**, $\mathbf{CFL_e}$, $\mathbf{CFL_{ew}}$, $\mathbf{CFL_{ec}}$, *and* **LK**.

3.2. Disjunction property and variable sharing property

A logic **L** has the *disjunction property* when for all α and β if $\alpha \vee \beta$ is provable in **L** then either α or β is provable in it. While classical logic doesn't have the disjunction property, intuitionistic logic does. This is usually attributed to constructive features of the latter. In fact the disjunction property of intuitionistic logic is an immediate consequence of the cut elimination theorem for **LJ**. The proof goes as follows. Suppose that the sequent $\Rightarrow \alpha \vee \beta$ is provable in **LJ**. Consider any cut-free proof of it. Since the sequent is not an initial sequent, it is obtained by applying some rule I. Then it is easily seen that I must be either $(\Rightarrow \vee 1)$ or $(\Rightarrow \vee 2)$. Therefore, the upper sequent must be either $\Rightarrow \alpha$ or $\Rightarrow \beta$.

This argument doesn't work for **LK**, since the last rule may be the right contraction rule $(\Rightarrow c)$ as the following example shows:

$$\dfrac{\dfrac{\dfrac{\dfrac{p \Rightarrow p}{\Rightarrow p, \neg p}}{\Rightarrow p \vee \neg p, \neg p}}{\Rightarrow p \vee \neg p, p \vee \neg p}}{\Rightarrow p \vee \neg p}\ (\Rightarrow c)$$

From this observation the following theorem can be easily derived.

THEOREM 3.3. *A basic substructural logic except* $\mathbf{FL_c}$ *has the disjunction property if and only if its sequent system doesn't have the right contraction rule.*

As we mentioned before, weakening rules allow us to introduce redundant assumptions and conclusions. For instance, using the weakening rule, we can show that the sequent $p * \neg p \Rightarrow q$ is provable in $\mathbf{FL_{ew}}$ though in this sequent there is no relation between the antecedent and the succedent. In fact, there are no propositional variables common to formulas in the left-hand side and the right-hand side of the sequent. We say that a logic **L** has the *variable sharing property*, if such a case never happens. More precisely, **L** has the variable sharing property, if for any formula $\alpha \rightarrow \beta$ containing no propositional constants, $\alpha \rightarrow \beta$ is never provable in **L** whenever formulas α and β have no propositional variables in common. Then, we have the following.

THEOREM 3.4. *A basic substructural logic, except* $\mathbf{FL_c}$, *has the variable sharing property if and only if its standard sequent system doesn't have the weakening rules.*

It is trivial that a basic substructural logic doesn't have the variable sharing property whenever its sequent system has the weakening rules. On the other hand, we can show that $\mathbf{CFL_{ec}}$ has the variable sharing property. Since it is the strongest among basic substructural logics without weakening rules, it follows that any of $\mathbf{FL}$, $\mathbf{FL_e}$, $\mathbf{FL_{ec}}$ and $\mathbf{CFL_e}$ has the property too (see [33] for the proof). Both the disjunction property and the variable sharing property for $\mathbf{FL_c}$ remain open.

3.3. Decision problems

Another important consequence of the cut elimination theorem is the decidability. Let us take first classical propositional logic, and see how Gentzen in [12] derived the decidability as a consequence of the cut elimination theorem for **LK**. To get the decidability, we introduce a proof-search procedure for any given sequent. This procedure tries to find a proof of a given sequent $\Gamma \Rightarrow \Delta$ when it is provable. Moreover, it terminates the search for proofs in a finite number of steps and tells us that the sequent is not provable when it fails to find a proof.

Suppose that a given sequent $\Gamma \Rightarrow \Delta$ is provable. By the cut elimination theorem and the subformula property, it has a cut-free proof in which every sequent is composed only of subformulas of formulas in $\Gamma \Rightarrow \Delta$. But if we count proofs with redundancies, the total number of proofs will be infinite. Here, we say that a proof contains a redundancy, if there exists a *branch* in the proof such that sequents of the same form appear more than once. So, we need to exclude proofs with redundancies.

Therefore, to show that a given sequent $\Gamma \Rightarrow \Delta$ is provable, it suffices to find a cut-free proof of $\Gamma \Rightarrow \Delta$ with no redundancies (in which every sequent is composed only of subformulas of formulas in $\Gamma \Rightarrow \Delta$). When we fail to find such a proof by an exhaustive search among all *possible proofs*, we can say that $\Gamma \Rightarrow \Delta$ is not provable. Here, by a possible proof we mean a proof-like figure (containing neither applications of the cut rule nor redundancies) in which every application of a rule is carried out in a correct way but some of the top sequents may not be initial ones.

Then, is the total number of all possible proofs finite? Unfortunately not, because of contraction rules. For example, suppose that $\alpha, \Sigma \Rightarrow \Pi$ is a top sequent in a possible proof P. As this top sequent may be obtained by applying contraction rule, the figure obtained from P by putting the sequent $\alpha, \alpha, \Sigma \Rightarrow \Pi$ over $\alpha, \Sigma \Rightarrow \Pi$ will be another possible proof. As this argument can be repeated, infinitely many possible proofs will be produced.

To settle this problem, Gentzen introduced the notion of *reduced* sequents. For $k > 0$, a sequent $\Sigma \Rightarrow \Pi$ is *k-reduced* if each formula in Σ (Π) occurs at most k times in Σ (and Π, respectively). A sequent $\Sigma' \Rightarrow \Pi'$ is a *1-reduced contraction* of a sequent $\Sigma \Rightarrow \Pi$ if $\Sigma' \Rightarrow \Pi'$ is 1-reduced, and Σ' and Π' are equal to Σ and Π, respectively, as sets of formulas. For instance, $\alpha, \beta \Rightarrow \gamma$ is a 1-reduced contraction of $\alpha, \beta, \alpha, \alpha \Rightarrow \gamma, \gamma$. With the help of the structural rules, we can easily show that a sequent $\Sigma \Rightarrow \Pi$ is provable if and only if any one of its 1-reduced contractions is provable. Therefore, to see whether a given sequent is provable or not, it is enough to take any one of its 1-reduced contractions and check whether it is provable or not. Now, by observing the form of each rule of **LK** and using the argument mentioned above, we can show the following lemma by using the length of proofs.

LEMMA 3.5. *For a given sequent $\Gamma \Rightarrow \Delta$, let $\Gamma' \Rightarrow \Delta'$ be any of its 1-reduced contraction. If $\Gamma \Rightarrow \Delta$ is provable in* **LK** *then $\Gamma' \Rightarrow \Delta'$ has a cut-free proof with no redundancies, in which only 3-reduced sequents appear.*

It is easily seen this time that for a given 1-reduced sequent $\Gamma' \Rightarrow \Delta'$, the total number of its possible proofs in which every sequent is 3-reduced and is composed only of subformulas of formulas in $\Gamma' \Rightarrow \Delta'$ is finite. Therefore, we have a proof-search procedure which searches for only such a proof.

As the above outline shows, the proof of the decidability relies much on structural rules. But, the necessity of the notion of reduced sequents comes from the contraction rules. In other words, if a sequent system under consideration has no contraction rules, then we can get a much simpler decision procedure. Let us consider $\mathbf{FL_{ew}}$ for example. In this case it is easily seen that in each rule except the cut rule, (each of) the upper sequent(s) is always simpler than the lower sequent. Thus, the proof-search procedure terminates always. This idea of proving the decidability can be extended also to sequent systems for basic substructural *predicate logics with function symbols.* Thus we have the following. (See e.g. [20] for the details. See also [32].)[4]

THEOREM 3.6. *Any of the sequent systems for basic substructural predicate logics without contraction rules, i.e. any of* **FL**, $\mathbf{FL_e}$, $\mathbf{FL_w}$, $\mathbf{FL_{ew}}$, $\mathbf{CFL_e}$ *and* $\mathbf{CFL_{ew}}$, *is decidable even when the language contains function symbols.*

[4]H. Wang noticed this fact already in his book [43] published in 1963.

On the other hand, the decision procedure becomes complicated for $\mathbf{FL_{ec}}$ and $\mathbf{CFL_{ec}}$, i.e. substructural logics with contraction rules but without weakening rules. To show the termination of the procedure, we need some combinatorial result like Kripke's lemma or Higman's theorem. The original proof is given essentially by Kripke in [23] of 1959. (See e.g. [33] for the details.) This complication leads us to the undecidability of their predicate extensions (see [20]).

THEOREM 3.7. *Both substructural propositional logics* $\mathbf{FL_{ec}}$ *and* $\mathbf{CFL_{ec}}$ *are decidable, while their predicate extensions are undecidable.*

The relevant logic **R** is obtained from $\mathbf{CFL_{ec}}$ by adding the distributive law (see §2.2). Urquhart proved the following quite nontrivial and important result in [42].

THEOREM 3.8. *The propositional logic* **R** *is undecidable.*

4. Residuated Lattices

In this section, we will introduce algebraic structures which serve us with semantics suitable for substructural logics. A key notion here is *residuation*, and the algebraic structures for substructural logics introduced here are *residuated lattices*, which have been already studied by algebraists in 1930s, e.g. Krull [24], and Ward and Dilworth [44] (see also [3]).

A residuated lattice consists of a lattice and a partially-ordered semigroup with residuation. By imposing some additional conditions on the lattice-part and/or the semigroup-part, we can get various interesting subclasses of residuated lattices, e.g. the class of MV-algebras and the class of BL-algebras. They correspond to Łukasiewicz's many-valued logics and extensions of the basic logic **BL** introduced by P. Hájek, respectively. For more information on recent results of residuated lattices, see [19] and also [22].

4.1. Residuated structures and residuation theory

In the following, to denote algebraic operations and constants, we use the same symbols as logical connectives and propositional constants corresponding to them, when no confusions may occur. A single exception is to use $\cdot$ for a semigroup operation, while we use $*$ for fusion.

A structure $\mathbf{P} = \langle P, \cdot, \leq \rangle$ is a *partially-ordered semigroup* if $\langle P, \cdot \rangle$ is a semigroup and $\leq$ is a partial order on P such that $\cdot$ is monotone increasing; i.e.

$$x \leq x' \text{ and } y \leq y' \text{ imply } x \cdot y \leq x' \cdot y'.$$

An algebra $\mathbf{P} = \langle P, \cdot, \backslash, /, \leq \rangle$ is a *residuated* partially-ordered semigroup $\mathbf{P}$ if $\langle P, \cdot \rangle$ is a semigroup, $\leq$ a partial order on P and moreover the following condition is satisfied by $\cdot, \backslash$ and $/$: For all $x, y, z \in P$

$$x \cdot y \leq z \text{ if and only if } y \leq x \backslash z \text{ if and only if } x \leq z/y.$$

This condition is sometimes called *the law of residuation*, and $\backslash$ and $/$ are called the right and left *residual* of $\cdot$, respectively. Any residuated partially-ordered semigroup is in fact a partially-ordered semigroup, though we don't assume the monotonicity of $\cdot$ in its definition. This can be shown by using the law of residuation as follows:

Suppose that both $x \leq x'$ and $y \leq y'$ hold. Since $x' \cdot y' \leq x' \cdot y'$, $y \leq y' \leq x' \backslash (x' \cdot y')$. Therefore $x' \cdot y \leq x' \cdot y'$. Thus, $x \leq x' \leq (x' \cdot y')/y$ and hence $x \cdot y \leq x' \cdot y'$. (This argument works in a more general setting. See Lemma 4.1.)

Any residuated partially-ordered semigroup $\mathbf{P}$ such that $\langle P, \leq \rangle$ forms a lattice and $\langle P, \cdot \rangle$ has a unit is called a *residuated lattice*. More precisely, an algebra $\mathbf{P} = \langle P, \wedge, \vee, \cdot, \backslash, /, 1 \rangle$ is a residuated lattice if

1. $\langle P, \wedge, \vee \rangle$ is a lattice,
2. $\langle P, \cdot, 1 \rangle$ is a monoid such that $\backslash$ and $/$ are the right and left residual of $\cdot$, respectively.

When $\cdot$ is commutative, we call $\mathbf{P}$ a commutative residuated lattice. In any commutative residuated lattice, $x \backslash y = y/x$ holds for all x, y. In such a case, we use the symbol $\rightarrow$ and write $x \rightarrow y$ instead of $x \backslash y$ (and of y/x). Also $\mathbf{P}$ is denoted by a sextuple $\langle P, \wedge, \vee, \cdot, \rightarrow, 1 \rangle$.

As an example, we show that any ℓ-group forms a residuated lattice. Recall that an algebra $\mathbf{G} = \langle G, \wedge, \vee, \cdot, ^{-1}, e \rangle$ is a lattice-ordered group (or an ℓ-group, for short) if

1. $\langle G, \wedge, \vee \rangle$ is a lattice with the lattice order $\leq$,
2. $\langle G, \cdot, ^{-1}, e \rangle$ is a group,

3. $\cdot$ is monotone with respect to $\leq$, i.e. $\langle G, \cdot, \leq \rangle$ is a partially-ordered semigroup.

Then $\cdot$ has both residuals $\backslash$ and $/$ that are defined by $x\backslash y = x^{-1} \cdot y$ and $y/x = y \cdot x^{-1}$, respectively.

Some of the basic properties of residuated lattices hold in a more general setting, and as *residuation theory* tells us, they come from consequences of residuated structures in general. To see this, we now take a brief look at some of basic facts in residuation theory. For general information on residuation theory, see [7].

Let $\mathbf{P}$ and $\mathbf{Q}$ be posets. A map f: $\mathbf{P} \to \mathbf{Q}$ is *residuated* if there exists a map g: $\mathbf{Q} \to \mathbf{P}$ such that the following holds for any $p \in P$ and any $q \in Q$:

$$f(p) \leq q \text{ if and only if } p \leq g(q).$$

If the above holds, we say that f and g form a *residuated pair*, and that g is a *residual* of f. We can show the following.

LEMMA 4.1. *If f and g form a residuated pair, both f and g are monotone increasing.*

It is easy to see that if f is residuated then its residual is determined uniquely by f. Hereafter, f^* denotes the residual of a map f when f is residuated. We can show conversely that f is determined uniquely by f^* when they form a residuated pair. In fact, these uniqueness results come from the following formulas:

1. $f^*(q) = \max\{p \in P : f(p) \leq q\}$,
2. $f(p) = \min\{q \in Q : p \leq f^*(q)\}$.

PROPOSITION 4.2. *Suppose that f and f^* form a residuated pair between posets $\mathbf{P}$ and $\mathbf{Q}$, and that $X \subseteq P$ and $Y \subseteq Q$. If the supremum $\bigvee X$ of X exists then $\bigvee f(X)$ $(= \bigvee\{f(p) : p \in X\})$ exists and $f(\bigvee X) = \bigvee f(X)$. Also if the infimum $\bigwedge Y$ of Y exists then $\bigwedge f^*(Y)$ exists and $f^*(\bigwedge Y) = \bigwedge f^*(Y)$.*

When both $\mathbf{P}$ and $\mathbf{Q}$ are lattices in the above proposition, we have the following.

COROLLARY 4.3. *Suppose that f and f^* form a residuated pair between lattices $\mathbf{P}$ and $\mathbf{Q}$. Then $f(x \vee y) = f(x) \vee f(y)$ for all $x, y \in \mathbf{P}$, and $f^*(u \wedge v) = f^*(u) \wedge f^*(v)$ for all $u, v \in \mathbf{Q}$.*

If both **P** and **Q** are moreover complete, the converse of Proposition 4.2 is true. More precisely, the following holds.

PROPOSITION 4.4. *Suppose that both* **P** *and* **Q** *are complete lattices and that* f *is a map from* **P** *to* **Q**. *Then,* f *is residuated if and only if* f *preserves all (possibly infinite) joins. Dually, a map* $f^* : \mathbf{Q} \to \mathbf{P}$ *is the residual of a map* $f : \mathbf{P} \to \mathbf{Q}$ *if and only if* f^* *preserves all (possibly infinite) meets.*

Now let **P** be a partially-ordered semigroup. For each fixed $u \in P$, define maps g_u and h_u from **P** to **P** by $g_u(x) = u \cdot x$ and $h_u(x) = x \cdot u$, respectively, for each $x \in P$. Then, **P** is a residuated partially-ordered semigroup if and only if both g_u and h_u are residuated maps for every $u \in P$. In this case, the residuals ${g_u}^*$ and ${h_u}^*$ are given by ${g_u}^*(y) = u\backslash y$ and ${h_u}^*(y) = y/u$, respectively.

Thus, applying Corollary 4.3 and Proposition 4.4 to residuated lattices, we have the following results.

COROLLARY 4.5. *The following equations hold in any residuated lattice. For all* x, y, z,
1. $(x \vee y) \cdot z = (x \cdot z) \vee (y \cdot z)$,
2. $z \cdot (x \vee y) = (z \cdot x) \vee (z \cdot y)$,
3. $(x \vee y)\backslash z = (x\backslash z) \wedge (y\backslash z)$,
4. $z/(x \vee y) = (z/x) \wedge (z/y)$,
5. $z\backslash(x \wedge y) = (z\backslash x) \wedge (z\backslash y)$,
6. $(x \wedge y)/z = (x/z) \wedge (y/z)$.

COROLLARY 4.6. *In any complete residuated lattice, i.e. any residuated lattice which is complete as a lattice, the following equations hold:*
1. $x\backslash z = \max\{y : x \cdot y \leq z\}$,
2. $z/y = \max\{x : x \cdot y \leq z\}$.

4.2. Residuated lattices as algebras for substructural logics

For each basic substructural logic **L**, there exists a class of residuated lattices which *characterizes* the logic **L**. To show this, we will introduce some classes of residuated lattices. A residuated lattice is *integral* if the unit 1 of the monoid is equal to the greatest element, denoted by $\top$, of the lattice. In any integral residuated lattice, both $x \cdot y \leq x$ and $x \cdot y \leq y$ hold, since by the monotonicity $x \cdot y \leq x \cdot \top = x \cdot 1 = x$ and similarly for the latter. A residuated lattice **P** is *increasing idempotent* if $x \leq x \cdot x$ for any $x \in P$. It is easy to see that a residuated lattice is both integral and increasing idempotent if and only if $x \cdot y = x \wedge y$ for all x, y.

By using residuated lattices, we can introduce an algebraic interpretation of each formula. To give an interpretation of negation in a given residuated lattice **P**, we need to introduce a element 0 which is an arbitrary element of P. Sometimes, $x \to 0$ is denoted as $\neg x$ in any commutative residuated lattice **P** with 0. If our language for describing formulas contains propositional constants $\top$ and $\bot$, we need to assume that **P** is bounded, i.e. it has a greatest element $\top$ and a least element $\bot$.

To simplify the naming of classes of residuated lattices corresponding to basic substructural logics and to avoid unnecessary complications, we will take the following approach.

Let us call any residuated lattice with a fixed element 0, an **FL**-algebra. A commutative (increasing idempotent, and commutative, increasing idempotent) **FL**-algebra is called an $\mathbf{FL_e}$- ($\mathbf{FL_c}$-, $\mathbf{FL_{ec}}$-, respectively) algebra. An integral **FL**-algebra *whose least element is 0*, which therefore is bounded, is called an $\mathbf{FL_w}$-algebra. Also, any commutative $\mathbf{FL_w}$-algebra is called an $\mathbf{FL_{ew}}$-algebra. If an $\mathbf{FL_e}$-algebra **P** satisfies $\neg\neg x \leq x$ for any x, it is called a $\mathbf{CFL_e}$-algebra. Similarly, we can define $\mathbf{CFL_{ec}}$- and $\mathbf{CFL_{ew}}$-algebras. Note that the equation $x \cdot y = \neg(x \to \neg y)$ holds always in any $\mathbf{CFL_e}$-algebra.

It is well-known that Heyting algebras are algebraic structures for intuitionistic logic. Recall that an algebra $\mathbf{P} = \langle P, \wedge, \vee, \to, 0\rangle$ is a *Heyting algebra* if

1. $\langle P, \wedge, \vee, 0\rangle$ is a lattice with least element 0, and
2. for all $x, y, z \in P$, $x \wedge y \leq z$ iff $x \leq (y \to z)$.

It is easy to see that any Heyting algebra is a bounded commutative residuated lattice with the greatest element 1 defined by $1 = 0 \to 0$, whose monoid operation is the meet $\wedge$, and in which $\wedge$ and $\to$ form a residuated pair. Therefore each Heyting algebra is moreover increasing idempotent and integral. Conversely, any increasing idempotent, integral commutative residuated lattice with least element 0 is a Heyting algebra. By Corollary 4.5, every Heyting algebra is distributive, i.e. $x \wedge (y \vee z) = (x \wedge y) \vee (x \wedge z)$ holds in it. Conversely, every finite distributive lattice is a Heyting algebra, by Proposition 4.4. When a Heyting algebra is complete, it satisfies the following infinite distributive law by Proposition 4.4

$$(\bigvee_i x_i) \wedge y = \bigvee_i (x_i \wedge y),$$

and conversely, any complete lattice with least element 0 satisfying the above law is a reduct of a Heyting algebra (see §4.3).

Also, residuated lattices related to relevant logics have been studied since the mid 1960s. The special feature of these residuated lattices is that they are increasing idempotent residuated lattices satisfying the distributive law between $\wedge$ and $\vee$. Both increasing idempotent, commutative residuated lattices with the distributive law and $\mathbf{CFL_{ec}}$-algebras with the distributive law are basic algebraic structures for relevant logics, which are called *Dunn monoids* and *De Morgan monoids*, respectively. See [1].

In the usual way, we can give an interpretation of formulas in an **FL**-algebra. For a given **FL**-algebra **P**, a *valuation* v on **P** is any mapping from the set of all propositional variables to the set P. Each valuation v is extended to a mapping from the set of all formulas to P inductively as follows. Here, recall that we use the same symbols for logical connectives (and constants) as those for corresponding algebraic operations (and constants, respectively).

1. $v(1) = 1$ and $v(0) = 0$,
2. $v(\top) = \top$ and $v(\bot) = \bot$, when the language contains $\top$ and $\bot$, and **P** is bounded,
3. $v(\alpha \wedge \beta) = v(\alpha) \wedge v(\beta)$,
4. $v(\alpha \vee \beta) = v(\alpha) \vee v(\beta)$,
5. $v(\alpha * \beta) = v(\alpha) \cdot v(\beta)$,
6. $v(\alpha \backslash \beta) = v(\alpha) \backslash v(\beta)$,
7. $v(\alpha / \beta) = v(\alpha) / v(\beta)$.

A formula α is *valid* in **P** if $v(\alpha) \geq 1$ holds for any valuation v on **P**. Also, a given sequent $\alpha_1, \ldots, \alpha_m \Rightarrow \beta$ is said to be valid in **P** if and only if the formula $(\alpha_1 * \cdots * \alpha_m) \to \beta$ is valid in it, or equivalently, $v(\alpha_1) \cdots v(\alpha_m) \leq v(\beta)$ holds for any valuation v on **P**. Here, we assume $v(\alpha_1) \cdots v(\alpha_m) = 1$ when $m = 0$, and $v(\beta) = 0$ when β is empty. For a commutative **P**, a sequent of the form $\alpha_1, \ldots, \alpha_m \Rightarrow \beta_1, \ldots, \beta_n$ is valid in **P** if and only if $v(\alpha_1) \cdots v(\alpha_m) \leq v(\beta_1) + \ldots + v(\beta_n)$ holds for any valuation v on it, where $x + y$ is defined by $x + y = \neg(\neg x \cdot \neg y)$.

We can show the following completeness theorem for basic substructural logics, by using the standard argument on Lindembaum algebras.

THEOREM 4.7. *For any sequent $\mathcal{S}$, $\mathcal{S}$ is provable in* **FL** *if and only if it is valid in all* **FL**-*algebras. This holds also for other basic substructural logics and corresponding classes of* **FL**-*algebras.*

Note that when the language contains constants $\top$ and $\bot$ it is necessary to replace the word "**FL**-algebras" in the above theorem by "bounded **FL**-algebras".

Until now, we were mainly concerned with basic substructural logics that are defined by using sequent systems. We go a step further now and introduce the notion of substructural logics in a general sense. Let **L** be a set of formulas. We say that **L** is a *substructural logic* (over **FL**) if

1. every formula provable in **FL** belongs to **L**,
2. for all formulas α and β, if both α and $\alpha\backslash\beta$ are in **L** then β is also in it,
3. for all formulas α and β, if both α and β are in **L** then $\alpha \wedge \beta$ is also in it,
4. for all formulas α and β, if α is in **L** then both $(\alpha\backslash\beta)\backslash\beta$ and $\beta/(\beta/\alpha)$ are also in it.

Here are some comments. First, every substructural logic **L** is closed under uniform substitutions. Second, for any substructural logic **L**, the following holds:

> for all formulas α and β, if both α and β/α are in **L** then β is also in it.

In fact, suppose that α belongs to **L**. Then $(\beta/\alpha)\backslash\beta$ belongs also to **L**, since $\alpha\backslash((\beta/\alpha)\backslash\beta)$ is provable and hence belongs to **L**. Thus, if moreover β/α is in **L** then β belongs to it by the second condition. Lastly, we can replace the fourth condition by the following two: For all formulas α, β and γ,

- if both α and $\beta\backslash(\alpha\backslash\gamma)$ are in **L** then $\beta\backslash\gamma$ is also in it,
- if both α and $(\gamma/\alpha)/\beta$ are in **L** then γ/β is also in it.

Note that the third condition becomes redundant when $(\alpha * \beta)\backslash(\alpha \wedge \beta) \in \mathbf{L}$ for all α, β, e.g. when the weakening rule holds in it. Also, the fourth condition becomes redundant when $(\alpha * \beta)\backslash(\beta * \alpha) \in \mathbf{L}$ for all α, β, i.e. when the exchange rule holds in **L**.

It is obvious that every basic substructural logic (as a set of formulas) is a substructural logic in our sense. We say that a substructural logic $\mathbf{L}$ is *characterized* by a set $\{\mathbf{P}_i : i \in I\}$ of $\mathbf{FL}$-algebras if $\mathbf{L} = \bigcap\{\mathbf{L}(\mathbf{P}_i) : i \in I\}$ holds. In particular, when $\mathbf{L}$ is characterized by a singleton set $\{\mathbf{P}\}$, we say simply that $\mathbf{L}$ is characterized by $\mathbf{P}$. It is easily seen that for any set $\{\mathbf{P}_i : i \in I\}$ of $\mathbf{FL}$-algebras, the set of formulas $\bigcap\{\mathbf{L}(\mathbf{P}_i) : i \in I\}$ is a substructural logic. We can show the following.

THEOREM 4.8. *Any substructural logic* $\mathbf{L}$ *is characterized by a single* $\mathbf{FL}$-*algebra.*

Here is an outline of the proof. Let Ψ be the set of all formulas. Define a binary relation $\approx$ on Ψ by

$\alpha \approx \beta$ if and only if both $\alpha\backslash\beta$ and $\beta\backslash\alpha$ are in $\mathbf{L}$.

Then, we can show that $\approx$ is a congruence relation on Ψ. Moreover the quotient set $\Psi/\approx$ forms a $\mathbf{FL}$-algebra, say $\mathbf{P}$, and $\mathbf{L} = \mathbf{L}(\mathbf{P})$ holds.

4.3. Quantales and completions of residuated lattices

A structure $\mathbf{Q} = \langle Q, \wedge, \vee, \cdot\rangle$ is a *quantale* if

1. $\langle Q, \wedge, \vee\rangle$ is a complete lattice (and hence is bounded),
2. $\langle Q, \cdot\rangle$ is a semigroup which satisfies that for $x_i, y \in Q$
   a) $(\bigvee_i x_i) \cdot y = \bigvee_i (x_i \cdot y)$,
   b) $y \cdot (\bigvee_i x_i) = \bigvee_i (y \cdot x_i)$.

By Proposition 4.4, $\mathbf{Q}$ is a residuated lattice whose residuals are defined by the equations in Corollary 4.6. Conversely, any complete residuated lattice, i.e. a residuated lattice whose lattice reduct is complete, is a quantale. Therefore, quantales are essentially equal to complete residuated lattices. For general information on quantales, see [40]. When $\langle Q, \cdot\rangle$ is moreover a commutative monoid, $\mathbf{Q}$ becomes a complete commutative residuated lattices. By adding 0 to it, we get a complete $\mathbf{FL_e}$-algebra, which sometimes is called an *intuitionistic phase structure.*

Theorem 4.7 says that for any basic substructural logic $\mathbf{L}$, a formula α is provable in $\mathbf{L}$ if and only if it is valid in all $\mathbf{L}$-algebras. We can strengthen this in the following way.

THEOREM 4.9. *Let* $\mathbf{L}$ *be any one of the basic substructural logics. Then, for any formula* α, α *is provable in* $\mathbf{L}$ *if and only if it is valid in all complete* $\mathbf{L}$-*algebras.*

To show this, it suffices to prove the if-part by Theorem 4.7. This is shown as follows. Let α be a formula not provable in **L**. By Theorem 4.7 α is not valid in a **L**-algebra **P**. We use the following lemma.

LEMMA 4.10. *Let* **L** *be any one of the basic substructural logics. Each* **L**-*algebra can be embedded into a complete* **L**-*algebra.*

By using this, **P** is embedded into a complete **L**-algebra $\mathbf{P}^{\dagger}$, and it is obvious that α is not valid in $\mathbf{P}^{\dagger}$. To get such $\mathbf{P}^{\dagger}$, it is enough to take the *Dedekind-MacNeille completion* of **P**. For the details of the proof of this completeness result, see [31]. For more information on completions of residuated lattices and complete embeddings, see [34] and [35].

5. A Prelude to Algebraic Study of Substructural Logics

The algebraic study of substructural logics is a rapidly growing research field, which attracts both logicians and algebraists. In this section we will touch some topics of the study briefly. These results tell us that the algebraic approach to substructural logics is quite useful and promising. For more information on the study in this direction, see [18].

5.1. Varieties of residuated lattices

We discuss further relations between substructural logics (in our sense) and classes of residuated lattices, using some basic notions and results from universal algebra. In the previous section, we have shown that each substructural logic is characterized by an **FL**-algebra and vice versa. Slightly changing our viewpoint, instead of taking a single **FL**-algebra we consider a class $\mathcal{V}(\mathbf{L})$ of **FL**-algebras for each substructural logic **L**, which is defined by:

$$\mathcal{V}(\mathbf{L}) = \{\mathbf{P} : \mathbf{L} \subseteq \mathbf{L}(\mathbf{P})\}.$$

Then, we can show that the class $\mathcal{V}(\mathbf{L})$ is closed under homomorphic images, subalgebras and direct products. In universal algebra, such a class is called a *variety*. By a fundamental result due to G. Birkhoff, a class $\mathcal{V}$ of algebras is a variety if and only if it is an *equational class*, i.e. it is defined by a set of equations. More precisely, a class $\mathcal{V}$ of algebras is an equational class if there exists a set Σ of equations such that

$$\mathcal{V} = \{\mathbf{P} : s = t \text{ holds in } \mathbf{P} \text{ for any } s = t \in \Sigma\}.$$

We explain here how the class of all residuated lattices is defined by a set of equations. It is well-known that both the class of (bounded) lattices and the class of monoids are defined by certain sets of equations. So it is enough to express the law of residuation by equations. Here, we consider the following.

$x \cdot y \leq z$ if and only if $y \leq x \backslash z$

We note first that each inequality $s \leq t$ can be expressed in a lattice by the equation $s \wedge t = s$. Thus, it suffices to show that the above condition can be expressed by using inequalities. In fact, it is expressed by the following two inequalities.

1. $y \leq x \backslash ((x \cdot y) \vee z)$,
2. $x \cdot (y \wedge (x \backslash z)) \leq z$.

THEOREM 5.1. *The class $\mathcal{V}(\mathbf{L})$ of* **FL**-*algebras is a variety and hence is an equational class, for each substructural logic* **L**.

Conversely, suppose that an equational class $\mathcal{V}$ of **FL**-algebras is given. We assume here that a set Σ of equations defines $\mathcal{V}$. Now let $\mathbf{L} = \{\alpha :$ the inequality $t_\alpha \geq 1$ follows from equations in $\Sigma\}$, where t_α is the term expression of a given formula α. In other words, $\mathbf{L} = \{\alpha :$ the inequality $t_\alpha \geq 1$ holds in every $\mathbf{P}$ in $\mathcal{V}\}$. Then it is easily seen that **L** is a substructural logic. Let us denote this **L** as $\mathbf{L}_\mathcal{V}$.

THEOREM 5.2. *For each variety $\mathcal{V}$ of* **FL**-*algebras, the set $\mathbf{L}_\mathcal{V}$ of formulas is a substructural logic. Moreover, $\mathcal{V}(\mathbf{L}_\mathcal{V}) = \mathcal{V}$ holds.*

Let $\mathcal{V}$ be a variety of **FL**-algebras, and W be a nonempty subclass of $\mathcal{V}$. We say that W *generates* the variety $\mathcal{V}$ if $\mathcal{V}$ is the smallest variety containing W. We can verify easily that a substructural logic **L** is characterized by a set $\{\mathbf{P}_i : i \in I\}$ of **FL**-algebras if and only if it generates the variety $\mathcal{V}(\mathbf{L})$.

Let us denote the variety of all **FL**-algebras by $\mathcal{FL}$, which is equal to $\mathcal{V}(\mathbf{FL})$. It is shown that all subvarieties of $\mathcal{FL}$ as well as all substructural logics over **FL** form a complete lattice. The above two theorems say that there exists a one-to-one correspondence φ between the lattice of all substructural logics and the lattice of all subvarieties of $\mathcal{FL}$, where φ is defined by $\varphi(\mathbf{L}) = \mathcal{V}(\mathbf{L})$. This φ is indeed a dual complete lattice isomorphism. This implies that we can understand the lattice structure of all substructural logics by studying the lattice of all subvarieties of $\mathcal{FL}$, instead. It follows also that for a given substructural logic $\mathbf{L}$, the restriction of φ to the lattice of all extensions of $\mathbf{L}$ is a dual complete lattice isomorphism from it to the lattice of all subvarieties of $\mathcal{V}(\mathbf{L})$. Based on this dual lattice isomorphism and using methods and results from universal algebra, we have many important results on the lattice structure of extensions of the logic $\mathbf{FL_{ew}}$.

Since the variety $\mathcal{HA}$ of Heyting algebras, which is one of the subvarieties of $\mathcal{V}(\mathbf{FL_{ew}})$, has been studied quite well, it is interesting to make a comparison of the lattice structure of subvarieties of $\mathcal{HA}$ with that of $\mathcal{V}(\mathbf{FL_{ew}})$. In fact, there is a big difference between them as the following example shows.

It is easy to see that there is only one *minimal* variety $\mathcal{BA}$ of $\mathbf{FL_{ew}}$-algebras, which is in fact the variety of Boolean algebras, since any nontrivial $\mathbf{FL_{ew}}$-algebra has the two-element Boolaen algebra as a subalgebra. Also, among subvarieties of $\mathcal{HA}$, there is only one *almost minimal variety*, i.e. the variety which covers $\mathcal{BA}$ in the lattice of subvarieties of $\mathcal{HA}$. This is the variety generated by the three-element Heyting algebra. On the other hand, T. Kowalski and M. Ueda obtained the following. See [22] and [18] for the details.

THEOREM 5.3. *There exist uncountably many almost minimal varieties of* $\mathbf{FL_{ew}}$*-algebras.*

For more information on extensions of the logic $\mathbf{FL_{ew}}$ and on subvarieties of the variety of $\mathbf{FL_{ew}}$-algebras, see e.g. [37], [22] and [36].

5.2. Algebras for many-valued logics and fuzzy logics

We will take here special subvarieties of the variety of $\mathbf{FL_{ew}}$-algebras that correspond to Łukasiewicz's many-valued logics and fuzzy logics in the narrow sense (see [9] and [15]).

Algebraic structures for Łukasiewicz's many-valued logics are originally introduced as follows. Let R be either the set $\{0, 1/n, 2/n, \ldots, (n-1)/n, 1\}$ (for $(n+1)$-valued logic) or the unit interval $[0,1]$ (for infinite-valued logic). In either case, R is totally ordered by the natural order on the reals. Define two operations $\rightarrow$ and $\cdot$ on R by

- $x \rightarrow y = \min\{1, 1 - x + y\}$,
- $x \cdot y = \max\{0, x + y - 1\}$.

Then the algebra $\mathbf{R} = \langle R, \min, \max, \cdot, \rightarrow, 1\rangle$ forms a bounded commutative residuated lattice with least element 0, and is in fact an $\mathbf{FL_{ew}}$-algebra. We note that when R denotes a set of the form $\{0, 1/n, 2/n, \ldots, (n-1)/n, 1\}$, $\mathbf{R}$ is isomorphic to an $\mathbf{FL_{ew}}$-algebra with the underlying set $\{a^n, \ldots, a^2, a, a^0\}$ such that $a^n = 0$, whose total order $<$ satisfies $a^{i+1} < a^i$ for each $i = 0, \ldots, n-1$. Here, a^j is defined by $a^0 = 1$ and $a^j = a * a^{j-1}$ for each $j > 0$, where $*$ is a semigroup operation on R. In this case, the residual $\rightarrow$ is defined by $a^k \rightarrow a^m = a^{\max\{0, m-k\}}$.

We can show that the following holds always in any algebra $\mathbf{R}$ under consideration:

$$(x \rightarrow y) \rightarrow y = (y \rightarrow x) \rightarrow x \quad \text{for all } x, y \qquad (1).$$

Now, let us generalize this. We say that a $\mathbf{FL_{ew}}$-algebra $\mathbf{P}$ is an *MV-algebra* if it satisfies the above condition (1). It is known that they serve as algebraic structures for many-valued logics.[5] We note that any MV-algebra is in fact a $\mathbf{CFL_{ew}}$-algebra, and moreover the following two equations hold in it: for all x, y $x \vee y = (x \rightarrow y) \rightarrow y$ and $x \wedge y = \neg(\neg x \wedge \neg y)$. The following completeness theorem holds. For further information on many-valued logics, see e.g. [9].

THEOREM 5.4. *The following three conditions are mutually equivalent for any formula α.*
1. *α is provable in the logic obtained from $\mathbf{FL_{ew}}$ by adding $((\beta \rightarrow \gamma) \rightarrow \gamma) \rightarrow ((\gamma \rightarrow \beta) \rightarrow \beta)$,*
2. *α is valid in the Łukasiewicz's infinite-valued algebra, i.e. the MV-algebra determined by the unit interval* $[0, 1]$,
3. *α is valid in any MV-algebra.*

A map T from $[0, 1]^2$ to $[0, 1]$ is a *triangular norm* (or, simply, a t-norm) if $\langle [0, 1], \circ, 1\rangle$ is a partially ordered commutative monoid where $\circ$ is defined by $x \circ y = T(x, y)$. It is obvious that $\langle [0, 1], \min, \max, 0, 1\rangle$ forms a complete lattice. Typical examples of t-norms are $\min\{x, y\}$, $x \times y$ (the multiplication of real numbers), and $\max\{x+y-1, 0\}$. The last one is the monoid operation used in Łukasiewicz's many-valued algebra.

Now, a question is when these partially ordered commutative monoids become residuated. Proposition 4.4 tells us that a partially ordered monoid defined by a t-norm T is residuated if and only if $\circ$ satisfies

$$(\bigvee x_i) \circ y = \bigvee (x_i \circ y).$$

---

[5]Usually, MV-algebras are defined in a different way. But, the definition given here determines essentially the same algebras as those defined in the standard way.

Since $\circ$ is defined on an interval $[0, 1]$ of reals, this can be expressed also as

$$T(x, y) = \lim_{z \to x-0} T(z, y) = \lim_{w \to y-0} T(x, w),$$

which means that a map T is *left-continuous*. Thus, we can see that every left-continuous t-norm determines a commutative, integral residuated lattice, and then this residuated lattice characterizes a substructural logic over $\mathbf{FL_{ew}}$. For instance, t-norms $\min\{x, y\}$, $x \times y$ and $\max\{x + y - 1, 0\}$ determine a superintuitionistic logic called Dummett-Gödel logic, the *product* logic introduced by Hájek (see [15]), and Łukasiewicz's infinite-valued logic, respectively.

When a left-continuous t-norm $\circ$ satisfies moreover that

$$(\bigwedge x_i) \circ y = \bigwedge (x_i \circ y),$$

it is *continuous*. We introduce here two extensions of $\mathbf{FL_{ew}}$ in an axiomatic way. Define **MTL** (monoidal t-norm logic) to be the logic obtained from $\mathbf{FL_{ew}}$ by adding $(\alpha \to \beta) \vee (\beta \to \alpha)$ as the axiom, and **BL** (basic logic) to be the logic obtained from **MTL** by adding $(\alpha \wedge \beta) \to (\alpha * (\alpha \to \beta))$ as the additional axiom. They are introduced by Esteva and Godo in [11] and Hájek in [15], respectively. Then the following results, called the *standard completeness theorems* for **MTL** and **BL**, are obtained in [17] and [8], respectively. Note here that Łukasiewicz's infinite-valued logic is shown to be equal also to the logic obtained from **BL** by adding $\neg\neg\alpha \to \alpha$ as the axiom.

THEOREM 5.5. *1. The logic* **MTL** *is complete with respect to the class of all* $\mathbf{FL_{ew}}$*-algebras determined by left-continuous t-norms.*
2. The logic **BL** *is complete with respect to the class of all* $\mathbf{FL_{ew}}$*-algebras determined by continuous t-norms.*

5.3. Finite model property and finite embeddability property

A substructural logic **L** has the *finite model propery* if there exists a set $\{\mathbf{P}_i : i \in I\}$ of finite **FL**-algebras such that $\mathbf{L} = \bigcap\{\mathbf{L}(\mathbf{P}_i) : i \in I\}$ holds. This is equivalent to saying that the variety $\mathcal{V}(\mathbf{L})$ is generated by its finite members. The finite model property is a useful property in showing the decidability of a given logic, as Harrop's theorem says that a logic is decidable if it has the finite model property and is finitely axiomatizable.

In modal logic, proving the finite model property with respect to Kripke models is a standard and most powerful technique of showing decidability. On the other hand, since Kripke frames for substructural logics, e.g. those introduced in [37], are hardly manageable, the finite model property of basic substructural logics had remained open until R.K. Meyer and the present author gave a positive answer to the implicational fragments of both $\mathbf{FL_{ew}}$ and $\mathbf{FL_{ec}}$ (see [28]). This makes an interesting contrast with the fact that the decidability of basic substructural logics except $\mathbf{FL_c}$ is an easy consequence of the cut elimination theorem, as discussed already in §3. Then, Y. Lafont [25] succeeded to show the finite model property of both $\mathbf{CFL_{ew}}$ and $\mathbf{CFL_{ec}}$. The methods used there were then extended and elaborated in the paper by M. Okada and K. Terui [29], who succeeded to show the finite model property of other basic substructural logics. Interestingly enough, in [25] and [29] the cut elimination theorem is used in an essential way in proving the finite model property of some of logics. An algebraic presentation of the method used in them is given in [5].

Extending an idea of [29], Blok and van Alten [6] have developed a method of proving the finite model property using the *finite embeddability property.* The method can be regarded as an algebraic substitute of the filtration method for Kripke frames.

We say that the class $\mathcal{K}$ of algebras has the *finite embeddability property* when for a given finite *partial* subalgebra $\mathbf{R}$ of an algebra $\mathbf{P}$ in $\mathcal{K}$, there exists a *finite* algebra $\mathbf{Q}$ in $\mathcal{K}$ into which $\mathbf{R}$ can be embedded. It is well-known that the class $\mathcal{HA}$ of Heyting algebras has the finite embeddability property. For, if $\mathbf{R}$ is a finite partial subalgebra of a Heyting algebra $\mathbf{P}$, then the sublattice $\mathbf{Q}$ generated by the underlying set of $\mathbf{R}$ becomes a finite distributive lattice and hence is a finite Heyting algebra, into which $\mathbf{R}$ is embedded.

Suppose that a formula α is not provable in $\mathbf{L}$. Then there exist an algebra $\mathbf{P}$ in $\mathcal{V}(\mathbf{L})$ and a valuation v of $\mathbf{P}$ such that $v(\alpha) \geq 1$ doesn't hold. The set $\{v(\delta) : \delta \text{ is a subformula of } \alpha\} \cup \{0, 1\}$ forms a finite partial subalgebra $\mathbf{R}$ of $\mathbf{P}$. We assume now that the variety $\mathcal{V}(\mathbf{L})$ has the finite embeddability property. Then $\mathbf{R}$ is embedded into a finite algebra $\mathbf{Q}$ in $\mathcal{V}(\mathbf{L})$. Therefore, α is not valid in $\mathbf{Q}$. Hence we have the finite model property of $\mathbf{L}$. Thus the finite embeddability property of $\mathcal{V}(\mathbf{L})$ implies the finite model property of $\mathbf{L}$.

For the class of $\mathbf{FL_{ew}}$-algebras, the proof of the finite embeddability property becomes much more complicated than that for $\mathcal{HA}$. We will give here a brief outline of the proof. Suppose that $\mathbf{R}$ is a partial subalgebra of a $\mathbf{FL_{ew}}$-algebra $\mathbf{P}$. Let $\mathbf{M} = \langle M, \cdot, 1, \leq \rangle$ be the partially-ordered submonoid generated by the domain R of $\mathbf{R}$. The set M is not necessarily finite even if R is finite. For each $u \in M$ and $r \in R$, we define a subset $(u \rightsquigarrow r]$ of M

by $(u \rightsquigarrow r] = \{w \in M : uw \leq r\}$. Let $\overline{D}$ be the set of all subsets of M of the form $(u \rightsquigarrow r]$ with $u \in M$ and $r \in R$. For each subset X of M, define $CX = \bigcap\{Z \in \overline{D} : X \subseteq Z\}$. Then C determines a closure operation on $\wp(\mathbf{M})$ and it is shown that the collection of all C-closed subsets of M forms a $\mathbf{FL_{ew}}$-algebra $\mathbf{Q}$, into which $\mathbf{R}$ can be embedded. Moreover, we can show that $\mathbf{Q}$ is finite when R is finite. Thus, we have the following (see [6]).

THEOREM 5.6. *The class of* $\mathbf{FL_{ew}}$*-algebras has the finite embeddability property. Therefore, the logic* $\mathbf{FL_{ew}}$ *has the finite model property. This holds also for* $\mathbf{FL_w}$.

We note that though $\mathbf{FL_e}$ has the finite model property, the class of $\mathbf{FL_e}$-algebras doesn't have the finite embeddability property. Thus, the finite model property of $\mathbf{L}$ doesn't always imply the finite embeddability property of $\mathcal{V}(\mathbf{L})$.

6. Final Remarks

More than ten years have already passed since a study of nonclassical logics under the name "substructural logics" started. The study seems to be quite promising, as so many different kinds of nonclassical logics studied so far can be discussed within this framework. On the other hand, it seems to be not clear yet why it works so well, or more precisely why restrictions on "structural rules" play such a key role when logics are formalized as sequent systems. Here we will try to give some explanations for this by using the notion of residuation.

Most of people will agree that the "implication" is the most important logical connective. The meanings and purpose of implication have been argued from various philosophical and mathematical points of view. Let us suppose here that our implication satisfies the following relation in a given sequent system $\mathbf{L}$:

> $\Gamma, \alpha \Rightarrow \beta$ is provable in $\mathbf{L}$ if and only if $\Gamma \Rightarrow \alpha \rightarrow \beta$ is provable in it.

This relation can be shown by the help of the cut rule as long as we have the standard sequent rules for $\rightarrow$. We introduce now an auxiliary logical connective $*$, called fusion, to represent commas (in the left-hand side) of sequents, as we have done in §2. Then the above relation, which we call the *residuation relation*, expresses the fact that implication is the residual of fusion. Fusion is regarded as a monoid operation from the mathematical point of view, that is much more manageable than implication. Thus, once we admit the residuation relation between them, we can shift our attention from implication to fusion, and study the latter in order to know properties of the former. This shift is justified by residuation theory, which tells us that each fusion determines uniquely an implication and vice versa.

Some of the basic properties of fusions as monoid operations can be described by structural rules in sequent systems, such as the commutativity of fusions is described by exchange rules, for instance. This means that by controling structural rules we can deal with logics having various kinds of implications. This, we think, will explain the reason why the framework of substructural logics works well.

Though there is already much literature on substructural logics, we have no common understanding of the definition of substructural logics. In the present paper, we have given a definition of substructural logics *over* $\mathbf{FL_{ew}}$. But there are still many other logics that should be counted substructural logics. For example, let us consider the implicational logic **BCK** which is the implicational fragment of $\mathbf{FL_{ew}}$. This logic is not an extension of **FL**, but should be regarded as a substructural logic. Then, what is a suitable definition of substructural logics in general? Based on discussions in the present paper, it will be quite reasonable, we think, to understand substructural logics in the following way:

Substructural logics are logics of residuated structures.

We need to add a comment on the above proposal. While **BCK** has no fusion, it can be extended conservatively to a logic having the law of residuation, in fact to the logic $\mathbf{FL_{ew}}$. Because of this, we may regard **BCK** as a substructural logic in our sense. Thus, by a "logic of residuated structures" we mean a logic which can be extended conservatively to a logic in which the law of residuation holds.

If we adopt this definition, logics like the "basic logic" **B** which was introduced in [4] and its extensions without the *deductive face*, i.e. logics without the law of residuation, will not be within the scope of substructural logics in our sense, though they are of their own importance and interest.

As a matter of fact, most nonclassical logics except linear logic and relevant logics don't contain fusion explicitly as a logical connective. On the other hand, since each sequent system contains usually comma as an auxiliary symbol, which behaves as a monoid operation, we may say that it includes already fusion in an implicit way. For example, consider a cut-free sequent system of **BCK** introduced by Y. Komori in [21]. By simply adding rules for fusion to it, we get a cut-free sequent system for an extension of **BCK** with fusion, in which the residuation relation holds between fusion and implication. Note that the fact that the extended sequent system is cut-free ensures that it is a conservative extension of the original **BCK**. This shows that sequent systems are quite useful in extracting residuation relations in logics. In other words, such residuation relations will hardly be observed as long as we take Hilbert-style formalization. These observations can be summarized as follows.

> *Formalizing logics in cut-free sequent systems will reveal hidden residuation relations in them. Therefore, sequent systems are suitable for describing residuated structures.*

The structure of the partially-ordered monoid in a given residuated lattice corresponds to structural rules while the lattice structure corresponds to rules for conjunction and disjunction. These two parts are combined by implication, through the law of residuation.

As we have shown, algebraic study of substructural logics is important not only from a technical point of view but also from a conceptual one.

References

[1] ANDERSON, A. R. and N. D. BELNAP JR., *Entailment: The Logic of Relevance and Necessity I*, Princeton University Press, 1975.

[2] ANDERSON, A. R. and N. D. BELNAP JR., and J. M.DUNN, *Entailment: The Logic of Relevance and Necessity II*, Princeton University Press, 1992.

[3] BALBES, R. and Ph. DWINGER, *Distributive Lattices*, University of Missouri Press, 1974.

[4] BATTILOTTI, G. and G. SAMBIN, 'Basic logic and the cube of its extensions', in: A. Cantini, E. Casari and P. Minari (eds.): *Logic and Foundations of Mathematics*, Kluwer Academic Publishers, pp.165-186, 1999.

[5] BELARDINELLI, F., P. JIPSEN and H. ONO, 'An algebraic aspects of cut elimination theorem', draft, 2003.

[6] BLOK, W. J. and C. J.VAN ALTEN, 'The finite embeddability property for residuated lattices, pocrims and BCK-algebras', *Algebra Universalis* 48:253-271, 2002.

[7] BLYTH, T. S. and M. F.JANOWITZ, *Residuation Theory*, International Series of Monographs in Pure and Applied Mathematics vol.102, Pergamon Press, 1972.

[8] CIGNOLI, R., F. ESTEVA, L. GODO and A. TORRENS, 'Basic Fuzzy Logic is the logic of continuous t-norms, and their residua', *Soft Computing* 4:106-112, 2000.

[9] CIGNOLI, R., D. MUNDICI and I. M. L. D'OTTAVIANO, *Algebraic Foundations of Many-Valued Reasoning*, Kluwer Academic Publishers, 2000.

[10] DOŠEN, K. and P. SCHROEDER-HEISTER eds., *Substructural Logics*, Oxford University Press, 1993.

[11] ESTEVA, F.and L. GODO, 'Monoidal t-norm based logic: towards a logic for left-continuous t-norms', *Fuzzy Sets and Systems* 124:271-288, 2001.

[12] GENTZEN, G., 'Untersuchungen über das logische Schliessen', *Mathematische Zeitschrift* 39:176-210, 405-413, 1934.

[13] GIRARD, J.-Y., 'Linear logic', *Theoretical Computer Science* 50:1-102, 1987.

[14] GRIŠIN, V., 'Predicate and set-theoretic calculi based on logic without contraction, *Math. USSR Izvestiya* 18:41-59, 1982.

[15] HÁJEK, P., *Metamathematics of Fuzzy Logic*, Kluwer Academic Publishers, 1998.

[16] HÖHLE, U., 'Commutative residuated monoids', in: U. Höhle and P. Klement (eds.): *Non-classical Logics and Their Applications to Fuzzy Subsets*, Kluwer Academic Publishers, pp.53-106, 1995.

[17] JENEI, S. and F. MONTAGNA, 'A proof of standard completeness for Esteva and Godo's Logic MTL', *Studia Logica* 70:183-192, 2002.

[18] JIPSEN, J., T. KOWALSKI and H. ONO, *Residuated Lattices: An algebraic glimpse at substructural logics*, in preparation.

[19] JIPSEN, J. and C. TSINAKIS, 'A survey of residuated lattices', in: J. Martinez (ed.): *Ordered Algebraic Structures*, Kluwer Academic Publishers, pp.19-56, 2002.

[20] KIRIYAMA, E. and H. ONO, 'The contraction rule and decision problems for logics without structural rules', *Studia Logica* 50:299-319, 1991.

[21] KOMORI, Y., 'The class of BCC–algebras is not a variety', *Mathematica Japonica*, 29:391-394, 1984.

[22] KOWALSKI, T. and H. ONO, *Residuated Lattices: An algebraic glimpse at logics without contraction*, monograph, March, 2001.

[23] KRIPKE, S. A., 'The problem of entailment (abstract)', *The Journal of Symbolic Logic* 24:324, 1959.

[24] KRULL, W., 'Axiomatische Begründung der allgemeinen Idealtheorie', *Sitzungsberichte der physikalisch medizinischen Societät der Erlangen* 56:47-63, 1924.

[25] LAFONT, Y., 'The finite model property for various fragments of linear logic', *The Journal of Symbolic Logic* 62:1202-1208, 1997.

[26] LAMBEK, J., 'The mathematics of sentence structure', *American Mathematical Monthly* 12:166-178, 1958.

[27] MEYER, R. K., *Topics in modal and many-valued logic*, Ph.D. dissertation, University of Pittsburgh, 1966.

[28] MEYER, R. K. and H. ONO, 'The finite model property for BCK and BCIW', *Studia Logica* 53:107-118, 1994.

[29] OKADA, M. and K. TERUI, 'The finite model property for various fragments of intuitionistic linear logic', *The Journal of Symbolic Logic* 64:790-802, 1999.

[30] ONO, H., 'Structural rules and a logical hierarchy', in: P.P. Petokov (ed.): *Mathematical Logic, Proceedings of the Summer School and the Conference 'Heyting '88'*, Plenum Press, pp.95-104, 1990.

[31] ONO, H., 'Semantics for substructural logics', in: K. Došen and P. Schroeder-Heister (eds.): *Substructural Logics*, Oxford University Press, pp.259-291, 1993.

[32] ONO, H., 'Decidability and the finite model property of substructural logics', in: J. Ginzburg et.al (eds.): *Tbilisi Symposium on Logic, Language and Computation: Selected Papers*, Studies in Logic, Language and Information, CSLI, pp.263-274, 1998.

[33] ONO, H., 'Proof-theoretic methods for nonclassical logic — an introduction', in: M. Takahashi, M. Okada and M. Dezani-Ciancaglini, (eds.): *Theories of Types and Proofs*, MSJ Memoirs vol.2, Mathematical Society of Japan, pp.207-254, 1998.

[34] ONO, H., 'Closure operators and complete embeddings of residuated lattices', to appear in *Studia Logica*.

[35] ONO, H., 'Completions of algebras and completeness of modal and substructural logics', to appear in *Advances in Modal Logic vol.4*.

[36] ONO, H., 'Logics without contraction rule and residuated lattices I', in E. Mares edited, *XXX*, CSLI, 2003.

[37] ONO, H. and Y. KOMORI, 'Logics without the contraction rule', *The Journal of Symbolic Logic* 50:169-201, 1985.

[38] PAOLI, F., *Substructural Logics: A Primer*, Trends in Logic vol.13 – Studia Logica Library, Kluwer Academic Publishers, 2002.

[39] RESTALL, G., *An Introduction to Substructural Logics*, Routledge, 2000.

[40] ROSENTHAL, K. I., *Quantales and Their Applications*, Pitman Research Notes in Mathematics 234, Longman, 1990.

[41] TROELSTRA, A. S., *Lectures on Linear Logic*, CSLI Lecture Notes No.29, Stanford University, 1992.

[42] URQUHART, A., 'The undecidability of entailment and relevant implication', *The Journal of Symbolic Logic* 49:1059-1073, 1984.

[43] WANG, H., *A Survey of Mathematical Logic*, Studies in Logic and the Foundations of Mathematics, North-Holland, 1963.

[44] WARD, M. and R. P. DILWORTH, 'Residuated lattices', *Transactions of the American Mathematical Society* 45:335-354, 1939.

HIROAKIRA ONO
School of Information Science
Japan Advanced Institute of Science and Technology
1-1 Asahidai, Tatsunokuchi
Ishikawa, 923-1292, Japan
ono@jaist.ac.jp

M. L. DALLA CHIARA
R. GIUNTINI
R. LEPORINI

Quantum Computational Logics: A Survey

Abstract. Quantum computation has suggested new forms of quantum logic, called *quantum computational logics* ([2]). The basic semantic idea is the following: the meaning of a sentence is identified with a *quregister*, a system of *qubits*, representing a possible *pure state* of a compound quantum system. The generalization to *mixed states*, which might be useful to analyse *entanglement-phenomena*, is due to Gudder ([7]). Quantum computational logics represent non standard examples of *unsharp quantum logic*, where the non-contradiction principle is violated, while conjunctions and disjunctions are strongly non-idempotent. In this framework, any sentence α of the language gives rise to a *quantum tree*: a kind of *quantum circuit* that transforms the quregister associated to the atomic subformulas of α into the quregister associated to α.

Keywords: quantum computation, quantum logic.

1. Introduction

Quantum computation has suggested new forms of quantum logic that have been called *quantum computational logics.* The main difference between orthodox quantum logic (first proposed by Birkhoff and von Neumann [1]) and quantum computational logics concerns a basic semantic question: how to represent the *meanings* of the sentences of a given language? The answer given by Birkhoff and von Neumann is the following: the meanings of the elementary experimental sentences of quantum theory (QT) have to be regarded as determined by convenient sets of *states* of quantum objects. Since these sets should satisfy some special closure conditions, it turns out that, in the framework of orthodox quantum logic, sentences can be adequately interpreted as *closed subspaces* of the Hilbert space associated to the physical systems under investigation. The answer given in the framework of quantum computational logics is quite different: meanings of sentences are represented by *information quantities*, a kind of abstract objects that are described in the framework of *quantum information theory.*

V. F. Hendricks and *J. Malinowski* (eds.),
Trends in Logic: 50 Years of Studia Logica

Trends in Logic **21**: 229–271, 2003.

2. From classical to quantum information

As is well known, the unit of measurement in classical information theory is the *bit*: one *bit* measures the information quantity that can be either transmitted or received whenever one chooses one element from a set consisting of two distinct elements (say, from the set $\{0, 1\}$). From an intuitive point of view, both the objects 0 and 1 can be imagined as a well determined *state* of a classical physical system (for instance, the state of a tape cell in a given machine).

Let us now refer to a quantum computational context, where information is supposed to be elaborated and transmitted by means of a quantum system. According to the standard axiomatization of QT, the *pure states* of our system are mathematically represented by unit vectors in a convenient Hilbert space $\mathcal{H}$. Let us refer to the simplest situation, where our Hilbert space $\mathcal{H}$ has dimension 2; hence $\mathcal{H} = \mathbb{C}^2$. In such a case $\mathcal{H}$ will have a *basis* consisting of two unit elements, and any vector of the space will be representable as a *superposition* of the two basis-elements. In quantum computation, it is customary to use Dirac's notation. Accordingly, the vectors of $\mathcal{H}$ are indicated by $|\psi\rangle$, $|\varphi\rangle$,... ; while the basis-elements are denoted by $|0\rangle$, $|1\rangle$. As a consequence, for any unit vector $|\psi\rangle$ we will have:

$$|\psi\rangle = a_0|0\rangle + a_1|1\rangle,$$

where the coefficients a_0, a_1 are complex numbers (also called amplitudes) such that:

$$|a_0|^2 + |a_1|^2 = 1.$$

Let us now try and apply such a formalism to a quantum information theory. The basic idea is to generalize the concept of *bit*, by introducing the notion of *qubit* or *quantum bit*. A *qubit* is a unit vector in the Hilbert space $\mathbb{C}^2$. Thus any *qubit* will have the form $|\psi\rangle = a_0|0\rangle + a_1|1\rangle$. The interpretation is determined by an axiom of QT that is usually called the *Born rule*. Suppose that (like in the classical case) the pure states $|0\rangle$ and $|1\rangle$ represent two *maximal* (and *precise*) pieces of information. Then the superposition-state $|\psi\rangle = a_0|0\rangle + a_1|1\rangle$ will represent an information that involves a certain degree of *uncertainty*. In particular, the number $|a_0|^2$ will correspond to the probability-value of the information described by the basic state $|0\rangle$; while $|a_1|^2$ will correspond to the probability-value of the information described by the basic state $|1\rangle$.

In this context, it makes sense to imagine $|\psi\rangle$ as an *epistemic state* that stocks two precise pieces of information *in parallel*: the information $|0\rangle$ and the information $|1\rangle$.

Let us now consider a situation characterized by many *bits* or *qubits*. As is well known, in classical information theory, a system consisting of n *bits* is naturally represented by a sequence of n elements belonging to the set $\{0,1\}$ (i.e. as an element of the set $\{0,1\}^n$). In the framework of quantum computation, it is convenient to adopt the tensor product formalism, which is used in quantum theory in order to represent compound physical systems. Suppose a two-particle quantum system:

$$\mathfrak{S} = \mathfrak{S}_1 + \mathfrak{S}_2.$$

For instance, $\mathfrak{S}_1$ and $\mathfrak{S}_2$ might correspond respectively to the two electrons in a given helium atom. In such a case, the Hilbert space $\mathcal{H}^{\mathfrak{S}}$ associated to $\mathfrak{S}$ will be the tensor product $\mathcal{H}^{\mathfrak{S}_1} \otimes \mathcal{H}^{\mathfrak{S}_2}$ of the two Hilbert spaces $\mathcal{H}^{\mathfrak{S}_1}$ and $\mathcal{H}^{\mathfrak{S}_2}$, that are associated to $\mathfrak{S}_1$ and $\mathfrak{S}_2$, respectively. Thus any pure state of $\mathfrak{S}$ will be a unit vector in the space $\mathcal{H}^{\mathfrak{S}}$.

A particularly interesting case is represented by those vectors $|\psi\rangle$ of $\mathcal{H}^{\mathfrak{S}}$ that can be expressed as the tensor product of two vectors $|\psi_1\rangle$ and $|\psi_2\rangle$, belonging to $\mathcal{H}^{\mathfrak{S}_1}$ and to $\mathcal{H}^{\mathfrak{S}_2}$, respectively. In other words:

$$|\psi\rangle = |\psi_1\rangle \otimes |\psi_2\rangle.$$

In such cases, one usually speak of *factorized states*. It is worthwhile recalling that not all vectors of $\mathcal{H}^{\mathfrak{S}}$ can be expressed in such a simple form.

How to represent, in this framework, a system consisting of n *qubits*? It seems quite natural to describe our system as the pure state of a compound physical system consisting of n quantum objects. On this basis our n-*qubit* system can be identified with a unit vector of the product space

$$\underbrace{\mathbb{C}^2 \otimes \ldots \otimes \mathbb{C}^2}_{n-times}.$$

Instead of $\underbrace{\mathbb{C}^2 \otimes \ldots \otimes \mathbb{C}^2}_{n-times}$ we will also write: $\otimes^n\mathbb{C}^2$. Particularly interesting examples will be represented by the vectors of $\otimes^n\mathbb{C}^2$ whose form is:

$$|x_1\rangle \otimes \ldots \otimes |x_n\rangle,$$

where each $|x_i\rangle$ is an element of the basis of $\mathbb{C}^2$ (i.e., $|x_i\rangle = |0\rangle$ or $|x_i\rangle = |1\rangle$). One can prove that the set of all vectors having this form represents a basis for the product space $\otimes^n\mathbb{C}^2$.

How can we deal with the concept of *quantum computation*, in this framework? The basic idea is to describe a computation by means of that kind of process that corresponds to the dynamic evolution of a quantum system. Suppose a physical system $\mathfrak{S}$, whose pure state at time t_0 is the vector $|\psi(t_0)\rangle$ (where $|\psi(t_0)\rangle$ belongs to the Hilbert space $\mathcal{H}^{\mathfrak{S}}$ associated to $\mathfrak{S}$). Owing to Schrödinger's equation, for any time t (where either $t \leq t_0$ or $t_0 \leq t$), there exists an operator $U_{[t_0,t]}$ that determines the state of the system at time t as a function of the state of the system at time t_0. In other words:

$$|\psi(t)\rangle = U_{[t_0,t]}|\psi(t_0\rangle.$$

Any operator $U_{[t_0,t]}$ of this kind is *unitary*. Hence, our operator preserves the length of the vectors and the orthogonality relation. Further it is *reversible*: one can go from $|\psi(t_0)\rangle$ to $|\psi(t)\rangle$ and viceversa, without any dissipation of information.

On this ground, it makes sense to represent a quantum computation by means of convenient unitary operators assuming arguments and values in particular sets of *qubit* systems. Since *qubits* are generally superposition-states, one obtains some typical *parallel configurations*.

3. Qubits, Quregisters and Qumixs

As we have seen, qubits and qubit-systems (also called *quregisters*) correspond to pure states, which are *maximal* pieces of information of the observer about the quantum system under investigation. In other words, one is dealing with a kind of information that cannot be consistently extended to a richer knowledge (expressed in the same language). In many concrete situations it may be interesting to consider also *mixed states* (or *mixtures*), describing pieces of information that are not generally maximal. According to the standard axiomatization of QT such states are mathematically represented by *density operators* ρ of the Hilbert space $\mathcal{H}$ (associated to the system). Any pure state $|\psi\rangle$ corresponds to a limit-case of a density operator: the projection $P_{|\psi\rangle}$ onto the one-dimensional closed subspace determined by the vector $|\psi\rangle$. Representing quantum information by density operators turns out to be important in order to deal with *entanglement-phenomena*, which play a fundamental role in teleportation and in quantum cryptography.

We will now sum up some basic formal definitions of quantum computation. Consider the two–dimensional Hilbert space $\mathbb{C}^2$ (where any vector $|\psi\rangle$ is represented by a pair of complex numbers). Let $\mathcal{B}^{(1)} = \{|0\rangle, |1\rangle\}$ be the canonical orthonormal basis for $\mathbb{C}^2$, where $|0\rangle = (1,0)$ and $|1\rangle = (0,1)$.

DEFINITION 3.1. (Qubit).
A *qubit* is a unit vector $|\psi\rangle$ of the Hilbert space $\mathbb{C}^2$.

As we have seen, from an intuitive point of view, any qubit $|\psi\rangle = a_0|0\rangle + a_1|1\rangle$ (with $|a_0|^2 + |a_1|^2 = 1$) can be regarded as an *uncertain piece of information*, where the answer *NO* has probability $|a_0|^2$, while the answer *YES* has probability $|a_1|^2$. The two basis-elements $|0\rangle$ and $|1\rangle$ are taken as encoding the classical bit-values 0 and 1, respectively. From a semantic point of view, they can be also regarded as the classical truth-values *Falsity* and *Truth.*

An n-qubit system (or n-*quregister*) is represented by a unit vector in the n-fold tensor product Hilbert space $\otimes^n\mathbb{C}^2 := \underbrace{\mathbb{C}^2 \otimes \ldots \otimes \mathbb{C}^2}_{n-times}$ (where $\otimes^1\mathbb{C}^2 := \mathbb{C}^2$). We will use $x, y, \ldots$ as variables ranging over the set $\{0, 1\}$. At the same time, $|x\rangle, |y\rangle, \ldots$ will range over the basis $\mathcal{B}^{(1)}$. Any factorized unit vector $|x_1\rangle \otimes \ldots \otimes |x_n\rangle$ of the space $\otimes^n\mathbb{C}^2$ will be called an n–*configuration* (which can be regarded as a quantum realization of a classical bit sequence of length n). Instead of $|x_1\rangle \otimes \ldots \otimes |x_n\rangle$ we will simply write $|x_1, \ldots, x_n\rangle$. Recall that the dimension of $\otimes^n\mathbb{C}^2$ is 2^n, while the set of all n–configurations $\mathcal{B}^{(n)} = \{|x_1, \ldots, x_n\rangle : x_i \in \{0, 1\}\}$ is an orthonormal basis for the space $\otimes^n\mathbb{C}^2$. We will call this set a *computational basis* for the n–quregisters. Since any string $x_1, \ldots, x_n$ represents a natural number $j \in [0, 2^n - 1]$ (where $j = 2^{n-1}x_1 + 2^{n-2}x_2 + \ldots + x_n$), any unit vector of $\otimes^n\mathbb{C}^2$ can be shortly expressed in the following form: $\sum_{j=0}^{2^n-1} c_j \, \|j\rangle\rangle$, where $c_j \in \mathbb{C}$, $\|j\rangle\rangle$ is the n-configuration corresponding to the number j and $\sum_{j=0}^{2^n-1} |c_j|^2 = 1$.

We will indicate by $\mathfrak{R}(\otimes^n\mathbb{C}^2)$ the set of all quregisters of $\otimes^n\mathbb{C}^2$, while $\mathfrak{R}$ will represent the set $\bigcup_{n=1}^{\infty} \mathfrak{R}(\otimes^n\mathbb{C}^2)$. The set $\mathfrak{R}(\otimes^1\mathbb{C}^2)$ of all qubits will be shortly indicated by $\mathfrak{Q}$.

Consider now the two following sets of natural numbers:

$$C_1^{(n)} := \{i : \|i\rangle\rangle = |x_1, \ldots, x_n\rangle \text{and } x_n = 1\}$$

and

$$C_0^{(n)} := \{i : \|i\rangle\rangle = |x_1, \ldots, x_n\rangle \text{and } x_n = 0\}.$$

Let us refer to a generic unit vector of the space $\otimes^n\mathbb{C}^2$:

$$|\psi\rangle = \sum_{i=0}^{2^n-1} a_i \, \|i\rangle\rangle.$$

We obtain:

$$|\psi\rangle = \sum_{i \in C_0^{(n)}} a_i \, \|i\rangle\!\rangle + \sum_{j \in C_1^{(n)}} a_j \, \|j\rangle\!\rangle.$$

Let $P_1^{(n)}$ and $P_0^{(n)}$ be the projections onto the span of $\left\{\|i\rangle\!\rangle : i \in C_1^{(n)}\right\}$ and $\left\{\|i\rangle\!\rangle : i \in C_0^{(n)}\right\}$, respectively. Clearly, $P_1^{(n)} + P_0^{(n)} = I^{(n)}$, where $I^{(n)}$ is the identity operator of $\otimes^n\mathbb{C}^2$. Apparently, $P_1^{(n)}$ and $P_0^{(n)}$ are density operators iff $n = 1$. Let $k_n = \frac{1}{2^{n-1}}$ be the normalization coefficient such that $k_n P_1^{(n)}$ and $k_n P_0^{(n)}$ are density operators. From an intuitive point of view, $k_n P_1^{(n)}$ can be regarded as a privileged information corresponding to the *Truth*, while $k_n P_0^{(n)}$ corresponds to the *Falsity*. In particular, $P_1^{(1)}$ represents the bit $|1\rangle$, while $P_0^{(1)}$ represents the bit $|0\rangle$. Let $\mathfrak{D}(\otimes^n\mathbb{C}^2)$ be the set of all density operators of $\otimes^n\mathbb{C}^2$ and let $\mathfrak{D} := \bigcup_{n=1}^{\infty} \mathfrak{D}(\otimes^n\mathbb{C}^2)$.

DEFINITION 3.2. (Qumix).
A *qumix* is a density operator in $\mathfrak{D}$.

Needless to say, quregisters correspond to particular qumixs that are *pure states* (i.e. projections onto one-dimensional closed subspaces of a given $\otimes^n\mathbb{C}^n$). For any quregister $|\psi\rangle$, we will indicate by $P_{|\psi\rangle}$ the *pure density operator* represented by the projection onto the one-dimensional subspace spanned by the vector $|\psi\rangle$. The set of all pure density operators will be indicated by $\mathfrak{D}_{\mathfrak{R}}$.

Recalling the Born rule, we can now define the *probability-value* of any qumix.

DEFINITION 3.3. (Probability of a qumix).
For any qumix $\rho \in \mathfrak{D}(\otimes^n\mathbb{C}^n)$:

$$\mathtt{p}(\rho) = \mathtt{tr}(P_1^{(n)}\rho).$$

From an intuitive point of view, $\mathtt{p}(\rho)$ represents the probability that the information stocked by the qumix ρ is true. In the particular case where ρ is a pure density operator $P_{|\psi\rangle}$, determined by the qubit $|\psi\rangle = a_0|0\rangle + a_1|1\rangle$, we obtain that $\mathtt{p}(\rho) = |a_1|^2$.

For any quregister $|\psi\rangle$, we will write $\mathtt{p}(|\psi\rangle)$ instead of $\mathtt{p}(P_{|\psi\rangle})$.

4. Quantum logical gates

We will now introduce some examples of *quantum logical gates.* Generally, a quantum logical gate can be described as a unitary operator, assuming arguments and values in a product-Hilbert space $\otimes^n\mathbb{C}^2$. First of all we will study the so called *Petri-Toffoli gate* ([8] and [9]). It will be expedient to start by analysing the simplest case, where our Hilbert space has the form:

$$\otimes^3\mathbb{C}^2 = \mathbb{C}^2 \otimes \mathbb{C}^2 \otimes \mathbb{C}^2.$$

In such a case, the Petri-Toffoli gate transforms the vectors of $\otimes^3\mathbb{C}^2$ into vectors of $\otimes^3\mathbb{C}^2$. In order to stress that our operator is defined on the product space $\otimes^3\mathbb{C}^2$, we will indicate it by $T^{(1,1,1)}$. Since we want to define a unitary operator, it will be sufficient to determine its behaviour for the elements of the basis, having the form $|x\rangle \otimes |y\rangle \otimes |z\rangle$ (where $x, y, z \in \{0,1\}$).

DEFINITION 4.1. (The Petri-Toffoli gate $T^{(1,1,1)}$).
The *Petri-Toffoli gate* $T^{(1,1,1)}$ is the linear operator $T^{(1,1,1)} : \otimes^3\mathbb{C}^2 \to \otimes^3\mathbb{C}^2$ that is defined for any element $|x\rangle \otimes |y\rangle \otimes |z\rangle$ of the basis as follows:

$$T^{(1,1,1)}(|x\rangle \otimes |y\rangle \otimes |z\rangle) = |x\rangle \otimes |y\rangle \otimes |xy \oplus z\rangle,$$

where $\oplus$ represents the sum modulo 2.

From an intuitive point of view, it seems quite natural to "see" the gate $T^{(1,1,1)}$ as a kind of self-reversible "truth-table" that transforms triples of *zeros* and *ones* into triples of *zeros* and *ones*. The "table" we obtain is the following:

$$\begin{array}{ccc}
|0,0,0\rangle & \mapsto & |0,0,0\rangle \\
|0,0,1\rangle & \mapsto & |0,0,1\rangle \\
|0,1,0\rangle & \mapsto & |0,1,0\rangle \\
|0,1,1\rangle & \mapsto & |0,1,1\rangle \\
|1,0,0\rangle & \mapsto & |1,0,0\rangle \\
|1,0,1\rangle & \mapsto & |1,0,1\rangle \\
|1,1,0\rangle & \mapsto & |1,1,1\rangle \\
|1,1,1\rangle & \mapsto & |1,1,0\rangle
\end{array}$$

In the first six cases, $T^{(1,1,1)}$ behaves like the identity operator; in the last two cases, instead, our gate transforms the last element of the triple into the opposite element (0 is transformed into 1 and 1 transformed into 0).

One can easily show that $T^{(1,1,1)}$ has been well defined for our aims: one is dealing with an operator that is not only linear but also unitary.

By using $T^{(1,1,1)}$, we can introduce a convenient notion of *conjunction*. Our conjunction, which will be indicated by **And**, is characterized as a function whose arguments are pairs of vectors in $\mathbb{C}^2$ and whose values are vectors of the product space $\otimes^3\mathbb{C}^2$.

DEFINITION 4.2. (**And**).
For any $|\varphi\rangle \in \mathbb{C}^2$ and any $|\psi\rangle \in \mathbb{C}^2$:

$$\mathtt{And}\,(|\varphi\rangle, |\psi\rangle) := T^{(1,1,1)}\,(|\varphi\rangle \otimes |\psi\rangle \otimes |0\rangle)\,.$$

Clearly, the qubit $|0\rangle$ behaves here as an "ancilla".

Let us check that **And** represents a good generalization of the corresponding classical truth-function. For the arguments $|0\rangle$ and $|1\rangle$ we will obtain the following "truth-table":

$$\begin{array}{lcl}
(|0\rangle, |0\rangle) & \mapsto & T^{(1,1,1)}(|0\rangle \otimes |0\rangle \otimes |0\rangle) = |0\rangle \otimes |0\rangle \otimes |0\rangle \\
(|0\rangle, |1\rangle) & \mapsto & T^{(1,1,1)}(|0\rangle \otimes |1\rangle \otimes |0\rangle) = |0\rangle \otimes |1\rangle \otimes |0\rangle \\
(|1\rangle, |0\rangle) & \mapsto & T^{(1,1,1)}(|1\rangle \otimes |0\rangle \otimes |0\rangle) = |1\rangle \otimes |0\rangle \otimes |0\rangle \\
(|1\rangle, |1\rangle) & \mapsto & T^{(1,1,1)}(|1\rangle \otimes |1\rangle \otimes |0\rangle) = |1\rangle \otimes |1\rangle \otimes |1\rangle
\end{array}$$

One immediately realizes the difference with respect to the classical case. The classical truth-table represents a typical irreversible transformation:

$$\begin{array}{lcl}
(0,0) & \mapsto & 0 \\
(0,1) & \mapsto & 0 \\
(1,0) & \mapsto & 0 \\
(1,1) & \mapsto & 1
\end{array}$$

The arguments of the function determine the value, but not the other way around. As is well known, irreversibility generally brings about dissipation of information. Mathematically, however, any Boolean function $f : \{0,1\}^n \to \{0,1\}^m$ can be transformed into a *reversible* function $\hat{f} : \{0,1\}^n \times \{0,1\}^m \to \{0,1\}^n \times \{0,1\}^m$ in the following way:

$$\forall u \in \{0,1\}^n\, \forall v \in \{0,1\}^m : \hat{f}(u,v) = (u, v \oplus f(u)),$$

where $\oplus$ is the sum modulo 2 pointwise defined. The function that is obtained by making reversible the irreversible classical "and" corresponds to the Petri-Toffoli gate. The classical "and" is then recovered by fixing the third input bit to 0.

Accordingly, the three arguments $(0,0)$, $(0,1)$, $(1,0)$ turn out to correspond to three distinct values, represented by the triples $(0,0,0)$, $(0,1,0)$, $(1,0,0)$. The price we have paid in order to obtain a reversible situation is the increasing of the complexity of our Hilbert space. The function **And** associates to pairs of arguments, belonging to the two-dimensional space $\mathbb{C}^2$, values belonging to the space $\otimes^3\mathbb{C}^2$ (whose dimension is 2^3).

All this happens in the simplest situation, when one is only dealing with elements of the basis (in other words, with precise pieces of information). Let us examine the case where the function **And** is applied to arguments that are superpositions of the basis-elements in the space $\mathbb{C}^2$. Consider the following *qubit* pair:

$$|\psi\rangle = a_0|0\rangle + a_1|1\rangle\,,\ |\varphi\rangle = b_0|0\rangle + b_1|1\rangle.$$

By applying the definitions of **And** and of $T^{(1,1,1)}$, we obtain:

$$\mathtt{And}(|\psi\rangle, |\varphi\rangle) = a_1b_1|1,1,1\rangle + a_1b_0|1,0,0\rangle + a_0b_1|0,1,0\rangle + a_0b_0|0,0,0\rangle.$$

This result suggests a quite natural logical interpretation. The four basis-elements that occur in our superposition-vector correspond to the four cases of the truth-table for the classical conjunction:

$$(1,1,1), (1,0,0), (0,1,0), (0,0,0).$$

However here, unlike the classical situation, each case is accompanied by a complex number, which represents a characteristic quantum *amplitude*. By applying the "Born rule" we will obtain the following interpretation: $|a_1b_1|^2$ represents the probability-value that both the *qubit*-arguments are equal to $|1\rangle$, and consequently their conjunction is $|1\rangle$. Similarly in the other three cases.

So far we have considered a very special situation, characterized by a Hilbert space having the form $\otimes^3\mathbb{C}^2$. However, our procedure can be easily generalized. The Petri-Toffoli gate can be defined in any Hilbert space having the form:

$$(\otimes^n\mathbb{C}^2) \otimes (\otimes^m\mathbb{C}^2) \otimes \mathbb{C}^2 (= \otimes^{n+m+1}\mathbb{C}^2).$$

DEFINITION 4.3. (The Petri-Toffoli gate $T^{(n,m,1)}$).
The *Petri-Toffoli gate* $T^{(n,m,1)}$ is the linear operator

$$T^{(n,m,1)} : (\otimes^n\mathbb{C}^2) \otimes (\otimes^m\mathbb{C}^2) \otimes \mathbb{C}^2 \;\rightarrow\; (\otimes^n\mathbb{C}^2) \otimes (\otimes^m\mathbb{C}^2) \otimes \mathbb{C}^2,$$

that is defined for any element $|x_1, \ldots, x_n\rangle \otimes |y_1, \ldots, y_m\rangle \otimes |z\rangle$ of the computational basis of $\otimes^{n+m+1}\mathbb{C}^2$ as follows:

$$\begin{aligned} &T^{(n,m,1)}(|x_1, \ldots, x_n\rangle \otimes |y_1, \ldots, y_m\rangle \otimes |z\rangle) \\ &= |x_1, \ldots, x_n\rangle \otimes |y_1, \ldots, y_m\rangle \otimes |x_n y_m \oplus z\rangle, \end{aligned}$$

where $\oplus$ represents the sum modulo 2.

On this basis one can immediately generalize our definition of **And**.

DEFINITION 4.4. (**And**).
For any $|\varphi\rangle \in \otimes^n\mathbb{C}^2$ and any $|\psi\rangle \in \otimes^m\mathbb{C}^2$:

$$\mathtt{And}\,(|\varphi\rangle, |\psi\rangle) := T^{(n,m,1)}\,(|\varphi\rangle \otimes |\psi\rangle \otimes |0\rangle)\,.$$

How to deal in this context with the concept of negation? A characteristic of quantum computation is the possibility of defining a plurality of negation-operations: some of them represent good generalizations of the classical negation. We will first consider the operator $\mathtt{Not}^{(\mathtt{n})}$ that simply inverts the value of the last element of any configuration of the space $\otimes^n\mathbb{C}^2$. Thus, if $|x_1, \ldots, x_n\rangle$ is a vector of the computational basis $\mathcal{B}^{(n)}$, the result of the application of $\mathtt{Not}^{(\mathtt{n})}$ to $|x_1, \ldots, x_n\rangle$ will be $|x_1, \ldots, x_{n-1}, 1 - x_n\rangle$.

DEFINITION 4.5. ($\mathtt{Not}^{(n)}$).
The negation-gate is the linear operator $\mathtt{Not}^{(n)}$ that is defined for any element $|x_1, \ldots, x_n\rangle$ of the computational basis of $\otimes^n\mathbb{C}^2$ as follows:

$$\mathtt{Not}^{(n)}(|x_1, \ldots, x_n\rangle) = |x_1, \ldots, x_{n-1}, 1 - x_n\rangle.$$

One can immediately check that $\mathtt{Not}^{(n)}$ represents a good generalization of the classical truth-table. Consider the basis-elements $|0\rangle$ and $|1\rangle$ of the space $\mathbb{C}^2$. In such a case we obtain:

$$\begin{aligned} &\mathtt{Not}^{(1)}(|1\rangle) = |0\rangle; \\ &\mathtt{Not}^{(1)}(|0\rangle) = |1\rangle. \end{aligned}$$

The matrix corresponding to $\mathtt{Not}^{(1)}$ is:

$$\begin{pmatrix} 0 & 1 \\ 1 & 0 \end{pmatrix}$$

Both the negation-gate and the Petri-Toffoli gate can be uniformly defined on the set $\mathfrak{R}$ of all quregisters in the expected way:

$$\texttt{Not}(|\psi\rangle) := \texttt{Not}^{(n)}(|\psi\rangle), \qquad \text{if } |\psi\rangle \in \otimes^n\mathbb{C}^2;$$
$$T(|\psi\rangle \otimes |\varphi\rangle \otimes |\chi\rangle) := T^{(n,m,1)}(|\psi\rangle \otimes |\varphi\rangle \otimes |\chi\rangle),$$
$$\text{if } |\psi\rangle \in \otimes^n\mathbb{C}^2,\ |\varphi\rangle \in \otimes^m\mathbb{C}^2 \text{ and } |\chi\rangle \in \mathbb{C}^2.$$

Finally, how to introduce a reasonable disjunction? A gate `Or` can be naturally defined in terms of `And` and `Not` via de Morgan.

DEFINITION 4.6. (`Or`).
For any quregisters $|\varphi\rangle$ and $|\psi\rangle$:

$$\texttt{Or}(|\varphi\rangle, |\psi\rangle) = \texttt{Not}\,(\texttt{And}\,(\texttt{Not}(|\varphi\rangle), \texttt{Not}(|\psi\rangle)))\,.$$

At first sight, `And` and `Or` may look as *irreversible transformations*. However, it is important to recall that, in this framework, $\texttt{And}(|\psi\rangle, |\varphi\rangle)$ should be regarded as a mere metalinguistic abbreviation for $T(|\psi\rangle \otimes |\varphi\rangle \otimes |0\rangle)$ (where T is reversible). Similarly `Or`.

The quantum logical gates we have considered so far are, in a sense, "semiclassical". A quantum logical behaviour only emerges in the case where our gates are applied to superpositions. When restricted to classical registers, our gates turn out to behave as classical truth-functions. We will now investigate *genuine quantum gates* that may transform classical registers into quregisters that are superpositions.

One of the most significant genuine quantum gates is the *square root of the negation*, which will be indicated by $\sqrt{\texttt{Not}}$. As suggested by the name, the characteristic property of the gate $\sqrt{\texttt{Not}}$ is the following: for any quregister $|\psi\rangle$,

$$\sqrt{\texttt{Not}}(\sqrt{\texttt{Not}}(|\psi\rangle)) = \texttt{Not}(|\psi\rangle).$$

In other words: applying twice the square root of the negation "means" negating.

Interestingly enough, the gate $\sqrt{\texttt{Not}}$ has some natural physical models (and implementations). As an example, consider an idealized atom with a single electron and two energy levels: a *ground state* (identified with $|0\rangle$) and an *excited state* (identified with $|1\rangle$). By shining a pulse of light of appropriate intensity, duration and wavelength, it is possible to force the electron to change energy level. As a consequence, the state (bit) $|0\rangle$ is transformed into the state (bit) $|1\rangle$, and viceversa:

$$|0\rangle \mapsto |1\rangle;\ |1\rangle \mapsto |0\rangle.$$

We have obtained a typical physical model for the gate $\mathtt{Not}$. Now, by using a light pulse of half the duration as the one needed to perform the $\mathtt{Not}$ operation, we effect a half-flip between the two logical states. The state of the atom after the half pulse is neither $|0\rangle$ nor $|1\rangle$, but rather a superposition of both states. As observed by Deutsch, Ekert, Lupacchini ([6]):

> Logicians are now entitled to propose a new logical operation $\sqrt{\mathtt{Not}}$. Why? Because a faithful physical model for it exists in nature.

The physical models of the gate $\sqrt{\mathtt{Not}}$ naturally suggest the following logical interpretation: $\sqrt{\mathtt{Not}}$ represents a kind of "tentative negation". By applying twice our "attempt" to negate, we obtain a full negation.

Interestingly enough, the gate $\sqrt{\mathtt{Not}}$ seems to have also some linguistic "models". For instance, consider the French language. Put:

$$\sqrt{\mathtt{Not}} = \text{“ne”} = \text{“pas”}.$$

We obtain:

$$\sqrt{\mathtt{Not}}\sqrt{\mathtt{Not}} = \text{“ne...pas”} = \mathtt{Not}.$$

Needless to observe, our linguistic example is only a partial model of the gate $\sqrt{\mathtt{Not}}$. In French, neither the expression "il ne pleut" nor the expression "il pleut pas" are grammatically correct sentences. And in the spoken language "il pleut pas" is simply used as an abbreviation for the correct "il ne pleut pas". In quantum computation, instead, for any quregister $|\psi\rangle$, the vector $\sqrt{\mathtt{Not}}(|\psi\rangle)$ is a quregister that is essentially different from the quregister $\mathtt{Not}(|\psi\rangle)$.

Let us now give the mathematical definition of $\sqrt{\mathtt{Not}}$.

DEFINITION 4.7. (The square root of the negation).
The square root of the negation on $\otimes^n\mathbb{C}^2$ is the linear operator $\sqrt{\mathtt{Not}}^{(n)}$ such that for every element $|x_1,\ldots,x_n\rangle$ of the computational basis $\mathcal{B}^{(n)}$:

$$\sqrt{\mathtt{Not}}^{(n)}(|x_1,\ldots,x_n\rangle) = |x_1,\ldots,x_{n-1}\rangle \otimes (\frac{1+i}{2}|x_n\rangle + \frac{1-i}{2}|1-x_n\rangle)$$

(where i is the imaginary unit).

In other words, $\sqrt{\mathtt{Not}}^{(n)}$ transforms the last element x_n of any configuration $|x_1,\ldots,x_n\rangle$ into the element $\frac{1+i}{2}|x_n\rangle + \frac{1-i}{2}|1-x_n\rangle$. As a consequence, for the two bits $|0\rangle$ and $|1\rangle$ ("living in the space $\mathbb{C}^2$) we obtain:

$$\sqrt{\mathtt{Not}}^{(1)}(|0\rangle) = \frac{1+i}{2}|0\rangle + \frac{1-i}{2}|1\rangle;$$

$$\sqrt{\mathtt{Not}}^{(1)}(|1\rangle) = \frac{1-i}{2}|0\rangle + \frac{1+i}{2}|1\rangle.$$

One can easily show that $\sqrt{\mathtt{Not}}^{(n)}$ is a unitary operator, which satisfies the following condition:

$$\text{for any } |\psi\rangle \in \otimes^n\mathbb{C}^2,\ \sqrt{\mathtt{Not}}^{(n)}(\sqrt{\mathtt{Not}}^{(n)}(|\psi\rangle)) = \mathtt{Not}^{(n)}(|\psi\rangle).$$

In other words, applying twice the square root of the negation means negating.

It turns out that the matrix associated to $\sqrt{\mathtt{Not}}^{(1)}$ is

$$\begin{pmatrix} \frac{1+i}{2} & \frac{1-i}{2} \\ \frac{1-i}{2} & \frac{1+i}{2} \end{pmatrix}$$

Like the negation, also the square root of the negation can be uniformly defined on the set $\mathfrak{R}$ of all quregisters:

$$\sqrt{\mathtt{Not}}(|\psi\rangle) := \sqrt{\mathtt{Not}}^{(n)}(|\psi\rangle), \qquad \text{if } |\psi\rangle \in \otimes^n\mathbb{C}^2.$$

As expected, the square root of the negation has no Boolean counterpart.

LEMMA 4.8. *There is no function $f : \{0,1\} \to \{0,1\}$ such that for any $x \in \{0,1\} : f(f(x)) = 1 - x$.*

PROOF. Suppose, by contradiction, that such a function f exists. Two cases are possible: (i) $f(0) = 0$; (ii) $f(0) = 1$.
(i) By hypothesis, $f(0) = 0$. Thus, $1 = f(f(0)) = f(0) = 0$, contradiction.
(ii) By hypothesis, $f(0) = 1$. Thus, $1 = f(f(0)) = f(1)$. Hence, $f(0) = f(1)$. Therefore, $1 = f(f(0)) = f(f(1)) = 0$, contradiction. ■

Interestingly enough, $\sqrt{\mathtt{Not}}$ does not even have any fuzzy counterpart, represented by a continuous function ([5]).

LEMMA 4.9. *There is no continuous function $f : [0,1] \to [0,1]$ such that for any $x \in [0,1] : f(f(x)) = 1 - x$.*

PROOF. Suppose, by contradiction, that such a function f exists. First, we prove that $f(\frac{1}{2}) = \frac{1}{2}$. By hypothesis, $f(f(\frac{1}{2})) = 1 - \frac{1}{2} = \frac{1}{2}$. Hence, $f(f(f(\frac{1}{2}))) = f(\frac{1}{2})$. Thus, $1 - f(\frac{1}{2}) = f(\frac{1}{2})$. Therefore, $f(\frac{1}{2}) = \frac{1}{2}$. Consider now $f(0)$. One can easily show: $f(0) \neq 0$ and $f(0) \neq 1$. Clearly, $f(0) \neq \frac{1}{2}$ since otherwise we would obtain $1 = f(f(0)) = f(\frac{1}{2}) = \frac{1}{2}$. Thus, only two cases are possible: (i) $0 < f(0) < \frac{1}{2}$; (ii) $\frac{1}{2} < f(0) < 1$.
(i) By hypothesis, $0 < f(0) < \frac{1}{2} < 1 = f(f(0))$. Consequently, by continuity, $\exists x \in (0, f(0))$ such that $\frac{1}{2} = f(x)$. Accordingly, $\frac{1}{2} = f(\frac{1}{2}) = f(f(x)) = 1-x$. Hence, $x = \frac{1}{2}$, which contradicts $x < f(0) < \frac{1}{2}$.
(ii) By hypothesis, $f(\frac{1}{2}) = \frac{1}{2} < f(0) < 1 = f(f(0))$. By continuity, $\exists x \in (\frac{1}{2}, f(0))$ such that $f(x) = f(0)$. Thus, $1 - x = f(f(x)) = f(f(0)) = 1$. Hence, $x = 0$, which contradicts $x > \frac{1}{2}$. ■

The gates considered so far can be naturally generalized to qumixs. When our gates will be applied to density operators, we will write: $\mathtt{NOT}$, $\sqrt{\mathtt{NOT}}$, $\mathtt{AND}$, $\mathtt{OR}$ (instead of $\mathtt{Not}$, $\sqrt{\mathtt{Not}}$, $\mathtt{And}$, $\mathtt{Or}$).

DEFINITION 4.10. (The negation).
For any qumix $\rho \in \mathfrak{D}(\otimes^n \mathbb{C}^2)$,

$$\mathtt{NOT}^{(n)}\rho = \mathtt{Not}^{(n)}\rho\mathtt{Not}^{(n)}.$$

DEFINITION 4.11. (The square root of the negation).
For any qumix $\rho \in \mathfrak{D}(\otimes^n \mathbb{C}^2)$,

$$\sqrt{\mathtt{NOT}}^{(n)}\rho = \sqrt{\mathtt{Not}}^{(n)}\rho\sqrt{\mathtt{Not}}^{(n)\,*}$$

(where $\sqrt{\mathtt{Not}}^{(n)\,*}$ is the adjoint of $\sqrt{\mathtt{Not}}^{(n)}$).

It is easy to see that for any $n \in \mathbb{N}^+$, both $\mathtt{NOT}^{(n)}(\rho)$ and $\sqrt{\mathtt{NOT}}^{(n)}(\rho)$ are qumixs of $\mathfrak{D}(\otimes^n \mathbb{C}^2)$. Further: $\mathtt{NOT}^{(n)}\mathtt{NOT}^{(n)} = I^{(n)}$.

DEFINITION 4.12. (The conjunction).
Let $\rho \in \mathfrak{D}(\otimes^n \mathbb{C}^2)$ and $\sigma \in \mathfrak{D}(\otimes^m \mathbb{C}^2)$.

$$\mathtt{AND}(\rho, \sigma) = T^{(n,m,1)}(\rho \otimes \sigma \otimes P_0^{(1)})T^{(n,m,1)}.$$

Like in the quregister-case, the gates $\mathtt{NOT}$, $\sqrt{\mathtt{NOT}}$, $\mathtt{AND}$, $\mathtt{OR}$ can be uniformly defined on the set $\mathfrak{D}$ of all qumixs.

The following theorem sums up some basic properties of our gates:

THEOREM 4.13.

(i) $\sqrt{\texttt{NOT}}\,\sqrt{\texttt{NOT}}\,\rho = \texttt{NOT}\rho$;

(ii) $\texttt{p}(\texttt{NOT}\,\rho) = 1 - \texttt{p}(\rho)$;

(iii) $\texttt{p}(\sqrt{\texttt{NOT}}\,\texttt{NOT}\,\rho) = \texttt{p}(\texttt{NOT}\,\sqrt{\texttt{NOT}}\,\rho)$;

(iv) $\texttt{p}(\texttt{AND}(\rho, \sigma)) = \texttt{p}(\rho)\texttt{p}(\sigma)$;

(v) $\texttt{p}(\sqrt{\texttt{NOT}}\,\texttt{AND}(\rho, \sigma)) = \frac{1}{2}$.

PROOF. [7] and [3] ■

5. Reversible and irreversible quantum computational structures

An interesting feature of the qumix system is the following: any real number $\lambda \in [0,1] \subset \mathbb{R}$ uniquely determines a qumix $\rho_\lambda^{(n)}$ (for any $n \in \mathbb{N}^+$):

$$\rho_\lambda^{(n)} := (1-\lambda)k_n P_0^{(n)} + \lambda k_n P_1^{(n)}. \tag{i}$$

Clearly, $\rho_\lambda^{(n)} \in \mathfrak{D}(\otimes^n \mathbb{C}^2)$. From an intuitive point of view, $\rho_\lambda^{(n)}$ represents a *mixture of pieces of information* that might correspond to the *Truth* with probability λ.

From a physical point of view, $\rho_\lambda^{(n)}$ corresponds to a particular preparation of the system such that the quantum system might be in the state $k_n P_0^{(n)}$ with probability $1-\lambda$ and in the state $k_n P_1^{(n)}$ with probability λ. It is worthwhile recalling that the random polarized states of the photon are represented by the density operator $\rho_{1/2}^{(1)} = \dfrac{1}{2} I^{(1)}$.

Two important properties of the qumix $\rho_\lambda^{(n)}$ are described by the following lemma:

LEMMA 5.1.

(i) $\forall n \in \mathbb{N}^+ \, \forall \lambda \in [0,1]$: $\texttt{p}(\rho_\lambda^{(n)}) = \lambda$;

(ii) $\texttt{p}(\sqrt{\texttt{NOT}}\rho_\lambda^{(n)}) = \dfrac{1}{2}$.

PROOF. [3] ■

We will now introduce two interesting relations that can be defined on the set of all qumixs. Both of them turn out to be a preorder-relation. We will speak of *weak* and of *strong preorder*, respectively.

DEFINITION 5.2. (Weak preorder).

$$\rho \leq \sigma \text{ iff } \mathtt{p}(\rho) \leq \mathtt{p}(\sigma).$$

DEFINITION 5.3. (Strong preorder).
$\rho \preceq \sigma$ iff the following conditions hold:

(i) $\mathtt{p}(\rho) \leq \mathtt{p}(\sigma)$;

(ii) $\mathtt{p}(\sqrt{\mathtt{NOT}}\sigma) \leq \mathtt{p}(\sqrt{\mathtt{NOT}}\rho)$.

Clearly, $\rho \preceq \sigma$ implies $\rho \leq \sigma$, but not the other way around. One immediately shows that both $\leq$ and $\preceq$ are reflexive and transitive, but not antisymmetric. Counterexamples can be easily found in $\mathfrak{D}(\mathbb{C}^2)$.

Consider now the following structure:

$$\left(\mathfrak{D}\,, \preceq, \mathtt{AND}\,, \mathtt{NOT}\,, \sqrt{\mathtt{NOT}}\,, P_0^{(1)}, P_1^{(1)}\,, \rho_{1/2}^{(1)}\right). \tag{ii}$$

We will call such a structure *the standard reversible quantum computational structure* (shortly *the RQC-structure*).

In the following we will generally write I, P_0, P_1 and $\rho_{1/2}$ instead of $I^{(1)}$, $P_0^{(1)}$,$P_1^{(1)}$, $\rho_{1/2}^{(1)}$. From an intuitive point of view, P_0, P_1 and $\rho_{1/2}$ represent privileged pieces of information that are *true*, *false*, *indeterminate*, respectively. Generally, our qumixs fail to satisfy *Duns Scotus law*: P_0 and P_1 are not the *minimum* and the *maximum* element of the RQC-structure. Hence, in this situation, it is interesting to isolate the elements that have a *Scotian* behaviour.

DEFINITION 5.4. (Down and up scotian qumixs).
Let ρ be a qumix of $\mathfrak{D}$.

(i) ρ is *down Scotian* iff $P_0 \preceq \rho$;

(ii) ρ is *up Scotian* iff $\rho \preceq P_1$;

(iii) ρ is *Scotian* iff ρ is both down and up Scotian.

LEMMA 5.5.

(i) $\rho \preceq \sqrt{\mathtt{NOT}}\, P_1$ *iff* $\mathtt{p}(\rho) \leq \frac{1}{2}$;

(ii) $\sqrt{\mathtt{NOT}}\, P_0 \preceq \rho$ *iff* $\mathtt{p}(\rho) \geq \frac{1}{2}$.

PROOF. [3] ∎

THEOREM 5.6.

(i) ρ *is down Scotian iff* $\mathtt{p}(\sqrt{\mathtt{NOT}}\rho) \leq \frac{1}{2}$ *iff* $\sqrt{\mathtt{NOT}}\rho \preceq \sqrt{\mathtt{NOT}}\, P_1$;

(ii) ρ *is up Scotian iff* $\frac{1}{2} \leq \mathtt{p}(\sqrt{\mathtt{NOT}}\rho)$ *iff* $\sqrt{\mathtt{NOT}} P_0 \preceq \sqrt{\mathtt{NOT}}\rho$;

(iii) ρ *is Scotian iff* $\mathtt{p}(\sqrt{\mathtt{NOT}}\rho) = \frac{1}{2}$;

(iv) $\forall n \in \mathbb{N}^+$: $k_n P_0^{(n)}, k_n P_1^{(n)}, \rho_{1/2}^{(n)}$ *are Scotian;*

(v) *For any* $\in \mathbb{N}^+$, *the set* $\mathfrak{D}(\otimes^n \mathbb{C}^2)$ *contains uncountably many Scotian density operators.*

PROOF. [3] ∎

The gates we have considered so far represent typical *reversible* logical operations. From a logical point of view, it might be interesting to consider also some *irreversible* operations. An important example is represented by a Łukasiewicz-like disjunction.

DEFINITION 5.7. (The Łukasiewicz disjunction).
Let $\tau \in \mathfrak{D}(\otimes^n \mathbb{C}^2)$ and $\sigma \in \mathfrak{D}(\otimes^m \mathbb{C}^2)$.

$$\tau \oplus \sigma := \rho^{(1)}_{\mathtt{p}(\tau)\oplus\mathtt{p}(\sigma)},$$

where $\oplus$ in $\mathtt{p}(\tau) \oplus \mathtt{p}(\sigma)$ is the Łukasiewicz "truncated sum" defined on the real interval $[0,1]$ (i.e. $\mathtt{p}(\tau) \oplus \mathtt{p}(\sigma) = \min\{1, \mathtt{p}(\tau) + \mathtt{p}(\sigma)\}$) ([10]).

The following lemmas sum up some basic properties of the Łukasiewicz disjunction:

LEMMA 5.8.

(i)

$$\tau \oplus \sigma = \begin{cases} \rho^{(1)}_{\mathtt{p}(\tau)+\mathtt{p}(\sigma)}, & \textit{if } \mathtt{p}(\tau) + \mathtt{p}(\sigma) \leq 1; \\ P_1^{(1)}, & \textit{otherwise;} \end{cases}$$

(ii) $\mathtt{p}(\tau \oplus \sigma) = \mathtt{p}(\tau) \oplus \mathtt{p}(\sigma)$;

(iii) $\mathtt{p}(\sqrt{\mathtt{NOT}}(\tau \oplus \sigma)) = \frac{1}{2}$.

PROOF. [3] ■

LEMMA 5.9. *Let* $\rho \in \mathfrak{D}(\otimes^n \mathbb{C}^2)$.

(i) $\forall n \in \mathbb{N}^+$: $\rho \oplus k_n P_1^{(n)} = P_1^{(1)}$;

(ii) $\forall n \in \mathbb{N}^+$: $\rho \oplus k_n P_0^{(n)} = \rho^{(1)}_{\mathtt{p}(\rho)}$;

(iii) $\rho \oplus \mathtt{NOT}\rho = P_1^{(1)}$.

PROOF. Straightforward. ■

From Lemma 5.9 it follows that $\mathtt{p}(\rho \oplus k_n P_1^{(n)}) = 1$, $\mathtt{p}(\rho \oplus k_n P_0^{(n)}) = \mathtt{p}(\rho)$ and $\mathtt{p}(\rho \oplus \mathtt{NOT}\rho) = 1$.

The preorder $\preceq$ permits us to define on the set of all qumixs an equivalence relation in the expected way.

DEFINITION 5.10. (The strong equivalence relation).

$$\rho \approxeq \sigma \text{ iff } \rho \preceq \sigma \text{ and } \sigma \preceq \rho.$$

Clearly, $\approxeq$ is an equivalence relation. Let

$$[\mathfrak{D}]_{\approxeq} := \{[\rho]_{\approxeq} \, : \, \rho \in \mathfrak{D}\}.$$

We will omit $\approxeq$ in $[\rho]_{\approxeq}$ if no confusion is possible.

Unlike the qumixs (which are only preordered by $\preceq$), the equivalence-classes of $[\mathfrak{D}]_{\approxeq}$ can be partially ordered in a natural way.

DEFINITION 5.11.

$$[\rho] \preceq [\sigma] \text{ iff } \rho \preceq \sigma.$$

The relation $\preceq$ (which is well defined) is a partial order.

LEMMA 5.12.

(i) $\forall n \in \mathbb{N}^+$: $[P_1] = \left[k_n P_1^{(n)}\right]$;

(ii) $\forall n \in \mathbb{N}^+$: $[P_0] = \left[k_n P_0^{(n)}\right]$;

(iii) $\forall n \in \mathbb{N}^+ \forall \lambda \in [0,1]$: $\left[\rho_\lambda^{(1)}\right] = \left[\rho_\lambda^{(n)}\right]$.

PROOF. [3] ■

On this basis, one can naturally define on the set $[\mathfrak{D}]_{\approx}$ a conjunction, a negation, the square root of the negation, a Łukasiewicz disjunction:

DEFINITION 5.13. Let $\rho \in \mathfrak{D}(\otimes^n\mathbb{C}^2)$ and $\sigma \in \mathfrak{D}(\otimes^m\mathbb{C}^2)$.

(i) $[\rho]\mathtt{AND}[\sigma] = [\mathtt{AND}(\rho, \sigma)]$;

(ii) $\mathtt{NOT}[\rho] = [\mathtt{NOT}\rho]$;

(iii) $\sqrt{\mathtt{NOT}}[\rho] = [\sqrt{\mathtt{NOT}}\rho]$;

(iv) $[\rho] \oplus [\sigma] = [\rho \oplus \sigma]$.

LEMMA 5.14. *The operations of Definition 5.13 are well defined.*

PROOF. [3] ■

LEMMA 5.15.

(i) The operation AND *is associative and commutative;*

(ii) The operation $\oplus$ *is associative and commutative;*

(iii) $\mathtt{NOT}\,\mathtt{NOT}[\rho] = [\rho]$;

(iv) $\sqrt{\mathtt{NOT}}\,\sqrt{\mathtt{NOT}}[\rho] = \mathtt{NOT}[\rho]$;

(v) $\sqrt{\mathtt{NOT}}\,\mathtt{NOT}[\rho] = \mathtt{NOT}\sqrt{\mathtt{NOT}}[\rho]$.

PROOF. Straightforward. ■

Consider now the structure

$$\left([\mathfrak{D}]_{\approx}, \mathtt{AND}, \oplus, \mathtt{NOT}, \sqrt{\mathtt{NOT}}, [P_0]_{\approx}, [P_1]_{\approx}, [\rho_{1/2}]\right). \qquad \text{(iii)}$$

We will call such a structure *the standard irreversible quantum computational algebra* (shortly *the IQC-algebra*).

As happens in the case of $\preceq$, also the weak preorder $\leq$ permits us to define an equivalence relation, which will be called *weak equivalence relation.*

DEFINITION 5.16. (Weak equivalence relation).

$$\rho \equiv \sigma \text{ iff } \rho \leq \sigma \text{ and } \sigma \leq \rho.$$

Clearly, $\equiv$ is an equivalence relation. Let

$$[\mathfrak{D}]_{\equiv} := \{[\rho]_{\equiv} : \rho \in \mathfrak{D}\}.$$

Also $[\mathfrak{D}]_{\equiv}$ can be partially ordered in a natural way.

DEFINITION 5.17.

$$[\rho]_{\equiv} \leq [\sigma]_{\equiv} \text{ iff } \rho \leq \sigma.$$

One can easily show that the relation $\leq$ (which is well defined) is a partial order.

A conjunction, a Łukasiewicz disjunction, a negation (but not the square root of the negation!) can be naturally defined on $[\mathfrak{D}]_{\equiv}$.

DEFINITION 5.18. Let $\rho \in \mathfrak{D}(\otimes^n \mathbb{C}^2)$ and $\sigma \in \mathfrak{D}(\otimes^m \mathbb{C}^2)$.

(i) $[\rho]_{\equiv}\mathtt{AND}[\sigma]_{\equiv} = [\mathtt{AND}(\rho, \sigma)]_{\equiv}$;

(ii) $\mathtt{NOT}[\rho]_{\equiv} = [\mathtt{NOT}\rho]_{\equiv}$;

(iii) $[\rho]_{\equiv} \oplus [\sigma]_{\equiv} = [\rho \oplus \sigma]_{\equiv}$.

LEMMA 5.19. *The operations of Definition 5.18 are well defined.*

PROOF. [3] ∎

Unlike $\approxeq$, the relation $\equiv$ is not a congruence with respect to $\sqrt{\mathtt{NOT}}$. In fact, the following situation is possible: $[\rho]_{\equiv} = [\sigma]_{\equiv}$ and $[\sqrt{\mathtt{NOT}}\,\rho]_{\equiv} \neq [\sqrt{\mathtt{NOT}}\,\sigma]_{\equiv}$. Consider for example the following unit vectors of $\mathbb{C}^2$: $|\psi\rangle := \frac{\sqrt{2}}{2}|0\rangle + \frac{\sqrt{2}}{2}|1\rangle$ and $|\varphi\rangle := \frac{\sqrt{2}}{2}|0\rangle + \frac{1+i}{2}|1\rangle$.

Let $P_{|\psi\rangle}$ and $P_{|\varphi\rangle}$ be the projections onto the unidimensional spaces spanned by $|\psi\rangle$ and $|\varphi\rangle$, respectively. It turns out that $\mathtt{p}(P_{|\psi\rangle}) = \mathtt{p}(P_{|\varphi\rangle}) = \frac{1}{2}$. Accordingly, $[P_{|\psi\rangle}]_{\equiv} = [P_{|\varphi\rangle}]_{\equiv}$. However, $\mathtt{p}(\sqrt{\mathtt{NOT}}P_{|\psi\rangle}) = \frac{1}{2}$ and $\mathtt{p}(\sqrt{\mathtt{NOT}}P_{|\varphi\rangle})$ $= \frac{1}{2} - \frac{\sqrt{2}}{4} \approx 0.146447$. Consequently, $[P_{|\psi\rangle}]_{\approxeq} \neq [P_{|\varphi\rangle}]_{\approxeq}$.

An interesting relation between the weak and the strong preorder is described by the following theorem.

THEOREM 5.20. *For any* $\rho, \sigma \in \mathfrak{D}$*:*

$$[\rho]_{\equiv} \leq [\sigma]_{\equiv} \textit{ iff } [\rho]_{\approxeq}\, \mathtt{AND}\, [P_1]_{\approxeq} \preceq [\sigma]_{\approxeq}\, \mathtt{AND}\, [P_1]_{\approxeq}.$$

PROOF. [3] ∎

6. The Poincaré quantum computational structures

We will now restrict our analysis to the qumixs living in the two-dimensional space $\mathbb{C}^2$. As is well known, every density operator of $\mathfrak{D}(\mathbb{C}^2)$ has the following matrix representation:

$$\frac{1}{2}\left(I + r_1 X + r_2 Y + r_3 Z\right), \tag{iv}$$

where r_1, r_2, r_3 are real numbers such that $r_1^2 + r_2^2 + r_3^2 \leq 1$ and X, Y, Z are the Pauli matrices:

$$X = \begin{pmatrix} 0 & 1 \\ 1 & 0 \end{pmatrix} \quad Y = \begin{pmatrix} 0 & -i \\ i & 0 \end{pmatrix} \quad Z = \begin{pmatrix} 1 & 0 \\ 0 & -1 \end{pmatrix}.$$

It turns out that a density operator $\frac{1}{2}\left(I + r_1 X + r_2 Y + r_3 Z\right)$ is pure iff $r_1^2 + r_2^2 + r_3^2 = 1$. Consequently,

- Pure density operators are in 1 : 1 correspondence with the points of the surface of the Poincaré sphere;

- Proper mixtures are in 1 : 1 correspondence with the inner points of the Poincaré sphere.

Let ρ be a density operator of $\mathfrak{D}(\mathbb{C}^2)$. We will denote by $\bar{\rho}$ the point of the Poincaré sphere that is univocally associated to ρ.

Let (r_1, r_2, r_3) be a point of the Poincaré sphere. We will denote by $\widehat{(r_1, r_2, r_3)}$ the density operator univocally associated to (r_1, r_2, r_3).

LEMMA 6.1. *Let $\rho \in \mathfrak{D}(\mathbb{C}^2)$ such that $\bar{\rho} = (r_1, r_2, r_3)$. The following conditions hold:*

(i) $\mathtt{p}(\rho) = \frac{1 - r_3}{2}$ *and* $\mathtt{p}(\sqrt{\mathtt{NOT}}\,\rho) = \frac{1 - r_2}{2}$;

(ii) $0 < \mathtt{p}(\rho) < 1$ *and* $0 < \mathtt{p}(\sqrt{\mathtt{NOT}}\rho) < 1$, *whenever ρ is a proper mixture.*

PROOF.

(i) Easy computation;

(ii) Since proper mixtures are in 1:1 correspondence with inner points of the Poincaré sphere, we have: $r_1^2 + r_2^2 + r_3^2 < 1$. Hence: $r_2^2, r_3^2 < 1$ and $-1 < r_2, r_3 < 1$. Consequently: $0 < \mathtt{p}(\rho) = \frac{1 - r_3}{2} < 1$ and $0 < \mathtt{p}(\sqrt{\mathtt{NOT}}\rho) = \frac{1 - r_2}{2} < 1$. ■

An irreversible conjunction can be now naturally defined on the set of all qumixs of $\mathfrak{D}(\mathbb{C}^2)$.

DEFINITION 6.2. (The irreversible conjunction).
Let $\tau, \sigma \in \mathfrak{D}(\mathbb{C}^2)$.

$$\mathtt{IAND}(\tau, \sigma) := \rho^{(1)}_{\mathtt{p}(\tau)\mathtt{p}(\sigma)} \tag{v}$$

Interestingly enough, the density operator $\mathtt{IAND}(\tau, \sigma)$ can be described in terms of the *partial trace*. Suppose we have a compound physical system consisting of three subsystems, and let

$$\mathcal{H} = (\otimes^n \mathbb{C}^2) \otimes (\otimes^m \mathbb{C}^2) \otimes (\otimes^r \mathbb{C}^2)$$

be the Hilbert space associated to our system. Then, for any density operator ρ of $\mathcal{H}$, there is a unique density operator $\mathtt{tr}_{1,2}(\rho)$ that represents the *partial trace* of ρ on the space $\otimes^r \mathbb{C}^2$ (associated to the third subsystem). The two operators ρ and $\mathtt{tr}_{1,2}(\rho)$ are statistically equivalent with respect to the third subsystem. In other words, for any self-adjoint operator $A^{(r)}$ of $\otimes^r \mathbb{C}^2$:

$$\mathtt{tr}(\mathtt{tr}_{1,2}(\rho)\, A^{(r)}) = \mathtt{tr}(\rho\, (I^{(n)} \otimes I^{(m)} \otimes A^{(r)})).$$

The density operator $\mathtt{tr}_{1,2}(\rho)$, obtained by "tracing out" the first and the second subsystem, is also called the *reduced state* of ρ on the third subsystem.

One can prove that:

$$\mathtt{IAND}(\tau, \sigma) = \mathtt{tr}_{1,2}(\mathtt{AND}(\tau, \sigma)).$$

In other words, $\mathtt{IAND}(\tau, \sigma)$ represents the reduced state of $\mathtt{AND}(\tau, \sigma)$ on the third subsystem.

An interesting situation arises when both τ and σ are pure states. For instance, suppose that:

$$\tau = P_{|\psi\rangle} \text{ and } \sigma = P_{|\varphi\rangle},$$

where $|\psi\rangle$ and $|\varphi\rangle$ are proper qubits. Then,

$$\mathtt{AND}(\tau, \sigma) = P_{T^{(1,1,1)}(|\psi\rangle \otimes |\varphi\rangle \otimes |0\rangle)},$$

which is a pure state. At the same time, we have:

$$\mathtt{IAND}(\tau, \sigma) = \mathtt{tr}_{1,2}(P_{T^{(1,1,1)}(|\psi\rangle \otimes |\varphi\rangle \otimes |0\rangle)}),$$

which is a proper mixture. Apparently, when considering only the properties of the third subsystem, we loose some information. As a consequence, we obtain a final state that does not represent a maximal knowledge. As is well known, situations where the state of a compound system represents a maximal knowledge, while the states of the subsystems are proper mixtures, play an important role in the framework of entanglement-phenomena.

LEMMA 6.3.

(i) $\mathtt{IAND}$ *is associative and commutative;*

(ii) $\mathtt{IAND}(\rho, P_0) = P_0$;

(iii) $\mathtt{IAND}(\rho, P_1) = \rho_{\mathtt{p}(\rho)}$;

(iv) $\mathtt{p}(\mathtt{IAND}(\rho, \sigma)) = \mathtt{p}(\rho)\mathtt{p}(\sigma)$;

(v) $\mathtt{p}(\sqrt{\mathtt{NOT}}\,\mathtt{IAND}(\rho, \sigma)) = \frac{1}{2}$.

PROOF. Easy. ■

Consider now the structure

$$\left(\mathfrak{D}(\mathbb{C}^2)\,, \mathtt{IAND}\,, \oplus\,, \mathtt{NOT}\,, \sqrt{\mathtt{NOT}}\,, P_0, P_1\,, \rho_{1/2}\right). \qquad \text{(vi)}$$

We will call such a structure *the Poincaré irreversible quantum computational algebra* (shortly *the Poincaré IQC-algebra*).

We can refer to the relation $\restriction\approxeq$, representing the restriction of $\approxeq$ to $\mathfrak{D}(\mathbb{C}^2)$. For any $\rho \in \mathfrak{D}(\mathbb{C}^2)$, let

$$[\rho]_{\restriction\approxeq} := \left\{\sigma \in \mathfrak{D}(\mathbb{C}^2) \,:\, \rho \approxeq \sigma\right\}. \qquad \text{(vii)}$$

Further define

$$[\mathfrak{D}(\mathbb{C}^2)]_{\restriction\approxeq} := \left\{[\rho]_{\restriction\approxeq} \,:\, \rho \in \mathfrak{D}(\mathbb{C}^2)\right\}. \qquad \text{(viii)}$$

The operations $\mathtt{IAND}\,, \oplus\,, \mathtt{NOT}\,, \sqrt{\mathtt{NOT}}$ and the relation $\preceq$ can be defined on $[D(\mathbb{C}^2)]_{\restriction\approxeq}$ in the expected way.

Consider now the quotient-structure

$$\left([\mathfrak{D}(\mathbb{C}^2)]_{\restriction\approxeq}\,, \mathtt{IAND}\,, \oplus\,, \mathtt{NOT}\,, \sqrt{\mathtt{NOT}}\,, [P_0]_{\restriction\approxeq}\,, [P_1]_{\restriction\approxeq}\,, [\rho_{1/2}]_{\restriction\approxeq}\right).$$

We will call such a structure *the contracted Poincaré irreversible quantum computational algebra* (shortly *the contracted Poincaré IQC-algebra*).

THEOREM 6.4. *The contracted Poincaré IQC-algebra is isomorphic to the IQC-algebra, via the map* $g : [\mathfrak{D}(\mathbb{C}^2)]_{\restriction\approx} \to [\mathfrak{D}]_{\approx}$ *such that* $\forall \rho \in \mathfrak{D}(\mathbb{C}^2)$:

$$g([\rho]_{\restriction\approx}) = [\rho]_{\approx}. \tag{ix}$$

Further, for any $\rho, \sigma \in \mathfrak{D}(\mathbb{C}^2)$: $[\rho]_{\restriction\approx} \preceq [\sigma]_{\restriction\approx}$ *iff* $g([\rho]_{\restriction\approx}) \preceq g([\sigma]_{\restriction\approx})$.

PROOF. [3] ■

One can prove that any density operator ρ in $\mathfrak{D}(\mathbb{C}^2)$ is associated to a qubit $|\psi_\rho\rangle$ that is "statistically equivalent" to ρ. In a sense, $|\psi_\rho\rangle$ represents a "purification" of ρ.

LEMMA 6.5. *For any* $\rho \in \mathfrak{D}(\mathbb{C}^2)$ *such that* $\bar{\rho} = (r_1, r_2, r_3)$, *there exists a qubit* $|\psi_\rho\rangle$ *that satisfies the following conditions:*

(i) $\mathtt{p}(\rho) = \mathtt{p}(|\psi_\rho\rangle)$;

(ii) $\mathtt{p}(\sqrt{\mathtt{NOT}}\rho) = \mathtt{p}(\sqrt{\mathtt{Not}}(|\psi_\rho\rangle))$.

PROOF. Let $\rho \in \mathfrak{D}(\mathbb{C}^2)$ such that $\bar{\rho} = (r_1, r_2, r_3)$. Consider the vector

$$|\psi_\rho\rangle = \frac{\sqrt{1 - r_2^2 - r_3^2} - ir_2}{\sqrt{2(1 - r_3)}}|0\rangle + \sqrt{\frac{1 - r_3}{2}}|1\rangle,$$

which turns out to be a qubit. An easy computation shows that

$$\mathtt{p}(|\psi_\rho\rangle) = \frac{1 - r_3}{2} \quad \text{and} \quad \mathtt{p}(\sqrt{\mathtt{Not}}|\psi_\rho\rangle) = \frac{1 - r_2}{2}.$$

Thus by Lemma 6.1 (i), we can conclude that

$$\mathtt{p}(|\psi_\rho\rangle) = \mathtt{p}(\rho) \quad \text{and} \quad \mathtt{p}(\sqrt{\mathtt{Not}}(|\psi_\rho\rangle)) = \mathtt{p}(\sqrt{\mathtt{NOT}}\rho).$$

■

As an interesting application of Lemma 6.5 consider a density operator whose form is: $\rho_\lambda = (1 - \lambda)P_0 + \lambda P_1$. Then, by Lemma 6.5, there exists a qubit $|\psi_{\rho_\lambda}\rangle$ such that $\mathtt{p}(|\psi_{\rho_\lambda}\rangle) = \lambda$. It turns out that

$$|\psi_{\rho_\lambda}\rangle = \sqrt{1 - \lambda}|0\rangle + \sqrt{\lambda}|1\rangle.$$

THEOREM 6.6. *Let* $f : \mathfrak{D}^n \to \mathfrak{D}(\mathbb{C}^2)$. *Consider the set* $\mathfrak{Q}$ *of all qubits. Then, there exists a map*

$$f_{\mathfrak{Q}} : \mathfrak{Q}^n \to \mathfrak{Q}$$

such that for any qubits $|\psi_1\rangle, \ldots, |\psi_n\rangle$ *the following conditions hold:*

(i) $\mathtt{p}(f_{\mathfrak{Q}}(|\psi_1\rangle, \ldots, |\psi_n\rangle)) = \mathtt{p}(f(P_{|\psi_1\rangle}, \ldots, P_{|\psi_n\rangle}));$

(ii) $\mathtt{p}(\sqrt{\mathtt{Not}}(f_{\mathfrak{Q}}(|\psi_1\rangle, \ldots, |\psi_n\rangle))) = \mathtt{p}(\sqrt{\mathtt{NOT}} f(P_{|\psi_1\rangle}, \ldots, P_{|\psi_n\rangle})).$

PROOF. Let $|\psi_1\rangle, \ldots, |\psi_n\rangle \in \mathfrak{Q}$. Then $P_{|\psi_1\rangle}, \ldots, P_{|\psi_n\rangle} \in \mathfrak{D}$ and $f(P_{|\psi_1\rangle}, \ldots, P_{|\psi_n\rangle}) \in \mathfrak{D}(\mathbb{C}^2)$. By lemma 6.5, there exists a qubit $|\psi_{f(P_{|\psi_1\rangle}, \ldots, P_{|\psi_n\rangle})}\rangle$ such that $\mathtt{p}(f(P_{|\psi_1\rangle}, \ldots, P_{|\psi_n\rangle})) = \mathtt{p}(|\psi_{f(P_{|\psi_1\rangle}, \ldots, P_{|\psi_n\rangle})}\rangle)$ and $\mathtt{p}(\sqrt{\mathtt{NOT}} f(P_{|\psi_1\rangle}, \ldots, P_{|\psi_n\rangle})) = \mathtt{p}(\sqrt{\mathtt{Not}}(|\psi_{f(P_{|\psi_1\rangle}, \ldots, P_{|\psi_n\rangle})}\rangle))$. Thus, we can put $f_{\mathfrak{Q}}(|\psi_1\rangle, \ldots, |\psi_n\rangle) := |\psi_{f(P_{|\psi_1\rangle}, \ldots, P_{|\psi_n\rangle})}\rangle$. ■

As a significant application of Theorem 6.6, we obtain that a Łukasiewicz disjunction $\oplus_{\mathfrak{Q}}$ and an irreversible conjunction $\mathtt{IAnd}_{\mathfrak{Q}}$ can be naturally defined for any qubits $|\varphi\rangle = a_0|0\rangle + a_1|1\rangle$ and $|\chi\rangle = b_0|0\rangle + b_1|1\rangle$:

$$|\varphi\rangle \oplus_{\mathfrak{Q}} |\chi\rangle := \begin{cases} \sqrt{1 - |a_1|^2 - |b_1|^2}|0\rangle + \sqrt{|a_1|^2 + |b_1|^2}|1\rangle, & \text{if } |a_1|^2 + |b_1|^2 \leq 1; \\ |1\rangle, & \text{otherwise;} \end{cases}$$

$$\mathtt{IAnd}_{\mathfrak{Q}}(|\varphi\rangle, |\chi\rangle) := \sqrt{1 - |a_1 b_1|^2}|0\rangle + |a_1 b_1||1\rangle.$$

From an intuitive point of view, it is interesting to compare $\mathtt{IAnd}_{\mathfrak{Q}}(|\varphi\rangle, |\chi\rangle)$ with $\mathtt{IAND}(P_{|\varphi\rangle}, P_{|\chi\rangle})$ and $\mathtt{And}(|\varphi\rangle, |\chi\rangle)$. As we already know, $\mathtt{And}(|\varphi\rangle, |\chi\rangle)$ represents a pure state of a compound physical system (living in the space $\otimes^3\mathbb{C}^2$). Hence, one is dealing with a *maximal knowledge*, that also includes a maximal knowledge about the component systems (described by the pure states $|\varphi\rangle$ and $|\chi\rangle$, respectively). Further, the transformation $(|\varphi\rangle, |\chi\rangle) \mapsto \mathtt{And}(|\varphi\rangle, |\chi\rangle)$ is reversible. The state $\mathtt{IAND}(P_{|\varphi\rangle}, P_{|\chi\rangle})$, instead, is generally a proper mixture: a non-maximal knowledge about a (non-decomposed) system, representing the output of a computation, where the original information about the component systems (the inputs) has been lost. The transformation $(P_{|\varphi\rangle}, P_{|\chi\rangle}) \mapsto \mathtt{IAnd}(P_{|\varphi\rangle}, P_{|\chi\rangle})$ is typically irreversible. The state $\mathtt{IAnd}_{\mathfrak{Q}}(|\varphi\rangle, |\chi\rangle)$ represents a "purification" of $\mathtt{IAND}(P_{|\varphi\rangle}, P_{|\chi\rangle})$: one is dealing with a maximal knowledge about the output, that does not preserve the original information about the inputs.

7. Quantum computational logics

The quantum computational structures we have investigated suggest a natural semantics, based on the following intuitive idea: any sentence α of the language is interpreted as a convenient qumix, that generally depends on the logical form of α; at the same time, the logical connectives are interpreted as operations that either are *gates* or can be conveniently simulated by gates. We will consider a *minimal (sentential) quantum computational language* $\mathcal{L}$ that contains a privileged atomic sentence $\mathbf{f}$ (whose intended interpretation is the truth-value *Falsity*) and the following primitive connectives: the *negation* ($\neg$), the *square root of the negation* ($\sqrt{\neg}$), the *conjunction* ($\wedge$). Let $Form^{\mathcal{L}}$ be the set of all sentences of $\mathcal{L}$. We will use the following metavariables: $\mathbf{q}, \mathbf{r} \dots$ for atomic sentences and $\alpha, \beta, \dots$ for sentences. The connective disjunction ($\vee$) is supposed to be defined via de Morgan ($\alpha \vee \beta := \neg(\neg\alpha \wedge \neg\beta)$), while the privileged sentence $\mathbf{t}$ representing the *Truth* is defined as the negation of $\mathbf{f}$ ($\mathbf{t} := \neg\mathbf{f}$). This minimal quantum computational language can be extended to richer languages containing other primitive connectives (for instance, a connective corresponding to the Łukasiewicz irreversible disjunction $\oplus$) that we will not consider here.

We will first introduce the notion of *reversible quantum computational model* (shortly, *RQC-model*).

DEFINITION 7.1. (RQC-model).
A *RQC-model* of $\mathcal{L}$ is a function $\mathtt{Qum} : Form^{\mathcal{L}} \to \mathfrak{D}$ (which associates to any sentence α of the language a qumix):

$$\mathtt{Qum}(\alpha) := \begin{cases} \text{a density operator of } \mathfrak{D}(\mathbb{C}^2) & \text{if } \alpha \text{ is an atomic sentence;} \\ P_0 & \text{if } \alpha = \mathbf{f}; \\ \mathtt{NOT}\,\mathtt{Qum}(\beta) & \text{if } \alpha = \neg\beta; \\ \sqrt{\mathtt{NOT}}\,\mathtt{Qum}(\beta) & \text{if } \alpha = \sqrt{\neg}\beta; \\ \mathtt{AND}(\mathtt{Qum}(\beta), \mathtt{Qum}(\gamma)) & \text{if } \alpha = \beta \wedge \gamma. \end{cases}$$

The concept of RQC-model seems to have a "quasi intensional" feature: the meaning $\mathtt{Qum}(\alpha)$ of the sentence α partially reflects the logical form of α. In fact, the dimension of the Hilbert space where $\mathtt{Qum}(\alpha)$ "lives" depends on the number of occurrences of atomic sentences in α. According to the intended physical interpretation, $\mathtt{Qum}$ will associate to each occurrence of one and the same atomic subformula $\mathbf{q}$ of α the state ρ, that corresponds to an *identical preparation* of the quantum system.

DEFINITION 7.2. (The atomic complexity of α).

$$At(\alpha) = \begin{cases} 1 & \text{if } \alpha \text{ is an atomic sentence;} \\ At(\beta) & \text{if } \alpha = \neg\beta \text{ or } \alpha = \sqrt{\neg}\beta; \\ At(\beta) + At(\gamma) + 1 & \text{if } \alpha = \beta \wedge \gamma. \end{cases}$$

(Recall that: $\mathtt{Qum}(\beta \wedge \gamma) = T^{(n,m,1)}(\mathtt{Qum}(\beta) \otimes \mathtt{Qum}(\gamma) \otimes \mathtt{Qum}(\mathbf{f}))T^{(n,m,1)}$, if $\mathtt{Qum}(\beta) \in \otimes^n\mathbb{C}^2$ and $\mathtt{Qum}(\gamma) \in \otimes^m\mathbb{C}^2$).

LEMMA 7.3. *If $At(\alpha) = n$, then $\mathtt{Qum}(\alpha) \in \mathfrak{D}(\otimes^n\mathbb{C}^2)$.*

PROOF. Straightforward. ∎

Given a reversible quantum computational model $\mathtt{Qum}$, any sentence α has a natural probability-value, which can be also regarded as its *extensional meaning* with respect to $\mathtt{Qum}$.

DEFINITION 7.4. (The probability-value of α in a model $\mathtt{Qum}$).
$\mathtt{p}_{\mathtt{Qum}}(\alpha) := \mathtt{p}(\mathtt{Qum}(\alpha))$.

As we already know, qumixs are naturally preordered by two basic relations: the strong preorder $\preceq$ and the weak preorder $\leq$. This suggests to introduce two different consequence relations: the *strong* and the *weak consequence.*

DEFINITION 7.5. (Strong and weak consequence in a model $\mathtt{Qum}$).

1. A sentence β is a *strong consequence in a model* $\mathtt{Qum}$ of a sentence α $(\alpha \Vdash_{\mathtt{Qum}} \beta)$ iff $\mathtt{Qum}(\alpha) \preceq \mathtt{Qum}(\beta)$;

2. A sentence β is a *weak consequence in a model* $\mathtt{Qum}$ of a sentence α $(\alpha \Vvdash_{\mathtt{Qum}} \beta)$ iff $\mathtt{Qum}(\alpha) \leq \mathtt{Qum}(\beta)$.

The notions of *strong* and *weak truth*, *strong* and *weak logical consequence*, *strong* and *weak logical truth* can be now defined in the expected way.

DEFINITION 7.6. (Strong and weak truth in a model $\mathtt{Qum}$).

1. A sentence α is *strongly true in a model* $\mathtt{Qum}$ iff $\mathbf{t} \Vdash_{\mathtt{Qum}} \alpha$;

2. A sentence α is *weakly true in a model* $\mathtt{Qum}$ iff $\mathbf{t} \Vvdash_{\mathtt{Qum}} \alpha$.

DEFINITION 7.7. (Strong and weak logical consequence).

1. A sentence β is a *strong logical consequence of a sentence* α $(\alpha \Vdash \beta)$ iff for any model $\mathtt{Qum}$, $\alpha \Vdash_{\mathtt{Qum}} \beta$;

2. A sentence β is a *weak logical consequence* of a sentence α $(\alpha \Vvdash \beta)$ iff for any model $\mathtt{Qum}$, $\alpha \Vvdash_{\mathtt{Qum}} \beta$.

DEFINITION 7.8. (Strong and weak logical truth).

1. A sentence α is a *strong logical truth* iff for any model $\mathtt{Qum}$, α is strongly true in $\mathtt{Qum}$;

2. A sentence α is a *weak logical truth* iff for any model $\mathtt{Qum}$, α is weakly true in $\mathtt{Qum}$.

The strong and the weak logical consequence relations ($\Vdash$ and $\Vvdash$) permit us to characterize semantically two different forms of *quantum computational logic*. We will indicate by $\sqrt{\neg}$**QCL** the logic that is semantically characterized by the strong logical consequence relation $\models$. At the same time, the logic that is characterized by the weak consequence relation will be indicated by **QCL**. In other words, we have:

- β is a logical consequence of α in the logic $\sqrt{\neg}$**QCL** ($\alpha \models_{\sqrt{\neg}\mathbf{QCL}} \beta$) iff β is a strong logical consequence of α;
- β is a logical consequence of α in the logic **QCL** ($\alpha \models_{\mathbf{QCL}} \beta$) iff β is a weak logical consequence of α.

Clearly, $\sqrt{\neg}$**QCL** is a sublogic of **QCL**. For:

$$\alpha \Vdash \beta \text{ implies } \alpha \Vvdash \beta.$$

But not the other way around!

An interesting relation between the two logics $\sqrt{\neg}$**QCL** and **QCL** is described by the following theorem:

THEOREM 7.9. $\alpha \models_{\mathbf{QCL}} \beta$ *iff* $\alpha \wedge \mathbf{t} \models_{\sqrt{\neg}\mathbf{QCL}} \beta \wedge \mathbf{t}$.

PROOF. The theorem is a direct consequence of the definition of $\sqrt{\neg}$**QCL** and **QCL** and of Theorem 5.20. ∎

Let us now turn to the concept of *irreversible quantum computational model* (shortly, *IQC-model*), where the "quasi-intensional" character of reversible models is lost. In fact, the interpretation of a sentence in an irreversible model does not generally reflect the logical form of our sentence: the meaning of the *whole* does not include the meanings of the *parts*. In spite of this, we will prove that reversible and irreversible models turn out to characterize the same logic.

DEFINITION 7.10. (IQC-model).
An *IQC-model* of $\mathcal{L}$ is a function $\mathtt{Qum}^{\mathbb{C}^2} : Form^{\mathcal{L}} \to \mathfrak{D}(\mathbb{C}^2)$ (which associates to any sentence α of the language a qumix of $\mathbb{C}^2$):

$$\mathtt{Qum}^{\mathbb{C}^2}(\alpha) := \begin{cases} P_0 & \text{if } \alpha = \mathbf{f}; \\ \mathtt{NOT}\,\mathtt{Qum}^{\mathbb{C}^2}(\beta) & \text{if } \alpha = \neg\beta; \\ \sqrt{\mathtt{NOT}}\,\mathtt{Qum}^{\mathbb{C}^2}(\beta) & \text{if } \alpha = \sqrt{\neg}\beta; \\ \mathtt{IAND}(\mathtt{Qum}^{\mathbb{C}^2}(\beta), \mathtt{Qum}^{\mathbb{C}^2}(\gamma)) & \text{if } \alpha = \beta \wedge \gamma. \end{cases}$$

The (strong and weak) notions of *consequence, truth, logical consequence, logical truth* are defined like in the reversible case, *mutatis mutandis.* We will shortly speak of *strong irreversible logical consequence* and of *weak irreversible logical consequence.* The logic that is determined by the strong irreversible logical consequence will be indicated by $^{\sqrt{\neg}}$**IQCL**, while **IQCL** will represent the logic determined by the weak irreversible logical consequence.

We will now prove that $^{\sqrt{\neg}}$**QCL** and $^{\sqrt{\neg}}$**IQCL** are the same logic.

LEMMA 7.11. *Let* $\mathtt{Qum}$ *be a RQC-model and let* $\mathtt{Qum}^{\mathbb{C}^2}$ *be an IQC-model such that for any atomic sentence* $\mathbf{q}$*:* $\mathtt{Qum}(\mathbf{q}) = \mathtt{Qum}^{\mathbb{C}^2}(\mathbf{q})$*. Then, for any sentence* $\alpha \in Form^{\mathcal{L}}$*:*

$$\mathtt{p}(\mathtt{Qum}(\alpha)) = \mathtt{p}(\mathtt{Qum}^{\mathbb{C}^2}(\alpha)).$$

PROOF. The proof is by induction on the *length* (i.e. the number of connectives) of α.

(i) $\alpha = \mathbf{q}$. Trivial.

(ii) $\alpha = \neg\beta$.

$$\begin{aligned}
\mathtt{p}(\mathtt{Qum}(\alpha)) &= \mathtt{p}(\mathtt{Qum}(\neg\beta)) \\
&= \mathtt{p}(\mathtt{NOT}\,\mathtt{Qum}(\beta)) \\
&= 1 - \mathtt{p}(\mathtt{Qum}(\beta)) && \text{(Theorem 4.13(ii))} \\
&= 1 - \mathtt{p}(\mathtt{Qum}^{\mathbb{C}^2}(\beta)) && \text{(Induction hypothesis)} \\
&= \mathtt{p}(\mathtt{NOT}\,\mathtt{Qum}^{\mathbb{C}^2}(\beta)) \\
&= \mathtt{p}(\mathtt{Qum}^{\mathbb{C}^2}(\neg\beta)).
\end{aligned}$$

(iii) $\alpha = \sqrt{\neg}\beta$. The following subcases are possible: (iiia) $\beta = \mathbf{q}$; (iiib) $\beta = \gamma \wedge \delta$; (iiic) $\beta = \sqrt{\neg}\gamma$; (iiid) $\beta = \neg\gamma$.

(iiia) $\beta = \mathbf{q}$. The proof follows from the assumption $\mathtt{Qum}(\mathbf{q}) = \mathtt{Qum}^{\mathbb{C}^2}(\mathbf{q})$.

(iiib)

$$\begin{aligned}
\mathtt{p}(\mathtt{Qum}(\alpha)) &= \mathtt{p}(\mathtt{Qum}(\sqrt{\neg}\beta) \\
&= \mathtt{p}(\sqrt{\mathtt{NOT}}\,\mathtt{Qum}(\gamma \wedge \delta)) \\
&= \mathtt{p}(\sqrt{\mathtt{NOT}}\,\mathtt{AND}(\mathtt{Qum}(\gamma), \mathtt{Qum}(\delta))) \\
&= \frac{1}{2} && \text{(Theorem 4.13(v))}
\end{aligned}$$

By induction hypothesis and by Lemma 6.3(v), we have: $\mathtt{p}(\mathtt{Qum}^{\mathbb{C}^2}(\alpha)) = \mathtt{p}(\mathtt{Qum}^{\mathbb{C}^2}(\sqrt{\neg}(\gamma \wedge \delta))) = \mathtt{p}(\sqrt{\mathtt{NOT}}\,\mathtt{IAND}(\mathtt{Qum}^{\mathbb{C}^2}(\gamma), \mathtt{Qum}^{\mathbb{C}^2}(\delta))) = \frac{1}{2} = \mathtt{p}(\mathtt{Qum}(\alpha))$.
(iiic)

$$\begin{aligned}
\mathtt{p}(\mathtt{Qum}(\alpha)) &= \mathtt{p}(\mathtt{Qum}(\sqrt{\neg}\sqrt{\neg}\gamma)) \\
&= \mathtt{p}(\mathtt{NOT}\,\mathtt{Qum}(\gamma)) \\
&= 1 - \mathtt{p}(\mathtt{Qum}(\gamma)) && \text{(Theorem 4.13(ii))} \\
&= 1 - \mathtt{p}(\mathtt{Qum}^{\mathbb{C}^2}(\gamma)) && \text{(Induction hypothesis)} \\
&= \mathtt{p}(\mathtt{NOT}\,\mathtt{Qum}^{\mathbb{C}^2}(\gamma)) \\
&= \mathtt{p}(\sqrt{\mathtt{NOT}}\,\sqrt{\mathtt{NOT}}\,\mathtt{Qum}^{\mathbb{C}^2}(\gamma)) \\
&= \mathtt{p}(\mathtt{Qum}^{\mathbb{C}^2}(\sqrt{\neg}\sqrt{\neg}\gamma)) \\
&= \mathtt{p}(\mathtt{Qum}^{\mathbb{C}^2}(\alpha)).
\end{aligned}$$

(iiid) The proof follows from induction hypothesis and Theorem 4.13(iii).
(iv) $\alpha = \beta \wedge \gamma$.

$$\begin{aligned}
\mathtt{p}(\mathtt{Qum}(\alpha)) &= \mathtt{p}(\mathtt{Qum}(\beta \wedge \gamma)) \\
&= \mathtt{p}(\mathtt{Qum}(\beta))\mathtt{p}(\mathtt{Qum}(\gamma)) && \text{(Theorem 4.13 (iv))} \\
&= \mathtt{p}(\mathtt{Qum}^{\mathbb{C}^2}(\beta))\mathtt{p}(\mathtt{Qum}^{\mathbb{C}^2}(\gamma)) && \text{(Induction hypothesis)} \\
&= \mathtt{p}(\mathtt{IAND}(\mathtt{Qum}^{\mathbb{C}^2}(\beta), \mathtt{Qum}^{\mathbb{C}^2}(\gamma))) && \text{(Lemma 6.3 (iv))} \\
&= \mathtt{p}(\mathtt{Qum}^{\mathbb{C}^2}(\beta \wedge \gamma)).
\end{aligned}$$

■

COROLLARY 7.12.

(i) For any RQC-model Qum, *there exists an IQC-model* $\mathtt{Qum}^{\mathbb{C}^2}$ *such that for any* $\alpha \in Form^{\mathcal{L}}$*:*

$$\mathtt{p}(\mathtt{Qum}(\alpha)) = \mathtt{p}(\mathtt{Qum}^{\mathbb{C}^2}(\alpha));$$

(ii) For any IQC-model $\mathtt{Qum}^{\mathbb{C}^2}$ *there exists a RQC-model* Qum *such that for any* $\alpha \in Form^{\mathcal{L}}$*:*

$$\mathtt{p}(\mathtt{Qum}^{\mathbb{C}^2}(\alpha)) = \mathtt{p}(\mathtt{Qum}(\alpha)).$$

THEOREM 7.13. $\alpha \models_{\sqrt{\neg}\mathbf{QCL}} \beta$ *iff* $\alpha \models_{\sqrt{\neg}\mathbf{IQCL}} \beta$.

PROOF. The theorem is a direct consequence of Corollary 7.12. ■

Hence, $\sqrt{\neg}$**QCL** and $\sqrt{\neg}$**IQCL** are the same logic. Similarly one can prove that **QCL** and **IQCL** are the same logic.

So far we have considered (reversible and irreversible) models, where the meaning of any sentence is represented by a qumix. A natural question arises: do density operators have an essential role in characterizing the logics $\sqrt{\neg}$**QCL** and **QCL**? This question has a negative answer. In fact, one can prove that quregisters are sufficient for our logical aims in the case of the *minimal* quantum computational language $\mathcal{L}$.

Let us first introduce the notion of (reversible) *qubit-model* (which is the basic concept of the qubit-semantics described in [2] and [5]).

DEFINITION 7.14. (Reversible qubit-model).
A *reversible qubit-model* of $\mathcal{L}$ is a function $\texttt{Qub} : Form^{\mathcal{L}} \to \mathfrak{R}$ (which associates to any sentence α of the language a quregister):

$$\texttt{Qub}(\alpha) := \begin{cases} \text{a qubit in } \mathbb{C}^2 & \text{if } \alpha \text{ is an atomic sentence;} \\ |0\rangle & \text{if } \alpha = \mathbf{f}; \\ \texttt{Not}(\texttt{Qub}(\beta)) & \text{if } \alpha = \neg\beta; \\ \sqrt{\texttt{Not}}(\texttt{Qub}(\beta)) & \text{if } \alpha = \sqrt{\neg}\beta; \\ \texttt{And}(\texttt{Qub}(\beta), \texttt{Qub}(\gamma)) & \text{if } \alpha = \beta \wedge \gamma. \end{cases}$$

The notions of (weak and strong) *consequence, truth, logical consequence, logical truth* are defined like in the case of reversible qumix models, *mutatis mutandis*. We will write $\alpha \models^{\texttt{Qub}}_{\sqrt{\neg}\mathbf{QCL}} \beta$, when β is a strong logical consequence of α in the qubit-semantics. Similarly, we will write $\alpha \models^{\texttt{Qub}}_{\mathbf{QCL}} \beta$ when β is a weak logical consequence in the same semantics.

Instead of the class $\mathfrak{R}$ of all quregisters, we could equivalently refer to the class $\mathfrak{D}_{\mathfrak{R}}$ of all *pure density operators* having the form $P_{|\psi\rangle}$, where $|\psi\rangle$ is a quregister. One can easily show that $\mathfrak{D}_{\mathfrak{R}}$ is closed under the gates $\texttt{NOT}, \sqrt{\texttt{NOT}}, \texttt{AND}$. At the same time, $\mathfrak{D}_{\mathfrak{R}}$ is not closed under $\texttt{IAND}$, because (as we have seen) $\texttt{IAND}(P_{|\psi\rangle}, P_{|\varphi\rangle})$ is, generally, a proper mixture.

LEMMA 7.15. *Consider a reversible qubit-model* Qub *and let* Qum *be a RQC-model such that for any atomic sentence* $\mathbf{q}$, $\texttt{Qum}(\mathbf{q}) = P_{\texttt{Qub}(\mathbf{q})}$. *Then, for any sentences* α*:*

$$\texttt{Qum}(\alpha) \approxeq P_{\texttt{Qub}(\alpha)}.$$

PROOF. Easy. ■

On this basis we can prove that the qubit-semantics and the qumix-semantics characterize the same logics.

THEOREM 7.16.

(1) $\alpha \models_{\sqrt{\neg}\mathbf{QCL}} \beta$ *iff* $\alpha \models^{\mathtt{Qub}}_{\sqrt{\neg}\mathbf{QCL}} \beta$;

(2) $\alpha \models_{\mathbf{QCL}} \beta$ *iff* $\alpha \models^{\mathtt{Qub}}_{\mathbf{QCL}} \beta$.

PROOF.

(1)

(1.1) Suppose that $\alpha \models_{\sqrt{\neg}\mathbf{QCL}} \beta$. Then for any RQC-model $\mathtt{Qum}$: $\mathtt{Qum}(\alpha) \preceq \mathtt{Qum}(\beta)$. Hence, for any $\mathtt{Qum}$ such that $\mathtt{Qum}(\alpha)$ and $\mathtt{Qum}(\beta)$ are pure density operators: $\mathtt{Qum}(\alpha) \preceq \mathtt{Qum}(\beta)$.
Consequently, by Lemma 7.15, for any qubit-model $\mathtt{Qub}$: $\mathtt{Qub}(\alpha) \preceq \mathtt{Qub}(\beta)$.

(1.2) Suppose, by contradiction, that $\alpha \models^{\mathtt{Qub}}_{\sqrt{\neg}\mathbf{QCL}} \beta$ and $\alpha \not\models_{\sqrt{\neg}\mathbf{QCL}} \beta$. Then, by Theorem 7.13 there exists an irreversible model $\mathbf{Qum}^{\mathbb{C}^2}$ such that $\mathbf{Qum}^{\mathbb{C}^2}(\alpha) \not\preceq \mathbf{Qum}^{\mathbb{C}^2}(\beta)$. By Lemma 6.5, there exists a qubit-model $\mathtt{Qub}$ such that for any sentential letter $\mathbf{q}$: $\mathtt{p}(\mathtt{Qub}(\mathbf{q})) = \mathtt{p}(\mathbf{Qum}^{\mathbb{C}^2}(\mathbf{q}))$ and $\mathtt{p}(\sqrt{\mathtt{Not}}(\mathtt{Qub}(\mathbf{q}))) = \mathtt{p}(\sqrt{\mathtt{NOT}}\mathbf{Qum}^{\mathbb{C}^2}(\mathbf{q}))$.One can easily prove that for any α,$\mathtt{p}(\mathtt{Qub}(\alpha)) = \mathtt{p}(\mathbf{Qum}^{\mathbb{C}^2}(\alpha))$ and $\mathtt{p}(\sqrt{\mathtt{Not}}(\mathtt{Qub}(\alpha))) = \mathtt{p}(\sqrt{\mathtt{NOT}}\mathbf{Qum}^{\mathbb{C}^2}(\alpha))$ (by induction on the length of α).
Consequently, $\alpha \not\models^{\mathtt{Qub}}_{\sqrt{\neg}\mathbf{QCL}} \beta$, contradiction.

(2) Similarly.

■

Needless to observe, Theorem 7.16 does not imply that the qumix-semantics is useless. First of all, qubit-models and qumix-models might characterize different logics for languages that are richer than $\mathcal{L}$. At the same time, even in the case of our minimal language $\mathcal{L}$, qumixs represent an important tool in order to describe entanglement-phenomena.

A remarkable property of the logics $\sqrt{\neg}\mathbf{QCL}$ and $\mathbf{QCL}$ is the following: our logics do not admit any "genuine" logical truth. In other words, any sentence α, that does not contain the atomic sentence $\mathbf{f}$, cannot be a logical truth. By Theorem 7.16, is is sufficient to prove that no "genuine" logical truths exist in the framework of the qubit-semantics.

Let us first prove the following theorem ([5]):

THEOREM 7.17. *Let* $\mathtt{Qub}$ *be a reversible qubit-model and let* α *be any sentence. If* $\mathtt{p}(\mathtt{Qub}(\alpha)) \in \{0,1\}$, *then there is an atomic subformula* $\mathbf{q}$ *of* α *such that* $\mathtt{p}(\mathtt{Qub}(\mathbf{q})) \in \{0, \frac{1}{2}, 1\}$.

PROOF. Suppose that $\mathtt{p}(\mathtt{Qub}(\alpha)) \in \{0,1\}$. The proof is by induction on the length of α.
(i) α is an atomic sentence. The proof is trivial.
(ii) $\alpha = \neg\beta$. By Theorem 4.13(ii), $\mathtt{p}(\mathtt{Qub}(\alpha)) = 1 - \mathtt{p}(\mathtt{Qub}(\beta)) \in \{0,1\}$. The conclusion follows by induction hypothesis.
(iii) $\alpha = \sqrt{\neg}\beta$. By hypothesis and by Theorem 4.13(v), β cannot be a conjunction. Consequently, only the following cases are possible: (iiia) $\beta = \mathbf{q}$; (iiib) $\beta = \neg\gamma$; (iiic) $\beta = \sqrt{\neg}\gamma$.
(iiia) $\beta = \mathbf{q}$. By hypothesis, $\mathtt{p}(\sqrt{\neg}\beta) \in \{0,1\}$. Hence, $\sqrt{\mathtt{Not}}(\mathtt{Qub}(q)) = c|x\rangle$, where $|x\rangle \in \{|0\rangle, |1\rangle\}$ and $|c| = 1$. We have:
$\mathtt{Not}(\mathtt{Qub}(\mathbf{q})) = \sqrt{\mathtt{Not}}(\sqrt{\mathtt{Not}}(\mathtt{Qub}(\mathbf{q}))) = \sqrt{\mathtt{Not}}(c|x\rangle)$. One can easily show that $\mathtt{p}(\sqrt{\mathtt{Not}}(c|x\rangle)) = \frac{1}{2}$. As a consequence, $\mathtt{p}(\mathtt{Qub}(\neg\mathbf{q})) = \frac{1}{2} = \mathtt{p}(\mathtt{Qub}(\mathbf{q}))$.
(iiib) $\beta = \neg\gamma$. By Theorem 4.13(iii), $\mathtt{p}(\mathtt{Qub}(\sqrt{\neg}\neg\gamma)) = \mathtt{p}(\mathtt{Qub}(\neg\sqrt{\neg}\gamma)) = 1 - \mathtt{p}(\mathtt{Qub}(\sqrt{\neg}\gamma))$. The conclusion follows by induction hypothesis.
(iiic) $\beta = \sqrt{\neg}\gamma$. Then $\mathtt{p}(\mathtt{Qub}(\sqrt{\neg}\sqrt{\neg}\gamma)) = \mathtt{p}(\mathtt{Qub}(\neg\gamma)) = 1 - \mathtt{p}(\mathtt{Qub}(\gamma))$. The conclusion follows by induction hypothesis.
(iv) $\alpha = \beta \wedge \gamma$. By Theorem 4.13(iv), $\mathtt{p}(\mathtt{Qub}(\beta\wedge\gamma)) = \mathtt{p}(\mathtt{Qub}(\beta))\mathtt{p}(\mathtt{Qub}(\gamma)) \in \{0,1\}$. The conclusion follows by induction hypothesis. ■

As a consequence, we immediately obtain the following Corollary.

COROLLARY 7.18. *If* α *does not contain* $\mathbf{f}$, *then* α *is not a logical truth either of* $^{\sqrt{\neg}}$**QCL** *or of* **QCL**.

PROOF. Suppose, by contradiction, that α is a logical truth either of $^{\sqrt{\neg}}$**QCL** or of **QCL**. Then, in both cases, we obtain that: $\mathtt{p}(\alpha) = 1$. Let $\mathbf{q}_1, \ldots, \mathbf{q}_n$ be the atomic sentences occurring in α. Since α does not contain $\mathbf{f}$, there exists a qubit-model $\mathtt{Qub}$ such that for any i $(1 \le i \le n)$, $\mathtt{p}(\mathtt{Qub}(\mathbf{q}_i)) \notin \{0, \frac{1}{2}, 1\}$. Then, by Theorem 7.17, $\mathtt{p}(\mathtt{Qub}(\alpha)) \notin \{0,1\}$, contradiction. ■

We will now list some interesting logical consequences and rules that hold for the logics $^{\sqrt{\neg}}$**QCL** and **QCL**.We will indicate by $\alpha \models \beta$ the logical consequence relation that refers either to $^{\sqrt{\neg}}$**QCL** or to **QCL**. According to the usual notation we will write:

$$\frac{\alpha_1 \models \beta_1, \ldots, \alpha_n \models \beta_n}{\gamma \models \delta},$$

to be read as: if $\alpha_1 \models \beta_1, \ldots, \alpha_n \models \beta_n$, then $\gamma \models \delta$. We will also write $\alpha \equiv \beta$ as an abbreviation for: $\alpha \models \beta$ and $\beta \models \alpha$.

Since $\sqrt{\neg}$**QCL** is a sublogic of **QCL**, any logical consequence that holds in $\sqrt{\neg}$**QCL** will also hold in **QCL**. At the same time, some rules that hold in $\sqrt{\neg}$**QCL** may be violated in **QCL** (and, of course, viceversa).

THEOREM 7.19 (Logical consequences and rules of both $\sqrt{\neg}$**QCL** and **QCL**).

(1) $\alpha \models \alpha$;
(identity)

(2) $\frac{\alpha \models \beta,\ \beta \models \gamma}{\alpha \models \gamma}$;
(transitivity)

(3) $\alpha \equiv \neg\neg\alpha$;
(double negation)

(4) $\frac{\alpha \models \beta}{\neg\beta \models \neg\alpha}$;
(contraposition for the negation)

(5) $\sqrt{\neg}\sqrt{\neg}\alpha \equiv \neg\alpha$;
(the double square root of the negation principle)

(6) $\neg\sqrt{\neg}\alpha \equiv \sqrt{\neg}\neg\alpha$;
(permutation of the negations)

(7) $\sqrt{\neg}\mathbf{f} \models \sqrt{\neg}\mathbf{t}$;
(a "tentative negation" of the falsity implies a "tentative negation" of the truth)

(8) $\alpha \wedge \beta \equiv \beta \wedge \alpha, \quad \alpha \vee \beta \equiv \beta \vee \alpha$;
(commutativity)

(9) $\alpha \wedge (\beta \wedge \gamma) \equiv (\alpha \wedge \beta) \wedge \gamma, \quad \alpha \vee (\beta \vee \gamma) \equiv (\alpha \vee \beta) \vee \gamma;$
(associativity)

(10) $\neg(\alpha \wedge \beta) \equiv \neg\alpha \vee \neg\beta, \quad \neg(\alpha \vee \beta) \equiv \neg\alpha \wedge \neg\beta;$
(de Morgan)

(11) $\alpha \wedge (\beta \vee \gamma) \models (\alpha \wedge \beta) \vee (\alpha \wedge \gamma), \quad (\alpha \vee \beta) \wedge (\alpha \vee \gamma) \models \alpha \vee (\beta \wedge \gamma);$
(distributivity 1)

(12) $\mathbf{f} \wedge \mathbf{f} \equiv \mathbf{f}, \quad \mathbf{t} \wedge \mathbf{t} \equiv \mathbf{t};$
(idempotence for the truth and the falsity)

(13) $\mathbf{f} \wedge \mathbf{t} \equiv \mathbf{f}, \quad \mathbf{f} \vee \mathbf{t} \equiv \mathbf{t};$

(14) $\frac{\alpha \equiv \beta}{\neg\alpha \equiv \neg\beta};$
(logical equivalence is a congruence for the negation)

(15) $\frac{\alpha \equiv \gamma, \ \beta \equiv \delta}{\alpha\wedge\beta \equiv \gamma\wedge\delta};$
(logical equivalence is a congruence for the conjunction)

(16) $\sqrt{\neg}(\alpha \wedge \beta) \models \sqrt{\neg}\mathbf{t};$

(17) $\frac{\sqrt{\neg}\alpha \models \sqrt{\neg}\mathbf{t}}{\alpha\wedge\beta \models \alpha}, \quad \frac{\sqrt{\neg}\beta \models \sqrt{\neg}\mathbf{t}}{\alpha\wedge\beta \models \beta};$

(18) $\frac{\alpha \models \sqrt{\neg}\mathbf{t}}{\mathbf{f}\models\alpha}.$
(Weak Duns Scotus)

PROOF. Easy. ■

Let us now consider examples of logical consequences and rules that hold in **QCL** and are violated in $\sqrt{\neg}$**QCL**.

THEOREM 7.20 (Logical consequences and rules of **QCL** that fail in $\sqrt{\neg}$**QCL**).

(1) $\alpha \wedge \beta \models_{\mathbf{QCL}} \alpha, \quad \alpha \wedge \beta \models_{\mathbf{QCL}} \beta;$

(2) $\alpha \models_{\mathbf{QCL}} \alpha \vee \beta, \quad \beta \models_{\mathbf{QCL}} \alpha \vee \beta;$

(3) $\alpha \wedge \alpha \models_{\mathbf{QCL}} \alpha, \quad \alpha \models_{\mathbf{QCL}} \alpha \vee \alpha;$
(semiidempotence 1)

(4) $\mathbf{f} \models_{\mathbf{QCL}} \alpha.$
(Duns Scotus)

PROOF. Easy. ■

THEOREM 7.21 (A rule that holds in $\sqrt{\neg}$**QCL** and fails in **QCL**).

$$\frac{\alpha \equiv \beta}{\sqrt{\neg}\alpha \equiv \sqrt{\neg}\beta}.$$

PROOF. Easy. ■

In other words, logical equivalence is a congruence for the square root of the negation.

THEOREM 7.22 (Logical consequences that fail both in **QCL** and $\sqrt{\neg}$**QCL**).

(1) $\alpha \not\models \alpha \wedge \alpha;$
(semiidempotence 2)

(2) $\mathbf{t} \not\models \alpha \vee \neg\alpha;$
(excluded middle)

(3) $\mathbf{t} \not\models \neg(\alpha \wedge \neg\alpha);$
(non contradiction)

(4) $(\alpha \wedge \beta) \vee (\alpha \wedge \gamma) \not\models \alpha \wedge (\beta \vee \gamma)$, $\quad \alpha \vee (\beta \wedge \gamma) \not\models (\alpha \vee \beta) \wedge (\alpha \vee \gamma)$.
(distributivity 2)

PROOF. Easy. ∎

Apparently, the logics **QCL** and $\sqrt{\neg}$**QCL** turn out to be non standard forms of quantum logic. Conjunction and disjunction do not correspond to lattice operations, because they are not generally idempotent. Unlike Birkhoff and von Neumann's quantum logic, the weak distributivity principle $((\alpha \wedge \beta) \vee (\alpha \wedge \gamma) \models \alpha \wedge (\beta \vee \gamma))$ breaks down. At the same time, the strong distributivity $(\alpha \wedge (\beta \vee \gamma) \models (\alpha \wedge \beta) \vee (\alpha \wedge \gamma))$, that is violated in orthodox quantum logic, is here valid. Both the excluded middle and the non contradiction principles are violated.As a consequence, one can say that the logics arising from quantum computation represent, in a sense, new examples of *fuzzy logics.*

The axiomatizability of **QCL** and $\sqrt{\neg}$**QCL** is an open problem.

8. Quantum trees

An interesting feature of the quantum computational semantics is the following: the *meaning* and the probability-value of any molecular sentence α can be naturally described (and calculated) by means of a convenient *quantum tree*, that illustrates a kind of reversible transformation of the atomic subformulas of α. By theorem 7.16, we know that we can refer to the qubit-semantics (instead of the qumix-semantics), without any loss of generality. For the sake of technical simplicity, we will first slightly modify our language. The new language $\mathcal{L}^{\bigwedge}$ contains, besides the atomic sentence $\mathbf{f}$ and the two negations ($\neg$ and $\sqrt{\neg}$), a ternary conjunction $\bigwedge$ (whose semantic behaviour is "close" to the Petri-Toffoli gate). For any sentences α and β, the expression $\bigwedge(\alpha, \beta, f)$ is a sentence of $\mathcal{L}^{\bigwedge}$. In this framework, the usual conjunction $\alpha \wedge \beta$ is dealt with as metalinguistic abbreviation for the ternary conjunction $\bigwedge(\alpha, \beta, \mathbf{f})$. The occurrence of $\mathbf{f}$ as the third element in the formula $\bigwedge(\alpha, \beta, \mathbf{f})$ is called a *non-genuine occurrence* of $\mathbf{f}$. The semantic definition of *qubit-model* of the language $\mathcal{L}^{\bigwedge}$ is then modified in the expected way. Besides the old conditions concerning the interpretation of $\mathbf{f}$ and of the two negations ($\neg$, $\sqrt{\neg}$), we require that for any $\mathtt{Qub}$:

$$\mathtt{Qub}(\bigwedge(\alpha, \beta, \mathbf{f})) = T(\mathtt{Qub}(\alpha) \otimes \mathtt{Qub}(\beta) \otimes \mathtt{Qub}(\mathbf{f})).$$

Needless to stress, the logics **QCL** and $\sqrt{\neg}$**QCL** can be equivalently formalized either in the language $\mathcal{L}$ or in $\mathcal{L}^{\bigwedge}$. In case where the language is

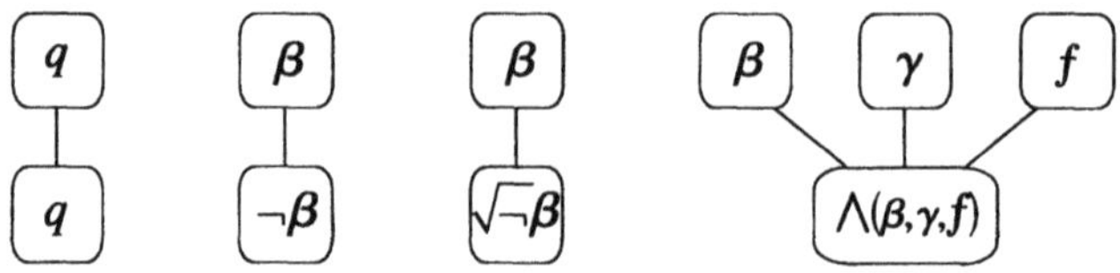

Figure 1. Branching rules for the construction of syntactical trees.

$\mathcal{L}^{\wedge}$, Corollary 7.18 shall be formulated as follows: *if* α *does not contain any genuine occurrence of* $\mathbf{f}$, *then* α *is not a logical truth either of* $\sqrt{\neg}$**QCL** *or of* **QCL**.

Before dealing with quantum trees, we will first introduce the notion of *syntactical tree* of a sentence α (abbreviated as $STree^{\alpha}$). Consider all subformulas of α.

Any subformula may be:

- an *atomic* sentence $\mathbf{q}$ (possibly $\mathbf{f}$);
- a *negated* sentence $\neg\beta$;
- a *square-root negated* sentence $\sqrt{\neg}\beta$;
- a *conjunction* $\bigwedge(\beta, \gamma, \mathbf{f})$.

The intuitive idea of *syntactical tree* can be illustrated as follows. Every occurrence of a subformula of α gives rise to a *node* of $STree^{\alpha}$. The tree consists of a finite number of *levels* and each level is represented by a sequence of subformulas of α:

$$Level_k(\alpha)$$

$$\vdots$$

$$Level_1(\alpha).$$

The *root-level* (denoted by $Level_1(\alpha)$) consists of α. From each node of the tree at most 3 edges may branch according to the *branching-rule* (Figure 1).

The second level ($Level_2(\alpha)$) is the sequence of subformulas of α that is obtained by applying the branching-rule to α. The third level ($Level_3(\alpha)$) is obtained by applying the branching-rule to each element (node) of $Level_2(\alpha)$, and so on. Finally, one obtains a level represented by the sequence of all atomic occurrences of α. This represents the *last level* of $STree^{\alpha}$. The *height* of $Stree^{\alpha}$ (denoted by $Height(\alpha)$) is then defined as the number of levels of $STree^{\alpha}$.

A more formal definition of *syntactical tree* can be given by using some standard graph-theoretical notions.

EXAMPLE 8.1. *The syntactical tree of* $\alpha = \neg\mathbf{q} \wedge (\mathbf{r} \wedge \sqrt{\neg}\mathbf{q})$ *is the following: Clearly the height of* $Stree^{\alpha}$ *is 4.*

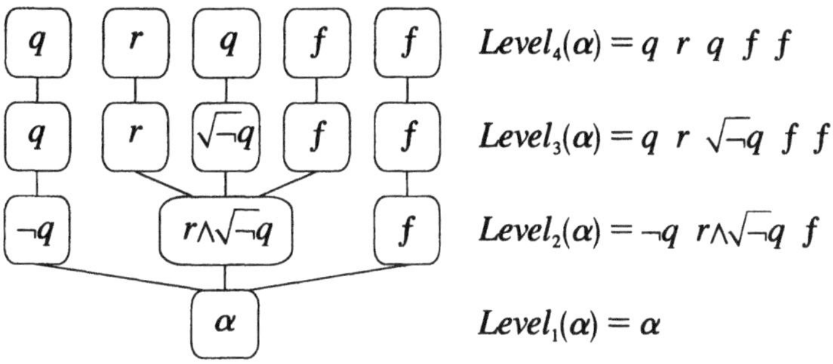

For any choice of a qubit-model $\mathtt{Qub}$, the syntactical tree of α determines a corresponding sequence of quregisters. Consider a sentence α with n atomic occurrences $(\mathbf{q}_1, \ldots, \mathbf{q}_n)$. Then $\mathtt{Qub}(\alpha) \in \otimes^n\mathbb{C}^2$. We can associate a quregister $|\psi_i\rangle$ to each $Level_i(\alpha)$ of $Stree^{\alpha}$ in the following way. Suppose that:

$$Level_i(\alpha) = (\beta_1, \ldots, \beta_r).$$

Then:

$$|\psi_i\rangle = \mathtt{Qub}(\beta_1) \otimes \ldots \otimes \mathtt{Qub}(\beta_r).$$

Hence:

$$\begin{cases} |\psi_1\rangle = \mathtt{Qub}(\alpha) \\ \vdots \\ |\psi_{Height(\alpha)}\rangle = \mathtt{Qub}(\mathbf{q}_1) \otimes \ldots \otimes \mathtt{Qub}(\mathbf{q}_n) \end{cases}$$

where all $|\psi_i\rangle$ belong to the same space $\otimes^n\mathbb{C}^2$.

From an intuitive point of view, $|\psi_{Height(\alpha)}\rangle$ can be regarded as a kind of *epistemic state*, corresponding to the input of a computation, while $|\psi_1\rangle$ represents the output.

We obtain the following correspondence:

$$Level_{Height(\alpha)}(\alpha) \leftrightsquigarrow |\psi_{Height(\alpha)}\rangle\text{: the input}$$
$$\ldots \leftrightsquigarrow \ldots$$
$$Level_1(\alpha) \leftrightsquigarrow |\psi_1\rangle\text{: the output}$$

The notion of *quantum tree* of a sentence α ($QTree^\alpha$) can be now defined as a particular sequence of unitary operators that is uniquely determined by the syntactical tree of α. As we already know, each $Level_i(\alpha)$ of $STree^\alpha$ is a sequence of subformulas of α. Let $Level_i^j(\alpha)$ represent the j-th element of $Level_i(\alpha)$. Each node $Level_i^j(\alpha)$ (where $1 \leq i < Height(\alpha)$) can be naturally associated to a unitary operator Op_i^j, according to the following *operator-rule*:

$$Op_i^j := \begin{cases} I^{(1)} & \text{if } Level_i^j(\alpha) \text{ is an atomic sentence;} \\ \mathtt{Not}^{(r)} & \text{if } Level_i^j(\alpha) = \neg\beta \text{ and } \mathtt{Qub}(\beta) \in \otimes^r\mathbb{C}^2; \\ \sqrt{\mathtt{Not}}^{(r)} & \text{if } Level_i^j(\alpha) = \sqrt{\neg}\beta \text{ and } \mathtt{Qub}(\beta) \in \otimes^r\mathbb{C}^2; \\ T^{(r,s,1)} & \text{if } Level_i^j(\alpha) = \bigwedge(\beta, \gamma, \mathbf{f}),\ \mathtt{Qub}(\beta) \in \otimes^r\mathbb{C}^2 \text{ and} \\ & \mathtt{Qub}(\gamma) \in \otimes^s\mathbb{C}^2. \end{cases}$$

On this basis, one can associate an operator U_i to each $Level_i(\alpha)$ (such that $1 \leq i < Height(\alpha)$):

$$U_i := \bigotimes_{j=1}^{|Level_i(\alpha)|} Op_i^j,$$

where $|Level_i(\alpha)|$ is the length of the sequence $Level_i(\alpha)$.

Being the tensor product of unitary operators, every U_i turns out to be a unitary operator. One can easily show that all U_i are defined in the same space $\otimes^n\mathbb{C}^2$, where n is the atomic complexity of α.

The notion of *quantum tree* of a sentence can be now defined as follows.

DEFINITION 8.2. (The quantum tree of α).
The *quantum tree* of α (denoted by $QTree^\alpha$) is the operator-sequence

$$(U_1, \ldots, U_{Height(\alpha)-1})$$

that is uniquely determined by the syntactical tree of α.

As an example, consider the following sentence: $\alpha = \mathbf{q} \wedge \neg \mathbf{q} = \bigwedge(\mathbf{q}, \neg \mathbf{q}, \mathbf{f})$. The syntactical tree of α is the following:

$$\begin{aligned} Level_1(\alpha) &= \bigwedge(\mathbf{q}, \neg \mathbf{q}, \mathbf{f}); \\ Level_2(\alpha) &= (\mathbf{q}, \neg \mathbf{q}, \mathbf{f}); \\ Level_3(\alpha) &= (\mathbf{q}, \mathbf{q}, \mathbf{f}). \end{aligned}$$

In order to construct the quantum tree of α, let us first determine the operators Op_i^j corresponding to each node of $Stree^\alpha$. We will obtain:

- $Op_1^1 = T^{(1,1,1)}$, because $\bigwedge(\mathbf{q}, \neg \mathbf{q}, \mathbf{f})$ is connected with $(\mathbf{q}, \neg \mathbf{q}, \mathbf{f})$ (at $Level_2(\alpha)$);
- $Op_2^1 = I^{(1)}$, because $\mathbf{q}$ is connected with $\mathbf{q}$ (at $Level_3(\alpha)$);
- $Op_2^2 = \mathtt{Not}^{(1)}$, because $\neg \mathbf{q}$ is connected with $\mathbf{q}$ (at $Level_3(\alpha)$);
- $Op_2^3 = I^{(1)}$, because $\mathbf{f}$ is connected with $\mathbf{f}$ (at $Level_3(\alpha)$).

The quantum tree of α is represented by the operator-sequence (U_1, U_2), where:

$$\begin{aligned} U_1 &= Op_1^1 = T^{(1,1,1)}; \\ U_2 &= Op_2^1 \otimes Op_2^2 \otimes Op_2^3 = I^{(1)} \otimes \mathtt{Not}^{(1)} \otimes I^{(1)}. \end{aligned}$$

Apparently, $QTree^\alpha$ is independent of the choice of $\mathtt{Qub}$.

THEOREM 8.3. *Let α be a sentence whose quantum tree is the operator-sequence $(U_1, \ldots, U_{Height(\alpha)-1})$. Given a quantum computational model* $\mathtt{Qub}$, *consider the quregister-sequence $(|\psi_1\rangle, \ldots, |\psi_{Height(\alpha)}\rangle)$ that is determined by* $\mathtt{Qub}$ *and by the syntactical tree of α. Then, $U_i(|\psi_{i+1}\rangle) = |\psi_i\rangle$ (for any i such that $1 \leq i < Height(\alpha)$).*

PROOF. Straightforward. ∎

The quantum tree of α can be naturally regarded as a *quantum circuit* that computes the output $\mathtt{Qub}(\alpha)$, given the input $\mathtt{Qub}(\mathbf{q}_1), \ldots, \mathtt{Qub}(\mathbf{q}_n)$ (where $\mathbf{q}_1, \ldots, \mathbf{q}_n$ are the atomic occurrences of α). In this framework, each U_i is the unitary operator that describes the computation performed by the i-th layer of the circuit.

References

[1] BIRKHOFF, G., and J. VON NEUMANN, *The logic of quantum mechanics*, Annals of Mathematics **37** (1936), 823–843.

[2] CATTANEO, G., M. L. DALLA CHIARA, R. GIUNTINI and R. LEPORINI, *An unsharp logic from quantum computation*, e-print: http://arxiv.org/abs/quant-ph/0201013.

[3] CATTANEO, G., M. L. DALLA CHIARA, R. GIUNTINI and R. LEPORINI, *Quantum computational structures*, preprint.

[4] DALLA CHIARA, M. L., and R. GIUNTINI, *Quantum logics*, in G. GABBAY and F. GUENTHNER (eds.), Handbook of Philosophical Logic, vol. VI, Kluwer, Dordrecht, 2002, pp. 129–228.

[5] DALLA CHIARA, M. L., R. GIUNTINI, A. LEPORATI and R. LEPORINI, *Qubit semantics and quantum trees*, e-print: http://arxiv.org/abs/quant-ph/0211190.

[6] DEUTSCH, D., A. EKERT, and R. LUPACCHINI, *Machines, logic and quantum physics*, Bulletin of Symbolic Logic, **3**, 2000, pp. 265–283.

[7] GUDDER, S., *Quantum computational logic*, preprint.

[8] PETRI, C. A., *Gründsatzliches zur Beschreibung diskreter Prozesse*, in Proceedings of the 3rd Colloquium über Automatentheorie (Hannover, 1965), Birkhäuser Verlag, Basel, 1967, pp. 121–140. English version: *Fundamentals of the Representation of Discrete Processes*, ISF Report 82.04 (1982), translated by H. J. GENRICH and P. S. THIAGARAJAN.

[9] TOFFOLI, T., *Reversible computing*, in J. W. DE BAKKER, J. VAN LEEUWEN (eds.), Automata, Languages and Programming, Springer, 1980, pp. 632–644. Also available as TechnicalMemo MIT/LCS/TM-151, MIT Laboratory for Computer Science, February 1980.

[10] ZAWIRSKI, Z., *Relation of many-valued logic to probability calculus*, (in Polish, original title: *Stosunek logiki wielowartościowej do rachunku prawdopodobieństwa*), Poznańskie Towarzystwo Przyjaciół Nauk, 1934.

MARIA LUISA DALLA CHIARA
Dipartimento di Filosofia,
Università di Firenze,
via Bolognese 52,
I-50139 Firenze, Italy
`dallachiara@unifi.it`

ROBERTO GIUNTINI
Dipartimento di Scienze Pedagogiche e Filosofiche,
Università di Cagliari,
via Is Mirrionis 1,
I-09123 Cagliari, Italy
`giuntini@unica.it`

ROBERTO LEPORINI
Dipartimento di Informatica, Sistemistica e Comunicazione,
Università degli Studi di Milano–Bicocca,
via Bicocca degli Arcimboldi 8,
I-20126 Milano, Italy
`leporini@disco.unimib.it`

Graham Priest

Inconsistent Arithmetics: Issues Technical and Philosophical

The study of inconsistent arithmetics is relatively young, going back about 25 years. It has, however, already occasioned a number of both interesting technical results and philosophical controversies. There are still, moreover, important technical questions to be answered, and philosophical issues to be debated. In this paper I will review the area and discuss some of these issues. In the first part of the paper I will survey the relevant technical material, ending with a number of open problems; in the second part, I will review some of the philosophical material, ending with a discussion of one central debate in the area.

1. Technical Issues

1.1. Historical Overview

The first person to construct an inconsistent arithmetic (as far as I know) was Nelson (1959), who used a realisability semantics to produce an inconsistent arithmetic, based on a paraconsistent logic of an intuitionist kind. Current developments in the subject, however, trace back, not to this, but to Meyer's paper, 'Relevant Arithmetic'. This paper, in an incomplete form, was circulated amongst relevant logicians, and was abstracted as Meyer (1976); sadly, the full version of the paper has never appeared, as far as I know. Meyer's concern was relevant Peano Arithmetic, that is, essentially, the axiomatic arithmetic in which one takes the Peano Axioms, replaces the conditionals employed with a relevant conditional, and then uses an underlying relevant logic, *R* in Meyer's case, to prove things about numbers. In investigating the properties of this theory, Meyer noticed that it could be given a finitary consistency proof—showing that Gödel's Second Incompleteness Theorem

V. F. Hendricks and *J. Malinowski* (eds.),
Trends in Logic: 50 Years of Studia Logica

Trends in Logic **21**: 273–299, 2003.

may not apply once one jettisons classical logic. Specifically, there are models with a two-element domain which verify all the theorems. The models were also models of the three-valued logic $RM3$, and they did a lot more than verify all the theorems of relevant Peano Arithmetic: they verified an inconsistent set of sentences.

In Meyer and Mortensen (1984) generalisations of Meyer's model were investigated.[1] Specifically, different finite sizes of the domain were employed, as were different many-valued semantics for the conditional. It thus became clear that there was a substantial family of inconsistent arithmetics. The models were constructed, in effect, by deploying an equivalence on the natural numbers which is a congruence with respect to successor, addition, and multiplication. Mortensen realised that similar techniques could be applied to the numbers in a non-standard model of arithmetic. In Mortensen (1986), he constructed many inconsistent arithmetics using this technique. In (1988) he investigated a number of the properties of this family of arithmetics. He also observed that the techniques in question could be applied, equally, to give inconsistent theories of other sorts of mathematical theory, such as fields and rings. In (1990) he deployed these ideas to produce inconsistent models of the differential calculus. Mortensen's work is nicely summarised in his (1995). Priest (1995), Part 4, Technical Appendix, used similar techniques to construct inconsistent set-theories with various properties. (In what follows, I will restrict myself to considering just the inconsistent natural-number arithmetics.)

Two things had become clear by this time. The first is that the inconsistent arithmetics are very powerful. Specifically, they can be made to contain all of the sentences true in the standard model of arithmetic—as expressed using just the classical propositional connectives, $\wedge$, $\vee$, $\neg$ and $\supset$, where $\alpha \supset \beta$ is defined in the usual way as $\neg\alpha \vee \beta$. The second was that, although a lot of the initial interest in these arithmetics was occasioned by an interest in a non-material conditional, and, specifically, in what could be proved using such a conditional, once one moved to a model-theoretic perspective, the non-material conditional was not playing a large role: all of the truths of the standard model came for free anyway. This meant that one could simply forget about the non-material conditional, and investigate the structure of the theories, as expressed in the classical vocabulary (though the underlying logic could not, of course, be classical, since the interpretations model inconsistent sets of sentences). This, in turn, allowed inconsistent models to

[1]In Mortensen and Meyer (1985) there is also an application of the inconsistent models to arithmetic based on a non-distributive quantum logic.

be constructed by a simple yet powerful model-theoretic construction now called the Collapsing Lemma. A form of this had already been established by Dunn in (1979); a version which allowed it to be immediately applicable to the construction of inconsistent models was given in Priest (1991).

All the tools were now at hand for circumscribing an important class of inconsistent arithmetics, and investigating their structure. The class was simple enough to be natural, and complex enough to be mathematically interesting. The analysis of the finite case was given in Priest (1997), and of the general case in Priest (2000). I will explain the details in what follows.

1.2. *LP* and *FDE*

The underlying logic for the arithmetics is First Degree Entailment (*FDE*) or *LP*. It does not make a difference which, as we will see in a moment. Let me start by specifying the semantics of the logic.[2]

The vocabulary of the languages in question is that of first-order logic with identity. (We take $\supset$ to be defined in the usual way.) For simplicity, I suppose that there are no free variables. An *FDE* interpretation, $\mathcal{I}$, for the language is a pair, $\langle D, d\rangle$, where D is the (non-empty) domain of quantification, and d is a function that maps every constant to an object in the domain, every n-place function symbol to an n-place function on the domain, and every n-place predicate, P, to a pair, $\langle E_P, A_P\rangle$, each member of which is a subset of the set of n-tuples of D, D^n. E_P is the *extension* of P; A_P is the *anti-extension*. The extension of the identity predicate, $E_=$, is, as usual, $\{\langle d, d\rangle : d \in D\}$. For an *LP* interpretation, we require, in addition, that for every n-place predicate, P, $E_P \cup A_P = D^n$. (This is the only difference between *FDE* and *LP*.)

Every term, t, is assigned a denotation, $d(t)$, in the usual recursive fashion. Every sentence, α, is assigned a truth value, $\nu(\alpha)$, which is a subset of $\{1, 0\}$, non-empty in the case of *LP*. For atomic sentences, the truth/falsity conditions are:

$$1 \in \nu(Pt_1...t_n) \text{ iff } \langle d(t_1), ..., d(t_n)\rangle \in E_P$$
$$0 \in \nu(Pt_1...t_n) \text{ iff } \langle d(t_1), ..., d(t_n)\rangle \in A_P$$

Truth/falsity conditions for the connectives are:

$$1 \in \nu(\neg\alpha) \text{ iff } 0 \in \nu(\alpha)$$
$$0 \in \nu(\neg\alpha) \text{ iff } 1 \in \nu(\alpha)$$

$$1 \in \nu(\alpha \wedge \beta) \text{ iff } 1 \in \nu(\alpha) \text{ and } 1 \in \nu(\beta)$$

---

[2]For further details, see Priest (2001), ch. 8.

$$0 \in \nu(\alpha \wedge \beta) \text{ iff } 0 \in \nu(\alpha) \text{ or } 0 \in \nu(\beta)$$
$$1 \in \nu(\alpha \vee \beta) \text{ iff } 1 \in \nu(\alpha) \text{ or } 1 \in \nu(\beta)$$
$$0 \in \nu(\alpha \wedge \beta) \text{ iff } 0 \in \nu(\alpha) \text{ and } 0 \in \nu(\beta)$$

For the quantifiers, we assume that the language has been augmented, if necessary, by names, in such a way that every member, d, of D has a name, $\mathbf{d}$.

$$1 \in \nu(\forall x\alpha(x)) \text{ iff for every } d \in D,\ 1 \in \nu(\alpha(\mathbf{d}))$$
$$0 \in \nu(\forall x\alpha(x)) \text{ iff for some } d \in D,\ 0 \in \nu(\alpha(\mathbf{d}))$$

$$1 \in \nu(\exists x\alpha(x)) \text{ iff for some } d \in D,\ 1 \in \nu(\alpha(\mathbf{d}))$$
$$0 \in \nu(\exists x\alpha(x)) \text{ iff for every } d \in D,\ 0 \in \nu(\alpha(\mathbf{d}))$$

An interpretation is a *model* of α iff, in that interpretation, $1 \in \nu(\alpha)$; it is a model of a set of sentences, Σ, iff it is a model of every member of Σ; and an inference is *valid* ($\models$) iff every model of the premises is a model of the conclusion.

It should be noted that if, for every predicate, P, E_P and A_P are exclusive and exhaustive, then we have, in effect, an interpretation of classical first order logic. All classical interpretations are therefore FDE (and LP) interpretations.

1.3. The Collapsing Lemma

We can now state the Collapsing Lemma. Let $\mathcal{I} = \langle D, d\rangle$ be any interpretation. Let $\sim$ be an equivalence relation on D, which is also a congruence relation on the denotations of the function symbols in the language (i.e., if g is such a denotation, and $d_i \sim e_i$ for all $1 \leq i \leq n$, then $g(d_1, ..., d_n) \sim g(e_1, ..., e_n)$). If $d \in D$ let $[d]$ be the equivalence class of d under $\sim$. Define an interpretation, $\mathcal{I}_\sim = \langle D_\sim, d_\sim\rangle$, to be called the *collapsed interpretation.* $D_\sim = \{[d]; d \in D\}$; if c is a constant, $d_\sim(c) = [d(c)]$; if f is an n-place function symbol:

$$d_\sim(f)([d_1], ..., [d_n]) = [d(f)(d_1, ..., d_n)]$$

(this is well defined, since $\sim$ is a congruence relation); and if P is an n-place predicate, its extension and anti-extension in $\mathcal{I}_\sim$, $E_P^\sim$ and $A_P^\sim$, are defined by:

$$\langle [d_1], ..., [d_n]\rangle \in E_P^\sim \text{ iff for all } 1 \leq i \leq n,\ \exists e_i \sim d_i,\ \langle e_1, ..., e_n\rangle \in E_P$$
$$\langle [d_1], ..., [d_n]\rangle \in A_P^\sim \text{ iff for all } 1 \leq i \leq n,\ \exists e_i \sim d_i,\ \langle e_1, ..., e_n\rangle \in A_P$$

where E_P and A_P are the extension and anti-extension of P in $\mathcal{I}$. It is easy to check that $E_=^\sim$ is $\{\langle [d], [d]\rangle\,; d \in D\}$, as required.

The collapsed interpretation, in effect, identifies all members of an equivalence class to produce a composite individual which has the properties of all of its members. This may, of course, be inconsistent, even if its members are not.

A swift induction confirms that for any term, t, $d_{\sim}(t) = [d(t)]$. Hence:

$$\begin{aligned} 1 \in \nu(Pt_1...t_n) &\Rightarrow \langle d(t_1), ..., d(t_n)\rangle \in E_P \\ &\Rightarrow \langle [d(t_1)], ..., [d(t_n)]\rangle \in E_P^{\sim} \\ &\Rightarrow \langle d_{\sim}(t_1), ..., d_{\sim}(t_n)\rangle \in E_P^{\sim} \\ &\Rightarrow 1 \in \nu_{\sim}(Pt_1...t_n) \end{aligned}$$

Similarly for 0 and anti-extensions. A routine argument by induction then establishes that this condition obtains for all formulas; i.e., for all α, $\nu(\alpha) \subseteq \nu_{\sim}(\alpha)$. This is the *Collapsing Lemma.*[3]

The Collapsing Lemma assures us that if an interpretation is a model of some set of sentences, then any interpretation obtained by collapsing it will also be a model. This gives us an important way of constructing inconsistent models of a theory. Start with any model of the theory, possibly a classical model, and collapse. Any collapsed model will be a model of the original theory, though, in general, it will be inconsistent. I will illustrate by constructing some inconsistent models of arithmetic.

1.4. Collapsed Models of Arithmetic

Let us start with a definition. Let L be the standard language of first-order arithmetic: one constant, $\mathbf{0}$, function symbols for successor, addition, and multiplication ($'$, $+$, and $\times$, respectively), and one predicate symbol, $=$. Let $\mathcal{N}$ be the standard (classical) interpretation of this language; and if $\mathcal{M}$ is an interpretation, let $Th(\mathcal{M})$ be the set of sentences true in $\mathcal{M}$. A *model of arithmetic* is any FDE or LP interpretation of L which is a model of $Th(\mathcal{N})$. Note that since $Th(\mathcal{N})$ is complete (i.e., for all α, either $\alpha \in Th(\mathcal{N})$ or $\neg\alpha \in Th(\mathcal{N})$), an FDE model is also an LP model. Note, also, that, as well as $\mathcal{N}$, any classical non-standard model of arithmetic is a model of arithmetic in the sense I will use the word here. But there are many more. In particular, as we will see, there are models of arithmetic, $\mathcal{M}$, such that $Th(\mathcal{M})$ is inconsistent. I will call such models, naturally enough, inconsistent models of arithmetic.

Now, let $\mathcal{M} = \langle M, d\rangle$ be any classical model of $Th(\mathcal{N})$. Let $\sim$ be an equivalence relation on M which is also a congruence relation with respect to the interpretations of the function symbols. Then we may construct the

---

[3]For details of the proof, see Priest (1991).

collapsed interpretation, $\mathcal{M}_\sim$. By the Collapsing Lemma, $\mathcal{M}_\sim$ is a model of arithmetic. Provided that $\sim$ is not the trivial equivalence relation, that relates each thing to only itself, then $\mathcal{M}_\sim$ will model inconsistencies. For suppose that $\sim$ relates the distinct members of M, n and m; then in $\mathcal{M}_\sim$, $[n] = [m]$ and so $\langle [n], [m] \rangle$ is in the extension of $=$. But since $n \neq m$ in $\mathcal{M}$, $\langle [n], [m] \rangle$ is in the anti-extension too. Thus, $\exists x(x = x \wedge x \neq x)$ holds in $\mathcal{M}_\sim$. Let me give a couple of simple examples of this.

Example one. Let $\mathcal{M}$ be the standard model of arithmetic. n, $p \in M$ and $p > 0$. Define a relation, $\sim$, on M, thus:

$$x \sim y \text{ iff } (x,\ y < n \text{ and } x = y) \text{ or } (x,\ y \geq n \text{ and } x = y \pmod p)$$

It is easy to check that $\sim$ is a congruence relation on M. Let $\mathcal{M}_n^p$ be the model obtained by collapsing with respect to this. The Collapsing Lemma assures us that it is a model of arithmetic. It is finite; it has an initial tail of length n that behaves consistently. The other numbers form a cycle of period p. The successor graph can be depicted as follows:

$$\begin{array}{ccccccccc} 0 & \to & 1 & \to & \ldots & \to & n & \to & n+1 \\ & & & & & & \uparrow & & \downarrow \\ & & & & & & n+p-1 & \leftarrow & \cdots \end{array}$$

Example two. Let $\mathcal{M}$ be any non-standard classical model of arithmetic. Define the relation $\sim$ as follows:

$$x \sim y \text{ iff } (x,\ y \text{ are standard numbers and } x = y) \text{ or } (x,\ y \text{ are non-standard})$$

Again, it is easy to check that $\sim$ is an equivalence relation which is also a congruence on the arithmetic operators. The model obtained by collapsing with respect to this equivalence relation contains the standard interpretation, plus an inconsistent "point at infinity". The successor graph can be depicted as follows:

$$\begin{array}{cccccc} 0 & \to & 1 & \to & \ldots & \Omega \\ & & & & & \circlearrowleft \end{array}$$

1.5. Inconsistent Models of Arithmetic

We now turn to the question of the general structure of models of arithmetic.[4] Let $\mathcal{M} = \langle M, d \rangle$ be any such model. I will refer to the denotations of $'$, $+$,

[4]The material in this and the next section is reproduced with minor revisions from sections 9.3, 9.4 of Priest (2002). I am grateful for permission to reuse the material. The contents of these sections are covered in more detail in Priest (2000).

and $\times$ as the arithmetic operations of $\mathcal{M}$; and since no confusion is likely, use the same signs for them. I will call the denotations of the numerals *regular* numbers.

Let $x \leq y$ be defined, in the usual way, as $\exists z\ x + z = y$. It is easy to check that $\leq$ is transitive. For if $i \leq j \leq k$, then for some x, y, $i + x = j$ and $j + y = k$. Hence $(i + x) + y = k$. But $(i + x) + y = i + (x + y)$ (since we are dealing with a model of arithmetic). The result follows.

If $i \in M$, let $N(i)$ (the *nucleus* of i) be $\{x \in M; i \leq x \leq i\}$. In a classical model, $N(i) = \{i\}$, but this need not be the case in an inconsistent model. For example, in any $\mathcal{M}_n^p$ the members of the cycle constitute a nucleus. If $j \in N(i)$ then $N(i) = N(j)$. For if $x \in N(j)$ then $i \leq j \leq x \leq j \leq i$, so $x \in N(i)$, and similarly in the other direction. Thus, every member of a nucleus defines the same nucleus.

Now, if N_1 and N_2 are nuclei, define $N_1 \preceq N_2$ to mean that for some (or all, it makes no difference) $i \in N_1$ and $j \in N_2$, $i \leq j$. It is not difficult to check that $\preceq$ is a partial ordering. Moreover, since for any i and j, $i \leq j$ or $j \leq i$, it is a linear ordering. The least member of the ordering is $N(0)$. If $N(1)$ is distinct from this, it is the next (since for any x, $x \leq 0 \vee x \geq 1$), and so on for all regular numbers.

Say that $i \in M$ has *period* $p \in M$ iff $i + p = i$. In a classical model every number has period 0 and only 0. But again, this need not be the case in an inconsistent model, as the $\mathcal{M}_n^p$ demonstrate. If $i \leq j$ and i has period p, so does j. For $j = i + x$, so $p + j = p + i + x = i + x = j$. In particular, if p is a period of some member of a nucleus, it is a period of every member. We may thus say that p is a period of the nucleus itself. It also follows that if $N_1 \preceq N_2$ and p is a period of N_1 it is a period of N_2.

If a nucleus has a regular non-zero period, m, then it must have a minimum (in the usual sense) non-zero period, since the sequence $0, 1, 2, ..., m$ is finite. If $N_1 \preceq N_2$ and N_1 has minimum regular non-zero period, p, then p is a period of N_2. Moreover, the minimum non-zero period of N_2, q, must be a divisor (in the usual sense) of p. For suppose that $q < p$, and that q is not a divisor of p. For some $0 < k < q$, p is some finite multiple of q plus k. So if $x \in N_2$, $x = x + q = x + p + ... + p + k$. Hence $x = x + k$, i.e., k is a period of N_2, which is impossible.

If a nucleus has period $p \geq 1$, I will call it *proper*. Every proper nucleus is closed under successors. For suppose that $j \in N$ with period p. Then $j \leq j' \leq j + p = j$. Hence, $j' \in N$. In an inconsistent model, a number may have more than one predecessor, i.e., there may be more than one x

such that $x' = j$.[5] (Although $(x' = y') \supset x = y$ holds in the model, we cannot necessarily detach to obtain $x = y$.) But if j is in a proper nucleus, N, it has a unique predecessor in N. For let the period of N be q'. Then $(j + q)' = j + q' = j$. Hence, $j + q$ is a predecessor of j; and $j \leq j + q \leq j + q' = j$. Hence, $j + q \in N$. Next, suppose that x and y are in the nucleus, and that $x' = y' = j$. We have that $x \leq y \vee y \leq x$. Suppose, without loss of generality, the first disjunct. Then for some z, $x + z = y$; so $j + z = j$, and z is a period of the nucleus. But then $x = x + z = y$. I will write the unique predecessor of j in the nucleus as $'j$.

Now let N be any proper nucleus, and $i \in N$. Consider the sequence $...,''i,'i, i, i', i''....$ Call this the *chromosome* of i. Note that if i, $j \in N$, the chromosomes of i and j are identical or disjoint. For if they have a common member, z, then all the finite successors of z are identical, as are all its finite predecessors (in N). Thus they are identical. Now consider the chromosome of i, and suppose that two members are identical. There must be members where the successor distance between them is a minimum. Let these be j and $j'^{...'}$ where there are n primes. Then $j = j + n$, and n is a period of the nucleus—in fact, its minimum non-zero period—and the chromosome of every member of the nucleus is a successor cycle of period n.

Hence, any proper nucleus is a collection of chromosomes, all of which are either successor cycles of the same finite period, or are sequences isomorphic to the integers (positive and negative). Both sorts are possible in an inconsistent model. Just consider the collapse of a non-standard model, of the kind given in 1.4, by an equivalence relation which leaves all the standard numbers alone and identifies all the others modulo p. If p is standard, the non-standard numbers collapse into a successor cycle; if it is non-standard, the nucleus generated is of the other kind.

To summarise so far, the general structure of a model is a liner sequence of nuclei. There are three segments (any of which may be empty). The first contains only improper nuclei. The second contains proper nuclei with linear chromosomes. The final segment contains proper nuclei with cyclical chromosomes of finite period. A period of any nucleus is a period of any subsequent nucleus; and in particular, if a nucleus in the third segment has minimum non-zero period, p, the minimum non-zero period of any subsequent nucleus is a divisor of p. Thus, we might depict the general structure

[5]In fact, it is not difficult to show that there is at most one number with multiple predecessors; and this can have only two.

of a model as follows (where $m+1$ is a multiple of $n+1$):

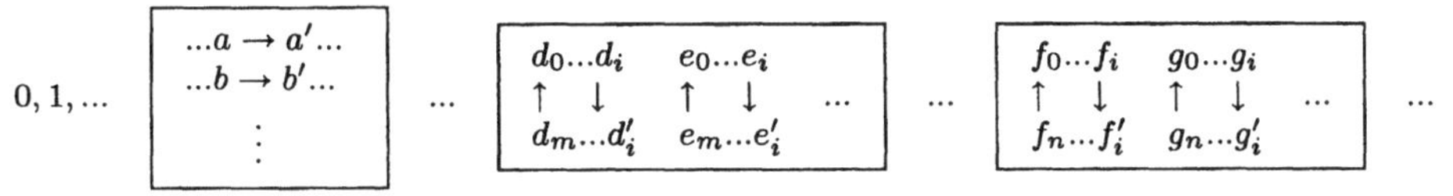

An obvious question at this point is what orderings the proper nuclei may have. For a start, they can have the order-type of any ordinal. To prove this, one establishes by transfinite induction that for any ordinal, α, there is a classical model of arithmetic in which the non-standard numbers can be partitioned into a collection of disjoint blocks with order-type α, closed under arithmetic operations. One then collapses this interpretation in such a way that each block collapses into a nucleus.

The proper nuclei need not be discretely ordered. They can also have the order-type of the rationals. To prove this, one considers a classical non-standard model of arithmetic, where the order-type of the non-standard numbers is that of the rationals. It is then possible to show that the numbers can be partitioned into a collection of disjoint blocks, closed under arithmetic operations, which themselves have the order-type of the rationals. One can then collapse this model in such a way that each of the blocks collapses into a proper nucleus, giving the result. This proof can be extended to show that any order-type which can be embedded in the rationals in a certain way can also be the order-type of the proper nuclei. This includes ω^* (the reverse of ω) and $\omega^* + \omega$, but *not* $\omega + \omega^*$.

1.6. Finite Models of Arithmetic

First-order arithmetic has many classical non-standard models, but none of them is finite. One of the intriguing features of inconsistent models of arithmetic is that they can be just that, e.g., the $\mathcal{M}^p_n$. For finite models, a complete characterisation is known.

Placing the constraint of finitude on the results of 1.5, we can infer as follows. The sequence of improper nuclei is either empty or is composed of the singletons of $0, 1, ..., n$, for some finite n. There must be a finite collection of proper nuclei, $N_1 \preceq ... \preceq N_m$; each N_i must comprise a finite collection of successor cycles of some minimum non-zero finite period, p_i. And if $1 \leq i \leq j \leq m$, p_j must be a divisor of p_i.[6]

Moreover, there are models of any structure of this form. To show this, take any non-standard classical model of arithmetic. This can be partitioned

[6]It is also possible to show that each nucleus is closed under addition and multiplication.

into the finite collection of blocks:

$$C_0, C_{1_0}, ..., C_{1_{k(1)}}, ...C_{i_0}, ..., C_{i_{k(i)}}, ..., C_{m_0}, ..., C_{m_{k(m)}}$$

where C_0 is either empty or is of the form $\{0, ..., n\}$, and each subsequent block is closed under arithmetic operations.[7] We now define a relation, $x \sim y$, as follows:

$(x, y \in C_0$ and $x = y)$ or
for some $1 \leq i \leq m$:
(for some $0 < j < k(i)$, $x, y \in C_{i_j}$, and $x = y \bmod p_i$) or
$(x, y \in C_{i_0} \cup C_{i_{k(i)}}$ and $x = y \bmod p_i)$

One can check that $\sim$ is an equivalence relation, and also that it is a congruence relation on the arithmetic operations. Hence we can construct the collapsed model. $\sim$ leaves all members of C_0 alone. For every i it collapses every C_{i_j} into a successor cycle of period p_i, and it identifies the blocks C_{i_0} and $C_{i_{k(i)}}$. Thus, the sequence $C_{i_0}, ..., C_{i_{k(i)}}$ collapses into a nucleus of period p_i with $k(i)$ chromosomes. The collapsed model therefore has exactly the required structure.

Finally in this section, note the following. If $\mathcal{M}$ is finite, $Th(\mathcal{M})$ is decidable. The truth value(s) in $\mathcal{M}$ of an atomic sentence can be computed, since the denotations of the functions and predicates, being on a finite domain, are computable. The truth values of propositional compounds are computed by *LP* (*FDE*) truth tables, and, since the domain is finite, quantified sentences are equivalent to finite conjunctions/disjunctions. (Thus, $\exists x\alpha(x)$ has the same truth value(s) as $\alpha(\mathbf{d}_0) \vee ... \vee \alpha(\mathbf{d}_n)$, where $D = \{d_0, ..., d_n\}$.)

1.7. Open Problems

There are many interesting questions about inconsistent models, even the finite ones, the answers to which are not known. I finish the technical part of this paper by listing some.

- Characterise the orderings of the proper cycles that may be realised in an inconsistent model.
- Can a nucleus have an infinitely descending sequence of periods?
- Given a model with a particular structure of cycles (nuclei and chromosomes), how many models of that structure are there? (The behaviour

[7]The existence of such a partition follows from a standard result in the study of classical models of arithmetic. See Kaye (1991), sect. 6.1.

of the successor function in a model does not determine the behavior of addition and multiplication, except in the tail.)

- Must nuclei always be closed under addition and multiplication?

- The set of sentences true in any finite model is decidable, and *a fortiori* axiomatisable. Are there any infinite models such that the set of sentences that hold in them is axiomatic?

Perhaps the most fundamental open question is as follows. Not all inconsistent models of arithmetic are collapses of classical models. Let $\mathcal{M}$ be any model of arithmetic; if $\mathcal{M}'$ is obtained from $\mathcal{M}$ by adding extra pairs to the anti-extension of $=$, call $\mathcal{M}'$ an *extension* in $\mathcal{M}$. If $\mathcal{M}'$ is an extension of $\mathcal{M}$, $\mathcal{M}'$ is itself a model of arithmetic (as may be shown by a simple inductive argument). Now, consider the extension of the standard model obtained by adding $\langle 0, 0\rangle$ to the anti-extension of $=$. This is not a collapsed model, since, if it were, 0, being inconsistent, would have to have been identified with some $x > 0$. But then 1 would have been identified with $x' > 1$. Hence, $\mathbf{0}' \neq \mathbf{0}'$ would also be true in the model, which it is not. Maybe, however, the following conjecture is true:

- Each inconsistent model is the extension of a collapsed classical model.

If this conjecture is correct, inconsistent models can be investigated via an analysis of the classical models of arithmetic and their congruence relations.

2. Philosophical Issues

2.1. Historical Overview

We now turn to some issues in the philosophy of arithmetic which are posed by the existence of inconsistent models. I start, as with the technical section, by giving a brief historical overview.

The first paper to argue for the inconsistency of arithmetic predates the technical investigations, and was Priest (1979). The argument appeals to Gödel's first Incompleteness Theorem. This was criticised by Chihara (1984). An extended defence of the argument appeared in Priest (1987), ch. 3.

The first person to deploy the technical material on inconsistent models in a philosophical context was van Bendegem (1993), (1994). He was particularly concerned with the finite models, and developed an argument for finitism on the basis of them. Priest (1994a) took up the idea, but used the finite models in defence of inconsistent arithmetic, rather than finitism. Where van Bendegem saw a greatest number, Priest saw a least inconsistent number. The idea of a least inconsistent number was discussed further in Priest (1994b). Priest (1994a) invokes various results about the finite models and metatheoretic notions, particularly provability. The relationship between finite models and provability is discussed further in Mortensen (1995). Priest and van Bendegem's deployment of inconsistent models is criticised by Batens (200a). Denyer (1995) is also a critique of Priest, to which Priest (1996a) is a reply.

Priest (1996b) invokes the inconsistent models of arithmetic in a quite different way: to argue the case for the possibility of arithmetic revision.[8] Most recently, Priest's (1987) argument for the inconsistency of arithmetic has been criticised by Shapiro (2002), though the technical material on inconsistent models is not deployed.

In the second half of this paper I will review and discuss some of these developments. At the end, I will address some of Shapiro's criticisms specifically in the light of the technical material on inconsistent arithmetics.

Let me start by setting the scene. If $\mathcal{M}$ is any model of arithmetic $Th(\mathcal{M})$ is a theory, that is, a set of arithmetic sentences closed under LP (and FDE) consequence, and contains $Th(\mathcal{N})$. If $\mathcal{M}$ is an inconsistent model of arithmetic, $Th(\mathcal{M})$ is also inconsistent.[9] I will call any such theory an *inconsistent arithmetic*. Now, when we count and perform arithmetic operations, which theory of arithmetic is right?

2.2. Arithmetic Revision

To answer this question we need to distinguish between pure arithmetic and applied arithmetic. A pure arithmetic is the set of truths about numbers themselves. An applied arithmetic is a pure arithmetic employed for the purposes of counting something or other. Now, if one askes what the correct pure arithmetic is, a natural answer is that the correct theory of arithmetic is $Th(\mathcal{N})$, the set of sentences true in the "standard model". Let us, for the

[8]This paper was due to appear in the proceedings of the conference at which it was given. These never, unfortunately, eventuated. It is worth noting that the idea that one might use a non-standard arithmetic to count appears as early as Gasking (1940).

[9]There are, in fact, LP theories that contain all of $Th(\mathcal{N})$, but that are not the theory of some collapsed model. This, for example, $\bigcap_{n\in\omega}\mathcal{M}^p_n$, being an intersection of theories, is a theory. But it contains the sentence $\exists x(x \neq x)$, whilst it contains nothing of the form $\mathbf{n} \neq \mathbf{n}$.

moment, agree that this is true.

It remains the case that, at least for some purposes, we might wish to use a different arithmetic to count some things. Compare the situation with that in geometry. Until the 19th century, the correct geometry for application to space was taken to be Euclidean geometry. But in the 20th century, this position was revised. The correct geometry is not Euclidean, but some non-Euclidean geometry. Could it be that we might want to revise our arithmetic in the same way?

It seems to me that we might. As fallibilists have argued, any theory that we employ may be revised under the pressure of recalcitrant evidence. Mathematical theories are no exception—as the history of geometry demonstrates. Whilst I have no situation to offer where the revision of arithmetic is currently plausible, it is easy enough to imagine the possibility of such things. Here, for example, is one where we might be inclined to revise our arithmetic in favour of $Th(\mathcal{M}_n^1)$ (where, for all $m \geq n$, $m = n$.)

Let us suppose that we come to predict a collision between an enormous star and a huge planet.[10] Using a standard technique, we compute their masses as x_1 and y_1, respectively. Since masses of this kind are, to within experimental error, the sum of the masses of the baryons (protons and neutrons) in them, it will be convenient to take a unit of measurement according to which a baryon has mass 1. In effect, therefore, these figures measure the numbers of baryons in the masses. After the collision, we measure the mass of the resulting (fused) body, and obtain the figure n, where n is much less than $x_1 + y_1$. Naturally, our results are subject to experimental error. But the difference is so large that it cannot possibly be explained by this. We check our instruments, suspecting a fault, but cannot find one; we check our computations for an error, but cannot find one. We have a puzzle. Some days later, we have the chance to record another collision. We record the masses before the collision. This time they are x_2 and y_2. Again, after the collision, the mass appears to be n (the same as before), less than $x_2 + y_2$. The first result was no aberration. We have an anomaly.

We investigate various ways of solving the anomaly. We might revise the theories on which our measuring devices depend, but there is no obvious way of doing this. We could say that some baryons disappeared in the collision; alternatively, we could suppose that under certain conditions the mass of a baryon decreases. But either of these options seems to amount to a rejection of the law of conservation of mass(-energy), which would seem to be a rather unattractive course of action.

---

[10]The following example comes, with minor revisions, from Priest (1996b).

Then someone, call them Einquine, fixes on the fact that the resultant masses of the two collisions were the same in both cases, n. This is odd. If mass has gone missing, why should this produce the same result in both cases? An idea occurs to Einquine. Maybe our arithmetic for counting baryons is wrong.[11] Maybe the appropriate arithmetic is $Th(\mathcal{M}_n^1)$. For in this arithmetic $x_1 + y_1 = x_2 + y_2 = n$, and our observations are explained without having to assume that the mass of baryons has changed, or that any are lost in the collisions! Einquine hypothesizes that n is a fundamental constant of the universe, just like the speed of light, or Planck's constant.[12]

While she is thus hypothesising, reports of the collisions start to come in from other parts of the galaxy. (The human race had colonised other planets some centuries before.) These reports all give the masses of the two new objects as the same, but all are different from each other. Some even measure them as greater that the sum of their parts. Einquine is about to give up her hypothesis, when she realises that this is quite compatible with it. Even if the observer measures the mass as m, provided only that $m \geq n$ then $m = n$, and their results are the same!

But this does leave a problem. Why do observers consistently *record* results that differ from each other? Analysing the data, Einquine sees that values of n (hers included), are related to the distance of the observer from the collision, d, by the (classical) equation $n = n_0 + kd$ (where n_0 and k are constants). In virtue of this, she revises her estimate of the fundamental constant to n_0, and hypothesizes that the effect of an inconsistent mass of baryons on a measuring device is a function of its distance from the mass. Further observational reports bear this hypothesis out; and Einquine starts to consider the mechanism involved in the distance-effect.

We could continue the story indefinitely, but it has gone far enough. For familiar reasons, there are likely to be theories other than Einquine's that could be offered to explain the data. Some of them might preserve orthodox arithmetic by jettisoning conservation laws, or by keeping these but varying some physical auxiliary hypotheses. Others might modify arithmetic in some other, but consistent, way. And each of these theories might become more or less plausible in the light of further experimentation, etc. But the point is made: it is quite possible that we might vary our arithmetic for empirical

---

[11]We already know that different sorts of fundamental particles satisfy different sorts of statistics.

[12]The revision of arithmetic envisaged here is a local one, in that it is only the counting of *baryons* that is changed. It would be interesting to speculate on what might happen which could motivate a global change, i.e., a move to a situation where everything is counted in the new way.

reasons. There can be alternative applied arithmetics, just as there are alternative applied geometries.

2.3. Consistent vs. Inconsistent Arithmetics

But now let us consider a stronger possibility—not just that we might want to apply an inconsistent arithmetic for some purpose, but that the correct pure arithmetic is one of the inconsistent ones. Which one? That is obviously an important question; but for the present, it does not need to be addressed.[13] The following considerations do not depend on which inconsistent arithmetic is at issue—or if they do, I will make this explicit.

The orthodox view is certainly that $Th(\mathcal{N})$ is the true arithmetic, not $Th(\mathcal{M})$, where $\mathcal{M}$ is some inconsistent model of arithmetic. Of course $Th(\mathcal{N})$ is true of $\mathcal{N}$, and $Th(\mathcal{M})$ is true of $\mathcal{M}$. That is not contentious. The question, then, is whether it is $\mathcal{N}$ or $\mathcal{M}$ that is the correct interpretation of the language. It might seem as though it is easy to resolve this issue, but it is not. A dispute between the proponent, A, of "standard arithmetic" and the proponent, B, of an inconsistent arithmetic is of a somewhat unusual kind. Anything (at least, anything arithmetic) that A endorses, B will endorse too. Thus, for example, A will insist that there is no greatest number ($\forall x \exists y\ y > x$); B will concur. The locus of disagreement will be in the fact that B will assert things that A will not wish to assent to. Why suppose A right and B wrong? A may point out that B's view of arithmetic is inconsistent; but unless they have some independent reason to suppose that inconsistency—or at least inconsistency in arithmetic—is a bad thing, this simply begs the question. A may, of course, attempt to mount a defence of consistency in general. I do not wish to enter into that debate here. Let me just say, for the record, that I am not aware of any very persuasive—and, in particular, non-question-begging—arguments for that conclusion.[14]

Are there any reasons, however, that push us towards endorsing an inconsistent arithmetic? One reason is that inconsistent arithmetics avoid some of the limitative results of the classical metatheory of arithmetic, and the unhappinesses associated with these.[15] Inconsistent arithmetics can do lots of things that consistent arithmetics cannot do. Thus, for example, as I have already noted, some inconsistent arithmetics are decidable. If one

---

[13]The question is discussed in the context of the finite models in Priest (1994b).

[14]See, e.g., Priest (1998).

[15]This matter is discussed further in Priest (1994a). I am not now happy with a number of the arguments used in that paper. Some reasons why are explained in Priest (1996a).

of these is the correct arithmetic then there is an algorithm for solving any arithmetical problem, which would certainly be very nice.

Another thing that inconsistent arithmetics can do is contain their own truth predicate; hence Tarski's theorem is avoided. Tarski's theorem shows that any theory that contains its own truth predicate is inconsistent; but this is obviously no problem in an inconsistent arithmetic! The language of arithmetic that we have been dealing with so far contains no truth predicate. However, it is well known[16] that any arithmetic based on LP can be extended conservatively with a truth predicate, T, satisfying the two way rule:

$$\frac{T\langle\alpha\rangle}{\alpha}$$

where $\langle\alpha\rangle$ is the numeral of the gödel number of α.[17]

Of course, since the extension of the language with a truth predicate is conservative, if we start with a consistent arithmetic, the purely arithmetic fragment of the theory with the truth predicate will also be consistent. So the inconsistency generated by the truth predicate gives no reason, as such, to suppose that the purely arithmetic fragment is inconsistent. But if one can have a truth predicate, excluding it from "pure arithmetic" is somewhat arbitrary. Truth has just as good a claim to be considered a logical predicate as the identity predicate. It should, therefore, be a part of all "pure theories".

2.4. Gödel's Theorems

Another thing that consistent arithmetic cannot do is provide a complete axiomatic theory. Inconsistent arithmetics can do this. As I have already noted, there are decidable complete inconsistent arithmetics; *a fortiori* they are axiomatic (and so, to point out the obvious, they can be specified by an axiom system in the usual way, quite independently of any consideration of collapsed models). In virtue of the methodological importance of axiomatisability in mathematics, this is a significant plus.

---

[16]See, e.g., Priest (2002), 8.1.

[17]It is worth noting also that any finite LP model of arithmetic will model all instances of the Induction Schema, however the language of arithmetic is extended. The Schema is of the form: $(\alpha(\mathbf{0}) \wedge \forall x(\alpha(x) \supset \alpha(x'))) \supset \forall x\alpha(x)$. With a little massaging, this can be seen to be equivalent to: $\neg\alpha(\mathbf{0}) \vee \exists x(\alpha(x) \wedge \neg\alpha(x')) \vee \forall x\alpha(x)$. Now, if the last disjunct is true, we are home. If not, there is some n such that $\alpha(\mathbf{n})$ fails, and since there is only a finite number of numbers in the domain, a least such n. Since $\alpha(\mathbf{n})$ fails, $\neg\alpha(\mathbf{n})$ holds. Thus, if $n = 0$ the first disjunct holds. If not, $n = m'$, and $\alpha(\mathbf{m})$, so the middle disjunct holds.

Naturally, it is worth asking what happens to the "Gödel undecidable sentence" in these arithmetics. Take any axiomatisable inconsistent arithmetic, Θ. Since it is axiomatisable, its membership can be represented in the theory of the standard model, $Th(\mathcal{N})$, by a formula of one free variable, $B(x)$. ($B(x)$ is of the form $\exists y\ Prov(y, x)$, where $Prov(y, x)$ represents the proof relation of Θ.) That is, for any sentence, α:[18]

If $\alpha \in \Theta$ then $B\langle\alpha\rangle \in Th(\mathcal{N})$.

If $\alpha \notin \Theta$ then $\neg B\langle\alpha\rangle \in Th(\mathcal{N})$.

Since $Th(\mathcal{N}) \subseteq \Theta$:

(1) If $\alpha \in \Theta$ then $B\langle\alpha\rangle \in \Theta$.

(2) If $\alpha \notin \Theta$ then $\neg B\langle\alpha\rangle \in \Theta$.

The undecidable sentence is a sentence, γ, of the form $\neg B\langle\gamma\rangle$. It is not difficult to see that both γ and $\neg\gamma$ are provable in Θ. For either $\gamma \in \Theta$ or $\gamma \notin \Theta$. In the first case, $B\langle\gamma\rangle = \neg\gamma \in \Theta$, by (1). In the second case, $\neg\gamma \in \Theta$, since Θ is complete, but $\neg B\langle\gamma\rangle = \gamma \in \Theta$, by (2). Either way, $\gamma \wedge \neg\gamma \in \Theta$. Note that, unlike the case of the contradiction connected with Tarski's Theorem, γ is a purely arithmetic sentence; that is, its vocabulary is just that of the pure language of arithmetic.

Given the inconsistency of the arithmetic in question, a consistency proof for it, and *a fortiori* a consistency proof within Θ, is not to be expected. Classically, of course, consistency and non-triviality go together; but in a paraconsistent context, this is not the case. In particular, though Θ is inconsistent, it is not trivial (unless it is produced by collapsing under the degenerate equivalence relation that relates everything to everything). And the non-triviality of Θ can be proved within Θ. In this sense, Gödel's second Incompleteness Theorem fails for inconsistent arithmetics. For take any unprovable sentence, α. Then since $\alpha \notin \Theta$, $\neg B\langle\alpha\rangle \in \Theta$, by (2). (Beware, however. This does not rule out $B\langle\alpha\rangle$ from being in Θ, too! We will return to this matter later.)

Finally, closely connected with Gödel's second Incompleteness Theorem is Löb's Theorem, to the effect that in classical arithmetics if $B\langle\alpha\rangle \supset \alpha$ is provable, so is α. It follows that not all instances of $B\langle\alpha\rangle \supset \alpha$ are provable. But this seems odd. All such sentences are clearly true; how is it that truths that seem as innocent as these *must* fail to be provable? In Θ, as one would

---

[18]I write $B\langle\alpha\rangle$ instead of $B(\langle\alpha\rangle)$, etc., for ease of readability.

expect, all instances are provable. For $\alpha \in \Theta$ or $\alpha \notin \Theta$. In the second case, $\neg B\langle\alpha\rangle \in \Theta$, by (2). In either case, $\neg B\langle\alpha\rangle \vee \alpha \in \Theta$. Note, also, that all instances of the converse are also provable. For if $\alpha \in \Theta$, then $B\langle\alpha\rangle \in \Theta$, by (1), so $\neg\alpha \vee B\langle\alpha\rangle \in \Theta$. And if $\alpha \notin \Theta$, $\neg\alpha \in \Theta$, since Θ is complete; hence $\neg\alpha \vee B\langle\alpha\rangle \in \Theta$. In a sense then, B is a truth predicate since $B\langle\alpha\rangle \equiv \alpha \in \Theta$ (though this does not necessarily mean that $B\langle\alpha\rangle$ and α have the same truth values in the collapsed model).

2.5. The Naive Notion of Proof

We see, then, that inconsistent arithmetics can do a lot of nice things, and can avoid a number of features that many have held to be problematic for consistent arithmetic. This does not demonstrate that true arithmetic is inconsistent, but it certainly moves us in this direction. There are considerations that drive us further.

As is clear to anyone who is familiar with Gödel's theorem, at it's heart there lies a paradox. Informally, the "undecidable" sentence is the sentence 'This sentence is not provable'. Suppose that it is provable; then since whatever is proved is true, it is not provable. Hence, it is not provable. But we have just proved this. So it is provable after all (as well).[19] Let us look at this paradox more closely.

When mathematicians establish things to be true, they give proofs. These are informal deductive arguments, appealing to things which have already been proved or, ultimately, from things that are obviously true, and so where no proof is required. I will call the notion of proof in question here the *naive* notion of proof. Let us restrict ourselves to what can be proved naively about natural numbers. The language of naive proof about numbers is standard mathematical English (or some other natural language), but it is natural to suppose that this can be regimented into a suitable formal language, so that sentences may be assigned gödel-codes. Let us write $B_N(x)$ as a predicate of natural numbers which expresses the fact that x is (the code of) a sentence that is naively proved. B_N satisfies the following principles:

(3) $\vdash B_N\langle\alpha\rangle \supset \alpha$

(4) If $\vdash \alpha$ then $\vdash B_N\langle\alpha\rangle$

where $\vdash$ records naive proof. For (3): it is analytic that whatever is naively proved is true. Naive proof just is that sort of mathematical argument that

[19] The paradox is of the same kind as the "Knower paradox"; see Priest (1995), 10.5.

establishes something as true. And since this is analytic, it is itself naively provable. (Whether it is axiomatic or is derivable from more fundamental principles, we do not need to go into here). For (4): if something is naively proved then this fact itself constitutes a proof that α is proved.

But from these two principles, we can show that $\vdash$ is inconsistent. By usual methods of self-reference, we can construct a sentence that says of itself that it is not provable, i.e., a sentence, γ, of the form $\neg B_N \langle\gamma\rangle$. Substituting in (3) gives us $\vdash B_N \langle\gamma\rangle \supset \neg B_N \langle\gamma\rangle$, i.e., $\vdash \neg B_N \langle\gamma\rangle \vee \neg B_N \langle\gamma\rangle$. Thus, $\vdash \neg B_N \langle\gamma\rangle$; that is, $\vdash \gamma$. By (2), $\vdash B_N \langle\gamma\rangle$ (i.e., $\vdash \neg\gamma$). Arithmetic is therefore inconsistent.

I have not assumed that B_N is itself a predicate that can be constructed from the usual arithmetic vocabulary ($'$, $+$, $\times$, $=$). But there are, in fact, reasons to suppose that it can be.[20] It is part of the very notion of proof that a proof should be effectively recognised as such. For the very point of a proof is that it gives us a way of settling whether something is true. It is, therefore, a proof only when it is recognised as such. Thus, Dummett, for example, has stressed the point: it is part of the very notion of proof, unlike truth, that we can recognise one when we see it—at least in principle.[21]

Moreover, proof of the kind in question is a human practice. It is one that must be taught and learned. The human brain is, presumably, some sort of finite-state machine. It could not grasp the notion of proof if this were not axiomatic; if it were not, it would transcend the abilities of such a machine. For similar reasons, one must suppose that the grammar of any speakable language must be generated by a decidable set of rules. It might be pointed out that standards of proof may change over time, and that there is no reason to suppose that the change, itself, must happen in a rule-governed way. Indeed so. But we may take naive proof to comprise the standards of proof that are in operation here and now.

If naive proof in this sense is, indeed, axiomatic, then we can find a Σ_1 sentence of the standard language of arithmetic that expresses B_N. That is, arithmetic, as expressed in the usual vocabulary, is itself inconsistent. Nor is this technically unfeasible. In 2.4, we have already seen how a pure arithmetic can contain its own proof predicate and the attendant contradiction concerning its Gödel sentence.

---

[20] This argument is developed further in Priest (1987), ch. 3.

[21] See Dummett (1975).

2.6. Shapiro's Criticisms

In (2002) Shapiro, following ideas of Priest (1987), constructs an axiomatic theory, PA^*, that can prove its own Gödel sentence. The language of the theory contains a truth predicate, which is involved in the proof of the sentence;[22] but the Gödel sentence itself, as Shapiro emphasises, is purely arithmetic, employing only the proof predicate for the theory, which, being axiomatisable, is expressible in terms of $'$, $+$, $\times$ and $=$. Actually, the exact details of PA^* are left somewhat under-determined; but we need not go into that here. What I want to discuss are the unpalatable consequences that Shapiro supposes to follow from the fact that this theory can prove its own Gödel sentence. The features that Shapiro points to are possessed just as much by the axiomatisable inconsistent arithmetics that we looked at in section 2.4. I will therefore discuss his objections in the context of these.[23]

In the inconsistent arithmetic, Θ, both γ and $\neg\gamma$ are provable, where γ is purely arithmetic, and is of the form $\neg B\langle\gamma\rangle$. Since γ is provable, there is some number, g, which is the code of its proof. Hence, $Prov(\mathbf{g}, \langle\gamma\rangle)$ is true in the standard model, and so is provable in Θ. But $\neg B\langle\gamma\rangle$ is $\neg\exists y\, Prov(y, \langle\gamma\rangle)$, i.e., $\forall y \neg Prov(y, \langle\gamma\rangle)$. Hence $\neg Prov(\mathbf{g}, \langle\gamma\rangle)$ is provable as well. Now, $Prov(x, y)$ expresses a primitive recursive relation. Hence, if Θ is the true arithmetic, we have to accept that there are inconsistencies concerning numbers that are of this very basic kind. Worse, consider the following biconditionals. From left to right, they are unproblematic. Suppose that we accept them from right to left too.

P+ m is the code of a proof with of formula with code n iff $Prov(\mathbf{m}, \mathbf{n}) \in \Theta$

P− m is the not code of a proof with of formula with code n iff $\neg Prov(\mathbf{m}, \mathbf{n}) \in \Theta$

Then we have to accept that some number both is and is not the code of a

[22] The argument appeals to the claim that whatever is provable is true: $\forall x(B_N(x) \supset Tx)$ (plus the T-schema for the truth predicate). Given the meaning of the naive notion of proof, this is certainly analytically true, though it may well be provable from more fundamental things. There is, however, a way that the truth predicate can, in fact, be dispensed with. All that we actually need are the instances of the schema $B_N\langle\alpha\rangle \supset \alpha$, which are, equally, analytic. In fact, given the conditionals of this sort, the proof of the "undecidable sentence", $\gamma = \neg B_N\langle\gamma\rangle$, reduces to a very simple form. Substituting γ in the schema gives: $\neg B_N\langle\gamma\rangle \vee \gamma$. That is, $\gamma \vee \gamma$, i.e., γ.

[23] The material in this section arose from a seminar at the University of St Andrews at the end of 2002. I am grateful to the participants, and particularly to Steward Shapiro, for their helpful comments.

proof, and, more generally, that something could be both provable and not provable. What could this mean?

Shapiro offers three responses to this situation:

A Reject the soundness of Θ, on the basis of the fact that primitive recursive relationships are consistent.

B Accept that Θ is sound, but reject the biconditionals **P+** and **P−**, and hence, on the assumption that Θ is the true arithmetic, the isomorphism between numbers with their operations and strings with theirs.

C Accept that Θ is sound, the biconditionals **P+** and **P−**, and hence that something can both provable and not provable.

All of these options, Shapiro argues, should be resisted. If one is to take seriously the idea that Θ is the true arithmetic, option **A** is obviously not the way to go. One has to accept that even primitive recursive relations may be inconsistent. But this is not news. In the finite models of arithmetic even numerical equations can be inconsistent; that is, there can be truths of the form $\mathbf{m} = \mathbf{n} \wedge \mathbf{m} \neq \mathbf{n}$. One also has to accept, more generally, that even the computational part of mathematics is inconsistent. But this is not a problem either. Θ itself tells us exactly what an inconsistent computation theory is like. The Δ_0 formulas (that is, the sentences obtainable from equations using connectives and bounded quantifiers) express the recursive properties/relations.

Option **B** certainly involves jettisoning a connection in terms of which logicians have become accustomed to thinking. This is certainly a loss, though I do not think it as devastating as Shapiro does. However, it seems to me that the simplest and most natural response is option **C**, so I will discuss this option at length. Shapiro marshals essentially two considerations against it. Let us consider these in turn.

2.7. The Inconsistency of Peano Arithmetic

Shapiro's first objection, and the quicker to deal with, is that if one holds that primitive recursive relations are inconsistent, it follows not just that Θ is inconsistent, but that Peano Arithmetic (PA) is inconsistent—which seems implausible. The reason is that all recursive relationships are known to be representable in PA.

The reply is simply that if the recursive relationships are as specified by Θ, they are *not* all representable in PA—just because it is consistent. Where does the proof of the fact that all recursive relationships are representable in PA break down, however? The answer depends on which proof we are talking about, and on which inconsistent theory of arithmetic is correct. But let us suppose, for the sake of argument, that Θ is $Th(\mathcal{M}^6_{10})$ (refer to the diagram of 1.4); and let us look at a direct proof to the effect that the formula $x = y'$ represents the successor relation in PA.[24] We need to show that if $i = j'$ then $\mathbf{i} = \mathbf{j}' \in PA$. This is proved by induction on j. Suppose that $j = 0$. Then if $i = 0'$, $i = 1$, and $\mathbf{1} = \mathbf{0}' \in PA$. Now suppose that the result holds for j, and show it for j'. So suppose that $i = j''$. Since i is not 0, there is a k such that $i = k'$. Hence, $k' = j''$, and $k = j'$. By induction, $\mathbf{k} = \mathbf{j}' \in PA$; so $\mathbf{i} = \mathbf{k}' = \mathbf{j}'' \in PA$.

The second part of this argument breaks down for $Th(\mathcal{M}^6_{10})$, since a number may have multiple predecessors, some of them *greater* than itself. Thus, suppose that j is 8. If $i = 8''$ then certainly for some k, $k' = 8''$; k can be 9 or 15. Now, $9' = 8''$, $9 = 8'$, $\mathbf{9} = \mathbf{8}' \in PA$ (by induction), and so $\mathbf{9}' = \mathbf{8}'' \in PA$. But, though $15' = 8''$, it does not follow that $15 = 8'$, so the argument breaks down. Indeed, $\mathbf{15} = \mathbf{8}' \notin PA$.

Thus, and in general, if you take Θ to provide the correct account of recursive relationships, then these will be representable (trivially) in Θ; but PA will be incomplete, since it captures only a consistent fragment of the truth. Dually, of course, if you take the usual classical line on recursive relationships, PA will be complete, but Θ will give *more* than the truth, because it is inconsistent. In other words, if you match up the formal arithmetic and the theory of recursive relations properly, then you will get representability. But if you mis-match these by taking one to be consistent and the other not, then things will go wrong.

2.8. The Incredulous Stare

Shapiro's other main objection amounts to a version of the incredulous stare. Let me put the start of it in his own words:[25]

> On all accounts—including the non-dialetheic perspective—we have that g is the code of a Θ-derivation of γ. This can be verified with a painstaking, but completely effective check. How can the dialetheist go on to maintain that, in addition, g is *not* the code of a Θ-derivation of γ? What does it mean to say this?

[24] Proofs of this kind can be found in Boolos and Jeffrey (1974), ch. 14, Part III.

[25] Shapiro (2002), p. 828. I have changed the notation to bring it in line with that used in this essay. The italics are original.

> Since $\neg Prov$ is recursive predicate, we can supposedly verify—at the same time, in almost exactly the same way—that g is not the code of a Θ-derivation of γ. How?

Shapiro asks how we can possibly verify a sentence expressing a recursive relation and its negation? What can this mean?

In principle, the answer is easy. Since we are endorsing **P+** and **P−**, we are now taking seriously the thought that metatheoretic sentences may be contradictory. If so, they must play by the same rules as those of Θ, and in particular, be based on the logic LP (or FDE). In any theory based on LP or FDE, α and $\neg\alpha$ are verified by *different* procedures. Thus, e.g., to determine whether $t_1 = t_2$ is true, we have to look to see whether $\langle d(t_1), d(t_2)\rangle \in E_=$. To determine whether $t_1 \neq t_2$ is true, we have to look to see whether $\langle d(t_1), d(t_2)\rangle \in A_=$. These are separate matters. Thus, in $Th(\mathcal{M}^6_{10})$, once we have checked to see whether $\mathbf{i} = \mathbf{j}$, the question of whether $\mathbf{i} \neq \mathbf{j}$ is a *further* question. $\mathbf{0} = \mathbf{0}$ is true, but $\mathbf{0} \neq \mathbf{0}$ is not. $\mathbf{10} = \mathbf{10}$ is true, but so is $\mathbf{10} \neq \mathbf{10}$.

Thus, to bring the matter to bear on proof explicitly, suppose that g is the code of a proof of γ. Suppose, for the sake of argument, that the code is 37. Then to say that g is the code is to say something equivalent to $g = 37$. What does it mean to say that it is also not the code of a proof of γ? It is to say that $g \neq 37$ as well. This is the case if $37 = 37 \wedge 37 \neq 37$, which it can be in an inconsistent arithemtic.

And what does it mean to say that γ is both provable and not? To say that it is provable is to say that $\exists x$(x is the code of a proof of γ), i.e., on the supposition at hand, $\exists x\, x = 37$. To say that it is not provable is to say that $\neg\exists x$(x is the code of a proof of γ) i.e., $\forall x \neg$(x is the code of a proof of γ), i.e., $\forall x \neg(x = 37)$, i.e., $0 \neq 37 \wedge 1 \neq 37 \wedge ... \wedge 37 \neq 37 \wedge ...$; which is, of course, true if $37 \neq 37$. In other words, to say that γ is not provable is to say that every number is distinct from a code of the proof of γ. This does not rule out there being a proof of γ. (In general, the truth of $\neg\alpha$ in a paraconsistent setting does not rule out the truth of α.) In particular, it will hold if the proof is distinct from itself. And how can a proof be distinct from itself? In the same way that a number can. After all, on option **C**, the one at issue, we are retaining the structural identity between strings and numbers. Both are, after all, abstract objects. And the inconsistent behaviour of strings is just as good or bad as the inconsistent behaviour of numbers.[26]

[26]It is worth noting that if numbers have inconsistent properties, then this will affect their behaviour whatever theory they are taken to be coding. Thus, the gödel codes

There are, or course, concrete objects whose behavior in some sense represents the behaviour of abstract objects. In this case, there are things such as marks of dried ink on paper that represent abstract proofs. But properties of abstract objects do not carry over of necessity to their physical representations. Thus, two tokens of one and the same proof type need not be identical. One would expect them, instead, to satisfy some sort of graphical relation, R. What of two proof tokens that represent different proof types? Arguably, they must satisfy some quite distinct relation, R', where both R and R' may consistently hold between tokens.

To illustrate, and for the sake of simplicity, consider not proofs, but letters. Let R_o be the relationship that holds between two tokens of 'o', R_c be the relationship that holds between two tokens of 'c', R be the relationship that holds between two tokens of the same letter-type, and R' be the relationship that holds between two tokens of different letter-types. It is natural to suppose that, whatever R and R' are, they satisfy at least the following conditions:

If xR_oy then xRy

If xR_oy and zR_cw then $xR'z$

Now let t_1, t_2 and t_3 be three token letters such that t_1 is a clear case of an 'o', that t_3 is a clear case of a 'c', but that t_2 is slightly ill-formed, and could be either an 'o' or a 'c'. Then we have $t_1R_ot_2$ (t_1 and t_2 are both 'o's) and $t_2R_ct_3$ (t_2 and t_3 are both 'c's). It follows that t_1Rt_2 and $t_1R't_2$.

We could also, of course, go down the dialetheic path: insist that identity is represented by the relation R, that difference is represented by its negation $\overline{R}$, but hold that, on some occasions at least, physical reality realises both xRy and $x\overline{R}y$. Thus, for example, t_2 might be a borderline case in a sorites sequence of tokens starting with clear 'o's and ending with things that are clearly not 'o's. If borderline cases realise truth value gluts, t_2 may be both an 'o' and not an 'o'. But this is not the place to pursue such metaphysical speculations.

---

of PA will behave just as inconsistenly as those of Θ. In other words, if numbers are inconsistent, we may expect things to be both provable and not provable in PA just as much as in Θ. This does not, of course, mean that PA is itself inconsistent. As to where the inconsistency of gödel codes arises, it might be only for numbers so large that they are larger than anything that is humanly meaningful. (See Priest (1994a).)

2.9. Conclusion

Shapiro's objections stem from being half-hearted about dialetheism. If one endorses an inconsistent arithmetic, but tries to hang on to either a consistent computational theory or a consistent metamathematics of proof, one is in for trouble. The solution to his problems is therefore not to be half-hearted, and to accept that these other things are inconsistent too. Indeed, the arithmetic itself shows us how to do this: the facts about computability and provability are simply read off from the arithmetic.[27]

Discussions in the philosophy of mathematics are always built on shaky foundations if they are not underpinned with the appropriate technical material. This is certainly true of discussions of the inconsistency of arithmetic. The inconsistent models show us exactly what can be done and how. That hardly settles many of the interesting philosophical questions. But it does put a firm skeleton below the philosophical flesh.

References

[1] Batens, D. (200a), 'The Demise of Rich Finitism; a Study in the Limitations of Paraconsistency' to appear.

[2] Boolos, G. and Jeffrey, R. (1974), *Computability and Logic*, Cambridge: Cambridge University Press.

[3] Chihara, C. (1984), 'Priest, the Liar and Gödel', *Journal of Philosophical Logic* 13, 117-24.

[4] Denyer, N. (1995), 'Priest's Paraconsistent Arithmetic', *Mind* 104, 567-75.

[5] Dummett, M. (1975), 'The Philosophical Basis of Intuitionist Logic', pp. 5-40 of H. Rose and J. Shepherdson (eds.), *Logic Colloquium '73*,

[27] The philosophical discussion has appealed to various metatheoretic properties of inconsistent arithmetics. How were these established? A natural assumption is that they were proved in a classical (consistent) metatheory, such as ZF. If we are now endorsing an inconsistent (meta-)arithmetic, we can no longer be working in ZF. What entitles us to be sure that we may still invoke those results? One answer goes essentially as follows. Start with a model of ZF, say (an initial segment of) the cumulative hierarchy. Then use the Collapsing Lemma to produce a collapsed model of ZF in which the structure of the numbers brings it into line with the inconsistent arithmetic we are envisaging. (For collapsed models of ZF, see Priest (1995), Part 3, Technical Appendix.) We can take the theory of that collapsed model to provide the metatheory in which we are working. And just as any theorem of standard arithmetic holds in the theory of a collapsed model of arithmetic, so any theorem of ZF holds in that theory.

Amsterdam: North Holland; reprinted as ch. 14 of Dummett's *Truth and Other Enigmas*, London: Duckworth (1978).

[6] Dunn, J. M. (1979), 'A Theorem of 3-Valued Model Theory with Connections to Number Theory, Type Theory and Relevance', *Studia Logica* 38, 149-69.

[7] Gasking, D. A. T. (1940), 'Mathematics and the World', *Australasian Journal of Philosophy* 18, 97-116; reprinted as pp. 390-403 of P. Benacerraf, and H. Putnam, (eds.), *Philosophy of Mathematics; Selected Readings*, Oxford: Oxford University Press, 1964.

[8] Kaye, R. (1991), *Models of Peano Arithmetic*, Oxford: Clarendon Press.

[9] Meyer, R. K. (1976), 'Relevant Arithmetic', *Bulletin of the Section of Logic, Polish Academy of Sciences* 5, 133-7.

[10] Meyer, R. K. and Mortensen, C. (1984), 'Inconsistent Models for Relevant Arithmetics', *Journal of Symbolic Logic* 49, 917-29.

[11] Mortensen, C. (1987), 'Inconsistent Nonstandard Arithmetic', *Journal of Symbolic Logic* 52, 512-8.

[12] Mortensen, C. (1988), 'Inconsistent Number Systems', *Notre Dame Journal of Formal Logic* 29, 45-60.

[13] Mortensen, C. (1990), 'Models for Inconsistent and Incomplete Differential Calculus', *Notre Dame Journal of Formal Logic* 31, 274-85.

[14] Mortensen, C. (1995), *Inconsistent Mathematics*, Dordrecht: Kluwer Academic Publishers.

[15] Mortensen, C. and Meyer, R. K. (1985), 'Relevant Quantum Arithmetic', pp. 211-26 of L. De Alcantara (ed.), *Mathematical Logic and Formal Systems*, New York, NY: Marcel Dekker.

[16] Nelson, D. (1959), 'Negation and Separation of Concepts in Constructive Systems', pp. 208-25 of A. Heyting (ed.), *Constructivity in Mathematics*, Amsterdam: North Holland Publishers.

[17] Priest, G. (1979), 'Logic of Paradox', *Journal of Philosophical Logic* 8, 219-41.

[18] Priest, G. (1987), *In Contradiction*, Dordrecht: Kluwer Academic Publishers.

[19] Priest, G. (1991), 'Minimally Inconsistent *LP*', *Studia Logica* 50, 321-31.

[20] Priest, G. (1994a), 'Is Arithmetic Consistent?', *Mind* 103, 321-31.

[21] Priest, G. (1994b), 'What Could the Least Inconsistent Number be?', *Logique et Analyse* 37, 3-12.

[22] Priest, G. (1995), *Beyond the Limits of Thought*, Cambridge: Cambridge University Press; second and revised edition, Oxford: Oxford University Press, 2002.

[23] Priest, G. (1996a), 'On Inconsistent Arithmetics: Reply to Denyer', *Mind* 105, 649-59.

[24] Priest, G. (1996b), 'On Alternative Geometries, Arithmetics and Logics: a Tribute to Łukasiewicz', a paper given at the conference *Łukasiewicz in Dublin*; *Studia Logica* 74 (2003), 441-68.

[25] Priest, G. (1997), 'Inconsistent Models of Arithmetic; I Finite Models', *Journal of Philosophical Logic* 26, 223-35.

[26] Priest, G. (1998), 'What's so Bad about Contradictions?', *Journal of Philosophy* 95, 410-26.

[27] Priest, G. (2000), 'Inconsistent Models of Arithmetic; Part II, the General Case', *Journal of Symbolic Logic* 65, 1519-29.

[28] Priest, G. (2001), *Introduction to Non-Classical Logic*, Cambridge: Cambridge University Press.

[29] Shapiro, S. (2002), 'Incompleteness and Inconsistency', *Mind* 111, 817-32.

[30] Van Bendegem, J.-P. (1993), 'Strict, Yet Rich Finitism', pp. 61-79 of Z. W. Wolkowski (ed.), *First International Symposium on Gödel's Theorems*, Singapore: World Scientific Press.

[31] Van Bendegem, J.-P. (1994), 'Strict Finitism as a Viable Alternative in the Foundations of Mathematics', *Logique et Analyse* 37, 23-40.

Serguei P. Odintsov*
Heinrich Wansing

Inconsistency-tolerant Description Logic: Motivation and Basic Systems

Abstract. In this paper, three systems of constructive and inconsistency-tolerant description logic are motivated and defined semantically. Moreover, sound and complete tableau calculi for these systems are presented.

Keywords: description logic, constructive logic, paraconsistent logic, modal logic, strong negation.

1. Introduction

Description logics are among the most prominent formalisms for knowledge representation. Many expressive but still decidable description logics have been investigated and implemented, see [3], [4], [9], [17]. There is some work on description logics using non-monotonic, many-valued, or fuzzy-logic, see [4, Chapter 6], however, the underlying logic of almost all systems of description logic is *classical*. One of the rare exceptions is the work [22] by Patel-Schneider. The author considers a four-valued semantics for description logic, but he does not introduce non-classical connectives into the object language. In the present paper, we shall replace the classical base logic by a *non-classical logic*, namely Nelson's four-valued constructive and paraconsistent logic N4, see, for example, [2], [11], [18], [19], [27], [28] [29]. In particular, we shall replace material implication by intuitionistic implication and Boolean negation by constructive, strong negation.[1]

It is well-known that the basic description logic $\mathcal{ALC}$ may be regarded as (is nothing but) the multi-modal extension of the smallest normal modal

---

* *S.P. Odintsov acknowledges support by the Alexander von Humboldt-Stiftung.*

[1]There are other paraconsistent logics that may be used to obtain inconsistency-tolerant description logics. Belnap's useful four-valued logic [5], for example, is a natural candidate. Note, however, that Belnap's four-valued logic has no genuine implication satisfying the Deduction Theorem.

V. F. Hendricks and *J. Malinowski* (eds.),
Trends in Logic: 50 Years of Studia Logica

Studia Logica **21**: 301–335, 2003.

propositional logic K. In the literature on intuitionistic modal logic, different proposals have been made for an intuitionistic counterpart of K, see, for example, [7], [12], [13], [15], [20], [21], [30], and in particular the discussion in [23]. One basic system is Sotirov's $\mathsf{IK}_0(\Box\Diamond)$ [24]. In the relational semantics for $\mathsf{IK}_0(\Box\Diamond)$, three accessibility relations are used: one for intuitionistic implication, one for $\Box$ and another one for $\Diamond$. Other systems are Božić's and Došen's $\mathsf{HK}\Box$ and $\mathsf{HK}\Diamond$ [10], [14, Chapter 10], [30], [31]. In this approach, interpreting both $\Box$ and $\Diamond$ with respect to the same accessibility relation gives rise to some problems. The system $\mathsf{HK}\Box\Diamond$ in [10], for example, fails to inherit the disjunction property from intuitionistic logic. $\mathsf{IK}_0(\Box\Diamond)$ may be described as the fragment of *classical* first-order logic obtained by the standard translation of the modal intuitionistic language into the language of classical first-order logic with a unary predicate for every propositional variable (concept name), a binary relation symbol for every modal operator, and one binary relation symbol for intuitionistic implication. Every formula of this fragment of first-order logic has exactly one free variable (for the state of evaluation), and certain conditions are imposed on the interaction between the relations denoted by the relation symbols. We shall see that the use of strong, constructive negation instead of intuitionistic negation overcomes the problems encountered with $\mathsf{HK}\Box\Diamond$. The translation of constructive description logic with strong negation into classical predicate logic induces a semantics. Since in the semantics for classical first-order logic formulas are not evaluated at states or possible worlds, both the role restrictions and implication are interpreted with respect to a single set of entities, namely individual/state pairs in which the state is left implicit.

Another basic intuitionistic modal logic is Fischer Servi's system FS, [12], [13], [14], [15]. This system corresponds to the fragment of *intuitionistic* first-order logic obtained by the standard translation into the language of intuitionistic predicate logic. Hence, as a result, there exists more than one basic constructive version of $\mathcal{ALC}$ with intuitionistic implication and constructive, strong negation. We shall refer to the system obtained by the translation into classical predicate logic as $\mathcal{CALC}^{\mathsf{C}}$ (constructive classical $\mathcal{ALC}$). Another constructive description logic is obtained by a translation into quantified N4, QN4. This translation treats the role restrictions $\forall r.$ and $\exists r.$ as restricted universal and existential quantifiers, and the resulting logic will be called $\mathcal{CALC}^{\mathsf{N4}}$. In $\mathcal{CALC}^{\mathsf{N4}}$, the role restrictions $\forall r.$ and $\exists r.$ fail to be dual operators with respect to strong negation. Duality can be obtained by another translation into QN4 that treats $\forall r.$ and $\exists r.$ as unary operators. The resulting system will be called $\mathcal{CALC}^{\mathsf{N4}d}$.[2] In this paper we shall motivate and define the three distinct basic systems $\mathcal{CALC}^{\mathsf{C}}$, $\mathcal{CALC}^{\mathsf{N4}}$, and $\mathcal{CALC}^{\mathsf{N4}d}$. In Section 2, we recall the syntax and semantics of $\mathcal{ALC}$ and consider various examples that motivate paraconsistent constructive description logic. Each of the three systems of constructive description logic to be developed can be seen as a response to these examples. Section 3 is devoted to a presentation of $\mathcal{CALC}^{\mathsf{C}}$ and some of its properties. In Section 4, the first-order extension QN4 of Nelson's N4 is introduced. Completeness with respect to a relational semantics is proved in order to define the description logic $\mathcal{CALC}^{\mathsf{N4}}$. The system $\mathcal{CALC}^{\mathsf{N4}d}$, in which the universal and existential role restrictions are duals of each other, is presented in Section 5. In Section 6 we introduce sound and complete tableau calculi for the three constructive description logics. Finally, in Section 7 we conclude with some summarizing remarks and a brief outlook.

2. Restricted subsumption, inconsistent data, and incomplete information

2.1. $\mathcal{ALC}$

The language of the basic description logic $\mathcal{ALC}$ consists of a non-empty set N_C of concept names, a non-empty set N_r of role names and a certain set of concept constructors used to build complex concept descriptions. The syntax is defined as follows:

concept names:	$A \in N_C$
role names:	$r \in N_r$
concept descriptions:	$C \in Con(N_C, N_r)$

$$C ::= A \mid \neg C \mid (C \sqcap D) \mid (C \sqcup D) \mid \forall r.C \mid \exists r.C$$

We shall sometimes omit outermost brackets (also in the syntax of $\mathcal{CALC}^{\mathsf{C}}$, $\mathcal{CALC}^{\mathsf{N4}}$, and $\mathcal{CALC}^{\mathsf{N4}d}$). The semantics for $\mathcal{ALC}$ employs interpretations $\mathcal{I} = \langle \Delta^{\mathcal{I}}, \cdot^{\mathcal{I}} \rangle$, where $\Delta^{\mathcal{I}}$ is a non-empty set, the individual domain of $\mathcal{I}$. The interpretation function $\cdot^{\mathcal{I}}$ maps every $A \in N_C$ to a set $A^{\mathcal{I}} \subseteq \Delta^{\mathcal{I}}$ and every

[2]Duality between $\Box$ and $\Diamond$ is the main motivation for Akama [1] to introduce a modal extension of Nelson's three-valued system N3. Note that Akama uses Fischer Servi's truth definition for formulas $\Box\varphi$ but refers to Božić's and Došen's HK$\Box$ and HK$\Diamond$ instead of FS. Akama comments upon intuitionistic modal logics comprising both $\Box$ and $\Diamond$ but does not refer to Božić's and Došen's HK$\Box\Diamond$. Moreover, Akama's definition of a canonical model is erroneous, because the canonical model validates $p \vee \sim p$ for propositional variables p.

constructor	semantics
concept name A	$A^{\mathcal{I}} \subseteq \Delta^{\mathcal{I}}$
negation $\neg C$	$\Delta^{\mathcal{I}} \setminus C^{\mathcal{I}}$
conjunction $(C \sqcap D)$	$C^{\mathcal{I}} \cap D^{\mathcal{I}}$
disjunction $(C \sqcup D)$	$C^{\mathcal{I}} \cup D^{\mathcal{I}}$
universal restriction $\forall r.C$	$\{x \in \Delta^{\mathcal{I}} : \forall y\, (x r^{\mathcal{I}} y \Rightarrow y \in C^{\mathcal{I}})\}$
existential restr. $\exists r.C$	$\{x \in \Delta^{\mathcal{I}} : \exists y\, (x r^{\mathcal{I}} y \;\&\; y \in C^{\mathcal{I}})\}$

Table 1. The semantics of concept descriptions in $\mathcal{ALC}$

role name $r \in N_r$ to a binary relation $r^{\mathcal{I}}$ on $\Delta^{\mathcal{I}}$. The function is extended to arbitrary concept descriptions as stated in Table 1. This simple yet expressive language may be used to represent and reason about knowledge (or, perhaps better, information). Moreover, algorithms are available for testing whether certain concepts are non-empty or subsumed by other concepts.

EXAMPLE 1. *Suppose the following concept descriptions and corresponding sets of individuals are of interest:*

$\forall$`child.Male`
the set of all individuals whose children are male

`Mother`$\sqcup\forall$`child.Live-in-Europe`
the set of all individuals who are a mother or all whose children live in Europe

`Father`$\sqcap\exists$`child.`$\neg$`Live-in-Europe`
the set of all individuals who are a father and have a child that does not live in Europe

Possible queries are, for example:

Is there an $\mathcal{I}$ such that (`Mother`$\sqcup\forall$`child.Live-in-Europe`)$^{\mathcal{I}} \neq \emptyset$ *?*

Does it hold for all interpretations $\mathcal{I}$ that
(`Father`$\sqcap\exists$`child.`$\neg$`Lives-in-Europe`$\sqsubseteq$ $\forall$`child.Male`)$^{\mathcal{I}}$
($=_{def}$ ($\neg$(`Father`$\sqcap\exists$`child.`$\neg$`Lives-in-Europe`) $\sqcup$ $\forall$`child.Male`)$^{\mathcal{I}}$) $= \Delta^{\mathcal{I}}$ *?*
(Is every father who has a child that does not live in Europe an individual all whose children are male?)

DEFINITION 1. *Let $C, D \in Con(N_C, N_r)$, and let $(C \sqsubseteq D) =_{def} (\neg C \sqcup D)$. C is subsumed by D (in symbols $C \models D$) iff for every interpretation $\mathcal{I}$, $(C \sqsubseteq D)^{\mathcal{I}} = \Delta^{\mathcal{I}}$, i.e., $C^{\mathcal{I}} \subseteq D^{\mathcal{I}}$. C is said to be satisfiable iff there exists an interpretation $\mathcal{I}$ such that $C^{\mathcal{I}} \neq \emptyset$.*

The syntax of description logics is extended by a non-empty set of individual names N_I. This extension allows the formulation of simple assertions that may be collected in so-called ABoxes.

DEFINITION 2. *An ABox is a finite set of expressions of the form* $C(a)$ *or* $r(a,b)$*, where* $a, b \in N_I$*,* $r \in N_r$*, and* $C \in Con(N_C, N_r)$*. An expression* $C(a)$ *is called a concept assertion, and an expression* $r(a,b)$ *is called a role assertion. Concept assertions and role assertions are also called ABox statements.*

Note that, for example, the following expressions are not well-formed: $C(a) \sqcup C(b)$ (object language syntax error), $C(a) \models C(b)$ (metalanguage syntax error). The function $\cdot^{\mathcal{I}}$ of an interpretation $\mathcal{I} = \langle \Delta^{\mathcal{I}}, \cdot^{\mathcal{I}} \rangle$ may be extended to apply also to individual names such that $a^{\mathcal{I}} \in \Delta^{\mathcal{I}}$. Such an interpretation is a model of an ABox $\mathcal{A}$ ($\mathcal{I} \models \mathcal{A}$) iff for every $C(a) \in \mathcal{A}$, $a^{\mathcal{I}} \in C^{\mathcal{I}}$ and for every $r(a,b) \in \mathcal{A}$, $a^{\mathcal{I}} r^{\mathcal{I}} b^{\mathcal{I}}$. The ABox $\mathcal{A}$ is called satisfiable iff it has a model.

2.2. Restricted subsumption

In $\mathcal{ALC}$ and other description logics based on classical logic, subsumption of concepts $C \models D$ is defined using material implication $\neg C \sqcup D$. It may be argued, however, that postulating $C^{\mathcal{I}} \subseteq D^{\mathcal{I}}$, for all $\mathcal{I}$, overgenerates. In particular, we want to be able to conclude that if $C \models D$, then $\forall r.C \models \forall r.D$ and $\exists r.C \models \exists r.D$, but for this we do not need material implication. Conceive of an element of an individual domain $\Delta^{\mathcal{I}}$ as an *individual at a state*, where the state (understood as an information state) is left implicit. If x is an individual at a state and y is an individual at a state, then y is *informationally accessible* from x (in symbols $x \preccurlyeq y$), if y is the same as x, however, at a state that is a possible expansion of the state of x. For subsumption it suffices that for every $\mathcal{I}$, every $x \in \Delta^{\mathcal{I}}$, and every $y \in \Delta^{\mathcal{I}}$ informationally accessible from x, y falls under D if y falls under C. The intuitive understanding of $\preccurlyeq$ clearly is such that $\preccurlyeq$ is both reflexive and transitive. The restriction to informationally accessible individuals in fact leads to a decrease of the subsumption relation, because it results in intuitionistic implication (seen as strict material implication) instead of material implication. The above considerations motivate the treatment of implication in the constructive description logic $\mathcal{CALC}^{\mathsf{C}}$. We shall define $\mathcal{CALC}^{\mathsf{C}}$ in Section 3.

EXAMPLE 2. *Consider a description language with the concept names* C *(say, for cranes) and* G *(say, for geese). Then in* $\mathcal{ALC}$ *the following subsumption relation holds: the concept* $\neg(\neg C \sqcup G) \sqcup C$ *is subsumed by* C*. If* *we read the complex concept in terms of material implication, we obtain that every object that is a crane provided it is a goose if it is a crane, is a crane. This subsumption is ruled out in constructive description logic, because* $(((C \to G) \to C) \to C)$ *is an instance of Peirce's law, which is classically but not intuitionistically valid.*

The motivation is similar, if the states are made explicit. We are then faced with two sorts of objects: individuals and states, and an interpretation will comprise in addition to a set of information states W an individual domain D_t for every $t \in W$. We assume expanding domains: if $x \in D_t$ and $s \in W$ is an expansion of t (in symbols $t \sqsubseteq s$ or $s \sqsupseteq t$), then $x \in s$. The concept C is subsumed by concept D, if for every interpretation $\mathcal{I}$ and every state $t \in W$ from $\mathcal{I}$ the following holds: for every $s \in W$ such that $t \sqsubseteq s$ and every individual $x \in D_t$, if x falls under C at s, then x falls under D at s. Again, Peirce's law fails. This approach to constructivizing $\mathcal{ALC}$ is taken in the systems $\mathcal{CALC}^{\mathsf{N4}}$ and $\mathcal{CALC}^{\mathsf{N4}d}$. As we shall see, in these two logics the strong negation of existential role restrictions is treated differently.

2.3. Inconsistent data

The data that form the basis for knowledge representation may be and often are inconsistent. Using classical logic as a base logic is problematic, because it leads to an 'explosion' of the computable subsumption hierarchy.

EXAMPLE 3. *Consider the following very simple background theory or TBox (and assume that* $\texttt{Mother} \in N_C$*):*

$$\{\texttt{Male} \models (\texttt{Father} \sqcap \exists \texttt{child}.\neg\texttt{Lives-in-Europe}),$$

$$(\texttt{Father} \sqcap \exists \texttt{child}.\neg\texttt{Lives-in-Europe}) \models \neg(\texttt{Painter} \sqcup \texttt{Lives-in-Europe}),$$

$$(\texttt{Father} \sqcap \exists \texttt{child}.\neg\texttt{Lives-in-Europe}) \models (\texttt{Musician} \sqcap \texttt{Lives-in-Europe})\}$$

A model of the TBox is an interpretation $\mathcal{I}$ *such that for every* $C \models D$ *from the TBox,* $(C \sqsubseteq D)^{\mathcal{I}} = \Delta^{\mathcal{I}}$*. In* $\mathcal{ALC}$*, like in any other description logic based on classical logic, for all models* $\mathcal{I}$ *of the above TBox we obtain, for example, the unwanted:*

$$(\texttt{Male} \sqsubseteq \texttt{Mother})^{\mathcal{I}} = \Delta^{\mathcal{I}}$$

simply because

$$(\texttt{Male} \sqsubseteq (\texttt{Lives-in-Europe} \sqcap \neg\texttt{Lives-in-Europe}))^{\mathcal{I}} = \Delta^{\mathcal{I}}$$

and $(\texttt{Lives-in-Europe} \sqcap \neg\texttt{Lives-in-Europe})^{\mathcal{I}} = \emptyset$.

If one thinks of knowledge representation not as a representation of *true* beliefs, but as a representation of *information*, then it is reasonable to treat *positive* and *negative* atomic information as independent and having equal status. In particular, information about the truth and falsity of atomic propositions may be supplied by different, independent, and equally reliable sources. If B is a concept name, then $\sim B$ will be used to express its strong, constructive negation. In ABoxes, the information that it is true that a falls under B is represented by $B(a)$ and the information that it is false that a falls under B is represented by $\sim B(a)$. Since it may happen that both $B(a)$ and $\sim B(a)$ is available, $B(a)$ and $\sim B(a)$ need not exclude each other, and hence it is certainly not the case that for every interpretation $\mathcal{I}$ and every $C \in Con(N_C, N_r)$, $((B \sqcap \sim B) \sqsubseteq C)^{\mathcal{I}} = \Delta^{\mathcal{I}}$. Therefore, for concept names (atomic propositions), the strong constructive negation is a primitive notion, whereas the constructive negation of compound concept descriptions will be defined in a natural way.

2.4. Incomplete information

Since the underlying logic of the standard description logics is classical, all concepts are total predicates. Category mistakes, therefore, cannot be accounted for.

EXAMPLE 4. *Suppose we want to represent information about the Mount Everest, m, but also about certain human beings enjoying mountain climbing, using the concept name* `Enjoy`. *For every interpretation $\mathcal{I}$, we have $m^{\mathcal{I}} \in (\neg\mathtt{Enjoy}\sqcup\mathtt{Enjoy})^{\mathcal{I}}$, (because for every interpretation $\mathcal{I}$, $(\neg\mathtt{Enjoy}\sqcup\mathtt{Enjoy})^{\mathcal{I}} = \Delta^{\mathcal{I}}$), although it is a category mistake to assert that the Mount Everest enjoys mountain climbing or does not enjoy mountain climbing.*

What we have tried to motivate is using a paraconsistent partial logic with constructive negation and implication as a basis for description logics. The four-valued paraconsistent logic **N4** is exactly such a logic. It is the base logic for constructivizing $\mathcal{ALC}$.

3. The system $\mathcal{CALC}^{\mathsf{C}}$

3.1. Syntax and semantics of $\mathcal{CALC}^{\mathsf{C}}$

The syntax of the logic $\mathcal{CALC}^{\mathsf{C}}$ is defined like the syntax of $\mathcal{ALC}$, except that classical negation $\neg C$ is replaced by constructive negation $\sim C$ and implication $C \sqsubseteq D$ is now a *primitive*, non-defined operation. To obtain a semantics

$$
\begin{array}{rcc}
A & \stackrel{ST_x}{\longmapsto} & A(x) \\
\sim A & \stackrel{ST_x}{\longmapsto} & \sim A(x) \\
\sim\sim C & \stackrel{ST_x}{\longmapsto} & ST_x(C) \\
C \sqcap D & \stackrel{ST_x}{\longmapsto} & ST_x(C) \wedge ST_x(D) \\
\sim(C \sqcap D) & \stackrel{ST_x}{\longmapsto} & ST_x(\sim C) \vee ST_x(\sim D) \\
C \sqcup D & \stackrel{ST_x}{\longmapsto} & ST_x(C) \vee ST_x(D) \\
\sim(C \sqcup D) & \stackrel{ST_x}{\longmapsto} & ST_x(\sim C) \wedge ST_x(\sim D) \\
C \sqsubseteq D & \stackrel{ST_x}{\longmapsto} & \forall y(x \preccurlyeq y \Rightarrow (ST_y(C) \Rightarrow ST_y(D))) \\
\sim(C \sqsubseteq D) & \stackrel{ST_x}{\longmapsto} & ST_x(C) \wedge ST_x(\sim D) \\
\forall r.C & \stackrel{ST_x}{\longmapsto} & \forall y(r(x,y) \Rightarrow ST_y(C)) \\
\sim\forall r.C & \stackrel{ST_x}{\longmapsto} & \exists y(r(x,y) \wedge ST_y(\sim C)) \\
\exists r.C & \stackrel{ST_x}{\longmapsto} & \exists y(r(x,y) \wedge ST_y(C)) \\
\sim\exists r.C & \stackrel{ST_x}{\longmapsto} & \forall y(r(x,y) \Rightarrow ST_y(\sim C))
\end{array}
$$

Table 2. The translation ST_x into classical predicate logic

for $\mathcal{CALC}^{\mathsf{C}}$, we use a translation ST_x of the language of $\mathcal{CALC}^{\mathsf{C}}$ into the language of *classical* predicate logic. The first-order language contains a binary predicate $\preccurlyeq$, a binary predicate r, for every $r \in N_r$, and unary predicates A and $\sim A$, for every concept name A, see Table 2. In this table, $\Rightarrow$ denotes classical implication, and y is an individual variable distinct from x. (It can easily be seen that two variables suffice, since we may flip between x and y in defining two translations ST_x and ST_y, see [6, p. 87].) The semantics for $\mathcal{CALC}^{\mathsf{C}}$ uses interpretations $\mathcal{I} = \langle \Delta^{\mathcal{I}}, \preccurlyeq, \cdot^{\mathcal{I}} \rangle$, where $\Delta^{\mathcal{I}}$ is a non-empty set and $\preccurlyeq \subseteq (\Delta^{\mathcal{I}} \times \Delta^{\mathcal{I}})$ is a relation of informational accessibility, i.e., $\preccurlyeq$ is reflexive and transitive. The interpretation function $\cdot^{\mathcal{I}}$ maps every $A \in N_C$ and every $\sim A$ with $A \in N_C$ to a subset of $\Delta^{\mathcal{I}}$, and it maps every role name $r \in N_r$ to a binary relation $r^{\mathcal{I}}$ on $\Delta^{\mathcal{I}}$, and every $a \in N_I$ to an element of $\Delta^{\mathcal{I}}$. The function is extended to arbitrary concept descriptions as stated in Table 3. Moreover, we require

(P1) for all $x, y \in \Delta^{\mathcal{I}}$, $A \in N_C$: if $x \preccurlyeq y$ and $x \in A^{\mathcal{I}}$, then $y \in A^{\mathcal{I}}$;

(P2) for all $x, y \in \Delta^{\mathcal{I}}$, $A \in N_C$: if $x \preccurlyeq y$ and $x \in \sim A^{\mathcal{I}}$, then $y \in \sim A^{\mathcal{I}}$;

(P3) for all $r \in N_r$: $\preccurlyeq r^{\mathcal{I}} \subseteq r^{\mathcal{I}} \preccurlyeq$;

constructor	semantics
A	$A^{\mathcal{I}} \subseteq \Delta^{\mathcal{I}}$
$\sim A$	$\sim A^{\mathcal{I}} \subseteq \Delta^{\mathcal{I}}$
$(C \sqcap D)$	$C^{\mathcal{I}} \cap D^{\mathcal{I}}$
$\sim(C \sqcap D)$	$\sim C^{\mathcal{I}} \cup \sim D^{\mathcal{I}}$
$(C \sqcup D)$	$C^{\mathcal{I}} \cup D^{\mathcal{I}}$
$\sim(C \sqcup D)$	$\sim C^{\mathcal{I}} \cap \sim D^{\mathcal{I}}$
$(C \sqsubseteq D)$	$\{x \in \Delta^{\mathcal{I}} : \forall y\,(x \preccurlyeq y \Rightarrow (y \in C^{\mathcal{I}} \Rightarrow y \in D^{\mathcal{I}}))\}$
$\sim(C \sqsubseteq D)$	$C^{\mathcal{I}} \cap \sim D^{\mathcal{I}}$
$\forall r.C$	$\{x \in \Delta^{\mathcal{I}} : \forall y\,(xr^{\mathcal{I}}y \Rightarrow y \in C^{\mathcal{I}})\}$
$\sim\forall r.C$	$\{x \in \Delta^{\mathcal{I}} : \exists y\,(xr^{\mathcal{I}}y \;\&\; y \in \sim C^{\mathcal{I}})\}$
$\exists r.C$	$\{x \in \Delta^{\mathcal{I}} : \exists y\,(xr^{\mathcal{I}}y \;\&\; y \in C^{\mathcal{I}})\}$
$\sim\exists r.C$	$\{x \in \Delta^{\mathcal{I}} : \forall y\,(xr^{\mathcal{I}}y \Rightarrow y \in \sim C^{\mathcal{I}})\}$

Table 3. The semantics of concept descriptions in $\mathcal{CALC}^{\mathrm{C}}$

(P4) for all $r \in N_r$: $\preccurlyeq^{-1} r^{\mathcal{I}} \subseteq r^{\mathcal{I}} \preccurlyeq^{-1}$.

Conditions (P3) and (P4) are used to prove persistence of arbitrary concept descriptions, a property which is intuitively correct, given our informational interpretation of ABox statements.

PROPOSITION 1. *(Persistence) Let $\mathcal{I} = (\Delta^{\mathcal{I}}, \preccurlyeq, \cdot^{\mathcal{I}})$ be an interpretation. Then for every concept description C and $x, y \in \Delta^{\mathcal{I}}$,*

if $x \preccurlyeq y$ and $x \in C^{\mathcal{I}}$, then $y \in C^{\mathcal{I}}$.

PROOF. By simultaneous induction on the complexity of C and $\sim C$. We consider here two example cases.
(1) Let $C = \sim\forall r.D$, $x \preccurlyeq y$, and $x \in \sim\forall r.D^{\mathcal{I}}$. Then $\exists z\ (xr^{\mathcal{I}}z \,\&\, z \in \sim D^{\mathcal{I}})$. Suppose for reductio that $y \notin \sim\forall r.D^{\mathcal{I}}$. Then (*) $\forall u\ (yr^{\mathcal{I}}u \Rightarrow u \notin \sim D^{\mathcal{I}})$. Since $y \preccurlyeq^{-1} r^{\mathcal{I}} z$ and $\preccurlyeq^{-1} r^{\mathcal{I}} \subseteq r^{\mathcal{I}} \preccurlyeq^{-1}$, by (P4), $yr^{\mathcal{I}} \preccurlyeq^{-1} z$. Thus, $\exists u\ yr^{\mathcal{I}}u$ and $z \preccurlyeq u$. By (*), $u \notin \sim D^{\mathcal{I}}$. By the induction hypothesis, however, $\forall u$ $(z \preccurlyeq u \Rightarrow u \in \sim D^{\mathcal{I}})$, quod non.
(2) Let $C = \forall r.D$, $x \preccurlyeq y$, and $x \in \forall r.D^{\mathcal{I}}$. Then (*) $\forall z\ (xr^{\mathcal{I}}z \Rightarrow z \in D^{\mathcal{I}})$. Suppose $yr^{\mathcal{I}}u$. Since $x \preccurlyeq y$, we have $x \preccurlyeq yr^{\mathcal{I}}u$ and by (P3), $xr^{\mathcal{I}} \preccurlyeq u$. Therefore $\exists w\ xr^{\mathcal{I}}w \preccurlyeq u$. By (*), $w \in D^{\mathcal{I}}$. By $w \preccurlyeq u$ and the induction hypothesis, $u \in D^{\mathcal{I}}$. Hence $\forall u\ (yr^{\mathcal{I}}u \Rightarrow u \in D^{\mathcal{I}})$ and thus $y \in \forall r.D^{\mathcal{I}}$. ■

EXAMPLE 5. *We modify the simple TBox of Example 3 and replace classical negation $\neg$ with constructive negation $\sim$ to obtain the TBox:*

$$\{\texttt{Male} \models (\texttt{Father} \sqcap \exists \texttt{child}.\sim\texttt{Lives-in-Europe}),$$

$$(\texttt{Father} \sqcap \exists \texttt{child}.\sim\texttt{Lives-in-Europe}) \models \sim(\texttt{Painter} \sqcup \texttt{Lives-in-Europe}),$$

$$(\texttt{Father} \sqcap \exists \texttt{child}.\sim\texttt{Lives-in-Europe}) \models (\texttt{Musician} \sqcap \texttt{Lives-in-Europe})\}$$

There are models $\mathcal{I}$ of this TBox such that

$$(\texttt{Lives-in-Europe} \sqcap \sim \texttt{Lives-in-Europe})^{\mathcal{I}} \neq \emptyset, \quad \textit{and}$$
$$(\texttt{Male} \sqsubseteq \texttt{Mother})^{\mathcal{I}} \neq \Delta^{\mathcal{I}}$$

EXAMPLE 6. *We reconsider Example 4. There are interpretations $\mathcal{I}$ such that $(\sim\texttt{Enjoy} \sqcup \texttt{Enjoy})^{\mathcal{I}} \neq \Delta^{\mathcal{I}}$. It does hold that for all interpretations $\mathcal{I}$, $(\texttt{Enjoy} \sqsubseteq \texttt{Enjoy})^{\mathcal{I}} = \Delta^{\mathcal{I}}$, but this is unproblematic, although the well-formed expression $\texttt{Enjoy}(m)$ is categorially mistaken.*

DEFINITION 3. *$C \in Con(N_C, N_r)$ is said to be $\mathcal{CALC}^{\mathsf{C}}$-valid ($\models_{\mathcal{CALC}^{\mathsf{C}}} C$) iff for every interpretation $\mathcal{I} = \langle \Delta^{\mathcal{I}}, \preccurlyeq, \cdot^{\mathcal{I}} \rangle$, $C^{\mathcal{I}} = \Delta^{\mathcal{I}}$. C is subsumed by D ($C \models D$) iff $\models_{\mathcal{CALC}^{\mathsf{C}}} C \sqsubseteq D$. C and D are positively equivalent ($C =^{+} D$) iff $C \models D$ and $D \models C$. C and D are negatively equivalent ($C =^{-} D$) iff $\sim C \models \sim D$ and $\sim D \models \sim C$. C and D are fully equivalent ($C \equiv D$) iff C and D are both positively and negatively equivalent.*

Note that in this paraconsistent setting, *every* concept description is satisfiable. $C^{\mathcal{I}} \neq \emptyset$ means that there is an $x \in \Delta^{\mathcal{I}}$ such that if $a^{\mathcal{I}} = x$, the information that $C(a)$ is available. But this availability is always possible, because the relevant information may be supplied by some source of information.

DEFINITION 4. *A TBox is a finite set of expressions of the following shape: $(C \models D)$, $(C =^{+} D)$, $(C =^{-} D)$, or $(C \equiv D)$. The elements of a TBox are called TBox statements. An interpretation $\mathcal{I}$ is a model of $(C \models D)$ iff $(C \sqsubseteq D)^{\mathcal{I}} = \Delta^{\mathcal{I}}$; $\mathcal{I}$ is a model of $(C =^{+} D)$ (respectively $(C =^{-} D)$) iff $\mathcal{I}$ is a model of $(C \models D)$ and $(D \models C)$ (respectively $(\sim C \models \sim D)$ and $(\sim D \models \sim C)$). $\mathcal{I}$ is a model of $(C \equiv D)$ iff $\mathcal{I}$ is a model of $(C =^{+} D)$ and $(C =^{-} D)$. Finally, $\mathcal{I}$ is said to be a model of a TBox $\mathcal{T}$ ($\mathcal{I} \models \mathcal{T}$) iff $\mathcal{I}$ is a model of every $\alpha \in \mathcal{T}$. The TBox $\mathcal{T}$ logically implies the TBox statement α (in symbols $\mathcal{T} \models \alpha$) iff every model of $\mathcal{T}$ is a model of $\{\alpha\}$.*

DEFINITION 5. *A knowledge base Σ is a pair $(\mathcal{T}, \mathcal{A})$, where $\mathcal{T}$ is a TBox and $\mathcal{A}$ is an ABox. An interpretation $\mathcal{I}$ is a model for Σ iff $\mathcal{I}$ is a model for both $\mathcal{T}$ and $\mathcal{A}$. Let α be an ABox statement or a TBox statement. The knowledge base Σ logically implies α (in symbols $\Sigma \models \alpha$) iff every model of Σ is a model of $\{\alpha\}$.*

If α is an ABox or a TBox statement and $\mathcal{I}$ is a model of α, we shall write $\mathcal{I} \models \alpha$.

In the constructive description logic $\mathcal{CALC}^{\mathsf{C}}$ (and similarly also in the systems $\mathcal{CALC}^{\mathsf{N4}}$ and $\mathcal{CALC}^{\mathsf{N4}d}$) we may consider a number of queries, not all of which are available in classical description logic.

$(C \models D)$?	subsumption
$(C =^{+} D)$?	positive equivalence
$(C =^{-} D)$?	negative equivalence
$(C \equiv D)$?	strong (complete) equivalence

EXAMPLE 7. *The following is an example of a knowledge base Σ in which the concepts C and D are negatively but not positively equivalent. This means that with respect to Σ, C and D contain the same negative information but fail to contain the same positive information. Constructive description logic thus allows more fine-grained queries than $\mathcal{ALC}$.*

$$\Sigma = (\{E \models F,\, C \equiv (\sim E \sqcup F),\, D \equiv (E \sqsubseteq F)\}, \{E(a), \sim F(a)\})$$

Since for every interpretation $\mathcal{I}$, $(C =^{-} D)^{\mathcal{I}} = \Delta^{\mathcal{I}}$, $\Sigma \models (C =^{-} D)$. However, $\Sigma \not\models (C =^{+} D)$. Note also that $\Sigma \models F(a)$ and $\Sigma \models \sim F(a)$.

The motivation for constructive description logic stemming from Examples 5–7 carries over (mutatis mutandis) from $\mathcal{CALC}^{\mathsf{C}}$ to the systems $\mathcal{CALC}^{\mathsf{N4}}$ and $\mathcal{CALC}^{\mathsf{N4}d}$, which we shall define in Sections 4 and 5.

3.2. Some properties of $\mathcal{CALC}^{\mathsf{C}}$

PROPOSITION 2. *The system $\mathcal{CALC}^{\mathsf{C}}$ has the disjunction property and the constructible falsity property:*

$$\begin{array}{lll} (i) & \models_{\mathcal{CALC}^{\mathsf{C}}} (C \sqcup D) & \Rightarrow \models_{\mathcal{CALC}^{\mathsf{C}}} C \textit{ or } \models_{\mathcal{CALC}^{\mathsf{C}}} D \\ (ii) & \models_{\mathcal{CALC}^{\mathsf{C}}} \sim(C \sqcap D) & \Rightarrow \models_{\mathcal{CALC}^{\mathsf{C}}} \sim C \textit{ or } \models_{\mathcal{CALC}^{\mathsf{C}}} \sim D \end{array}$$

PROOF. Consider (ii); the proof of (i) is analogous. Suppose that for every interpretation $\mathcal{I}$, $\sim (C \sqcap D)^{\mathcal{I}} = \Delta^{\mathcal{I}}$ and that there are interpretations $\mathcal{I}_1, \mathcal{I}_2$ and individuals $x_1 \in \Delta^{\mathcal{I}_1}$, $x_2 \in \Delta^{\mathcal{I}_2}$, such that $x_1 \notin \sim C^{\mathcal{I}_1}$ and $x_2 \notin \sim D^{\mathcal{I}_2}$. Let $Lit = N_C \cup \{\sim A : A \in N_C\}$, and define $\mathcal{I}_3 = (\Delta^{\mathcal{I}_3}, \preccurlyeq^{\mathcal{I}_3}, \cdot^{\mathcal{I}_3})$ such that

- $\Delta^{\mathcal{I}_3} = \Delta^{\mathcal{I}_1} \cup \Delta^{\mathcal{I}_2} \cup \{x_0\}$, $x_0 \notin \Delta^{\mathcal{I}_1} \cup \Delta^{\mathcal{I}_2}$
- $\preccurlyeq^{\mathcal{I}_3} = \preccurlyeq^{\mathcal{I}_1} \cup \preccurlyeq^{\mathcal{I}_2} \cup \{(x_0, x_0)\} \cup \{(x_0, x) : x \in \Delta^{\mathcal{I}_1} \cup \Delta^{\mathcal{I}_2}\}$

- $\forall C \in Lit$, $x_0 \notin C^{\mathcal{I}_3}$ and $\forall x \in \Delta^{\mathcal{I}_i} (i = 1, 2)$, $x \in C^{\mathcal{I}_3}$ iff $x \in C^{\mathcal{I}_i}$

Since $\mathcal{I}_3$ is an interpretation, $x_0 \in \sim (C \sqcap D)^{\mathcal{I}_3}$. Hence $x_0 \in \sim C^{\mathcal{I}_3}$ or $x_0 \in \sim D^{\mathcal{I}_3}$. By persistence, $x_1 \in \sim C^{\mathcal{I}_3}$ or $x_2 \in \sim D^{\mathcal{I}_3}$. But then $x_1 \in \sim C^{\mathcal{I}_1}$ or $x_2 \in \sim D^{\mathcal{I}_2}$, a contradiction. ■

A concept description is in *negation normal form* if it contains $\sim$ only in front of concept names. Let $\forall r.^d = \exists r.$ and $\exists r.^d = \forall r.$ The following translation $\overline{(\cdot)}$ sends every concept description C to a concept description in negation normal form, where $A \in N_C$, $\diamond \in \{\sqcap, \sqcup, \sqsubseteq\}$ and $\sharp \in \{\forall r., \exists r.\}$:

$$
\begin{array}{rclrcl}
\overline{A} & = & A & \overline{\sim A} & = & \sim A \\
\overline{\sim\sim C} & = & \overline{C} & \overline{C \diamond D} & = & \overline{C} \diamond \overline{D} \\
\overline{\sim (C \sqcap D)} & = & \overline{\sim C} \sqcup \overline{\sim D} & \overline{\sim (C \sqcup D)} & = & \overline{\sim C} \sqcap \overline{\sim D} \\
\overline{\sim (C \sqsubseteq D)} & = & \overline{C} \sqcap \overline{\sim D} & & & \\
\overline{\sharp C} & = & \sharp \overline{C} & \overline{\sim \sharp C} & = & \sharp^d \overline{\sim C}
\end{array}
$$

The notion of negation normal form extends in a natural way to ABox and TBox statements. If $C(a)$ is an ABox statement, its negation normal form $\overline{C(a)}$ is defined as $\overline{C}(a)$. The negation normal form of a TBox statement is defined as follows:

$$
\begin{array}{lll}
\overline{C \models D} & =_{def} & \overline{C} \models \overline{D} \\
\overline{C =^{+} D} & =_{def} & \overline{C} =^{+} \overline{D} \\
\overline{C =^{-} D} & =_{def} & \overline{\sim C} =^{+} \overline{\sim D} \\
\overline{C \equiv D} & =_{def} & \{\overline{C =^{+} D}, \overline{C =^{-} D}\}
\end{array}
$$

PROPOSITION 3. *For every concept description C, and every interpretation $\mathcal{I}$, $\overline{C}$ is in negation normal form, and $(C =^{+} \overline{C})^{\mathcal{I}} = \Delta^{\mathcal{I}}$.*

4. The system $\mathcal{CALC}^{\mathsf{N4}}$

We now define a variant of description logic based on a certain translation of concept descriptions into first-order logic with strong negation.

4.1. Quantified N4

First of all, we define a first-order variant QN4 of the propositional logic N4. We will consider languages with only constant and predicate symbols. For an arbitrary first-order language $\mathcal{L}$, $T_{\mathcal{L}}$ ($CT_{\mathcal{L}}$) denotes the set of $\mathcal{L}$-terms (closed $\mathcal{L}$-terms, i.e., constants of $\mathcal{L}$), $At_{\mathcal{L}}$ the set of atomic formulas, finally, $For_{\mathcal{L}}$ denotes the set of $\mathcal{L}$-formulas. Consider a Hilbert-style deductive system for **QN4** with the following sets of axioms and rules.

Axiom schemes:

A1. $\varphi \to (\psi \to \varphi)$

A2. $(\varphi \to (\psi \to \chi)) \to ((\varphi \to \psi) \to (\varphi \to \chi))$

A3. $(\varphi \wedge \psi) \to \varphi$

A4. $(\varphi \wedge \psi) \to \psi$

A5. $(\varphi \to \psi) \to ((\varphi \to \chi) \to (\varphi \to (\psi \wedge \chi)))$

A6. $\varphi \to (\varphi \vee \psi)$

A7. $\psi \to (\varphi \vee \psi)$

A8. $(\varphi \to \chi) \to ((\psi \to \chi) \to ((\varphi \vee \psi) \to \chi))$

A9. $\sim\sim \varphi \leftrightarrow \varphi$

A10. $\sim (\varphi \vee \psi) \leftrightarrow (\sim \varphi \wedge \sim \psi)$

A11. $\sim (\varphi \wedge \psi) \leftrightarrow (\sim \varphi \vee \sim \psi)$

A12. $\sim (\varphi \to \psi) \leftrightarrow (\varphi \wedge \sim \psi)$

A13. $\sim \exists x \varphi \leftrightarrow \forall x \sim \varphi$

A14. $\sim \forall x \varphi \leftrightarrow \exists x \sim \varphi$

A15. $\varphi(t) \to \exists x \varphi(x)$ (t is free for x in φ)

A16. $\forall x \varphi(x) \to \varphi(t)$ (t is free for x in φ)

Rules:

$$\frac{\varphi, \varphi \to \psi}{\psi}; \quad \frac{\varphi \to \psi(x)}{\varphi \to \forall x \psi(x)} \ (x \text{ not free in } \varphi); \quad \frac{\varphi(x) \to \psi}{\exists x \varphi(x) \to \psi} \ (x \text{ not free in } \psi).$$

The notion of deducible formulas and the consequence relation $\vdash_{\mathsf{QN4}}$ are defined in the usual way.

In the literature one can find only a first-order variant of **N3**, a three-valued, explosive extension of **N4**, see [25] (with the restriction to constant

$\mathcal{M}, t \models^+ \varphi$	iff	$t \in v^+(\varphi)$, $\varphi \in At_{\mathcal{L}}$
$\mathcal{M}, t \models^- \varphi$	iff	$t \in v^-(\varphi)$, $\varphi \in At_{\mathcal{L}}$
$\mathcal{M}, t \models^+ \varphi \wedge \psi$	iff	$\mathcal{M}, t \models^+ \varphi$ and $\mathcal{M}, t \models^+ \psi$
$\mathcal{M}, t \models^- \varphi \wedge \psi$	iff	$\mathcal{M}, t \models^- \varphi$ or $\mathcal{M}, t \models^- \psi$
$\mathcal{M}, t \models^+ \varphi \vee \psi$	iff	$\mathcal{M}, t \models^+ \varphi$ or $\mathcal{M}, t \models^+ \psi$
$\mathcal{M}, t \models^- \varphi \vee \psi$	iff	$\mathcal{M}, t \models^- \varphi$ and $\mathcal{M}, t \models^- \psi$
$\mathcal{M}, t \models^+ \varphi \rightarrow \psi$	iff	$(\forall u \in W)$ if $t \preccurlyeq u$, then $\mathcal{M}, u \models^+ \varphi$ implies $\mathcal{M}, u \models^+ \psi$
$\mathcal{M}, t \models^- \varphi \rightarrow \psi$	iff	$\mathcal{M}, t \models^+ \varphi$ and $\mathcal{M}, t \models^- \psi$
$\mathcal{M}, t \models^+ \forall x \varphi(x)$	iff	$(\forall u \in W)$ if $t \preccurlyeq u$, then $(\forall c \in D_u)$ $\mathcal{M}, u \models^+ \varphi(c)$
$\mathcal{M}, t \models^- \forall x \varphi(x)$	iff	$(\exists c \in D_t)$ $\mathcal{M}, t \models^- \varphi(c)$
$\mathcal{M}, t \models^+ \exists x \varphi(x)$	iff	$(\exists c \in D_t)$ $\mathcal{M}, t \models^+ \varphi(c)$
$\mathcal{M}, t \models^- \exists x \varphi(x)$	iff	$(\forall u \in W)$ if $t \preccurlyeq u$, then $(\forall c \in D_u)$ $\mathcal{M}, u \models^- \varphi(c)$
$\mathcal{M}, t \models^+ {\sim} \varphi$	iff	$\mathcal{M}, t \models^- \varphi$
$\mathcal{M}, t \models^- {\sim} \varphi$	iff	$\mathcal{M}, t \models^+ \varphi$

Table 4. The forcing relations for **QN4**

domains) and [16]. Therefore, we introduce a semantics and give a short completeness proof for **QN4** by embedding this logic into positive intuitionistic predicate logic.

We define the notion of a first-order Nelson model with a view on information representation. Therefore, we consider only models with specific individual domains consisting of terms. A first-order *Nelson model* is a structure $\langle W, \preccurlyeq, \Delta, D, v^+, v^- \rangle$, where $\langle W, \preccurlyeq \rangle$ is a partial order; Δ is a set, $CT_{\mathcal{L}} \subseteq \Delta \subseteq T_{\mathcal{L}}$; D is a function from W to subsets of Δ such that 1) $CT_{\mathcal{L}} \subseteq D_u$ for all $u \in W$, 2) $D_u \subseteq D_t$ whenever $u \preccurlyeq t$. Finally, v^+ and v^- are functions from $At_{\mathcal{L}}$ to cones (upwards closed sets) of the partial order $\langle W, \preccurlyeq \rangle$ such that if $u \in v^{+(-)}(P(a_1, \ldots, a_n))$, then $a_1, \ldots, a_n \in D_u$. Function D assigns to every state $u \in W$ its domain D_u, and the valuation function v^+ (v^-) defines, in fact, a set of states, in which one or another atomic formula is verified (falsified).

Let $\mathcal{M} = \langle W, \preccurlyeq, \Delta, D, v^+, v^- \rangle$ be a Nelson model, $t \in W$ and $\varphi \in For_{\mathcal{L}}$. The notions $\mathcal{M}, t \models^+ \varphi$ (φ is verified at t in $\mathcal{M}$) and $\mathcal{M}, t \models^- \varphi$ (φ is falsified at t in $\mathcal{M}$), the 'forcing relations' for **QN4**, are inductively defined as shown in Table 4.

It is not hard to check that verification and falsification of arbitrary formulas are persistent with respect to $\preccurlyeq$. We say that a formula $\varphi \in For_{\mathcal{L}}$ is **QN4**-valid if for any Nelson model $\mathcal{M} = \langle W, \preccurlyeq, \Delta, D, v^+, v^- \rangle$ and $t \in W$, φ is verified at t in $\mathcal{M}$.

The correctness of QN4 with respect to the above semantics can be proved in the usual way by induction on the length of proofs.

PROPOSITION 4. *For any* $\varphi \in For_{\mathcal{L}}$, *if* QN4 $\vdash \varphi$, *then* φ *is* QN4*-valid.*

Completeness can be proved by embedding QN4 into positive first-order intuitionistic logic QInt^+ [8], which we will consider in the language $\mathcal{L}^+$, where the symbol $\sim$ is deleted. Let us recall that the Hilbert calculus for QInt^+ has the same inference rules as QN4 and axioms A1–A8, A15, and A16.
A QInt^+-model is a structure $\langle W, \preccurlyeq, \Delta, D, v\rangle$ defined like a Nelson model, except that it has only one valuation function v from $At_{\mathcal{L}}$ to cones of the partial order $\langle W, \preccurlyeq\rangle$ such that if $u \in v(P(a_1, \dots, a_n))$, then $a_1, \dots, a_n \in D_u$.
Let $\mathcal{M} = \langle W, \preccurlyeq, \Delta, D, v\rangle$ be a QInt^+-model, $t \in W$ and $\varphi \in For_{\mathcal{L}^+}$. The relation $\mathcal{M}, t \models \varphi$ (φ is true at t in $\mathcal{M}$) is defined exactly so as the relation $\mathcal{M}, t \models^+ \varphi$ was defined for positive connectives and quantifies (with v^+ replaced by v in case of atomic formulas). We say that a formula $\varphi \in For_{\mathcal{L}^+}$ is QInt^+-valid if for any QInt^+-model $\mathcal{M} = \langle W, \preccurlyeq, \Delta, D, v\rangle$ and $t \in W$, φ is true at t in $\mathcal{M}$.

PROPOSITION 5. *For any* $\varphi \in For_{\mathcal{L}^+}$, QInt $\vdash \varphi$ *iff* φ *is* QInt^+*-valid.*

PROOF. The proof is essentially the same as the completeness proof for full first-order intuitionistic logic [8], omitting the part concerning negation. ■

We say that a formula $\varphi \in For_{\mathcal{L}}$ is in *negation normal form*, if it contains strong negations only in front of atomic formulas, in other words, if $\sim \psi$ is a subformula of φ, then ψ is atomic.

DEFINITION 6. *We define the transformation* $\overline{(\cdot)}$ *on* $For_{\mathcal{L}}$ *as follows:*

1. $\overline{\psi} := \psi$, $\overline{\sim \psi} := \sim \psi$ *for* $\psi \in At_{\mathcal{L}}$.
2. $\overline{\sim\sim \varphi} := \overline{\varphi}$ *for a formula* φ.
3. $\overline{\varphi \diamond \psi} := \overline{\varphi} \diamond \overline{\psi}$, *where* φ *and* ψ *are formulas and* $\diamond \in \{\vee, \wedge, \rightarrow\}$.
4. *For any formula* φ *and* $Q \in \{\forall, \exists\}$, $\overline{Qx\varphi} := Qx\overline{\varphi}$.
5. *For any formulas* φ *and* ψ, $\overline{\sim (\varphi \vee \psi)} := \overline{\sim \varphi} \wedge \overline{\sim \psi}$, $\overline{\sim (\varphi \wedge \psi)} := \overline{\sim \varphi} \vee \overline{\sim \psi}$, *and* $\overline{\sim (\varphi \rightarrow \psi)} := \overline{\varphi} \wedge \overline{\sim \psi}$.
6. *For any formula* φ, $\overline{\sim \forall x\varphi} := \exists x\overline{\sim \varphi}$ *and* $\overline{\sim \exists x\varphi} := \forall x\overline{\sim \varphi}$.

PROPOSITION 6. *For any formula φ, $\overline{\varphi}$ is in negation normal form and* $\mathsf{QN4} \vdash \varphi \leftrightarrow \overline{\varphi}$.

Let $\mathcal{L}$ be a first-order language with strong negation. We define the positive intuitionistic first-order language $\mathcal{L}'^{+}$ by deleting $\sim$ from $\mathcal{L}$ and adding for any predicate symbol P of $\mathcal{L}$ a new predicate symbol $P^{\sim}$ of the same arity.

DEFINITION 7. *We define the transformation $(\cdot)^*$ of formulas from $For_{\mathcal{L}}$ into the language $\mathcal{L}'^{+}$ as follows:*

1. *For an atomic $P(\overline{a})$, $(P(\overline{a}))^* := P(\overline{a})$ and $(\sim P(\overline{a}))^* := P^{\sim}(\overline{a})$.*

2. *$(\varphi \circ \psi)^* := \varphi^* \circ \psi^*$, where φ and ψ are formulas in negation normal form and $\circ \in \{\vee, \wedge, \rightarrow\}$.*

3. *$(Qx\varphi)^* := Qx\varphi^*$, where φ is a formula in negation normal form and $Q \in \{\forall, \exists\}$.*

4. *$(\varphi)^* := (\overline{\varphi})^*$ for any formula φ not in negation normal form.*

PROPOSITION 7. *For any formula φ, $\mathsf{QN4} \vdash \varphi$ iff $\mathsf{QInt}^{+} \vdash \varphi^*$.*

PROOF. If $\mathsf{QN4} \vdash \varphi$, one can easily check by induction on the length of proofs that φ^* is provable in positive intuitionistic logic. If $\mathsf{QInt}^{+} \vdash \varphi^*$ and $\psi_0, \dots \psi_n = \varphi^*$ is a proof, then $\psi'_0, \dots \psi'_n = \overline{\varphi}$, where ψ'_i is the result of replacing any subformula $P^{\sim}(\overline{a})$ by $\sim P(\overline{a})$, is a proof of the negation normal form of φ in $\mathsf{QN4}$. By Proposition 6, we then have $\mathsf{QN4} \vdash \varphi$. ■

Now we can easily prove the completeness result for $\mathsf{QN4}$.

PROPOSITION 8. *For any $\varphi \in For_{\mathcal{L}}$, $\mathsf{QN4} \vdash \varphi$ iff φ is $\mathsf{QN4}$-valid.*

PROOF. Let φ_0 be a $\mathsf{QN4}$-valid formula. Assume that $\mathsf{QN4} \nvdash \varphi_0$. Then $\mathsf{QInt}^{+} \nvdash (\varphi_0)^*$ by Proposition 7. This means that there is a QInt^{+}-model $\mathcal{M} = \langle W, \preccurlyeq, \Delta, D, v\rangle$ of the language $\mathcal{L}'^{+}$ and $t_0 \in W$ such that $\mathcal{M}, t_0 \nvDash (\varphi_0)^*$. We define a Nelson model $\mathcal{M}' = \langle W, \preccurlyeq, \Delta, D, v^{+}, v^{-}\rangle$ with the same W, $\preccurlyeq$, and Δ. The assignment functions v^{+} and v^{-} are defined as follows:

$$v^{+}(P(\overline{a})) := v(P(\overline{a})) \text{ and } v^{-}(P(\overline{a})) := v(P^{\sim}(\overline{a})).$$

Using induction on the structure of a formula, we can prove that

$$\mathcal{M}, t \models \varphi^* \text{ if and only if } \mathcal{M}', t \models^{+} \varphi$$

A	$\stackrel{T_x}{\longmapsto}$	$A(x)$
$\sim C$	$\stackrel{T_x}{\longmapsto}$	$\sim T_x(C)$
$C \sqcap D$	$\stackrel{T_x}{\longmapsto}$	$T_x(C) \wedge T_x(D)$
$C \sqcup D$	$\stackrel{T_x}{\longmapsto}$	$T_x(C) \vee T_x(D)$
$C \sqsubseteq D$	$\stackrel{T_x}{\longmapsto}$	$T_x(C) \rightarrow T_x(D)$
$\forall r.C$	$\stackrel{T_x}{\longmapsto}$	$\forall y(r(x,y) \rightarrow T_y(C))$
$\exists r.C$	$\stackrel{T_x}{\longmapsto}$	$\exists y(r(x,y) \wedge T_y(C))$

Table 5. The translation T_x into constructive predicate logic

for any $t \in W$ and formula φ in negation normal form. Thus, $\mathcal{M}', t_0 \not\models^+ \overline{\varphi_0}$ and $\mathcal{M}', t_0 \not\models^+ \varphi_0$ by Proposition 6, which conflicts with the QN4-validity of φ_0. ■

For two finite sets of formulas Γ and Δ, $\Gamma \vdash_{\mathsf{QN4}} \Delta$ means that the disjunction of formulas from Δ can be obtained from formulas of Γ using axioms and inference rules of QN4, and $\Gamma \models_{\mathsf{QN4}} \Delta$ means that for any Nelson model $\mathcal{M} = \langle W, \preccurlyeq, \Delta, D, v^+, v^- \rangle$ and $t \in W$, if all formulas from Γ are verified at t, then one of the formulas of Δ is verified at t.

QN4 contains all axioms and inference rules of QInt^+. Therefore, the Deduction Theorem holds for sentences of QN4 without a restriction on the application of rules for quantifies, as it holds in QInt^+. Using the standard procedure of replacing variables by new constants, we obtain the following generalized completeness result for QN4.

PROPOSITION 9. *For any finite sets of formulas* Γ *and* Δ, $\Gamma \vdash_{\mathsf{QN4}} \Delta$ *iff* $\Gamma \models_{\mathsf{QN4}} \Delta$.

4.2. Syntax and semantics of $\mathcal{CALC}^{\mathsf{N4}}$

The syntax of the description logic $\mathcal{CALC}^{\mathsf{N4}}$ does not differ from the syntax of $\mathcal{CALC}^{\mathsf{C}}$. The semantics of $\mathcal{CALC}^{\mathsf{N4}}$ is induced by a translation T_x from the language of constructive description logic into the language of *constructive* predicate logic. This translation into QN4 is presented in Table 5. In the semantics induced by T_x, the interpretations of a concept description C and roles are relativized to a state. We will not assume that all individual names are interpreted in all states. Moreover, we assume that every interpretation domain is simply a subset of the set N_I. With this semantical assumption

constructor	semantics
A	$(A,t)^{\mathcal{I}} \subseteq D_t$
$\sim A$	$(\sim A,t)^{\mathcal{I}} \subseteq D_t$
$(C \sqcap D)$	$(C,t)^{\mathcal{I}} \cap (D,t)^{\mathcal{I}}$
$\sim (C \sqcap D)$	$(\sim C,t)^{\mathcal{I}} \cup (\sim D,t)^{\mathcal{I}}$
$(C \sqcup D)$	$(C,t)^{\mathcal{I}} \cup (D,t)^{\mathcal{I}}$
$\sim (C \sqcup D)$	$(\sim C,t)^{\mathcal{I}} \cap (\sim D,t)^{\mathcal{I}}$
$(C \sqsubseteq D)$	$\{x \in D_t : (\forall s \in I)\,(t \preccurlyeq s \Rightarrow (x \in (C,s)^{\mathcal{I}} \Rightarrow x \in (D,s)^{\mathcal{I}}))\}$
$\sim (C \sqsubseteq D)$	$(C,t)^{\mathcal{I}} \cap (\sim D,t)^{\mathcal{I}}$
$\forall r.C$	$\{x \in D_t : (\forall s \succcurlyeq t)\,(\forall y \in D_s)\,((x,y) \in (r,s)^{\mathcal{I}} \Rightarrow y \in (C,s)^{\mathcal{I}})\}$
$\sim \forall r.C$	$\{x \in D_t : (\exists y \in D_t)\,((x,y) \in (r,t)^{\mathcal{I}}\ \&\ y \in (\sim C,t)^{\mathcal{I}})\}$
$\exists r.C$	$\{x \in D_t : (\exists y \in D_t)\,((x,y) \in (r,t)^{\mathcal{I}}\ \&\ y \in (C,t)^{\mathcal{I}})\}$
$\sim \exists r.C$	$\{x \in D_t : (\forall s \succcurlyeq t)\,(\forall y \in D_s)\,((x,y) \in (\sim r,s)^{\mathcal{I}} \vee y \in (\sim C,s)^{\mathcal{I}})\}$

Table 6. The semantics of concept descriptions in $\mathcal{CALC}^{\mathsf{N4}}$

we emphasize the informational character of constructive description logics. They are not used to describe real situations, but information contained in a data base. An interpretation for $\mathcal{CALC}^{\mathsf{N4}}$ is a structure $\mathcal{I} = \langle W, \preccurlyeq, D, (\cdot,\cdot)^{\mathcal{I}} \rangle$, where $(W, \preccurlyeq)$ is a partial order, $D : W \rightarrow 2^{N_I}$ is a domain function such that $D_u \subseteq D_t$ whenever $u \preccurlyeq t$. Further, for every $t \in W$, every $A \in N_C$, and every $r \in N_r$, $(A,t)^{\mathcal{I}}$ and $(\sim A,t)^{\mathcal{I}}$ are subsets of D_t and $(r,t)^{\mathcal{I}}$ and $(\sim r,t)^{\mathcal{I}}$ are binary relations on D_t. Moreover, if $t, s \in W$ and $t \preccurlyeq s$, then $(A,t)^{\mathcal{I}} \subseteq (A,s)^{\mathcal{I}}$, $(\sim A,t)^{\mathcal{I}} \subseteq (\sim A,s)^{\mathcal{I}}$, $(r,t)^{\mathcal{I}} \subseteq (r,s)^{\mathcal{I}}$, and $(\sim r,t)^{\mathcal{I}} \subseteq (\sim r,s)^{\mathcal{I}}$. Intuitively, $(x,y) \in (\sim r,s)^{\mathcal{I}}$ iff at the state s the information is available that the ABox statement $r(x,y)$ is false. The interpretations of concept descriptions are presented in Table 6 in the style of Table 3. Note that the translation T_x treats the role restrictions as restricted universal and existential quantifiers rather than unary operators. This becomes important in the case of concept descriptions $\sim \exists r.C$. We have $T_x(\sim \exists r.C) = \sim T_x(\exists r.C)$ by definition and the following equivalence in **QN4**:

$$\sim T_x(\exists r.C) \;=\; \sim \exists y(r(x,y) \wedge T_y(C)) \leftrightarrow \forall y(\sim r(x,y) \vee \sim T_y(C)).$$

This equivalence explains the choice of interpretation for $\sim \exists r.C$ and why we need relations $(\sim r,s)^{\mathcal{I}}$, although there are no strongly negated role names in the language of $\mathcal{CALC}^{\mathsf{N4}}$.

DEFINITION 8. *$C \in Con(N_C, N_r)$ is said to be $\mathcal{CALC}^{\mathsf{N4}}$-valid ($\models_{\mathcal{CALC}^{\mathsf{N4}}} C$) iff for every interpretation $\mathcal{I} = \langle W, \preccurlyeq, D, (\cdot,\cdot)^{\mathcal{I}} \rangle$, and every $t \in W$, $(C,t)^{\mathcal{I}} = D_t$. C is subsumed by D ($C \models D$) iff $\models_{\mathcal{CALC}^{\mathsf{N4}}} C \sqsubseteq D$.*

EXAMPLE 8. *It can easily be seen that $\mathcal{CALC}^{\mathsf{C}}$ and $\mathcal{CALC}^{\mathsf{N4}}$ are distinct logics. In the semantics for $\mathcal{CALC}^{\mathsf{N4}}$, we have*

$$\models_{\mathcal{CALC}^{\mathsf{N4}}} (\exists r.C \sqsubseteq \forall r.D) \sqsubseteq \forall r.(C \sqsubseteq D)$$

which can be checked directly. However, the following situation is a counterexample to $\exists r.C \sqsubseteq \forall r.D \models \forall r.(C \sqsubseteq D)$ in the semantics for $\mathcal{CALC}^{\mathsf{C}}$:

$$\begin{array}{l}\Delta^{\mathcal{I}} = \{x, y\}, \;\; xr^{\mathcal{I}}y, \;\; \textit{both } r^{\mathcal{I}} \textit{ and } \preccurlyeq \textit{ are reflexive,}\\ x \notin C^{\mathcal{I}}, \;\; y \in C^{\mathcal{I}}, y \notin D^{\mathcal{I}}.\end{array}$$

Here, $x \in (\exists r.C \sqsubseteq \forall r.D)^{\mathcal{I}}$ but $x \notin \forall r.(C \sqsubseteq D)^{\mathcal{I}}$.

Now we extend the translation T_x to ABox and TBox statements. Unlike concept descriptions, these statements will be translated as *sentences*. We will use, however, the same notation for the extended translation. Note that any TBox statement is equivalent to a finite number of TBox statements $C \models D$; therefore, we consider only this case. Let $C, D \in Con(N_C, N_r)$ and $a \in N_I$:

$$\begin{array}{rcl} C(a) & \overset{T_x}{\longmapsto} & T_x(C)(a/x) \\ C \models D & \overset{T_x}{\longmapsto} & \forall x T_x(C \sqsubseteq D) \end{array}$$

We also introduce semantical notions for ABox and TBox statements.

DEFINITION 9. *Let $\mathcal{I} = \langle W, \preccurlyeq, D, (\cdot,\cdot)^{\mathcal{I}} \rangle$ be an interpretation and $t \in W$. We put*

$$\begin{array}{lll} \mathcal{I}, t \models^{+} C(a) & \textit{iff} & a \in (C, t)^{\mathcal{I}} \\ \mathcal{I}, t \models^{+} C \models D & \textit{iff} & (C \sqsubseteq D, t)^{\mathcal{I}} = D^{t} \end{array}$$

Let $\Sigma = (\mathcal{T}, \mathcal{A})$ be a knowledge base and α an ABox or TBox statement. The relation $\Sigma \models_{\mathcal{CALC}^{\mathsf{N4}}} \alpha$ holds if for any interpretation $\mathcal{I}$ and $t \in W$, $\mathcal{I}, t \models^{+} \beta$ for all $\beta \in \mathcal{T} \cup \mathcal{A}$ implies $\mathcal{I}, t \models^{+} \alpha$. We will write $\models_{\mathcal{CALC}^{\mathsf{N4}}} \alpha$ instead of $(\emptyset, \emptyset) \models_{\mathcal{CALC}^{\mathsf{N4}}} \alpha$.

The following proposition expresses in an exact form the correspondence between the semantics of $\mathsf{QN4}$ and $\mathcal{CALC}^{\mathsf{N4}}$.

PROPOSITION 10. *Let $\Sigma = (\mathcal{T}, \mathcal{A})$ be a knowledge base and α an ABox or TBox statement. The relation $\Sigma \models_{\mathcal{CALC}^{\mathsf{N4}}} \alpha$ holds if and only if $T_x(\Sigma) \models_{\mathsf{QN4}} T_x(\alpha)$, where $T_x(\Sigma) := \{T_x(\beta) \mid \beta \in \mathcal{T} \cup \mathcal{A}\}$.*

In analogy to the proof of the disjunction and the constructible falsity property of $\mathcal{CALC}^{\mathsf{C}}$ one can prove

constructor	semantics
A	$(A,t)^{\mathcal{I}} \subseteq D_t$
$\sim A$	$(\sim A,t)^{\mathcal{I}} \subseteq D_t$
$(C \sqcap D)$	$(C,t)^{\mathcal{I}} \cap (D,t)^{\mathcal{I}}$
$\sim(C \sqcap D)$	$(\sim C,t)^{\mathcal{I}} \cup (\sim D,t)^{\mathcal{I}}$
$(C \sqcup D)$	$(C,t)^{\mathcal{I}} \cup (D,t)^{\mathcal{I}}$
$\sim(C \sqcup D)$	$(\sim C,t)^{\mathcal{I}} \cap (\sim D,t)^{\mathcal{I}}$
$(C \sqsubseteq D)$	$\{x \in D_t : (\forall s \in I)\,(t \preccurlyeq s \Rightarrow (x \in (C,s)^{\mathcal{I}} \Rightarrow x \in (D,s)^{\mathcal{I}}))\}$
$\sim(C \sqsubseteq D)$	$(C,t)^{\mathcal{I}} \cap (\sim D,t)^{\mathcal{I}}$
$\forall r.C$	$\{x \in D_t : (\forall s \succcurlyeq t)\,(\forall y \in D_s)\,((x,y) \in (r,s)^{\mathcal{I}} \Rightarrow y \in (C,s)^{\mathcal{I}})\}$
$\sim \forall r.C$	$\{x \in D_t : (\exists y \in D_t)\,((x,y) \in (r,t)^{\mathcal{I}} \;\&\; y \in (\sim C,t)^{\mathcal{I}})\}$
$\exists r.C$	$\{x \in D_t : (\exists y \in D_t)\,((x,y) \in (r,t)^{\mathcal{I}} \;\&\; y \in (C,t)^{\mathcal{I}})\}$
$\sim \exists r.C$	$\{x \in D_t : (\forall s \succcurlyeq t)\,(\forall y \in D_s)\,((x,y) \in (r,s)^{\mathcal{I}} \Rightarrow y \in (\sim C,s)^{\mathcal{I}})\}$

Table 7. The semantics of concept descriptions in $\mathcal{CALC}^{\mathsf{N4}d}$

PROPOSITION 11. *The system $\mathcal{CALC}^{\mathsf{N4}}$ has the disjunction property and the constructible falsity property:*

$$\begin{array}{lllll} (i) & \models_{\mathcal{CALC}^{\mathsf{N4}}} (C \sqcup D) & \Rightarrow & \models_{\mathcal{CALC}^{\mathsf{N4}}} C \textit{ or } \models_{\mathcal{CALC}^{\mathsf{N4}}} D \\ (ii) & \models_{\mathcal{CALC}^{\mathsf{N4}}} \sim(C \sqcap D) & \Rightarrow & \models_{\mathcal{CALC}^{\mathsf{N4}}} \sim C \textit{ or } \models_{\mathcal{CALC}^{\mathsf{N4}}} \sim D \end{array}$$

5. The system $\mathcal{CALC}^{\mathsf{N4}d}$

5.1. Syntax and semantics of $\mathcal{CALC}^{\mathsf{N4}d}$

The syntax of $\mathcal{CALC}^{\mathsf{N4}d}$ is the same as the syntax of $\mathcal{CALC}^{\mathsf{N4}}$ and $\mathcal{CALC}^{\mathsf{C}}$. The semantics for $\mathcal{CALC}^{\mathsf{N4}d}$ is based on a translation T'_x into QN4 that differs from T_x only in the translation of concept descriptions $\sim \exists r.C$. The relevant clause is:

$$\sim \exists r.C \quad \overset{T'_x}{\longmapsto} \quad \forall y(r(x,y) \rightarrow T'_y(\sim C)).$$

The semantics for $\mathcal{CALC}^{\mathsf{N4}d}$ is given in Table 7.

In contrast to the semantics for $\mathcal{CALC}^{\mathsf{N4}}$, in the semantics for $\mathcal{CALC}^{\mathsf{N4}d}$ we do not need relations $(\sim r,t)^{\mathcal{I}}$. The reason for this is that

$$\mathsf{QN4} \vdash T'_x(\sim \exists r.C) \leftrightarrow T'_x(\forall r. \sim C),$$

which allows one to calculate $(\sim \exists r.C,t)^{\mathcal{I}}$ as $(\forall r. \sim C,t)^{\mathcal{I}}$.

We extend the translation T'_x to ABox and TBox statements and define the relation $\Sigma \models_{\mathcal{CALC}^{\mathsf{N4}d}} \alpha$ exactly as in the previous section for $\mathcal{CALC}^{\mathsf{N4}}$.

EXAMPLE 9. *To see that the logics $\mathcal{CALC}^{\mathsf{N4}d}$ and $\mathcal{CALC}^{\mathsf{C}}$ are distinct, we may consider the same countermodel as in Example 8. But also $\mathcal{CALC}^{\mathsf{N4}d}$ and $\mathcal{CALC}^{\mathsf{N4}}$ are distinct. In $\mathcal{CALC}^{\mathsf{N4}d}$ we have*

$$\models_{\mathcal{CALC}^{\mathsf{N4d}}} \sim \exists r.C =^{+} \forall r. \sim C.$$

whereas $\not\models_{\mathcal{CALC}^{\mathsf{N4}}} \sim \exists r.C =^{+} \forall r. \sim C.$

The property of $\mathcal{CALC}^{\mathsf{N4}d}$ noted in this example allows one to reduce any concept description to negation normal form. The notion of negation normal form $\overline{C}$ of a concept description C and of ABox and TBox statements is defined exactly as for the logic $\mathcal{CALC}^{\mathsf{C}}$. Like for $\mathcal{CALC}^{\mathsf{C}}$ we can prove

PROPOSITION 12. *For every concept description C, $\overline{C}$ is in negation normal form, and* $\models_{\mathcal{CALC}^{\mathsf{N4d}}} (C =^{+} \overline{C})$

This allows modifying the translation T'_x so as to obtain a translation of $\mathcal{CALC}^{\mathsf{N4}d}$ into QInt^{+}. We put

$$T''_x(C) := T'_x(\overline{C})$$

and consider expressions $\sim A$, $A \in N_C$ as new unary predicate symbols. The following analogue of Proposition 10 holds.

PROPOSITION 13. *Let $\Sigma = (\mathcal{T}, \mathcal{A})$ be a knowledge base and α an ABox or TBox statement. The relation $\Sigma \models_{\mathcal{CALC}^{\mathsf{N4d}}} \alpha$ holds iff $T''_x(\Sigma) \models_{\mathsf{QInt}^{+}} T''_x(\alpha)$, where $T''_x(\Sigma) := \{T''_x(\beta) \mid \beta \in \mathcal{T} \cup \mathcal{A}\}$.*

5.2. FS and the decidability of validity in $\mathcal{CALC}^{\mathsf{N4}d}$

Fischer Servi [12] formulated her logic FS in a language with two modalities, $\Box$ and $\Diamond$, and defined a Kripke-style semantics for this logic. In [15], Grefe proved that FS possesses the finite model property and, consequently, is decidable. The results of Fischer Servi and Grefe can be immediately generalized to the multi-modal case. Moreover, one can see that the positive fragment of multi-modal FS coincides, in fact, with the positive fragment of $\mathcal{CALC}^{\mathsf{N4}d}$. Combining this fact with the existence of a negation normal form for all concept descriptions in $\mathcal{CALC}^{\mathsf{N4}d}$, we can prove the decidability of validity in $\mathcal{CALC}^{\mathsf{N4}d}$.

Now we give a detailed presentation of the above considerations. First, we define the multi-modal variant FS_ω of Fischer Servi's modal logic. We will

$$
\begin{array}{lcl}
p_A & \stackrel{T_x^m}{\longmapsto} & A(x) \\
p_{\sim A} & \stackrel{T_x^m}{\longmapsto} & \sim A(x) \\
\bot & \stackrel{T_x^m}{\longmapsto} & \bot \\
\varphi \wedge \psi & \stackrel{T_x^m}{\longmapsto} & T_x^m(\varphi) \wedge T_x^m(\psi) \\
\varphi \vee \psi & \stackrel{T_x^m}{\longmapsto} & T_x^m(\varphi) \vee T_x^m(\psi) \\
\varphi \to \psi & \stackrel{T_x^m}{\longmapsto} & T_x^m(\varphi) \to T_x^m(\psi) \\
[r]\varphi & \stackrel{T_x^m}{\longmapsto} & \forall y(r(x,y) \to T_y^m(\varphi)) \\
\langle r \rangle \varphi & \stackrel{T_x^m}{\longmapsto} & \exists y(r(x,y) \wedge T_y^m(\varphi))
\end{array}
$$

Table 8. The translation of FS_ω into intuitionistic predicate logic

consider FS_ω in the language of intuitionistic propositional logic, enriched by the family of modalities indexed with the role names of $\mathcal{CALC}^{\mathsf{N4}d}$:

$$\langle \vee, \wedge, \to, \bot, \{[r], \langle r \rangle | \ r \in N_r\} \rangle.$$

Moreover, we assume that propositional variables are indexed with concept names and negated concept names, $Var = \{p_A, p_{\sim A} | \ A \in N_C\}$. The logic FS_ω is the least set of formulas in the above language containing the standard axioms for propositional intuitionistic logic, the modal axioms:

1. $[r]p \wedge [r]q \leftrightarrow [r](p \wedge q)$,
2. $\langle r \rangle p \vee \langle r \rangle q \leftrightarrow \langle r \rangle (p \vee q)$,
3. $\neg \langle r \rangle \bot$,
4. $\langle r \rangle (p \to q) \to ([r]p \to \langle r \rangle q)$,
5. $(\langle r \rangle p \to [r]q) \to [r](p \to q)$

for every $r \in N_r$, and is closed under the rules of substitution, modus ponens, and the monotonicity rules $p \to q/[r]p \to [r]q$ and $p \to q/\langle r \rangle p \to \langle r \rangle q$ for each pair of modalities.

We define the translation T_x^m of FS_ω into QInt with the set of unary predicates $\{A, \sim A | \ A \in N_C\}$ and the set of binary predicates N_r as it is presented in Table 8.

The next statement is a generalization of Theorem 3.1 in [15] to the multi-modal case, in the sense that its proof exactly follows the proof of Theorem 3.1 (repeating stages for different pairs of modalities).

$$\begin{array}{ccc}
A & \overset{(\cdot)^t}{\longmapsto} & p_A \\
\sim A & \overset{(\cdot)^t}{\longmapsto} & p_{\sim A} \\
C \sqcap D & \overset{(\cdot)^t}{\longmapsto} & (C)^t \wedge (D)^t \\
C \sqcup D & \overset{(\cdot)^t}{\longmapsto} & (C)^t \vee (D)^t \\
C \sqsubseteq D & \overset{(\cdot)^t}{\longmapsto} & (C)^t \rightarrow (D)^t \\
\forall r.C & \overset{(\cdot)^t}{\longmapsto} & [r](C)^t \\
\exists r.C & \overset{(\cdot)^t}{\longmapsto} & \langle r\rangle(C)^t
\end{array}$$

Table 9. The translation of $\mathcal{CALC}^{\mathsf{N4d}}$ into FS_ω

PROPOSITION 14. *A modal formula φ is a theorem of FS_ω if and only if $T_x^m(\varphi)$ is a theorem of intuitionistic first-order logic.*

In the same way we can generalize Corollary 5.13 of [15].

PROPOSITION 15. *FS_ω is decidable.*

Furthermore, we can see that the translation T_x^m is defined for positive formulas essentially as we defined the translation T_x' of $\mathcal{CALC}^{\mathsf{N4d}}$ into constructive first-order logic. More exactly, let us consider the translation $(\cdot)^t$ from the language of $\mathcal{CALC}^{\mathsf{N4d}}$ into the language of FS_ω defined in Table 9. From Propositions 13 and 14 we immediately obtain

PROPOSITION 16. *For any $C \in Con(N_C, N_r)$ in negation normal form, we have $\models_{\mathcal{CALC}^{\mathsf{N4d}}} C$ if and only if $(C)^t$ is a theorem of FS_ω.*

COROLLARY 1. *The validity problem for $\mathcal{CALC}^{\mathsf{N4d}}$ is decidable.*

Recall that for any $C \in Con(N_C, N_r)$, there was defined the negation normal form $\overline{C}$. Moreover, by Proposition 12 we have $\models_{\mathcal{CALC}^{\mathsf{N4d}}} (C =^+ \overline{C})$. In this way, $\models_{\mathcal{CALC}^{\mathsf{N4d}}} C$ if and only if $(\overline{C})^t$ is a theorem of FS_ω.

COROLLARY 2. *The system $\mathcal{CALC}^{\mathsf{N4d}}$ has the disjunction property and the constructible falsity property:*

$$\begin{array}{llll}
(i) & \models_{\mathcal{CALC}^{\mathsf{N4d}}} (C \sqcup D) & \Rightarrow & \models_{\mathcal{CALC}^{\mathsf{N4d}}} C \text{ or } \models_{\mathcal{CALC}^{\mathsf{N4d}}} D \\
(ii) & \models_{\mathcal{CALC}^{\mathsf{N4d}}} \sim(C \sqcap D) & \Rightarrow & \models_{\mathcal{CALC}^{\mathsf{N4d}}} \sim C \text{ or } \models_{\mathcal{CALC}^{\mathsf{N4d}}} \sim D
\end{array}$$

6. Tableau calculi for constructive description logic

In this section we shall present sound and complete tableau calculi for the description logics $\mathcal{CALC}^{\mathsf{C}}$, $\mathcal{CALC}^{\mathsf{N4}}$, and $\mathcal{CALC}^{\mathsf{N4}d}$. The tableau calculi use signed ABox and TBox statements, where the sign $+$ stands for "true" and the sign $-$ stands for "not true". In other words, $-A(a)$ is *not* a shorthand for $+\sim A(a)$. A set of signed ABox statements is called a signed ABox. Since every TBox statement is equivalent to a finite number of TBox statements $C \models D$, we shall consider only TBox statements of this shape.

6.1. An explicit tableau calculus for $\mathcal{CALC}^{\mathsf{C}}$

For heuristic reasons, we shall present a tableau calculus $\mathcal{TCALC}^{\mathsf{C}}$ for $\mathcal{CALC}^{\mathsf{C}}$ with explicit accessibility relations $\preccurlyeq$ and $r \in N_r$. This calculus closely follows the semantics of $\mathcal{CALC}^{\mathsf{C}}$ and is an adaptation of familiar tableau calculi for intuitionistic logic and multimodal K with explicit accessibility relations. The tableau rules for $\mathcal{CALC}^{\mathsf{C}}$ are given in Table 10. There are no rules for compound negations, because concept descriptions can be reduced to negation normal form. We mark with * parameters that must be new.

DEFINITION 10. *A node set is a finite set $\mathcal{A}$ such that every element of $\mathcal{A}$ is a signed ABox or TBox statement or an expression of the form $x \preccurlyeq y$. $\mathcal{A}$ is said to be closed if there is an $a \in N_I$ and $C \in N_C$ such that $\{+C(a), -C(a)\} \subseteq \mathcal{A}$; if $\mathcal{A}$ is not closed it is called open. A tableau for $\mathcal{A}$ is a binary tree τ such that (i) every node of τ is a node set, (ii) $\mathcal{A}$ is the root of τ, (iii) every transition from a node to its successor or pair of successors is an instance of one of the tableau rules, and (iv) nodes of τ that contain a closed node set are leaves. If **b** is a branch of a tableau τ, **b** is called closed if it contains a closed set, otherwise **b** is called open. A tableau τ is said to be closed if all branches of τ are closed. A branch **b** is said to be complete if **b** is closed or all rules that can be applied to a node of **b** have been applied. The tableau τ is called complete if all branches of τ are complete.*

In order to check the validity of a concept description C, we consider tableaux for the signed ABox $\{-\overline{C}(a)\}$, in order to check $C \models D$ we consider tableaux for $\{+\overline{C}(a), -\overline{D}(a)\}$, in order to check $\Sigma \models \alpha$ for a knowledge base $\Sigma = (\mathcal{T}, \mathcal{A})$ and ABox or TBox statement α, we consider tableaux for the set $\{+\overline{\beta} : \beta \in \mathcal{T} \cup \mathcal{A}\} \cup \{-\overline{\alpha}\}$.

EXAMPLE 10. *The following is a simple closed tableau for the signed ABox $\{+(C \models D), -(\exists r.C \models \exists r.D)\}$. Note that we do not rewrite node sets but only register new elements:*

$(\sqcap+)$	$\dfrac{+(C \sqcap D)(a) \in \mathcal{A}}{\mathcal{A}, +C(a), +D(a)}$	$(\sqcap-)$	$\dfrac{-(C \sqcap D)(a) \in \mathcal{A}}{\mathcal{A}, -C(a) \mid \mathcal{A}, -D(a)}$
$(\sqcup+)$	$\dfrac{+(C \sqcup D)(a) \in \mathcal{A}}{\mathcal{A}, +C(a) \mid \mathcal{A}, +D(a)}$	$(\sqcup-)$	$\dfrac{-(C \sqcup D)(a) \in \mathcal{A}}{\mathcal{A}, -C(a), -D(a)}$
$(\sqsubseteq +)$	$\dfrac{+(C \sqsubseteq D)(a),\ a \preccurlyeq b \in \mathcal{A}}{\mathcal{A}, -C(b) \mid \mathcal{A}, +D(b)}$	$(\sqsubseteq -)$	$\dfrac{-(C \sqsubseteq D)(a) \in \mathcal{A}}{\mathcal{A}, +C(b^*), -D(b^*), a \preccurlyeq b^*}$
$(\forall+)$	$\dfrac{+\forall r.C(a),\ r(a,b) \in \mathcal{A}}{\mathcal{A}, +C(b)}$	$(\forall-)$	$\dfrac{-\forall r.C(a) \in \mathcal{A}}{\mathcal{A},\ r(a,b^*),\ -C(b^*)}$
$(\exists+)$	$\dfrac{+\exists r.C(a) \in \mathcal{A}}{\mathcal{A}, r(a,b^*), +C(b^*)}$	$(\exists-)$	$\dfrac{-\exists r.C(a),\ r(a,b) \in \mathcal{A}}{\mathcal{A}, -C(b)}$
$(per+)$	$\dfrac{+A(a),\ a \preccurlyeq b \in \mathcal{A}}{\mathcal{A}, +A(b)}$	$(per\ \sim)$	$\dfrac{+\!\sim A(a),\ a \preccurlyeq b \in \mathcal{A}}{\mathcal{A}, +\!\sim A(b)}$
$(\preccurlyeq r)$	$\dfrac{a \preccurlyeq c,\ r(c,b) \in \mathcal{A}}{\mathcal{A}, r(a,d^*), d^* \preccurlyeq b}$	$(\preccurlyeq^{-1} r)$	$\dfrac{c \preccurlyeq a,\ r(c,b) \in \mathcal{A}}{\mathcal{A}, r(a,d^*), b \preccurlyeq d^*}$
(ref)	$\dfrac{\mathcal{A}}{\mathcal{A}, a \preccurlyeq a}$ *for any a occurring in* $\mathcal{A}$	$(tran)$	$\dfrac{a \preccurlyeq b,\ b \preccurlyeq c \in \mathcal{A}}{\mathcal{A}, a \preccurlyeq c}$
$(\models +)$	$\dfrac{+(C \models D) \in \mathcal{A}}{\mathcal{A}, +(C \sqsubseteq D)(a)}$	$(\models -)$	$\dfrac{-(C \models D) \in \mathcal{A}}{\mathcal{A}, -(C \sqsubseteq D)(a^*)}$

Table 10. Tableau rules for $\mathcal{CALC}^{\mathsf{C}}$ with explicit accessibility relations

$$\cfrac{\cfrac{\cfrac{+(C \models D),\ -(\exists r.C \models \exists r.D)}{-(\exists r.C \sqsubseteq \exists r.D)(a)}}{+\exists r.C(b),\ -\exists r.D(b),\ a \preccurlyeq b}}{\cfrac{r(b,c),\ +C(c)}{\cfrac{-D(c)}{\cfrac{c \preccurlyeq c}{\cfrac{+(C \sqsubseteq D)(c)}{-C(c) \mid +D(c)}}}}}$$

DEFINITION 11. *Let $\mathcal{I} = \langle \Delta^{\mathcal{I}}, \preccurlyeq, \cdot^{\mathcal{I}} \rangle$ be an interpretation, let τ be a tableau, and let **b** be a branch of τ. Interpretation $\mathcal{I}$ is faithful to **b** iff there is a function f from a non-empty set N_I of individual names into $\Delta^{\mathcal{I}}$ such that*

1. *for every $+C(a)$ on **b**, $f(a) \in C^{\mathcal{I}}$;*
2. *for every $-C(a)$ on **b**, $f(a) \notin C^{\mathcal{I}}$;*
3. *for every $r(a,b)$ on **b**, $f(a)r^{\mathcal{I}}f(b)$;*
4. *for every $a \preccurlyeq b$ on **b**, $f(a) \preccurlyeq f(b)$;*
5. *for every $+(C \models D)$ on **b**, $f(a) \in (C \sqsubseteq D)^{\mathcal{I}}$ for every $a \in N_I$;*
6. *for every $-(C \models D)$ on **b**, $f(a) \notin (C \sqsubseteq D)^{\mathcal{I}}$ for some $a \in N_I$.*

LEMMA 1. *Let $\mathcal{I} = \langle \Delta^{\mathcal{I}}, \preccurlyeq, \cdot^{\mathcal{I}} \rangle$ be an interpretation, let τ be a tableau, and let **b** be a branch of τ. If $\mathcal{I}$ is faithful to **b** and a tableau rule is applied to **b**, then among the resulting branches, there is at least one branch **b′** to which $\mathcal{I}$ is faithful.*

PROOF. By induction on the construction of tableaux. By way of example, we here consider only two cases. (1) ($\models -$). Suppose $-(C \models D) \in \mathcal{A}$ and $\mathcal{I}$ is faithful to **b**. Then there is a function f with $f(a) \notin (C \sqsubseteq D)^{\mathcal{I}}$ for some a. Hence there exists $x \in \Delta^{\mathcal{I}}$ such that $f(a) \preccurlyeq x$, $x \in C^{\mathcal{I}}$, and $x \notin D^{\mathcal{I}}$. Define the function g to be exactly as f, except that $g(a^*) = x$ for some new a^*. Since a^* is a new individual name, g is faithful to **b**. By definition of g, $g(a) \preccurlyeq g(a^*)$, $g(a^*) \in C^{\mathcal{I}}$, and $g(a^*) \notin D^{\mathcal{I}}$. Hence g is faithful to the extended branch. (2) ($\preccurlyeq r$). Suppose $\mathcal{I}$ is faithful to **b** and $a \preccurlyeq c$, $r(c,b) \in \mathcal{A}$. Then there exists f with $f(a) \preccurlyeq f(c)$, and $f(c)r^{\mathcal{I}}f(b)$. By (P3), $(\exists x \in \Delta^{\mathcal{I}})$ $f(a)r^{\mathcal{I}}x$ and $x \preccurlyeq f(b)$. Define the function g exactly as f, except that $g(d^*)$ $= x$. Since d^* is new, g is faithful to **b** and to the extended branch. ■

DEFINITION 12. *Let τ be a tableau and let **b** be an open branch of τ. The structure induced by **b** is the structure $\mathcal{I}_{\mathbf{b}} = (\Delta^{\mathcal{I}_b}, \preccurlyeq, \cdot^{\mathcal{I}_b})$, where*

- $\Delta^{\mathcal{I}_b} = \{a : a$ *occurs on* $\mathbf{b}\}$
- $a \preccurlyeq b$ *iff* $a \preccurlyeq b$ *occurs on* $\mathbf{b}$;
- $ar^{\mathcal{I}_b}b$ *iff* $r(a,b)$ *occurs on* $\mathbf{b}$;
- $a \in A^{\mathcal{I}_b}$ *iff* $+A(a)$ *occurs on* $\mathbf{b}$;
- $a \in \sim A^{\mathcal{I}_b}$ *iff* $+\sim A(a)$ *occurs on* $\mathbf{b}$.

LEMMA 2. *Let τ be a tableau and let **b** be an open complete branch of τ. Then $\mathcal{I}_{\mathbf{b}}$ is an interpretation.*

PROOF. It must be shown that $\preccurlyeq$ is reflexive and transitive and that (P1) – (P4) hold. These properties follow easily by noticing that the respective tableau rules have been applied, because $\boldsymbol{b}$ is complete. Note that since $\boldsymbol{b}$ is open, for no $C \in N_c$ and $a \in N_I$, both $a^{\mathcal{I}_b} \in C^{\mathcal{I}_b}$ and $a^{\mathcal{I}_b} \notin C^{\mathcal{I}_b}$. ∎

LEMMA 3. *Let τ be a tableau and let **b** be an open complete branch of τ. Then for every ABox or TBox statement α,*

(i) $\mathcal{I}_{\mathbf{b}} \models \alpha$ *if* $+\alpha$ *occurs on* $\mathbf{b}$; *(ii)* $\mathcal{I}_{\mathbf{b}} \not\models \alpha$ *if* $-\alpha$ *occurs on* $\mathbf{b}$.

PROOF. (i) Suppose $+\alpha$ occurs on $\boldsymbol{b}$ and α is an ABox statement. If $\alpha \in$ *Lit*, the claim holds by definition. If α is a complex ABox statement, use induction on the structure of α. If α is a TBox statement $(C \models D)$ and $+(C \models D)$ occurs on $\boldsymbol{b}$, then for arbitrary $a \in \Delta^{\mathcal{I}_b}$, $+(C \sqsubseteq D)(a)$ occurs on $\boldsymbol{b}$, because $\boldsymbol{b}$ is complete. Since the claim holds for ABox statements, $a \in (C \models D)^{\mathcal{I}_b}$, for arbitrary a. Thus $(C \models D)^{\mathcal{I}_b} = \Delta^{\mathcal{I}_b}$. In other words, $\mathcal{I}_{\mathbf{b}} \models (C \models D)$. (ii) Suppose $-\alpha$ occurs on $\boldsymbol{b}$, and α is an ABox statement. If $\alpha \in Lit$, the claim holds by definition, since $\boldsymbol{b}$ is open. If α is a complex ABox statement, use induction on the structure of α. If α is a TBox statement $(C \models D)$ and $-(C \models D)$ occurs on $\boldsymbol{b}$, then for some $a \in \Delta^{\mathcal{I}_b}$, $-(C \sqsubseteq D)(a)$ occurs on $\boldsymbol{b}$, because $\boldsymbol{b}$ is complete. Since the claim holds for ABox statements, $a \notin (C \models D)^{\mathcal{I}_b}$. Thus $(C \models D)^{\mathcal{I}_b} \neq \Delta^{\mathcal{I}_b}$. In other words, $\mathcal{I}_{\mathbf{b}} \not\models (C \models D)$. ∎

PROPOSITION 17. *Let $\Sigma = (\mathcal{T}, \mathcal{A})$ be a knowledge base and α an ABox or TBox statement. Then $\Sigma \models \alpha$ iff there exists a closed tableau for $\mathcal{B} = \{+\overline{\beta} : \beta \in \mathcal{T} \cup \mathcal{A}\} \cup \{-\overline{\alpha}\}$.*

PROOF. $\Leftarrow$: Let $\Sigma \not\models \alpha$ and suppose that τ is a closed tableau for $\mathcal{B}$. Then there is an interpretation $\mathcal{I}$ such that $\mathcal{I} \models \mathcal{T}$, $\mathcal{I} \models \mathcal{A}$, but $\mathcal{I} \not\models \{\alpha\}$. Since $\mathcal{I}$ is faithful to the root of τ, by Lemma 1, there is a branch $\boldsymbol{b}$ of τ such that $\mathcal{I}$ is faithful to $\boldsymbol{b}$. Since $\boldsymbol{b}$ is closed, it contains $+C(a)$ and $-C(a)$ for some $a \in N_I$ and $C \in N_C$. But since $\mathcal{I}$ is faithful to $\boldsymbol{b}$, there is a function f such that $f(a) \in C^{\mathcal{I}}$ and $f(a) \notin C^{\mathcal{I}}$, a contradiction. $\Rightarrow$: Let τ be a complete but open tableau for $\mathcal{B}$, and let $\boldsymbol{b}$ be an open branch of τ. By Lemma 3, $\Sigma \not\models \alpha$. ∎

6.2. A tableau calculus for $\mathcal{CALC}^{\mathsf{N4}}$

A tableau calculus for QN4

From the Hilbert-style calculus for QN4 described in Subsection 4.1 one can easily pass to a tableau calculus, which we will call TQN4. As in the case of the tableau calculus for intuitionistic logic (see, for example, [26]), we will work with sets of formulas signed with + or −. Below, $\mathcal{A}$ denotes a set of signed formulas.

Axioms of TQN4 have the form $\{+\varphi, -\varphi\}$, where φ is an arbitrary formula. The *reduction rules* of the calculus are given in Table 11. In this table, * denotes the variable restriction that the parameter a introduced must not occur in any formula of $\mathcal{A}, \varphi(x)$. $\mathcal{A}^+$ stands for $\{+\psi \mid +\psi \in \mathcal{A}\}$. A set of signed formulas is *closed* if it contains $+\varphi$ and $-\varphi$ simultaneously for some φ. The notions of a tableau and closed tableau are defined in the usual way. A set $\mathcal{A}$ of signed formulas is said to be TQN4-consistent if there is a non-closed tableau for $\mathcal{A}$, otherwise, $\mathcal{A}$ is TQN4-inconsistent.

We can prove in a standard way a statement on the equivalence of the two calculi for QN4 we have defined.

PROPOSITION 18. *For any two finite sets of formulas* Γ *and* Δ, $\Gamma \vdash_{\mathsf{QN4}} \Delta$ *if and only if the set* $\{+\varphi, -\psi \mid \varphi \in \Gamma, \psi \in \Delta\}$ *is* TQN4*-inconsistent.*

6.2.1. The tableau calculus $\mathcal{TCALC}^{\mathsf{N4}}$

From the calculus TQN4 one can easily pass to a tableau calculus for the logic $\mathcal{CALC}^{\mathsf{N4}}$. We formulate the calculus $\mathcal{TCALC}^{\mathsf{N4}}$, which works with sets of signed ABox and TBox statements. Axioms of $\mathcal{TCALC}^{\mathsf{N4}}$ have the form

$\{+\alpha, -\alpha\}$, where α is an ABox or TBox statement.

The reduction rules of $\mathcal{TCALC}^{\mathsf{N4}}$ are given in Table 12, where $\mathcal{A}$ is a set of signed ABox and TBox statements, $C, D \in Con(N_C, N_r)$, $a, b \in N_I$,

$(\wedge +)$	$\dfrac{+(\varphi \wedge \psi) \in \mathcal{A}}{\mathcal{A}, +\varphi, +\psi}$	$(\wedge -)$	$\dfrac{-(\varphi \wedge \psi) \in \mathcal{A}}{\mathcal{A}, -\varphi \mid \mathcal{A}, -\psi}$
$(\vee +)$	$\dfrac{+(\varphi \vee \psi) \in \mathcal{A}}{\mathcal{A}, +\varphi \mid \mathcal{A}, +\psi}$	$(\vee -)$	$\dfrac{-(\varphi \vee \psi) \in \mathcal{A}}{\mathcal{A}, -\varphi, -\psi}$
$(\rightarrow +)$	$\dfrac{+(\varphi \rightarrow \psi) \in \mathcal{A}}{\mathcal{A}, -\varphi \mid \mathcal{A}, +\psi}$	$(\rightarrow -)$	$\dfrac{-(\varphi \rightarrow \psi) \in \mathcal{A}}{\mathcal{A}^{+}, +\varphi, -\psi}$
$(\sim \wedge +)$	$\dfrac{+\sim(\varphi \wedge \psi) \in \mathcal{A}}{\mathcal{A}, +\sim\varphi \mid \mathcal{A}, +\sim\psi}$	$(\sim \wedge -)$	$\dfrac{-\sim(\varphi \wedge \psi) \in \mathcal{A}}{\mathcal{A}, -\sim\varphi, -\sim\psi}$
$(\sim \vee +)$	$\dfrac{+\sim(\varphi \vee \psi) \in \mathcal{A}}{\mathcal{A}, +\sim\varphi, +\sim\psi}$	$(\sim \vee -)$	$\dfrac{-\sim(\varphi \vee \psi) \in \mathcal{A}}{\mathcal{A}, -\sim\varphi \mid \mathcal{A}, -\sim\psi}$
$(\sim\rightarrow +)$	$\dfrac{+\sim(\varphi \rightarrow \psi) \in \mathcal{A}}{\mathcal{A}, +\varphi, +\sim\psi}$	$(\sim\rightarrow -)$	$\dfrac{-\sim(\varphi \rightarrow \psi) \in \mathcal{A}}{\mathcal{A}, -\varphi \mid \mathcal{A}, -\sim\psi}$
$(\sim\sim +)$	$\dfrac{+\sim\sim\varphi \in \mathcal{A}}{\mathcal{A}, +\varphi}$	$(\sim\sim -)$	$\dfrac{-\sim\sim\varphi \in \mathcal{A}}{\mathcal{A}, -\varphi}$
$(\forall +)$	$\dfrac{+\forall x\varphi(x) \in \mathcal{A}}{\mathcal{A}, +\varphi(a)}$	$(\forall -)$	$\dfrac{-\forall x\varphi(x) \in \mathcal{A}}{\mathcal{A}^{+}, -\varphi(a^{*})}$
$(\exists +)$	$\dfrac{+\exists x\varphi(x) \in \mathcal{A}}{\mathcal{A}, +\varphi(a^{*})}$	$(\exists -)$	$\dfrac{\exists x\varphi(x) \in \mathcal{A}}{\mathcal{A}, -\varphi(a)}$
$(\sim \forall +)$	$\dfrac{+\sim\forall x\varphi(x) \in \mathcal{A}}{\mathcal{A}, +\sim\varphi(a^{*})}$	$(\sim \forall -)$	$\dfrac{-\sim\forall x\varphi(x) \in \mathcal{A}}{\mathcal{A}, -\sim\varphi(a)}$
$(\sim \exists +)$	$\dfrac{+\sim\exists x\varphi(x) \in \mathcal{A}}{\mathcal{A}, +\sim\varphi(a)}$	$(\sim \exists -)$	$\dfrac{-\sim\exists x\varphi(x) \in \mathcal{A}}{\mathcal{A}^{+}, -\sim\varphi(a^{*})}$

Table 11. Tableau rules for QN4

and $\mathcal{A}^+ = \{+\alpha| \ +\alpha \in \mathcal{A}\}$. The sign * denotes a parameter restriction. In a statement $C(b^*)$, the individual name b must be new. The notion of a $\mathcal{TCALC}^{\mathsf{N4}}$-inconsistent set of signed statements is defined exactly as that of a TQN4-inconsistent set of signed formulas.

Most of the rules of this calculus for $\mathcal{CALC}^{\mathsf{N4}}$ are obtained from the respective rules of TQN4 by rewording of antecedents and succedents with the help of T_x. In the case of role restrictions, the rules of $\mathcal{TCALC}^{\mathsf{N4}}$ are obtained by a combination of two rules of TQN4. For example, consider the rule $(\forall+)$. By definition, $T_x(\forall r.C(a)) = \forall y(r(a,y) \to T_y(C))$. Applying successively rules $(\forall+)$ and $(\to +)$ of TQN4 we obtain

$$\frac{\dfrac{+\forall y(r(a,y) \to T_y(C)) \in \mathcal{A}}{\mathcal{A},\ +(r(a,b) \to T_y(C)(b/y))}}{\mathcal{A},\ -r(a,b) \mid \mathcal{A},\ +T_y(C)(b/y)}$$

Taking into account that $T_x(C(b)) = T_y(C)(b/y)$, we have obtained the result of applying the rule $(\forall+)$ reworded with the help of T_x.

In this way, we infer from Propositions 9, 10, and 18 the following completeness result.

PROPOSITION 19. *Let* $\Sigma = (\mathcal{T}, \mathcal{A})$ *be a knowledge base and* α *an ABox or TBox statement. Then* $\Sigma \models_{\mathcal{CALC}^{\mathsf{N4}}} \alpha$ *if and only if the set* $\{+\beta|\ \beta \in \mathcal{T} \cup \mathcal{A}\} \cup \{-\alpha\}$ *is* $\mathcal{TCALC}^{\mathsf{N4}}$*-inconsistent.*

6.2.2. A tableau calculus for $\mathcal{CALC}^{\mathsf{N4}d}$

Due to the fact that any concept description of $\mathcal{CALC}^{\mathsf{N4}d}$ can be reduced to a negation normal form, we need no reduction rules for negated statements. The calculus $\mathcal{TCALC}^{\mathsf{N4}d}$ has axioms $\{+\alpha, -\alpha\}$, where α is an ABox or TBox statement in negation normal form. The reduction rules of $\mathcal{TCALC}^{\mathsf{N4}d}$ are presented in Table 13.

As in the case of $\mathcal{TCALC}^{\mathsf{N4}}$, the rules of the considered calculus are obtained by a combination of rules in the tableau calculus for QInt^+ [26]. As in the previous subsection, we can prove

PROPOSITION 20. *Let* $\Sigma = (\mathcal{T}, \mathcal{A})$ *be a knowledge base and* α *an ABox or TBox statement. Then* $\Sigma \models_{\mathcal{CALC}^{\mathsf{N4}d}} \alpha$ *iff the set* $\{+\overline{\beta}|\ \beta \in \mathcal{T} \cup \mathcal{A}\} \cup \{-\overline{\alpha}\}$ *is* $\mathcal{TCALC}^{\mathsf{N4}d}$*-inconsistent.*

$(\sqcap+)$	$\dfrac{+(C \sqcap D)(a) \in \mathcal{A}}{\mathcal{A}, +C(a), +D(a)}$	$(\sqcap-)$	$\dfrac{-(C \sqcap D)(a) \in \mathcal{A}}{\mathcal{A}, -C(a) \mid \mathcal{A}, -D(a)}$
$(\sqcup+)$	$\dfrac{+(C \sqcup D)(a) \in \mathcal{A}}{\mathcal{A}, +C(a) \mid \mathcal{A}, +D(a)}$	$(\sqcup-)$	$\dfrac{-(C \sqcup D)(a) \in \mathcal{A}}{\mathcal{A}, -C(a), -D(a)}$
$(\sqsubseteq +)$	$\dfrac{+(C \sqsubseteq D)(a) \in \mathcal{A}}{\mathcal{A}, -C(a) \mid \mathcal{A}, +D(a)}$	$(\sqsubseteq -)$	$\dfrac{-(C \sqsubseteq D)(a) \in \mathcal{A}}{\mathcal{A}^{+}, +C(a), -D(a)}$
$(\sim \sqcap+)$	$\dfrac{+ \sim (C \sqcap D)(a) \in \mathcal{A}}{\mathcal{A}, + \sim C(a) \mid \mathcal{A}, + \sim D(a)}$	$(\sim \sqcap-)$	$\dfrac{- \sim (C \sqcap D)(a) \in \mathcal{A}}{\mathcal{A}, - \sim C(a), - \sim D(a)}$
$(\sim \sqcup+)$	$\dfrac{+ \sim (C \sqcup D)(a) \in \mathcal{A}}{\mathcal{A}, + \sim C(a), + \sim D(a)}$	$(\sim \sqcup-)$	$\dfrac{- \sim ((C \sqcup D)(a) \in \mathcal{A}}{\mathcal{A}, - \sim C(a) \mid \mathcal{A}, - \sim D(a)}$
$(\sim\sqsubseteq +)$	$\dfrac{+ \sim (C \sqsubseteq D)(a) \in \mathcal{A}}{\mathcal{A}, +C(a), + \sim D(a)}$	$(\sim\sqsubseteq -)$	$\dfrac{- \sim (C \sqsubseteq D)(a) \in \mathcal{A}}{\mathcal{A}, -C(a) \mid \mathcal{A}, - \sim D(a)}$
$(\sim\sim +)$	$\dfrac{+ \sim\sim C(a) \in \mathcal{A}}{\mathcal{A}, +C(a)}$	$(\sim\sim -)$	$\dfrac{- \sim\sim C(a) \in \mathcal{A}}{\mathcal{A}, -C(a)}$
$(\forall+)$	$\dfrac{+\forall r.C(a) \in \mathcal{A}}{\mathcal{A}, -r(a,b) \mid \mathcal{A}, +C(b)}$	$(\forall-)$	$\dfrac{-\forall r.C(a) \in \mathcal{A}}{\mathcal{A}^{+}, +r(a,b^{*}), -C(b^{*})}$
$(\exists+)$	$\dfrac{+\exists r.C(a) \in \mathcal{A}}{\mathcal{A}, +r(a,b^{*}), +C(b^{*})}$	$(\exists-)$	$\dfrac{-\exists r.C(a) \in \mathcal{A}}{\mathcal{A}, -r(a,b) \mid \mathcal{A}, -C(b)}$
$(\sim \forall+)$	$\dfrac{+ \sim \forall r.C(a) \in \mathcal{A}}{\mathcal{A}, +r(a,b^{*}) \mid \mathcal{A}, + \sim C(b^{*})}$	$(\sim \forall-)$	$\dfrac{- \sim \forall r.C(a) \in \mathcal{A}}{\mathcal{A}, -r(a,b) \mid \mathcal{A}, - \sim C(b)}$
$(\sim \exists+)$	$\dfrac{+ \sim \exists r.C(a) \in \mathcal{A}}{\mathcal{A}, + \sim r(a,b) \mid \mathcal{A}, + \sim C(b)}$	$(\sim \exists-)$	$\dfrac{- \sim \exists r.C(a) \in \mathcal{A}}{\mathcal{A}^{+}, - \sim r(a,b^{*}), - \sim C(b^{*})}$
$(\models +)$	$\dfrac{+(C \models D) \in \mathcal{A}}{\mathcal{A}, +(C \sqsubseteq D)(b)}$	$(\models -)$	$\dfrac{-(C \models D) \in \mathcal{A}}{\mathcal{A}^{+}, -(C \sqsubseteq D)(b^{*})}$

Table 12. Tableau rules for $\mathcal{TCALC}^{\mathsf{N4}}$

$$(\sqcap+)\quad \frac{+(C \sqcap D)(a) \in \mathcal{A}}{\mathcal{A}, +C(a), +D(a)} \qquad (\sqcap-)\quad \frac{-(C \sqcap D)(a) \in \mathcal{A}}{\mathcal{A}, -C(a) \mid \mathcal{A}, -D(a)}$$

$$(\sqcup+)\quad \frac{+(C \sqcup D)(a) \in \mathcal{A}}{\mathcal{A}, +C(a) \mid \mathcal{A}, +D(a)} \qquad (\sqcup-)\quad \frac{-(C \sqcup D)(a) \in \mathcal{A}}{\mathcal{A}, -C(a), -D(a)}$$

$$(\sqsubseteq+)\quad \frac{+(C \sqsubseteq D)(a) \in \mathcal{A}}{\mathcal{A}, -C(a) \mid \mathcal{A}, +D(a)} \qquad (\sqsubseteq-)\quad \frac{-(C \sqsubseteq D)(a) \in \mathcal{A}}{\mathcal{A}^+, +C(a), -D(a)}$$

$$(\forall+)\quad \frac{+\forall r.C(a) \in \mathcal{A}}{\mathcal{A}, -r(a,b) \mid \mathcal{A}, +C(b)} \qquad (\forall-)\quad \frac{-\forall r.C(a) \in \mathcal{A}}{\mathcal{A}^+, +r(a,b^*), -C(b^*)}$$

$$(\exists+)\quad \frac{+\exists r.C(a) \in \mathcal{A}}{\mathcal{A}, +r(a,b^*), +C(b^*)} \qquad (\exists-)\quad \frac{-\exists r.C(a) \in \mathcal{A}}{\mathcal{A}, -r(a,b) \mid \mathcal{A}, -C(b)}$$

$$(\models+)\quad \frac{+(C \models D) \in \mathcal{A}}{\mathcal{A}, +(C \sqsubseteq D)(b)} \qquad (\models-)\quad \frac{-(C \models D) \in \mathcal{A}}{\mathcal{A}^+, -(C \sqsubseteq D)(b^*)}$$

Table 13. Tableau rules for $\mathcal{TCALC}^{\mathsf{N4d}}$

7. Conclusion and outlook

In the present paper, we have motivated the investigation and use of constructive paraconsistent description logics. We argued that subsumption of concepts need not be tied to material implication. Moreover, we pointed out that if knowledge representation is understood not as reflecting real situations but rather *information* about real situations, it is advantageous to work with a paraconsistent logic. By considering translations from the language of constructive description logic into the languages of classical and constructive first-order logic, we defined three basic paraconsistent systems of constructive description logic. Each of theses systems answers to the needs deriving from our motivating examples and replaces classical implication by intuitionistic implication and classical negation by strong, constructive negation. It should have become clear that constructive description logic is a rich and promising field offering, for example, reasoning tasks not available in the classical framework. Now that certain foundations have been laid, many problems and questions naturally arise. Although we presented sound and complete tableau calculi, in this paper we did not consider tableaux-based decision procedures. In particular, our tableau calculus for $\mathcal{CALC}^{\mathsf{C}}$ is not very well-suited for that purpose. Therefore, in a sequel to this paper we plan to consider tableau algorithms. Other directions of future research are also obvious, namely investigating extensions to constructive description logics with qualified number restrictions, complex roles, role restrictions, modal operators, and so forth. Another prominent issue, of course, is the complexity of tableau algorithms and their implementation.

References

[1] S. Akama, On Constructive Modality, in: S. Akama (ed.), *Logic, Language and Computation*, Kluwer Academic Publishrs, Dordrecht, 1997, 143–158.

[2] A. Almukdad and D. Nelson, Constructible Falsity and Inexact Predicates, *Journal of Symbolic Logic* 29 (1984), 231–233.

[3] F. Baader and U. Sattler, An Overview of Tableau Algorithms for Description Logics, *Studia Logica* 69 (2001), 5–40.

[4] F. Baader et al. (eds.), *The Description Logic Handbook*, Cambridge University Press, Cambridge, 2003.

[5] N. Belnap, A Useful Four-Valued Logic, in: J.M. Dunn and G. Epstein (eds.), *Modern Uses of Multiple-Valued Logic*, Reidel, Dordrecht, 8–37, 1977.

[6] P. Blackburn, M. de Rijke and Y. Venema, *Modal Logic*, Cambridge Tracts in Teoretical Computer Science 53, Cambridge University Press, Cambridge, 2001.

[7] R. Bull, MIPC as the Formalisation of an Intuitionistic Concept of Modality, *Journal of Symbolic Logic* 31 (1966), 609–616.

[8] D. van Dalen, Intuitionistic Logic, in: D. Gabbay and F. Guenthner (eds.), *Handbook of Philosophical Logic. Vol.III: Alternatives to Classical Logic*, Reidel, Dordrecht, 1986, 225–339.

[9] F. Donini, M. Lenzerini, D. Nardi, and A. Schaerf, Reasoning in Description Logics, in: G. Brewka (ed.), *Principles of Knowledge Representation*, CSLI Publications, Stanford, 1996, 191–236.

[10] M. Božić's and K. Došen, Models for Normal Intuitionistic Modal Logics, *Studia Logica* 43 (1984), 217–245.

[11] J.M. Dunn, Partiality and its Dual, *Studia Logica*, 66 (2000), 5–40.

[12] G. Fischer Servi, On Modal Logic with an Intuitionistic Base, *Studia Logica*, 36 (1977), 141–149.

[13] G. Fischer Servi, Axiomatizations for some Intuitionistic Modal Logics, *Rend. Sem. Mat. Univers. Polit.*, 42 (1984), 179–184.

[14] D. Gabbay, A. Kurucz, F. Wolter, and M. Zakharyaschev, *Many-Dimensional Modal Logics: Theory and Applications*, book manuscript, 2002.

[15] C. Grefe, Fischer Servi's Intuitionistic Modal Logic has the Finite Model Property, in: M. Kracht et al. (eds.), *Advances in Modal logic. Vol. 1.*, CSLI Lecture Notes 87, 1998, 85–98.

[16] Yu. Gurevich, Intuitionistic Logic with Strong Negation, *Studia Logica* 27 (1977), 49–49.

[17] C. Lutz and U. Sattler, The Complexity of Reasoning with Boolean Modal Logic, in: F. Wolter et al. (eds.), *Advances in Modal Logic. Volume 3*, World Scientific Publishers, Singapore, 2002, 329–348.

[18] S.P. Odintsov, Algebraic Semantics for Paraconsitent Nelson's Logic, 2002, to appear in: *Journal of Logic and Computation.*

[19] S.P. Odintsov, On the Embedding of Nelson's Logics, *Bulletin of the Section of Logic*, 31/4(2002), 241–248.

[20] H. Ono, On some Intuitionistic Modal Logics, *Publ. Res. Inst. Math. Sci. Kyoto Univ.* 13 (1977), 687-722.

[21] H. Ono and N.-Y. Suzuki, Relations between Intuitionistic Modal Logics and Intermediate Predicate Logics, *Reports on Mathematical Logic* 22 (1989), 65-87.

[22] P.F. Patel-Schneider, A Four-valued Semantics for Terminological Logics, *Artificial Intelligence* 38, No.3 (1989), 319–351.

[23] A.K. Simpson, *The Proof Theory and Semantics of Intuitionistic Modal Logic*, PhD Thesis, University of Edinburg, 1994.

[24] V. Sotirov, Modal Theories With Intuitionistic Logic, in: *Mathematical Logic. Proceedings of the Conference on Mathematical Logic, Dedicated to the Memory of A.A. Markov (1903-1979), Sofia, September 22-23, 1980*, Sofia, 1984, 139–171.

[25] R. Thomason, A Semantical Study of Constructive Falsity, *Zeitschrift für mathematische Logik und Grundlagen der Mathematik*, 15 (1969), 247–257.

[26] A. Waaler and L. Wallen, Tableaux for Intuitionistic Logic, in: M. D'Agostino et al. (eds.), *Handbook of Tableaux Methods*, Kluwer Academic Publishers, Dordrecht, 1999, 255–296.

[27] H. Wansing, *The Logic of Information Structures*, Lecture Notes in AI 681, Springer-Verlag, Berlin, 1993.

[28] H. Wansing, Tarskian Structured Consequence Relations and Functional Completeness, *Math. Logic Quarterly* 41 (1995), 73-92.

[29] H. Wansing, Negation, in: L. Goble (ed.), *The Blackwell Guide to Philosophical Logic*, Basil Blackwell Publishers, Cambridge/Ma., 2001, 415–436.

[30] F. Wolter and M. Zakharyaschev, Intuitionistic Modal Logic, in: A. Cantini et al. (eds.), *Logic and Foundations of Mathematics*, Kluwer Academic Publishers, Dordrecht, 1999, 227–238.

[31] F. Wolter and M. Zakharyaschev, Intuitionistic Modal Logics as Fragments of Classical Bimodal Logics, in: E. Orłowska (ed.), *Logic at Work*, Springer Verlag, Berlin, 1999, 168–186.

SERGUEI P. ODINTSOV
Institute of Mathematics
Siberian Division of the Russian Academy of Sciences
Novosibirsk, Russia
`odintsov@math.nsc.ru`

HEINRICH WANSING
Institute of Philosophy
Dresden University of Technology
01062 Dresden, Germany
`Heinrich.Wansing@mailbox.tu-dresden.de`

W. Buszkowski

Type Logics in Grammar

Introduction

Type logics are logics whose formulas are interpreted as types. For instance, $A \to B$ is a type of functions (procedures) which send inputs of type A to outputs of type B, and $A \otimes B$ is a type of pairs (f, g) such that f is of type A and g is of type B. The scope of possible realizations is huge: from constructivism in mathematics to logics of computation, from combinators and lambda calculus to linear logics, from type theories of Russell and Church to theories of syntax, proposed by Leśniewski, Ajdukiewicz, Chomsky, Curry and Lambek.

In this paper I try to illuminate the current state of the latter theories, often characterized as logical grammar or, more generally, formal grammar. Not every logician would be inclined to attach this region to *logic proper*. Fortunately, numerous contributions of last thirty years have lifted up the logic of grammar to a vivid research branch with a well-defined conceptual framework and promising connections with linguistics and computer science (also philosophy, cognitive science, semantics). From the standpoint of logic, formal systems appearing in this discipline belong to substructural logics, since their sequential formats abandon structural rules (Weakening, Contraction, Exchange); see [69].

Hiż [34] dubbed the term *grammar logicism* for the thesis: grammar reduces to logic. Not discussing all subtleties of this standpoint, I believe Hiż soundly characterizes the real content of so-called formal linguistics with its standard syntactic devices (automata, grammars, trees, transformations) which are conceptually and methodologically akin to logical entities (routines of computation, theories, proofs, deductive systems). It seems to be important that grammatical structures influence the creation and evolution of new logical formalisms, as e.g. the Lambek calculus and noncommutative linear logics.

Reduction of grammar to logic need not be restricted to the dimension of

---

V. F. Hendricks and *J. Malinowski* (eds.),
Trends in Logic: 50 Years of Studia Logica

Trends in Logic **21**: 337–382, 2003.

syntax. Since Montague [57], semantics of natural language has extensively been studied within philosophical logic. Semantical problems seem to be more sophisticated and solutions less definitive. I am inclined to remove any strict borderline between syntax and semantics; the former is a representation of the latter, the latter must be anchored in the former. Type grammars are especially close to the description of syntax according to general semantic paradigms. Connections of type grammars with Montague semantics are based on the Curry-Howard isomorphism between proofs in Natural Deduction format and lambda terms. For Lambek logics, main results are due to van Benthem [76, 78], Moortgat [58, 59] and their collaborators. In this survey, semantical aspects are briefly mentioned at the beginning of section 3.

Grammars based on logics of types are traditionally called *categorial grammars* (after Bar-Hillel et al. [5]). From the modern perspective, the term *type grammars* seems to be more adequate: all formal grammars classify language expressions into categories, but a characteristic feature of type grammars is the usage of (logical) types as carriers of grammatical information. The present paper mainly focuses on type grammars and their logics, which evidently reflects the author's predilection and experience. Hopingly, it shows a more general significance of some central ideas.

We begin with the simplest formalization of type grammar: Classical Categorial Grammar (CCG) of Ajdukiewicz [3] and Bar-Hillel [5]. Ajdukiewicz was strongly influenced by logical syntax, designed by Leśniewski [54], and the idea of universal grammar, due to Husserl [36]. The logic of CCG is a purely implicational propositional logic with modus ponens and no axioms; in the bidirectional format, stemming from Lambek [49] and Bar-Hillel [5], it admits two conditionals $A \rightarrow B$ and $B \leftarrow A$ and two MP-schemes. We discuss three major topics concerning CCG: (i) syntactic unambiguity (related to Polish notation), (ii) connections with Chomsky grammars and tree automata, (iii) learnability. The latter exhibits a new, promising direction of formal learning theory, employing logical unification to design successful algorithms for language acquisition.

Next, we consider full-fledged type logics, originating in Syntactic Calculus of Lambek [49]. They are closely connected with other logical formalisms, as e.g. linear logics, lambda calculus and combinators, labelled deductive systems. First, we present different versions of type logics and discuss their proof-theoretic and computational aspects. Second, we sketch a landscape of models of these logics. We also regard central problems related to type grammars based on these logics.

This survey is complementary to the author's [14]. Certain topics, largely discussed there, e.g. algebra of syntactic structures and modelling type logics in the lambda calculus, are almost completely omitted here. On the other hand, we present in more detail some special matters, characteristic of current research, as e.g. learning theory and computational complexity. In section 3, we mix algebraic models and computational problems, because they are mutually dependent. Algebraic structures influence the shape of formal systems and are useful for illuminating proof-theoretic properties, i.e. the subformula property.

We report results of many people, including the author, who apologizes for the great number of self-citations in comparing with others. The reason is obvious: the author focuses on problems he has been involved in and informs the reader where she may find proofs abandoned here. In compensation, let me emphasize a particular role of two persons. Joachim Lambek has originated some central ideas, and Johan van Benthem has shown their broader significance for logic.

Classical Categorial Grammar

Pr denotes a set of *primitive types* (propositional variables). *Types* (formulas) are formed out of two conditionals $\rightarrow$ (right arrow) and $\leftarrow$ (left arrow). We denote types by A, B, C, D. Γ, Δ denote finite sequences (strings) of types, and $\Gamma\Delta$ denotes the concatenation of Γ and Δ. *Sequents* are expressions of the form $\Gamma \vdash A$. The empty string is denoted by ϵ.

The logic AB (after Ajdukiewicz and Bar-Hillel) can be presented as a structured consequence relation (in the sense of Gabbay [28]), which is determined by MP-schemes:

$$\text{(MP-R) } A(A \rightarrow B) \vdash B, \text{ (MP-L) } (B \leftarrow A)A \vdash B,$$

and the following structural rules:

$$\text{(Id) } A \vdash A,$$

$$\text{(CUT) } \frac{\Gamma A \Gamma' \vdash B;\ \Delta \vdash A}{\Gamma \Delta \Gamma' \vdash B}.$$

Equivalently, AB can be presented as a rewriting system, based on reduction rules:

$$A, A \rightarrow B \Rightarrow B \text{ and } B \leftarrow A, A \Rightarrow B.$$

Then, $\Gamma \vdash_{AB} A$ iff Γ reduces to A by a finite number of applications of reduction rules. We write $\Gamma \vdash_{AB} A$ for: $\Gamma \vdash A$ is provable in AB, and similarly for other systems. Clearly, if $\Gamma \vdash_{AB} A$ then $\Gamma \neq \epsilon$.

Ajdukiewicz [3] introduced the $(\leftarrow)$–fragment of this logic; precisely, he regarded multiple-argument types $B \leftarrow A_1, \ldots, A_n$ with reduction rules:

$$(B \leftarrow A_1, \ldots, A_n), A_1, \ldots, A_n \Rightarrow B.$$

In Natural Deduction (ND) format, AB is axiomatized by (Id) and two conditional elimination rules:

$$(\rightarrow\mathrm{E})\ \frac{\Gamma \vdash A;\ \Delta \vdash A \rightarrow B}{\Gamma\Delta \vdash B},\quad (\leftarrow\mathrm{E})\ \frac{\Gamma \vdash B \leftarrow A;\ \Delta \vdash A}{\Gamma\Delta \vdash B}.$$

Proofs in the ND-format determine the so-called *functor-argument structures* (fa-structures) on antecedents of provable sequents. Types are atomic fa-structures. If X, Y are fa-structures, then $(XY)_1$ and $(XY)_2$ are fa-structures. X is the functor, and Y is the argument of $(XY)_1$, and the converse holds for $(XY)_2$. The logic ABs (i.e. AB in structure form) arises from AB (in ND-format) by replacing $\Gamma$with X, Δ with Y in premises of rules, and $\Gamma\Delta$ with $(XY)_2$ (resp. $(XY)_1$) in the conclusion of $(\rightarrow$E) (resp. $(\leftarrow$E)). Every proof of $\Gamma \vdash A$ in AB determines a unique fa-structure X on Γ such that $X \vdash_{ABs} A$. The same could be done for the original version of AB.

Syntactic unambiguity

It is well known that the Polish notation is unambigous, that means, the structure of a formula is uniquely determined by the string of symbols. This property is actually a property of the logic AB, restricted to one conditional. Since this a rather untypical property of a logical system, and it is in focus of Ajdukiewicz [3], we have decided to discuss the problem in some detail.

For any structure X, there is at most one type A such that $X \vdash_{ABs} A$. One computes this type by a successive replacement of every substructure of X of the form $(A(A \rightarrow B))_2$ or $((B \leftarrow A)A)_1$ with type B. This property will be referred to as *type unambiguity* of ABs. Also, every provable sequent $X \vdash A$ has a unique proof tree in the ND-format of ABs (*proof unambiguity*).

AB is both type and proof ambiguous. $(A \leftarrow A)A(A \rightarrow A) \vdash A$ has two proofs in AB which determine structures $(C(AD)_2)_1$ and $((CA)_1D)_2$, for $C = A \leftarrow A$, $D = A \rightarrow A$. For $\Gamma = ((A \leftarrow D) \leftarrow A)AD$, with D as above, we have both $\Gamma \vdash_{AB} A \leftarrow D$ and $\Gamma \vdash_{AB} A$.

One-directional fragments of AB are both type and proof unambigous, which is a nonstandard proof-theoretic property, clearly related to (generalized) Polish notation: a string of symbols supplied with one-directional types admits a unique syntactic analysis (see [13] for a discussion and the proof for the general case of multiple-argument types). We sketch the proof for the $(\leftarrow)$–fragment of AB.

In types $B \leftarrow A_1 \leftarrow \cdots \leftarrow A_n$ parentheses are grouped to the left. If D is of the above form, with B primitive, then $C < D$ means that C equals $B \leftarrow A_1 \leftarrow \cdots \leftarrow A_i$, for some $0 \leq i < n$. The relation $<$ is irreflexive, transitive and quasi-linear: if $A < C$ and $B < C$, then $A < B$ or $A = B$ or $B < A$. The following lemmas are crucial.

(L1) If $A\Gamma \vdash_{AB} B$ and $\Gamma \neq \epsilon$ then $B < A$.

(L2) If $B < A$ then $A\Gamma \vdash_{AB} B$, for some $\Gamma \neq \epsilon$.

(L3) If $\Gamma \vdash_{AB} A$ then there is no $\Delta \neq \epsilon$ such that $\Gamma\Delta \vdash_{AB} A$.

We prove (L3). We use the following property of the restricted AB: if $\Gamma \vdash_{AB} A$ and Γ has a length greater than 1, then $\Gamma = (A_2 \leftarrow A_1)\Gamma_1\Gamma_2$ with $\Gamma_1 \vdash_{AB} A_1$ and $A_2\Gamma_2 \vdash_{AB} A$. The proof of (L3) proceeds by induction on the length of Γ.

Let Γ be of length 1. Assume $\Gamma \vdash_{AB} A$ and $\Gamma\Delta \vdash_{AB} A$, for $\Delta \neq \epsilon$. Then, $\Gamma = A$, and consequently, $A < A$, by (L1). Contradiction.

Let Γ be of length greater than 1. Assume the same as above. Using the property, mentioned above, we get types A_1, A_2 and nonempty strings Γ_i, Δ_i, for $i = 1, 2$, such that:

1. $\Gamma = (A_2 \leftarrow A_1)\Gamma_1\Gamma_2$,
2. $\Gamma\Delta = (A_2 \leftarrow A_1)\Delta_1\Delta_2$,
3. $\Gamma_1 \vdash_{AB} A_1$, $\Delta_1 \vdash_{AB} A_1$,
4. $A_2\Gamma_2 \vdash_{AB} A$, $A_2\Delta_2 \vdash_{AB} A$.

Clearly, the length of Γ_1 is less than the length of Γ, and similarly for Γ_2. By the induction hypothesis, neither Γ_1 is a proper initial segment of Δ_1, nor Δ_1 is a proper initial segment of Γ_1. Consequently, $\Gamma_1 = \Delta_1$. Then, Γ_2 is a proper initial segment of Δ_2, hence $A_2\Gamma_2$ is a proper initial segment of $A_2\Delta_2$, and its length is less than that of Γ. This contradicts the induction hypothesis.

We prove type unambiguity. Assume $\Gamma \vdash_{AB} A$ and $\Gamma \vdash_{AB} B$, with $A \neq B$. Let C be the first type in Γ. Since $A \neq B$, then Γ is of length greater than 1. By (L1), $A < C$ and $B < C$. Consequently, $A < B$ or $B < A$. If $A < B$ then there exists $\Delta \neq \epsilon$ such that $B\Delta \vdash_{AB} A$, by (L2), which yields $\Gamma\Delta \vdash_{AB} A$, by (CUT). This contradicts (L3). The case $B < A$ is symmetrical.

We prove proof unambiguity. Precisely, we show that every provable sequent admits a unique ND proof tree in the restricted AB. Let d_1, d_2 be proofs of $\Gamma \vdash A$. We prove $d_1 = d_2$, by induction on the length of Γ.

Assume Γ is of length 1. Then, $\Gamma = A$, and d_1, d_2 consist of the only application of (Id). Assume Γ is of length greater than 1. Then, both d_1 and d_2 use ($\leftarrow$E). We consider the last application of this rule in either proof. For d_1, let the premises be $\Gamma_1 \vdash A \leftarrow C$, $\Delta_1 \vdash C$, and for d_2, let them be $\Gamma_2 \vdash A \leftarrow D$, $\Delta_2 \vdash D$. We show $C = D$. Assume $C \neq D$. Then, Γ_1 or Γ_2 is of length greater than 1, hence $A \leftarrow C < A \leftarrow D$ or $A \leftarrow D < A \leftarrow C$, by (L1) and properties of $<$. This is impossible, by the definition of $<$. Now, Γ_1 is an initial segment of Γ_2 or conversely. Neither of them can be a proper initial segment of the other, by (L3), which yields $\Gamma_1 = \Gamma_2$, and consequently, $\Delta_1 = \Delta_2$. By the induction hypothesis, subproofs of d_1, d_2 leading to the left premise are equal, and similarly for the right premise. Consequently, $d_1 = d_2$.

Ajdukiewicz [3] has designed a simple parsing procedure. Given a string Γ, find the left-most occurrence of $(B \leftarrow A)A$ and replace it with B. Then, $\Gamma \vdash_{AB} C$ iff Γ reduces to C by finitely many applications of left-most reduction. As shown in [13], this is not true for multiple-argument types (considered by Ajdukiewicz) but is true for one-argument types, considered here. The Ajdukiewicz procedure is fully deterministic and shows, in fact, that languages generated by rigid type grammars based on this fragment of AB are deterministic context-free languages, i.e. they can be recognized by deterministic push-down automata (rigid grammars are defined below).

CCG versus other formal grammars

A CCG can formally be defined as a triple $G = (V_G, I_G, S_G)$ such that V_G is a nonempty finite lexicon (alphabet), I_G is a finite relation between elements of V_G and types, and $S_G \in$Pr; V_G, I_G, S_G are called *the lexicon, the initial type assignment* and *the principal type*, respectively, of G. We write $G : v \mapsto A$ for $(v, A) \in I_G$. We say that G *assigns* type A to string $a = v_1 \ldots v_n$, $v_i \in V_G$, if there are types A_i such that $G : v_i \mapsto A_i$, for $i = 1, \ldots, n$, satisfying $A_1 \ldots A_n \vdash_{AB} A$; we write $a \mapsto_G A$. *The language* of G ($L(G)$) is the set of all strings a such that $a \mapsto_G S_G$. Other type grammars, considered later on, are defined in a similar way; only AB is to be replaced by a different type logic.

The fa-structures on V_G are defined as fa-structures of types except that atomic fa-structures are elements of V_G. By $F(V_G)$ we denote the set of fa-structures on V_G. We say that G assigns type A to structure $X \in F(V_G)$ if there is an fa-structure X' of types which arises from X by replacing each atom v by a type B such that $G : v \mapsto B$, satisfying $X' \vdash_{ABs} A$; we again write $X \mapsto_G A$. *The f-language* of G ($L^f(G)$) is the set of all $X \in F(V_G)$ such that $X \mapsto_G S_G$.

For instance, let G admit the following assignment:

$$\text{Joan} \mapsto A,\ \text{works} \mapsto B,\ \text{hardly} \mapsto C,$$

where A =PN (proper noun), $B = A \rightarrow$S, $C = B \rightarrow B$, and S_G =S (sentence). Then, $L^f(G)$ contains structures:

1. (Joan works)$_2$,

2. (Joan (works hardly)$_2$)$_2$,

3. (Joan ((works hardly)$_2$ hardly)$_2$)$_2$,

and so on. $L(G)$ contains the strings resulting from the above structures after one has dropped all structure markers.

The above grammar is a *rigid* CCG, i.e. it assigns at most one type to any lexical atom. Rigidness is characteristic of grammars designed for formal languages of logic and mathematics, while natural language usually requires several types to be assigned to one lexical atom. The negation 'not' is assigned type S$\leftarrow$S, as in the sentence 'not every man works' with structure (not ((every man) works)), but also type $C \leftarrow C$ (C defined as above), as in 'John works not hardly'.

We have used a structure without functor markers; such structures are called *phrase structures* (p-structures). For $X \in F(V_G)$, by $p(X)$ we denote the p-structure resulting from X after one has dropped all functor markers. *The phrase language* of G ($L^p(G)$) consists of all $p(X)$, for $X \in L^f(G)$.

Now, let us turn to some basic notions of Chomsky linguistics. *A context-free grammar* (CF-grammar) is a quadruple $G = (V_G, N_G, R_G, S_G)$ such that V_G, N_G are disjoint finite alphabets, $S_G \in N_G$, and R_G is a finite subset of $N_G \times (V_G \cup N_G)^\star$ ($W^\star$ denotes the set of all finite strings of elements of set W). Elements of V_G, N_G, R_G are called *terminal symbols*, *nonterminal symbols* and *production rules*, respectively, of G, and S_G is called *the initial symbol* of G. Production rules are written $A \mapsto a$ instead of (A, a). We say that string b is *directly derivable* from string a in G (write $a \Rightarrow_G b$) if there exist strings c, d, e and rule $A \mapsto e$ in R_G such that $a = cAd$, $b = ced$. We say that string b is *derivable* from string a in G (write $a \Rightarrow^\star_G b$) if there exists a sequence $(a_0, \ldots, a_n)$ such that $n \geq 0$, $a_0 = a$, $a_n = b$ and $a_{i-1} \Rightarrow_G a_i$, for all $i = 1, \ldots, n$. *The language* of G ($L(G)$) is the set of all $a \in V_G^\star$ such that $S_G \Rightarrow_G^\star a$.

Grammars G, G' are said to be (weakly) *equivalent* if $L(G) = L(G')$. It is well-known (see e.g. [72]) that every CF-grammar G such that $\epsilon \notin L(G)$ can be transformed into an equivalent CF-grammar G' in *the Chomsky normal form*: all production rules of G' are of the form $A \mapsto v$ or $A \mapsto BC$, where $A, B, C \in N_{G'}$, $v \in V_{G'}$.

The CCG exemplified above can be replaced with a CF-grammar (in the Chomsky normal form) whose production rules are as follows:

$$\mathrm{S} \mapsto AB, \; B \mapsto BC,$$

$$A \mapsto \text{Joan}, \; B \mapsto \text{works}, \; C \mapsto \text{hardly},$$

with $S_G =$S, $N_G = \{S, A, B, C\}$, and V_G consisting of lexical atoms 'Joan', 'works' and 'hardly'. A derivation of 'Joan works hardly' is:

$$\mathrm{S} \Rightarrow AB \Rightarrow ABC \Rightarrow \cdots \Rightarrow \text{Joan works hardly.}$$

Every CF-derivation determines a unique p-structure on the derived string; the above derivation leads to structure (Joan (works hardly)). *The phrase language* of CF-grammar G ($L^p(G)$) consists of all p-structures on strings from $L(G)$ which are determined by possible derivations of these strings.

A principal difference between CCG and CF-grammar is that the former is *lexical*, that means: all particular linguistic information is put in the initial assignment of types to lexical atoms, and the derivation procedure is based on universal rules, common for all languages, whereas the latter puts the linguistic information in the production rules which underly the derivation procedure. Lexicality is characteristic of all basic kinds of type grammar: the universal rules for derivation procedures are provable sequents of some logics, being independent of the particular language. In section 3, we shall show that lexicality is, actually, a restriction: larger classes of languages can only be described by type grammars avoiding lexicality.

The Gaifman theorem [5] establishes the (weak) equivalence of CCG's and CF-grammars (for $\epsilon-$free languages). It is easy to show that every CCG is equivalent to some CF-grammar. Let G be a CCG, and let T_G denote the set of all types appearing in I_G. By $T(G)$ we denote the set of all subtypes of types from T_G. Clearly, $T(G)$ is finite and contains all types assigned by

G to any strings. A CF-grammar G' is defined by: $V_{G'} = V_G$, $N_{G'} = T(G)$, $S_{G'} = S_G$, and $R_{G'}$ consists of all rules:

$$B \mapsto A(A \to B),\ B \mapsto (B \leftarrow A)A,$$

for $(A \to B), (B \leftarrow A) \in T(G)$, and all lexical rules $A \mapsto v$, for $(v, A) \in I_G$. One easily proves:

$$A_1 \ldots A_n \vdash_{AB} A \text{ iff } A \Rightarrow^{\star}_{G'} A_1 \ldots A_n,$$

for all $A_i, A \in T(G)$, and consequently, $L(G) = L(G')$.

The converse direction is more sophisticated. It is easier if one assumes that the CF-grammar G is in *the Greibach normal form*, that means: all production rules are of the form $A \mapsto v$, $A \mapsto vB$ or $A \mapsto vBC$, for $A, B, C \in N_G$, $v \in V_G$ (every CF-grammar G such that $\epsilon \notin L(G)$ is equivalent to a CF-grammar in the Greibach normal form [72]). We identify nonterminal symbols of G with primitive types. A CCG G' is defined by: $V_{G'} = V_G$, $S_{G'} = S_G$, and $I_{G'}$ consists of:

1. all pairs (v, A) such that $(A \mapsto v) \in R_G$,
2. all pairs $(v, A \leftarrow B)$ such that $(A \mapsto vB) \in R_G$,
3. all pairs $(v, A \leftarrow C \leftarrow B)$ such that $(A \mapsto vBC) \in R_G$.

By induction on n, one proves:

$$A \Rightarrow^{\star}_{G} v_1 \ldots v_n \text{ iff } v_1 \ldots v_n \mapsto_{G'} A,$$

for all $A \in N_G$, $v_i \in V_G$, which yields $L(G) = L(G')$.

Actually, the Gaifman theorem has been proven earlier than the Greibach normal form theorem. Since the CCG, constructed by Gaifman, is precisely of the above form, then his proof can be treated as the first proof of the Greibach normal form theorem for CF-grammars (see an analysis of this proof in [10]).

CCG's are not equivalent to CF-grammars on the level of p-structures. Let $P(V)$ denote the set of all p-structures on alphabet V. Let $L \subseteq P(V)$. We say that $X \in P(V)$ is equivalent to $Y \in P(V)$ with respect to L (write $X \sim_L Y$) if, for all $Z \in P(V)$, $Z[X] \in L$ iff $Z[Y] \in L$ (as usual in logic, $Z[X]$ denotes Z with a distinguished occurrence of substructure X, and $Z[Y]$ denotes the result of replacing X with Y in Z). By the classical theorem of Thatcher, $L = L^p(G)$, for some CF-grammar G, iff the relation $\sim_L$ is of finite index. By *the external degree* of $X \in P(V)$ (think of X as a tree) we mean the length of the shortest branch of X, and *the degree* of X is the maximal external degree of substructures of X. For instance, v is of degree 0, (vw) is of degree 1, and $((vw)(v'w'))$ is of degree 2. It has been shown in [9] that $L = L^p(G)$, for some CCG G, iff both $\sim_L$ is of finite index and all structures in L are of bounded degree. Accordingly, phrase languages of CCG's are a narrower class than those of CF-grammars. For instance, the CF-grammar given by rules S$\mapsto$SS, S$\mapsto$ v produces all possible p-structures on $\{v\}$, hence its phrase language is of unbounded degree, and consequently, it cannot be generated by any CCG.

Phrase languages generated by CCG's are regular tree languages in the sense of tree automata (see [30]); this follows from the above characterization of them as languages of finite index. Tiede [74] obtains analogous results for ND proof trees associated with CCG's and other kinds of categorial grammars. For grammars based on the nonassociative Lambek calculus, some related results have been proven by Kandulski [43].

A similar characterization can be given for functor languages generated by CCG's: for $L \subseteq F(V)$, there is a CCG G such that $L = L^f(G)$ iff both the relation $\sim_L$ is of finite index and all structures in L are of bounded functor degree (one counts the length of functor branches only; see [14]). Consequently, functor languages of CCGs are regular tree languages. Standard techniques of tree automata [30] yield the decidability of the emptiness problem, the inclusion problem and the equality problem for these languages; for a discussion, see [14]. In particular, the problem of whether $L^f(G) \subseteq L^f(G')$ is decidable, which is crucial for some learnability results in the next subsection.

Learnability

Formal learning theory, stemming from Gold [31], is a theory of language acquisition, based on formal grammars and recursion theory. It has evident links with logical theories of inductive inference and knowledge discovery. Technical results are mainly due to people working in theoretical computer science; see the survey [64].

In [20], the logical method of unification has been applied to design learning procedures for CCG's. Kanazawa [38, 39] uses these procedures to construct convergent learning functions for several classes of CCG's. Negative postulates are considered in [20, 56, 18, 26, 27]. Unification in type grammar has also been studied in e.g. [77, 75].

Due to lexicality (and the functional structure of types), type grammars seem to be especially suitable to experiment learning algorithms employing unification. Hopingly, similar techniques can be developed for dependency grammars and minimalist grammars (C. Retore, E. Stabler).

We shall describe fundamental lines of this approach for CCG's. First, we recall some basic notions of formal learning theory.

A grammar system is a triple (Ω, E, L) such that Ω and E are countably infinite sets, and L is a function from Ω into the powerset of E. Elements of Ω are called *grammars*, elements of E are called *expressions*, and $L(G)$, for $G \in \Omega$, is called *the language* of G.

W^+ denotes the set of all nonempty strings of elements of W. *A learning function* for the grammar system is a partial function from E^+ into Ω. Intuitively, the learning function assigns grammars to finite samples of a language. Let $(s_i)_{i \in \omega}$ be an infinite sequence of expressions. One says that a learning function φ *converges* on this sequence to $G \in \Omega$ if $\varphi((s_i)_{i<n})$ is defined and equals G, for all but finitely many $n \in \omega$.

Let $\mathcal{G} \subseteq \Omega$. We denote $L(\mathcal{G}) = \{L(G) : G \in \mathcal{G}\}$. A sequence $(s_i)_{i \in \omega}$ is called *a text* for a language $L \subseteq E$ if $L = \{s_i : i \in \omega\}$. One says that a learning function φ *learns* $\mathcal{G}$ if, for every $L \in L(\mathcal{G})$ and every text for L, there is $G \in \mathcal{G}$ such that $L = L(G)$ and φ converges to G on this text. A class $\mathcal{G}$ is said to be (effectively) *learnable* if there exists a computable learning function φ that learns $\mathcal{G}$.

Informally, $\mathcal{G}$ is learnable iff there exists an algorithmic procedure which hypothesizes a grammar on the basis of a finite language sample (it need not be defined on all samples); if the inputs are successive initial segments of any text for any language generated by a grammar from this class, then, after a finite number of steps, the outputs provide a fixed grammar from $\mathcal{G}$ which generates this language.

Let $\mathcal{L}$ be a class of subsets of E (languages on E). One says that $\mathcal{L}$ *admits a limit point* if there exists a stricly ascending chain $(L_n)_{n \in \omega}$ of languages from $\mathcal{L}$ such that the join of this chain belongs to $\mathcal{L}$. It is known that if $L(\mathcal{G})$ admits a limit point then $\mathcal{G}$ is not (even uneffectively) learnable. As a consequence, if $\mathcal{G}$ generates all finite languages and at least one infinite language, then $\mathcal{G}$ is not learnable. This holds for all standard classes of formal grammars, e.g. regular grammars, CF-grammars, CCG's and so on.

Accordingly, learnability can be gained for some restricted classes only, as e.g. context-sensitive grammars with at most k production rules [71]. Kanazawa [38, 39] shows that rigid CCG's and $k-$valued CCG's are learnable.

Wright [81] has defined the following property of a class $\mathcal{L}$ of languages: $\mathcal{L}$ has *finite elasticity* if there exists no pair $((s_i)_{i \in \omega}, (L_i)_{i \in \omega})$, $s_i \in E$, $L_i \in \mathcal{L}$, such that $s_i \notin L_i$ but $s_0, \ldots, s_i \in L_{i+1}$, for all $i \in \omega$. Kapur [44] has proven that finite elasticity entails the following condition:

(D) for every $L \in \mathcal{L}$, there exists a finite set $D_L \subseteq L$ such that L is the smallest language $L' \in \mathcal{L}$, satisfying $D_L \subseteq L'$.

We prove that finite elasticity is equivalent to a strengthening of (D):

(D') for every $L \subseteq E$, there exists a finite set $D_L \subseteq L$ such that, for every $L' \in \mathcal{L}$, if $D_L \subseteq L'$ then $L \subseteq L'$.

Assume that $\mathcal{L}$ does not satisfy (D'). Then, there exists $L \subseteq E$ such that, for any finite $D \subseteq L$, there is $L' \in \mathcal{L}$ such that $D \subseteq L'$ but $L \not\subseteq L'$. Take $D = \emptyset$. There is $L' \in \mathcal{L}$ such that $L \not\subseteq L'$. Put $s_0 \in L - L'$, $L_0 = L'$. Assume $s_i \in L$ and $L_i \in \mathcal{L}$, for $i = 0, \dots n$, have already been defined with $s_i \notin L_i$, for $0 \leq i \leq n$, and $s_0, \dots, s_i \in L_{i+1}$, for $0 \leq i < n$. There is $L' \in \mathcal{L}$ such that $s_0, \dots, s_n \in L'$ but $L \not\subseteq L'$. Put $s_{n+1} \in L - L'$, $L_{n+1} = L'$. In this way, we define a sequence $(s_i)_{i \in \omega}$ and a sequence $(L_i)_{i \in \omega}$, showing that $\mathcal{L}$ does not have finite elasticity.

Assume that $\mathcal{L}$ satisfies (D'). Let (s_i) be an infinite sequence of expressions. By (D'), there is $n \in \omega$ such that, for any $L' \in \mathcal{L}$, if $s_0, \dots, s_n \in L'$ then $s_i \in L'$, for all $i \in \omega$. Consequently, there exists no infinite sequence (L_i), of languages from $\mathcal{L}$, such that $s_0, \dots, s_n \in L_{n+1}$ but $s_{n+1} \notin L_{n+1}$, which proves the finite elasticity of $\mathcal{L}$.

This yields Kapur's theorem, since (D) follows from (D'). If $\mathcal{L}$ is a closure system, that means: $\mathcal{L}$ is closed under arbitrary meets, then $\mathcal{L}$ satisfies (D') iff it admits no infinite ascending chains, that means: there is no infinite sequence (L_i), of languages from $\mathcal{L}$, such that $L_i \subset L_{i+1}$, for all $i \in \omega$ (this observation and the equivalence of finite elasticity with (D') seem to have been overlooked in literature). We prove this fact.

Assume $\mathcal{L}$ satisfies (D'). Let (L_i) be an infinite sequence of languages from $\mathcal{L}$ such that $L_i \subseteq L_{i+1}$, for all $i \in \omega$. Let L be the join of this sequence. Take a finite set $D_L \subseteq L$, as in (D'). Then, $D_L \subseteq L_n$, for some $n \in \omega$. Consequently, $L \subseteq L_n$, hence $L_n = L_{n+1}$. Therefore, $\mathcal{L}$ admits no infinite ascending chains.

Assume that $\mathcal{L}$ is a closure system which does not satisfy (D'). Then, there exists $L \subseteq E$ such that, for every finite $D \subseteq L$, there is $L' \in L$, satisfying $D \subseteq L'$ but $L \not\subseteq L'$. Clearly, L is infinite. Let (s_i) be an infinite sequence of all expressions from L. For every $n \in \omega$, let L_n denote the smallest language $L' \in \mathcal{L}$ such that $s_0, \dots, s_n \in L'$; it exists, since $\mathcal{L}$ is a closure system. Clearly, $L_n \subseteq L_{n+1}$, for all $n \in \omega$. For every $n \in \omega$, there is $L' \in \mathcal{L}$ such that $s_0, \dots, s_n \in L'$ but $L \not\subseteq L'$, and consequently, there is

$m > n$ such that $s_m \notin L_n$, which yields $L_n \subset L_m$. Then, there exists an infinite, strictly ascending chain of languages from $\mathcal{L}$.

Let $\mathcal{G} \subseteq \Omega$. We always assume that $\mathcal{G}$ is recursively enumerable and the universal membership problem $s \in L(G)$, for $s \in E$, $G \in \mathcal{G}$, is decidable. We need one more condition:

(MIN) there exists a computable partial function χ from the set of finite subsets of E into $\mathcal{G}$ such that the domain of χ is recursive and, for any finite $D \subseteq E$, the family $\{L \in L(\mathcal{G}) : D \subseteq L\}$ is nonempty iff $\chi(D)$ is defined; if so, $L(\chi(D))$ is a $\subseteq$ −minimal language in this family.

We prove a basic theorem, due to Kapur [44]:

(T1) if $\mathcal{G}$ satisfies (MIN) and $L(\mathcal{G})$ has finite elasticity, then $\mathcal{G}$ is learnable.

Assume $\mathcal{G}$ satisfies (MIN), and $L(\mathcal{G})$ has finite elasticity. A learning function φ is defined as follows:

(10) $\varphi((s_0)) = \chi(\{s_0\})$,

(11) if $\varphi((s_0, \ldots, s_i)) = G$ is and $s_{i+1} \in L(G)$ then $\varphi((s_0, \ldots, s_{i+1})) = G$,

(12) $\varphi((s_0, \ldots, s_{i+1})) = \chi(\{s_0, \ldots, s_{i+1}\})$, otherwise.

The relation 'φ is defined on $(s_0, \ldots, s_i)$' is recursive, since both the domain of χ and the relation $s \in L(G)$, for $G \in \mathcal{G}$, are recursive. Consequently, φ is recursive. It is easy to show the following: if $\varphi((s_0, \ldots, s_i)) = G$ is defined, then $L(G)$ is a minimal language in the family of all $L \in L(\mathcal{G})$ such that $s_0, \ldots, s_i \in L$.

Let $L \in L(\mathcal{G})$, and let (s_i) be a text for L. Clearly, $\varphi((s_0, \ldots, s_i))$ is defined, for all $i \in \omega$. Let $D_L \subseteq L$ be as in (D). There exists $n \in \omega$ such that $D_L \subseteq \{s_0, \ldots, s_n\}$. Since L is the smallest language $L' \in \mathcal{L}$ such that $D_L \subseteq L'$, then $L = L(G)$, for $G = \varphi((s_0, \ldots, s_n))$. By (11), $G = \varphi((s_0, \ldots, s_m))$, for all $m \geq n$.

We have shown that φ learns $\mathcal{G}$. Since φ satisfies (11), it is *conservative* in the sense of formal learning theory. If conservativity is not required, then the finite elasticity of $L(\mathcal{G})$ is a sufficient condition for the learnability of $\mathcal{G}$ [64].

Now, we apply these methods to CCG. Let us consider the grammar system (Ω, E, L) such that Ω is the class of CCG's (with a fixed lexicon V), $E = F(V)$, and $L(G) = L^f(G)$, for $G \in \Omega$. We assume that all grammars in Ω have the same principal type S. Since V and S are fixed, a grammar G can be identified with I_G.

First, we recall a unification-based learning procedure from [20]. Let $D \subseteq F(V)$ be a nonempty, finite set. We define a CCG GF(D), called *the general form of a grammar* for D. S is treated as a constant, whereas other primitive types are treated as variables. We assign S to all structures from D and distinct variables to all occurrences of argument substructures of these structures. Then, we assign types to functor substructures of these structures according to the rules:

(fr) if $(XY)_2$ is assigned B and X is assigned A then Y is assigned $A \to B$,

(fl) if $(XY)_1$ is assigned B and Y is assigned A then X is assigned $B \leftarrow A$.

GF(D) contains all assignments $v \mapsto A$ obtained in this way, for $v \in V$; the principal type of GF(D) is S. For instance, if D consists of structures (Joan works)$_2$ and (Joan (works hardly)$_2$)$_2$, then GF(D) contains the assignments:

$$\text{Joan} \mapsto x, y, \text{ works} \mapsto z, \ x \to \mathrm{S}, \text{ hardly} \mapsto z \to (y \to \mathrm{S}).$$

A substitution is a mapping from variables to types; it is naturally extended to a mapping from types to types. If G is a CCG, and σ is a substitution, then $G\sigma$ is a (unique) CCG which assignments $v \mapsto A\sigma$, whenever $G : v \mapsto A$. We have $L^f(G) \subseteq L^f(G\sigma)$, for all G, σ. For $G, G' \in \Omega$, we write $G \subseteq G'$ if $I_G \subseteq I_{G'}$. The following lemma is crucial [20].

(L1) Let $D \subseteq F(V)$ be nonempty and finite. Then, for every $G \in \Omega$, $D \subseteq L^f(G)$ iff there exists σ such that $GF(D)\sigma \subseteq G$.

The following notions come from the logical theory of unification; we modify definitions from [20]. A substitution σ is called *a unifier* of $G \in \Omega$ if, for all $v \in V$ and types A, B such that $G : v \mapsto A$ and $G : v \mapsto B$, there holds $A\sigma = B\sigma$. It is called *a most general unifier* (m.g.u.) of G if it is a unifier of G and, for every unifier η of G, there exists a substitution γ such that $\eta = \sigma\gamma$. G is said to be *unifiable* if there exists a unifier of G. The standard algorithm of unification can be used to decide whether a given CCG G is unifiable and, if so, to produce an m.g.u. of G (it is unique up to variants). For rigid CCG's, and any nonempty, finite $D \subseteq F(V)$, there holds the following theorem [20].

(T2) There exists a rigid $G \in \Omega$ such that $D \subseteq L^f(G)$ iff GF(D) is unifiable. If σ is an m.g.u. of GF(D), then GF(D)σ is a rigid CCG from Ω whose functor language is the smallest language $L(G)$ such that $G \in \Omega$ is rigid and $D \subseteq L(G)$.

GF(D) from the above example is unifiable; an m.g.u. σ is the substitution: y/x, $z/x \rightarrow$ S, hence GF(D)σ is given by:

$$\text{Joan} \mapsto x, \text{ works} \mapsto x \rightarrow \text{S}, \text{ hardly} \mapsto (x \rightarrow \text{S}) \rightarrow (x \rightarrow \text{S}),$$

which agrees with our earlier analysis of these sentences.

We define RG(D) =GF(D)σ, if GF(D) is unifiable and σ is an m.g.u. of GF(D). Kanazawa [38, 39] proves that RG($\{s_0, \ldots, s_n\}$) is a conservative learning function for the class $\mathcal{G}_r$, of all rigid grammars from Ω. It is true if one identifies 'isomorphic' grammars. It is easier to show that, for $\chi(D)$ =RG(D), the learning function φ, defined by (10)-(12), learns $\mathcal{G}_r$ (now, 'isomorphic' grammars need not be identified); this function is conservative. By (T1), one needs (MIN) and finite elasticity of $L(\mathcal{G}_r)$. (MIN) holds, by (T2). To prove finite elasticity one may use Kanazawa's lemma on chains of rigid languages:

(L2) There exists no infinite sequence (G_i) of rigid CCG's from Ω such that $L^f(G_i) \subset L^f(G_{i+1})$, for all $i \in \omega$.

We omit a rather involved, combinatorial proof of (L2). Let $\mathcal{L}$ be the class $L(\mathcal{G}_r)$ enriched with the total language $F(V)$. Using (L2) and (T2), we prove that $\mathcal{L}$ is a closure system. Let $\{L_i\}_{i \in I}$ be a family of languages from $\mathcal{L}$, and let L be the meet of this family. If $I = \emptyset$, then $L = F(V)$, hence $L \in \mathcal{L}$. So, we may assume $I \neq \emptyset$ and $L \neq F(V)$; further, we may also assume $L_i \in L(\mathcal{G}_r)$, for all $i \in I$. We consider two cases. (1) L is finite. By (T2), RG(L) exists, and the functor language of RG(L) is the smallest language $L' \in L(\mathcal{G}_r)$ such that $L \subseteq L'$. Consequently, this functor language contains L and is contained in all L_i, for $i \in I$, hence it equals L. Thus, $L \in \mathcal{L}$. (2) L is infinite. Let (s_i) be a text for L. We define $D_n = \{s_0, \ldots, s_n\}$. By (T2), RG($D_n$) exists, for all $n \in \omega$. Let $L(n)$ denote the functor language of RG(D_n). Clearly, $L(n) \subseteq L(n+1)$ and $L(n) \subseteq L_i$, for all $n \in \omega$, $i \in I$. By (L2), there exists $n \in \omega$ such that $L(m) = L(n)$, for all $m \geq n$, and consequently, $L = L(n)$. Again, $L \in \mathcal{L}$.

So, $\mathcal{L}$ is a closure system which admits no infinite ascending chains. By the fact on closure systems, proven above, $\mathcal{L}$ has finite elasticity, and consequently, $L^f(\mathcal{G}_r)$ has finite elasticity (this property is preserved by subclasses). This yields the learnability of $\mathcal{G}_r$.

A CCG G is said to be $k-$valued if $\{A : (v, A) \in G\}$ is of cardinality at most k, for any $v \in V$. By $\mathcal{G}_k$ we denote the class of all $k-$valued grammars in Ω. Again, (T1) yields the learnability of $\mathcal{G}_k$.

To show the finite elasticity of $L^f(\mathcal{G}_k)$ Kanazawa uses the following lemma.

(L3) Let $R \subseteq E_1 \times E_2$ be a finite valued relation, that means, for any $s \in E_1$, there exist at most finitely many $t \in E_2$ such that $(s,t) \in R$. Let $\mathcal{L}$ be a family of languages on E_2 which has finite elasticity. Then, the family $\{R^{-1}[L] : L \in \mathcal{L}\}$ has finite elasticity.

The proof of (L3) is an application of the König lemma: every infinite, finitely branching tree has an infinite branch.

Each atom from V is associated with k copies $v^1, \ldots, v^k$. Let W denote the set of all copies of atoms from V. A mapping $f : F(W) \mapsto F(V)$ is defined by: (i) $f(v^i) = v$, (ii) $f((XY)_i) = (f(X)f(Y))_i$. We define a finite valued relation $R \subseteq F(V) \times F(W)$: $(X,Y) \in R$ iff $X = f(Y)$. For any grammar $G \in \mathcal{G}_k$, we define a rigid grammar G' with lexicon W in the following way: if $A_1, \ldots, A_i$ are all types assigned to v by G, then G' assigns A_j to v^j, for $j = 1, \ldots, i$. Clearly, $L^f(G) = R^{-1}[L^f(G')]$ and $L^f(\mathcal{G}_k) = R^{-1}[L^f(\mathcal{G}_r)]$. So, the latter class has finite elasticity.

To prove (MIN) one uses a modification of the construction of RG(D). Let GF$_k(D)$ denote the set of all grammars G with lexicon W, satisfying the conditions: for all $v \in V$ and types A, $GF(D) : v \mapsto A$ iff, for some i, $G : v^i \mapsto A$, (ii) if $i \neq j$ then G assigns no common type to both v^i and v^j. Clearly, GF$_k(D)$ is a finite set, efectively constructed from GF(D). Let GF$^k(D)$ be defined as follows: (i) construct all grammars $G\sigma$ such that $G \in$GF$_k(D)$, σ is an m.g.u. of G (if exists), (ii) return to lexicon V, by assigning to v all types $G\sigma$ assigns to copies of v. GF$^k(D)$ is an effectively constructed set of $k-$valued grammars, fulfilling an analogue of (L1): for any $k-$valued grammar G, $D \subseteq L^f(G)$ iff there exist $G' \in$GF$^k(D)$ and substitution σ such that $G'\sigma \subseteq G$. Consequently, among languages $L^f(G)$, $G \in$GF$^k(D)$, there appear all minimal languages L such that $L = L^f(G)$, for some $k-$valued grammar G with $D \subseteq L^f(G)$. Since the inclusion problem $L^f(G) \subseteq L^f(G')$ is decidable, then we can effectively find a grammar $G \in$GF$^k(D)$ such that $L^f(G)$ is minimal among all languages of grammars from GF$^k(D)$, and consequently, among all languages of $k-$valued grammars G such that $D \subseteq L^f(G)$. Accordingly, a function χ satisfying (MIN) exists.

Kanazawa also considers learnability from strings. Now, the set E consists of all nonempty strings on V, and $L(G)$ is the string language of G. For $X \in F(V)$, by $s(X)$ we denote the string on V resulting from X after one has dropped all structure markers. The relation R between strings and structures, given by: $(a, X) \in R$ iff $a = s(X)$, is finite valued, and consequently, if $L^f(\mathcal{G})$ has finite elasticity, then $L(\mathcal{G})$ has finite elasticity, by (L3). So, both $L(\mathcal{G}_r)$ and $L(\mathcal{G}_k)$ have finite elasticity, hence $\mathcal{G}_r$ and $\mathcal{G}_k$ are learnable from strings, by a general theorem of formal learning theory, mentioned above.

The corresponding learning functions need not be conservative. Kanazawa defines nonconservative learning functions for these classes which use the above constructions performed on all possible structures admissible for a given string.

We have discussed learnability from positive data. As observed by Gold [31], if one admits texts consisting of all expressions from E and informing which expressions belong and which expressions do not belong to the language, then every recursively enumerable class of grammars is learnable. The learning function simply picks the first grammar G_n (in the fixed recursive sequence of all grammars from the class) which is compatible with the sample.

In [20, 56], negative data have been handled in a different way. Samples consist of positive data only, while some negative information is employed to restrict the search space of admissible grammars. In [56], there are considered classes $\mathcal{B}$, of grammars, fulfilling the following conditions:

(N1) if $G \subseteq G'$ and $G' \in \mathcal{B}$, then $G \in \mathcal{B}$,

(N2) if $G\sigma \in \mathcal{B}$, then $G \in \mathcal{B}$.

The order of type A is defined by:

$$o(A) = 0, \text{ for primitive } A,\ o(A \to B) = o(B \leftarrow A) = \max(o(B), o(A) + 1).$$

Thus, $o(p \to q) = 1$ and $o((p \to q) \to r) = 2$. The order of G is the maximal order of types in G; we denote it by $o(G)$. Always $o(G) \leq o(G\sigma)$, hence the class of all CCG's of order at most n satisfies (N1) and (N2). Another example is the following. We admit Pr contains some other constant types besides S, say PN, N etc. Let $X_1, \ldots, X_n$ be structures in $F(V)$, and let $A_1, \ldots, A_n$ be variable-free types. Then, the class of all CCG's which do not assign A_i to X_i, for $i = 1, \ldots, n$, satisfies (N1), (N2).

A set S, of substitutions, is said to be *hereditary*, if: $\sigma\eta \in S$ entails $\sigma \in S$. If $\mathcal{B}$ satisfies (N2), then, for any G, the set of all σ such that $G\sigma \in \mathcal{B}$ is hereditary. If $\mathcal{B}$ is recursive, then the latter set is recursive. We say that G is unifiable in S if there exists a unifier of G which belongs to S. If S is hereditary, then G is unifiable in S iff the m.g.u. of G exists and belongs to S.

Using these facts, all results discussed above for classes $\mathcal{G}_r$ and $\mathcal{G}_k$ can also be obtained for these classes restricted to grammars from a recursive class $\mathcal{B}$, fulfilling (N1), (N2). Many results of that kind can be found in [56, 18, 26, 27].

Kanazawa considers similar problems for type grammars based on logics stronger than AB; Moortgat [60] proposes an extension to grammars based on multimodal logics. Recently, Foret has shown that string languages of rigid grammars based on Associative Lambek Calculus (see section 3) admit a limit point, hence this class of grammars is not learnable from strings. Rigidness is not a natural constraint for Lambek grammars, since the Lambek calculus transforms each type to infinitely many types. Consequently, no Lambek grammar is esentially rigid (or $k-$valued), and grammars of that kind seem to require some different notion of learnability. On the other hand, methods discussed above can be adapted to various grammar formalisms, different from CCG but interpretable in it, as e.g. finite state automata, CF-grammars in (generalized) Greibach normal form, planar dependency grammars, and others.

Type logics

The basic logic AB can be extended in various ways. Lambek [49] has introduced Syntactic Calculus, now called Associative Lambek Calculus (AL). Types are formed out of primitive types by $\otimes, \rightarrow, \leftarrow$. The axioms are (Id) and:

$$\text{(ASS)}\ (A \otimes B) \otimes C \vdash A \otimes (B \otimes C),\ A \otimes (B \otimes C) \vdash (A \otimes B) \otimes C,$$

and inference rules are:

$$\text{(RRES)}\ A \otimes B \vdash C \text{ iff } B \vdash A \rightarrow C,$$

$$\text{(LRES)}\ A \otimes B \vdash C \text{ iff } A \vdash C \leftarrow B,$$

$$\text{(CUT1) if } A \vdash B \text{ and } B \vdash C \text{ then } A \vdash C.$$

Theorems of AL (in this format) are simple sequents $A \vdash B$, interpreted as: every expression of type A is also of type B. By (Id), (RRES) and (LRES), one gets:

$$A \otimes (A \rightarrow B) \vdash B,\ (B \leftarrow A) \otimes A \vdash B,$$

which correspond to the reduction rules of AB. AL provides many theorems not derivable in AB, as e.g.:

(1) $A \vdash B \leftarrow (A \rightarrow B),\ A \vdash (B \leftarrow A) \rightarrow B,$

(2) $(A \rightarrow B) \otimes (B \rightarrow C) \vdash A \rightarrow C,\ (C \leftarrow B) \otimes (B \leftarrow A) \vdash C \leftarrow A,$

(3) $(A \rightarrow B) \leftarrow C \vdash A \rightarrow (B \leftarrow C)$, $A \rightarrow (B \leftarrow C) \vdash (A \rightarrow B) \leftarrow C$,

(4) $A \vdash B \rightarrow (B \otimes A)$, $A \vdash (A \otimes B) \leftarrow B$,

called expansion, composition, associativity and lifting, respectively.

According to (1), type PN (proper noun) can be expanded to NP (noun phrase) equal to S← (PN→S). As a consequence, noun phrase conjunctions of type (NP→NP)←NP can be used in contexts like 'Joan and some teacher', analysed as compound noun phrases. According to (2), the sentence negation 'not' of type S←S can be composed with a noun phrase to produce a negated noun phrase, e.g. 'not every teacher'. An application of (3) was observed by Lambek. Let NP' equal to (S←PN)→S be the type of noun phrases on the object position, as in 'likes every teacher'. (PN→S)←PN is the type of transitive verb phrases. If we assign NP' to 'him', then 'Joan likes him' gives rise to the sequent:

$$\text{NP (PN}\rightarrow\text{S)}\leftarrow\text{PN (S}\leftarrow\text{PN)}\rightarrow\text{S} \vdash \text{S},$$

which is derivable in AL but not in AB (in AL, use (3) to the middle type). The sequence $A_1 \ldots A_n$ in the antecedent is interpreted as $A_1 \otimes \cdots \otimes A_n$.

Type transformations provided by AL are justified by an algebra of types, which is assumed by Lambek but not by Ajdukiewicz and Bar-Hillel. In CCG, types of expressions are completely determined by the initial type assignment of a categorial grammar. If 'Joan' is assigned type PN, then PN is the only type of 'Joan' in all syntactic analyses. According to Lambek, 'Joan' has type PN, but also type NP, NP' and infinitely many other types A such that PN$\vdash$ A is derivable in AL. Types are interpreted as sets of nonempty strings on a lexicon V (languages on V). The logical constants $\otimes, \rightarrow, \leftarrow$ are interpreted by the following operations on languages:

(5) $L_1 \cdot L_2 = \{ab : a \in L_1,\ b \in L_2\}$,

(6) $L_1 \backslash L_2 = \{a \in V^+ : (\forall b \in L_1) ba \in L_2\}$,

(7) $L_1 / L_2 = \{a \in V^+ : (\forall b \in L_2) ab \in L_1\}$.

A language model is defined as a pair (V, μ) such that V is a nonempty finite set, and μ is a mapping from Pr into $P(V^+)$. One extends μ to a mapping from Tp (the set of types) into $P(V^+)$, by setting:

$$\mu(A \otimes B) = \mu(A) \cdot \mu(B),\ \mu(A \rightarrow B) = \mu(A) \backslash \mu(B),\ \mu(A \leftarrow B) = \mu(A)/\mu(B).$$

A sequent $A \vdash B$ is said to be true in this model, if $\mu(A) \subseteq \mu(B)$. It is easy to show that sequents provable in AL are true in all language models. Pentus

[66] proves the completeness: every sequent true in all language models is provable in AL (for product-free sequents, it has been shown in [7]).

Since AL is sound and complete with respect to language models, grammars based on AL cannot yield incorrect types of expressions provided that the initial type assignment is correct. Lambek [50] discusses the following example. Let us assign type PN←PN to adjectives. Then, AL yields type S to pseudo-sentence 'Joan likes poor him', if 'Joan', 'likes' and 'him' are typed as above. Lambek concludes that AL is too strong, and proposes to replace AL by its nonassociative variant without (ASS). Nonassociative Lambek Calculus (NL) is interesting for many reasons, but Lambek's conclusion seems to be misleading. Type PN←PN of 'poor' and type (PN→S)←PN of 'likes' are correct, and the sequent:

$$\text{(PN}\rightarrow\text{S)}\leftarrow\text{PN PN}\leftarrow\text{PN (S}\leftarrow\text{PN)}\rightarrow\text{S} \vdash \text{PN}\rightarrow\text{S}$$

is provable in AL (hence true in language models), but 'likes poor him' is not a syntactically correct expression of type PN→S (intransitive verb phrase). The conslusion must be: NP' is not a correct type of 'him'. On the other hand, it is difficult to find totally correct types of lexical atoms in natural language. Dropping associativity makes this task easier: now, 'likes poor' is not a syntactically correct expression of type (PN→S)←PN, hence NP' can correctly be assigned to 'him'. So, Lambek is right in his final solution. Language models of NL interpret types as phrase languages on V.

Type tranformations in AL correspond to semantic transformations definable in the lambda calculus. Primitive semantic types are t (truth value), e (entity), and possibly others. $a \rightarrow b$ is the type of functions which assign objects of type b to objects of type a. Syntactic types of AL correspond to semantic types: PN corresponds to e, S to t, and $A \rightarrow B$, $B \leftarrow A$ correspond to $a \rightarrow b$, where a corresponds to A and b to B. For instance, the semantic type of intransitive verb phrases is $e \rightarrow t$, of functions which assign truth values to entities; they can be identified with sets of entities. The semantic type corresponding to NP is $(e \rightarrow t) \rightarrow t$, of functions which assign truth values to objects of type $e \rightarrow t$; they can be identified with families of sets of entities. (The noun phrase 'every student' denotes the family of all sets W such that every student belongs to W.)

We say that a type transformation $a_1 \ldots a_n \Rightarrow a$ is lambda definable, if there exists a lambda term M of type a with free variables x_i of type a_i, for $i = 1, \ldots, n$. Let us consider composition (1) for A =PN, B =S. The corresponding type transformation $e \Rightarrow (e \rightarrow t) \rightarrow t$ is definable by the term:

$$\lambda x^{e\rightarrow t}.x^{e\rightarrow t}y^e,$$

where the superscripts indicate types of variables. For any given value u of y^e, this term denotes a mapping F_u such that $F_u(f) = f(u)$, for any function f of type $e \to t$. Equivalently, F_u is the family of all sets W such that $u \in W$. Less formally, expansion corresponds to a semantic transformation which sends each entity u to the family of all (monadic) predicates true on u (or: properties of u). In semantical terms, the phrase 'John and every student' can be interpreted as follows. First, we expand type e of 'John' to type $(e \to t) \to t$ of noun phrases. Now, 'John' and 'every student' are of the same type. The entity denoted by 'John' is transformed by the lambda term to the family of all properties of this entity. The conjunction 'and' is interpreted as the Boolean meet on families of sets. Consequently, 'John and every student' denotes the family of all sets W such that John and all students belong to W, as expected.

Interconnections between type logics and semantic transformations have been studied by van Benthem [76, 78], Moortgat [58, 59] and their collaborators. van Benthem proves a kind of lambda completeness of the commutative AL: sequents provable in this system are precisely those which are definable by lambda terms, fulfilling three constraints: (i) every subterm has a free variable, (ii) every variable occurs at most once in each subterm, (iii) each lambda binds a variable in its scope. This refines the well-known Curry-Howard isomorphism between intuitionistic proofs and lambda calculus. Along these lines, derivations in type logics have a sense from the standpoint of Montague semantics and may serve as a basis for 'semantic parsing'.

Type logics as substructural logics

Lambek [49] presents AL in the form of a Gentzen-style sequent system. Sequents are of the form $\Gamma \vdash A$, where $\Gamma \in \mathrm{Tp}^+$ and $A \in \mathrm{Tp}$. The axioms are (Id) (it suffices to restrict them to primitive types A), and the inference rules are:

$$(\otimes \mathrm{L})\ \frac{\Gamma A B \Gamma' \vdash C}{\Gamma(A \otimes B)\Gamma' \vdash C},\quad (\otimes \mathrm{R})\ \frac{\Gamma \vdash A;\ \Delta \vdash B}{\Gamma\Delta \vdash A \otimes B},$$

$$(\to \mathrm{L})\ \frac{\Gamma B \Gamma' \vdash C;\ \Delta \vdash A}{\Gamma\Delta(A \to B)\Gamma' \vdash C},\quad (\to \mathrm{R})\ \frac{A\Gamma \vdash B}{\Gamma \vdash A \to B},$$

$$(\leftarrow \mathrm{L})\ \frac{\Gamma B \Gamma' \vdash C;\ \Delta \vdash A}{\Gamma(B \leftarrow A)\Delta\Gamma' \vdash C},\quad (\leftarrow \mathrm{R})\ \frac{\Gamma A \vdash B}{\Gamma \vdash B \leftarrow A},$$

where $\Gamma \neq \epsilon$ in ($\to$R) and ($\leftarrow$R). Dropping the latter constraint yields a stronger system AL0 in which provable sequents may have empty antecedents. AL0 is not conservative over AL: $\vdash p \to p$ is provable in AL0, by (Id) and (→R), hence $(p \to p) \to p \vdash p$ is provable in AL0, by (Id) and (→L), but the latter is not provable in AL. AL0 is complete with respect to language models in which languages are subsets of $V^\star$ (the set of all strings over V).

Provability in AL and AL0 is decidable, since the above axiomatizations yield cut-free sequent systems in which the complexity of the conclusion of each rule is greater than the complexity of any premise of this rule. As shown by Lambek [49], (CUT) is admissible in AL, and the same holds for AL0. Using (CUT), it is easy to prove that the sequent system of AL is equivalent to the 'algebraic' system given at the beginning of this section. As concerns computational complexity, AL and AL0 are NP-TIME decidable; recently, Pentus [67] shows NP-completeness.

Nonassociative Lambek Calculus (NL) deals with sequents $X \vdash A$ such that X is a type structure and A is a type. *Type structures* are defined as follows: (i) all types are (atomic) type structures, (ii) if X, Y are type structures, then $(X \circ Y)$ is a type structure. So, up to notation, type structures amount to phrase structures on Tp in the sense of section 2. The axioms are (Id), and the inference rules are:

$$(\otimes\mathrm{L})\ \frac{X[A \circ B] \vdash C}{X[A \otimes B] \vdash C},\quad (\otimes\mathrm{R})\ \frac{X \vdash A;\ Y \vdash B}{X \circ Y \vdash A \otimes B},$$

$$(\to\mathrm{L})\ \frac{X[B] \vdash C;\ Y \vdash A}{X[Y \circ (A \to B)] \vdash C},\quad (\to\mathrm{R})\ \frac{A \circ X \vdash B}{X \vdash A \to B},$$

$$(\leftarrow\mathrm{L})\ \frac{X[B] \vdash C;\ Y \vdash A}{X[(B \leftarrow A) \circ Y] \vdash C},\quad (\leftarrow\mathrm{R})\ \frac{X \circ A \vdash B}{X \vdash B \leftarrow A}.$$

As usual, $X[Y]$ denotes structure X with a distinguished substructure Y, and in this context $X[Z]$ denotes the result of replacing Y with Z in X. NL0 admits the empty type structure 0, satisfying $0 \circ X = X$ and $X \circ 0 = X$, for any type structure X. The cut rule:

$$(\mathrm{NCUT})\ \frac{X[A] \vdash B;\ Y \vdash A}{X[Y] \vdash B}$$

is admissible in NL and NL0 [50]. As shown in [33], NL is P-TIME decidable. A different proof of this fact will be sketched below.

NL and NL0 are genuine substructural logics. In the Gentzen format, they employ logical rules, that means, rules for logical constants only. They employ no structural rules, manipulating type structures. AL (AL0) can be obtained from NL (NL0) by affixing the structural rule of associativity:

$$(\mathrm{RASS})\ X[(Y_1 \circ Y_2) \circ Y_3] \vdash A \text{ iff } X[Y_1 \circ (Y_2 \circ Y_3)] \vdash A;$$

as a result of (RASS), type structures can be identified with strings of types. Stronger systems arise by adding other structural rules, as e.g.

$$\text{(EXC)}\ \frac{\Gamma AB\Gamma' \vdash C}{\Gamma BA\Gamma' \vdash C},$$

$$\text{(WEA)}\ \frac{\Gamma\Gamma' \vdash B}{\Gamma A\Gamma' \vdash B},$$

$$\text{(CON)}\ \frac{\Gamma \vdash B}{\Delta \vdash B};$$

in (CON), Γ contains at least two occurrences of A, and Δ drops one of them. As a result of (EXC) (the rule of exchange), antecedents of sequents can be identified with multisets of types. For systems with (EXC), rules (WEA) (weakening) and (CON) (contraction) can be formulated in a more standard way:

$$\text{(WEA)}\ \frac{\Gamma \vdash B}{\Gamma A \vdash B},\ \text{(CON)}\ \frac{\Gamma AA \vdash B}{\Gamma A \vdash B}.$$

In the presence of (EXC), $\otimes$ is commutative, and $A \rightarrow B$ is equivalent to $B \leftarrow A$ (so, $\leftarrow$ can be abandoned). The product-free fragment of AL0 plus (EXC) amounts to the logic BCI (the system of types of combinators formed out of combinators B, C and I). A Hilbert-style system of BCI employs the axioms:

(B) $(B \rightarrow C) \rightarrow ((A \rightarrow B) \rightarrow (A \rightarrow C))$,

(C) $(A \rightarrow (B \rightarrow C)) \rightarrow (B \rightarrow (A \rightarrow C))$,

(I) $A \rightarrow A$,

and the only rule MP: from $A \rightarrow B$ and A infer B. In the tradition of substructural logics, $\otimes$ is called *fusion*. AL0 plus (EXC) amounts to BCI with fusion. Analogously, the product-free AL0 plus (EXC), (WEA) (resp. (CON)) amounts to BCK (resp. BCIW), and the system enriched with (EXC), (WEA), (CON) amounts to the logic of positive implication. The rule (CUT) is admissible for all systems, mentioned above.

Zielonka [84] proves that the product-free fragment of AL0 cannot be axiomatized by any finite number of axiom schemes with MP and its dual (from $B \leftarrow A$ and A infer B) as the only rules; much earlier [83], this author obtained similar results for AL. It holds for AL plus (EXC), but it remains open for NL0. Notice that, for systems like AL, (AL) plus (EXC), one takes axioms of the form $\Gamma \vdash A$ (closed under substitution), $\Gamma \neq \epsilon$, and (CUT) as the only rule.

Systems with (EXC) may be regarded as logics of semantic types. For instance, the product-free AL plus (EXC) has been studied by van Benthem [76] as a basic logic of semantics.

It is reasonable to consider further extensions of these systems. One direction is to admit lattice operations $\wedge$ and $\vee$. For associative systems, the rules for $\wedge$ are:

$$(\wedge\mathrm{L1})\ \frac{\Gamma A\Gamma' \vdash C}{\Gamma(A\wedge B)\Gamma' \vdash C},\quad (\wedge\mathrm{L2})\ \frac{\Gamma B\Gamma' \vdash C}{\Gamma(A\wedge B)\Gamma' \vdash C},$$

$$(\wedge\mathrm{R})\ \frac{\Gamma\vdash A;\ \Gamma\vdash B}{\Gamma\vdash A\wedge B},$$

and the rules for $\vee$ are:

$$(\vee\mathrm{L})\ \frac{\Gamma A\Gamma' \vdash C;\ \Gamma B\Gamma' \vdash C}{\Gamma(A\vee B)\Gamma' \vdash C},$$

$$(\vee\mathrm{R1})\ \frac{\Gamma\vdash A}{\Gamma\vdash A\vee B},\quad (\vee\mathrm{R2})\ \frac{\Gamma\vdash B}{\Gamma\vdash A\vee B}.$$

Again, (CUT) is admissible, and the resulting systems are decidable. Since AL with $\wedge, \vee$ is a conservative extension of AL, then the decision problem for the former is NP-complete, by the Pentus result for AL. The precise computational complexity of NL with $\wedge, \vee$ is not known.

Systems with $\wedge, \vee$ lack some important property, characteristic of systems like AL, AL0, NL, NL0 (also with (EXC)). Proof trees in the latter systems show no 'material' dependence of different branches. Primitive types appearing in the conclusion of ($\otimes$R) come from premises independently of each other, and similarly for other rules. As a consequence, every proof tree can be transformed into another proof tree in which each axiom $p\vdash p$ appears precisely once (we use different copies of p for different occurrences of $p\vdash p$). In sequents appearing in the modified proof tree, each primitive type has 0 or 2 occurrences, and the initial proof tree can be obtained from the latter by substitution (one substitutes p for every copy of p). Accordingly, every provable sequent is a substitution of a provable sequent in which each primitive type occurs twice. This property is called Multiplicativity. In the tradition of linear logic [32], constants $\otimes, \rightarrow, \leftarrow$ are said to be *multiplicative*, while $\wedge, \vee$ are *additive*. Actually, in rules ($\wedge$R) and ($\vee$L) the antecedent string Γ must be the same in both premises, hence the two branches of the proof tree are 'materially' dependent.

Multiplicative systems admit a strong form of Interpolation. For $p \in\mathrm{Pr}$, let $|A|_p$ denote the number of occurrences of p in A. $|\Gamma|_p$ is defined in a similar way. $|\Gamma|$ denotes the total number of occurrences of primitive types in Γ, and similarly for $|A|$. We state an interpolation lemma, proven by Roorda [70].

(INT) Let $\Gamma\Delta\Gamma' \vdash B$ be provable in AL with $\Delta \neq \epsilon$. Then, there exists type A (an interpolant of Δ), satisfying: (i) $\Delta \vdash A$ is provable in AL, (ii) $\Gamma A\Gamma' \vdash B$ is provable in AL, (iii) for any $p \in$Pr, $|A|_p$ is not greater than the minimum of $|\Delta|_p$, $|\Gamma\Gamma' B|_p$.

The proof proceeds by induction on proofs in AL. Let us demonstrate one case. Assume $\Gamma\Gamma' \vdash A \otimes B$ is derivable from $\Gamma \vdash A$ and $\Gamma' \vdash B$, by ($\otimes$R). Let $\Gamma = \Gamma_1\Gamma_2$, $\Gamma' = \Gamma'_1\Gamma'_2$, $\Delta = \Gamma_2\Gamma'_1$ with Γ_2 and Γ'_1 being nonempty. We find an interpolant C of Δ. By the induction hypothesis, there exists an interpolant C_1 of Γ_2 in $\Gamma \vdash A$, and there exists an interpolant C_2 of Γ'_1 in $\Gamma' \vdash B$. Set $C = C_1 \otimes C_2$. $\Delta \vdash C$ is provable, by ($\otimes$R). Since $\Gamma_1 C_1 \vdash A$ and $C_2\Gamma'_2 \vdash B$ are provable, then $\Gamma_1 C\Gamma'_2 \vdash A \otimes B$ is provable, by ($\otimes$R) and ($\otimes$L). We also have:

$$|C_1|_p \leq |\Gamma_2|_p, \; |C_2|_p \leq |\Gamma'_1|_p,$$

which yields $|C|_p \leq |\Delta|_p$ (add both inequalities). In a similar way, we show $|C|_p \leq |\Gamma_1\Gamma'_2(A \otimes B)|_p$.

Using multiplicativity and interpolation, Pentus [65] proves that type grammars based on AL generate context-free languages (the converse statement that all context-free languages are generated by these grammars follows from the Gaifman theorem and the fact that AL is equivalent to AB for sequents $A_1 \ldots A_n \vdash p$ such that $A_1, \ldots, A_n$ are types of order at most 1 and p is primitive). By multiplicativity, for every sequent $\Gamma \vdash A$, provable in AL, there exists a provable sequent $\Gamma' \vdash A'$ in which each primitive type occurs exactly twice, and $\Gamma \vdash A = (\Gamma' \vdash A')\sigma$, for some substitution σ. Let $\Gamma = A_1 \ldots A_n$, $\Gamma' = A'_1 \ldots A'_n$. For each $i = 1, \ldots, n$, one finds an interpolant B_i of A'_i in $\Gamma' \vdash A'$ and an interpolant B of Γ'. Then, $B_1 \ldots B_n \vdash B$ is provable in AL, each primitive type occurs twice in this sequent and at most once in every type of this sequent; such sequents are said to be *thin*. By a combinatorial argument, Pentus proves the following binary reduction lemma for thin sequents: if $B_1 \ldots B_n \vdash B$ is thin, $|B_i| \leq m$, $|B| \leq m$, then there exist $1 \leq j < n$ and type C such that $|C| \leq m$ and C is an interpolant of B_jB_{j+1} in $B_1 \ldots B_n \vdash B$. Applying σ, we prove the same binary reduction lemma for the initial sequent $\Gamma \vdash A$. Let G be a grammar based on AL, and let m be the maximal $|A|$, for types A appearing in G. All types in G employ only a finite collection of primitive types. By the binary reduction lemma, every sequent $A_1 \dots A_n \vdash A$ provable in AL and such that all types appear in G can be derived by (CUT) from provable sequents $C_1C_2 \vdash C_3$, $C_1 \vdash C_2$ such that $|C_i| \leq m$. By the decidability of AL, one can effectively determine all sequents of this form, and they yield production rules of a CF-grammar equivalent to the initial type grammar.

This proof cannot be adjusted to systems with $\wedge, \vee$. Grammars based on AL, NL with $\wedge$ generate languages which are not context-free [37], which follows from the fact that the intersection of two context-free languages need not be context-free.

For NL, NL0, the formulation of (INT) is the following. Let $X[Y] \vdash B$ be provable, $Y \neq 0$. Then, there exists type A, satisfying: (i) $Y \vdash A$ is provable, (ii) $X[A] \vdash B$ is provable, (iii) for any $p \in$Pr, $|A|_p$ is not greater than the minimum of $|Y|_p$, $|X[q]B|_p$, where $q \neq p$.

Interestingly, nonassociative systems admit another form of interpolation, not admissible in associative ones. Let T be a finite set of types, closed under subtypes. $T-$sequents are sequents using types from T only.

(NINT) Let $X[Y] \vdash B$ be a provable $T-$sequent. Then, there exists $A \in T$ such that (i) and (ii) hold true.

(NINT) can easily be proven by induction on proofs in NL, NL0 (also with (EXC)); a similar result has been announced by Jäger (unpublished) As a consequence, grammars based on these systems yield context-free languages. Other proofs, based on a normalization procedure for nonassociative systems, has been given in [9, 40, 41].

Other type logics, e.g. multimodal systems, pregroup systems, will be discussed later on.

Models and grammars

Abstract algebraic frames of NL are *residuated groupoids*, i.e. structures of the form $(M, \leq, \cdot, \backslash, /)$ such that $(M, \leq)$ is a nonempty poset, and $\cdot, \backslash, /$ are binary operations on M, satisfying the condition of residuation:

$$\text{(RES) } ab \leq c \text{ iff } b \leq a\backslash c \text{ iif } a \leq c/b,$$

for all $a, b, c \in M$. *A model* is a pair (M, μ) such that M is a residuated groupoid, and μ is a mapping from Pr into M, which is extended to all types as for the case of language models. It is also extended to type structures: $\mu(X \circ Y) = \mu(X)\mu(Y)$. A sequent $X \vdash A$ is said to be *true* in a model (M, μ), if $\mu(X) \leq \mu(A)$.

A standard argument, using a canonical model, yields the completeness of NL with respect to residuated groupoids. We can consider theories in NL, i.e. systems arising from NL by affixing a set Φ of new axioms; then, the rule (NCUT) is treated as a legal rule of the system. A system S enriched by a set Φ of new axioms will be denoted S(Φ). Caution: we do not assume new axioms are closed under substitution. The strong completeness also holds: the sequents provable in NL(Φ) are precisely those which are true in all models (M, μ) in which all sequents from Φ are true.

Frames for NL0 are residuated groupoids with unit 1, satisfying: $1a = a1 = a$, for all elements a. We set $\mu(0) = 1$. To construct a canonical model (from equivalence classes of types) it is expedient to add a propositional constant 1 to the language, with the following rules:

$$\text{(1L)}\ \frac{X[0] \vdash A}{X[1] \vdash A},\ \text{(1R)}\ 0 \vdash 1;$$

recall that $0 \circ Y = Y \circ 0 = Y$ in the metalanguage, hence $X[1 \circ Y] \vdash A$ can be infereed from $X[Y] \vdash A$. This system is denoted NL1. Cut elimination holds for NL1, and NL1 is conservative over NL0. The completeness of NL1 can be shown by the construction of a canonical model, and consequently, NL0 is also complete with respect to residuated groupoids with unit.

Residuated groupoids in which $\cdot$ is associative are called *residuated semigroups*. Residuated semigroups with unit are called *residuated monoids*. AL is strongly complete with respect to residuated semigroups, and AL1 is strongly complete with respect to residuated monoids (AL1 is NL1 plus (RASS)). Since AL1 is conservative over AL0, the latter system is complete with respect to residuated monoids. For systems with $\wedge, \vee$, the frames are residuated groupoids (semigroups) $(M, \leq, \cdot, \backslash, /)$ such that $(M, \leq)$ is a lattice; we call them lattice ordered (l.o.) residuated semigroups (monoids), but the term 'residuated lattice' is also used in literature. Completeness can be obtained, as above. Systems with (EXC) are complete with respect to commutative residuated semigroups (monoids, groupoids) in which $ab = ba$, for all elements a, b.

More special frames are powerset frames. Let $(M, \cdot)$ be a groupoid. For $L_1, L_2 \subseteq M$, one defines operations $\cdot, \backslash, /$, by (5), (6), (7) (replace V^+ by M). The structure $(P(M), \subseteq, \cdot, \backslash, /)$ is a residuated groupoid (*the powerset residuated groupoid* over $(M, \cdot)$). If $(M, \cdot)$ is a semigroup, then the powerset structure is a residuated semigroup (*the powerset residuated semigroup* over $(M, \cdot)$). If $(M, \cdot, 1)$ is a monoid (i.e. a semigroup with unit), then $\{1\}$ is the unit in the powerset structure, and similarly for groupoids. AL (resp. AL1) is strongly complete with respect to powerset residuated semigroups (resp. monoids) [8].

The proof given in [8] seems to be one of the first applications of labelled deductive systems, extensively studied by Gabbay [29]. Systems of that kind admit expressions of the form $t : A$ such that A is a logical formula, and t is a label. Rules of the system perform some logical operations on formulas and algebraic operations on labels. They are intended to formalize a part of metalanguage for the object language; for instance, $t : A$ may express the relation: A is true in state (world) t. In [8], $t : A$ means: the element t of M belongs to $\mu(A)$. Labels are defined by recursion: (i) all types are atomic labels, (ii) if $t_1, \ldots, t_n$, $n \geq 1$, are atomic labels, then $t_1 \ldots t_n$ is a label, (iii) if t is a label, and A, B are types, then $(t, A, B, 1)$ and $(t, A, B, 2)$ are atomic labels. One says that label t *reduces* to label t', if t' arises from t by a finite number of replacements of $(s, A, B, 1)(s, A, B, 2)$ by s. The axioms are (LID) $A : A$, and the rules are ND rules of elimination and introduction of $\otimes, \rightarrow, \leftarrow$.

(L$\otimes$E) from $t : A \otimes B$ infer both $(t, A, B, 1) : A$ and $(t, A, B, 2) : B$,

(L$\otimes$I) from $s : A$ and $t : B$ infer $st : A \otimes B$,

(L$\rightarrow$E) from $s : A$ and $t : A \rightarrow B$ infer $st : B$,

(L$\rightarrow$I) from $As : B$ infer $s : A \rightarrow B$,

and the dual rules (L$\leftarrow$E), (L$\leftarrow$I). We also need a reduction rule: from $t : A$ infer $t' : A$, whenever t reduces to t'. For any sequent $A_1 \ldots A_n \vdash B$, $n \geq 1$, we introduce a rule: from $t_i : A_i$, for $i = 1, \ldots, n$, infer $t_1 \ldots t_n : B$. The system LAL(Φ) admits the latter rule, for every sequent from Φ. To any label t one assigns a string of types $T(t)$ by the following recursion: (i) $T(A) = A$, (ii) $T(t_1 \ldots t_n) = T(t_1) \ldots T(t_n)$, (iii) $T((t, A, B, 1)) = A$, $T((t, A, B, 2)) = B$. One shows: $t : A$ is provable in LAL(Φ) iff $T(t) \vdash A$ is provable in AL(Φ). A label is said to be *irreducible* if it starts no proper reduction. We construct a canonical model. Let $(M, \cdot)$ be a semigroup in which M is the set of irreducible labels, and, for $s, t \in M$, $s \cdot t$ is the only irreducible label t' such that st reduces to t' (the uniqueness follows from the Church-Rosser property of reduction). We consider the powerset structure $P(M)$ and the canonical mapping μ, defined by: $\mu(p)$ equals the set of all $t \in M$ such that $t : p$ is provable in LAL(Φ). We prove: $\mu(A)$ equals the set of all $t \in M$ such that $t : A$ is provable in LAL(Φ). Now, if $A \vdash B$ is not provable in AL(Φ), then $A : B$ is not provable in LAL(Φ) (by $T(A) = A$), hence $A \in \mu(A)$ and $A \notin \mu(B)$. Consequently, $A \vdash B$ is not true in $(P(M), \mu)$. This yields the strong completeness of AL with respect to powerset frames. In a similar way, one can prove a representation theorem: every residuated semigroup is embeddable into a powerset residuated semigroup.

In [46], a similar argument shows the strong completeness of NL with respect to powerset residuated groupoids and the corresponding representation theorem: every residuated groupoid is embeddable into a powerset residuated groupoid. If K, K' are classes of frames such that $K \subseteq K'$, a propositional system is (strongly) complete with respect to K, and there holds the representation theorem: every frame from K is embeddable into a frame from K', then this system is (strongly) complete with respect to K'. So, the completeness of a system with respect to concrete frames follows from its completeness with respect to abstract frames and the embeddability of abstract frames into concrete frames.

Language models discussed at the beginning of this section are special powerset structures in which the underlying semigroup is the free semigroup generated by V, i.e. the set V^+ with concatenation. Pentus [66] shows the completeness of AL with respect to language models (the proof is very sophisticated). Strong completeness fails. Take $p \vdash p \otimes p$ as a new axiom. If $\mu(p) \subseteq \mu(p)\mu(p)$, then $\mu(p) = \emptyset$, hence $\mu(p) \subseteq \mu(q)$ in all language models, satisfying this axiom, but $p \vdash q$ is not provable in AL($p \vdash p \otimes p$) (it is easy to find a powerset model over a semigroup in which the axiom is true, but $p \vdash q$ is false). Consequently, the representation theorem also fails: not every residuated semigroup is embeddable into the powerset residuated semigroup over a free semigroup.

These completeness results are easier for product-free fragments of type logics. Consider the free semigroup generated by Tp. In the powerset structure we define $\mu(p)$ as the set of all Γ such that $\Gamma \vdash p$ is provable in AL. One proves by induction: for any product-free type A, $\mu(A)$ is the set of all Γ such that $\Gamma \vdash A$ is provable in AL. Now, if $A \vdash B$ is not provable in AL, then $A \in \mu(A)$ and $A \notin \mu(B)$, hence $A \vdash B$ is not true in the model. Similar arguments can be provided for product-free fragments of AL0, NL, NL0 and their variants with 1, $\wedge$ [8].

For systems with product, this simpler method can be applied, if powerset models are replaced with more general cone models. In a poset $(M, \leq)$, a set $L \subseteq M$ is called *a (lower) cone* if $a \leq b$ and $b \in L$ entail $a \in M$. By $C(M)$ we denote the set of cones in M. Let $(M, \leq, \cdot)$ be a partially ordered semigroup (that means: we assume: $a \leq b$ entails $ca \leq cb$ and $ac \leq bc$). One defines operations $\backslash, /$ on $C(M)$ as above and the product as follows:

(5') $L_1 \cdot L_2 = \{x : (\exists y \in L_1, z \in L_2) x \leq yz\}$.

Then, $(C(M), \subseteq, \cdot, \backslash, /)$ is a residuated semigroup. There holds the representation theorem: every residuated semigroup M is embeddable into the residuated semigroup $C(M)$. This yields the strong completeness of AL with respect to cone models. Došen [23] gives a direct proof of this fact, using a canonical model construction with cone models instead of powerset models; the underlying structure is Tp with $\otimes$ as the operation and the provable $\vdash$ as $\leq$ (it is a preorder, but see below).

A refinement of this construction can be used to prove the subformula property for theories (since cut elimination fails, this property cannot be shown in a standard way). In [8], it was proven with the aid of the labelled deductive system, mentioned above. Here, we sketch a proof using canonical cone models [17]. First, observe that, in the construction of $C(M)$, the underlying structure $(M, \leq, \cdot)$ may be a preordered semigroup, that means, $\leq$ is reflexive and transitive, $\cdot$ is monotone in both arguments, and associativity holds modulo the equivalence generated by $\leq$: $a \sim b$ iff both $a \leq b$ and $b \leq a$. Let T be a set of types, closed under subtypes, which contains all types appearing in Φ. Take $M = T^+$ with concatenation. We write $\Gamma \vdash_T A$ if there exists a proof of $\Gamma \vdash A$ in AL(Φ) which consists of T–sequents only (see the end of subsection 3.1). We define a preorder $\leq$ on M as the reflexive and transitive closure of the following relation: $\Gamma\Delta\Gamma' \leq \Gamma A \Gamma'$ if $\Delta \vdash_T A$, $\Gamma AB\Gamma' \leq \Gamma(A \otimes B)\Gamma'$ if $(A \otimes B) \in T$. Then, $(M, \leq)$ with concatenation is a preordered semigroup. We consider the residuated semigroup $C(M)$. A mapping μ is defined as follows: $\mu(p) = \{\Gamma \in M : \Gamma \vdash_T p\}$. One proves $\mu(A) = \{\Gamma \in M : \Gamma \vdash_T A\}$, for all $A \in T$. Now, let $\Gamma \vdash A$ be provable in AL(Φ). Let T be the set of all subtypes of the types appearing in Φ and $\Gamma \vdash A$. Axioms from Φ are true in $(C(M), \mu)$, hence $\mu(\Gamma) \subseteq \mu(A)$, by soundness. Since $\Gamma \in \mu(\Gamma)$, then $\Gamma \in \mu(A)$, and consequently, $\Gamma \vdash_T A$. Thus, $\Gamma \vdash A$ has a proof in AL(Φ) in which all types are in T. A similar argument can be given for NL. For product-free fragments, powerset frames can be used instead of cone frames.

Now, we can prove that the consequence relation '$X \vdash A$ follows from Φ in NL' (that means, $X \vdash A$ is provable in NL(Φ)) is P-TIME decidable (Φ is a finite set of sequents $C \vdash D$). Let T be the smallest set of types, being closed under subtypes and containing all types appearing in $X \vdash A$ and Φ. We can construct T from the data in polynomial time, and the size of T is not greater than the number of primitive types and logical constants in the data. Recall that a T–sequent is a sequent whose all types are in T. A sequent is said to be *basic* if it is a T–sequent of the form $C_1 \circ C_2 \vdash C_3$ or $C_1 \vdash C_2$. The number of basic sequents equals $n^2 + n^3$, where n is the size of T. We can effectively find all basic sequents provable in NL(Φ).

First, we construct a set S^T of provable basic sequents. S_0 consists of all sequents from Φ, T-sequents (Id), and T-sequents of the form:

$$A \circ B \vdash A \otimes B,\ A \circ (A \rightarrow B) \vdash B,\ (B \leftarrow A) \circ A \vdash B.$$

Assume S_n has already been defined. S_{n+1} is S_n enriched with sequents resulting from the following rules:

(S1) if $A \circ B \vdash C$ is in S_n and $A \otimes B$ is in T, then $A \otimes B \vdash C$ is in S_{n+1},

(S2) if $A \circ B \vdash C$ is in S_n and $A \rightarrow C$ is in T, then $B \vdash A \rightarrow C$ is in S_{n+1},

(S3) if $A \circ B \vdash C$ is in S_n and $C \leftarrow B$ is in T, then $A \vdash C \leftarrow B$ is in S_{n+1},

(S4) if $A \circ B \vdash C$ and $A' \vdash A$ are in S_n, then $A' \circ B \vdash C$ is in S_{n+1},

(S5) if $A \circ B \vdash C$ and $B' \vdash B$ are in S_n, then $A \circ B' \vdash C$ is in S_{n+1},

(S6) if $X \vdash A$ and $A \vdash B$ are in S_n, then $X \vdash B$ is in S_{n+1}.

We define S^T as the join of this chain. Clearly, $S^T = S_k$, for the least k such that $S_k = S_{k+1}$, and this k is not greater than the number of basic sequents. The construction of S_{n+1} from S_n can be done in a time polynomial in the number of basic sequents. Consequently, S^T can be constructed in a time polynomial in the size of the data. Clearly, S^T is closed under (NCUT) restricted to basic sequents. Let $S(T)$ denote the system whose axioms are all sequents from S^T and the only rule is (NCUT). It is easy to show that all basic sequents derivable in $S(T)$ belong to S^T (use induction on derivations). All derivations in $S(T)$ are actually derivations in a CF-grammar whose production rules are (reversed) sequents from S^T. By the known CYK algorithm, checking derivability in CF-grammars, the derivability in $S(T)$ is P-TIME decidable. Now, it suffices to prove that NL(Φ) is equivalent to $S(T)$ in the scope of T-sequents. It is easy to show that all sequents derivable in $S(T)$ are derivable in NL(Φ). To show the converse, we first observe that the interpolation lemma (NINT) evidently holds for $S(T)$. By the subformula property, T-sequents provable in NL(Φ) have proofs which consist of T-sequents only. All axioms of NL(Φ), being T-sequents, are axioms of $S(T)$. It suffices to show that $S(T)$ is closed under rules of NL(Φ), restricted to T-sequents. It is obvious for (NCUT). Let us consider ($\otimes$L). Assume $X[A \circ B] \vdash C$ is provable in $S(T)$. By (NINT) for $S(T)$, there is $D \in T$ such that both $A \circ B \vdash D$ and $X[D] \vdash C$ are provable in $S(T)$. $A \circ B \vdash D$ is a basic sequent, so it belongs to S^T. Since $A \otimes B$ is in T, then $A \otimes B \vdash D$ belongs to S^T, by (S1). Then, $X[A \otimes B] \vdash C$ is provable in $S(T)$, by (NCUT). Let us consider ($\otimes$R). Assume $X \vdash A$ and $Y \vdash B$ are provable in $S(T)$, and $A \otimes B$ is in T. Since $A \circ B \vdash A \otimes B$ is in S_0, then $X \circ Y \vdash A \otimes B$ is provable in $S(T)$, by (NCUT). The remaining rules are handled in a similar way.

The above construction yields a CF-grammar equivalent to the given type grammar based on NL(Φ). Accordingly, type grammars based on finite theories over NL generate context-free languages. These results strentgthen earlier results on P-TIME decidability of NL [33] and the equivalence of grammars based on NL with CF-grammars. Analogous facts can be proven for other nonassociative systems, e.g. NL1, NL with (EXC), multi-modal logics and GLC (see below) [17].

Nonlogical axioms can be of interest for linguistics for several reasons. For the case of AL, grammars based on finite theories generate all recursively enumerable languages. This fact has been proven in [7] for the product free fragment. With product, the proof is much easier: every rewriting system with rules of the form $pq \mapsto r$ and $p \mapsto qr$ can be simulated in AL enriched with axioms $p \otimes q \vdash r$ and $p \vdash q \otimes r$ (use the subformula property to show conservativity [17]). Thus, associative systems with nonlogical axioms surpass the context-free world. Nonlogical axioms can express subtyping; for instance, restrictive adjectives are a subcategory of adjectives. Further, it is known that not every linguistically sound transformation can be deduced in AL. For instance, 'and' of type (S→S)←S can also act as a verb phrase conjunction of type (VP→VP)←VP, e.g. in 'Joan sings and dances'. One cannot transform the former type to the latter in AL (it is possible in AL with (EXC) and (CON)). So, the corresponding sequent can be added to AL or NL as a nonlogical axiom. By enriching NL with finitely many new axioms, derivable in AL or not, we can improve its expressibility without lacking the nice computational simplicity.

In powerset frames and cone frames, $\wedge$ can be interpreted as the intersection of sets: $\mu(A \wedge B) = \mu(A) \cap \mu(B)$. In general, $\vee$ should not be modelled as the union of sets; this would entail the distributivity of $\wedge$ under $\vee$, which could not be derived in the systems. Adequate frames consist of ideals, i.e. cones closed under $\vee$. The join of two ideals is the smallest ideal containing them [69].

An interesting problem is to extend type logics to be able to express negative facts. We briefly discuss some attempts from [12].

A De Morgan negation $\neg$ satisfies Double Negation $\neg\neg a = a$ and Transposition: if $a \leq b$ then $\neg b \leq \neg a$ [24]. AL enriched with $\neg$ with additional axioms: $\neg\neg A \vdash A$, $A \vdash \neg\neg A$ and the rule:

$$\text{(TRA)}\ \frac{A \vdash B}{\neg B \vdash \neg A}$$

does not admit cut elimination. With (CUT), it is complete with respect to residuated semigroups with De Morgan negation. Its product free fragment is strongly complete with respect to powerset residuated semigroups with a quasi-Boolean complement in the sense of [6]: for a fixed involutive mapping $f : M \mapsto M$ and $P \subseteq M$, one defines $-P = M - f[P]$. The full system is strongly complete with respect to cone models over semigroups (the complement is quasi-Boolean). These facts follow from representation theorems: (i) for every residuated semigroup with De Morgan negation, its product free reduct is embeddable into some powerset residuated semigroup with quasi-Boolean complement, (ii) every residuated semigroup with De Morgan negation is embeddable into some residuated semigroup of cones (over some semigroup) with quasi-Boolen complement [12].

Unfortunately, AL with De Morgan negation is not complete with respect to linguistically intended models: powerset residuated semigroups with Boolean complement [12]. Set $C = B \rightarrow B$ and consider the sequent:

$$(C \rightarrow C) \rightarrow \neg A \vdash \neg((C \rightarrow C) \rightarrow A).$$

In powerset models, $\mu(C \rightarrow C)$ must be nonempty: if $\mu(C)$ is empty then $\mu(C \rightarrow C) = M$; if $\mu(C)$ is nonempty, then $\mu(C \rightarrow C)$ is nonempty, since $\mu(C) \subseteq \mu(C \rightarrow C)$. Consequently, the sequent is true in all powerset models in which $\neg$ is interpreted as the Boolean complement of sets. It is not provable in the system, since it is not true in the following model. Set $V = \{a, b\}$ and consider the free semigroup V^+. For any string $x \in V^+$, let $f(x)$ be the mirror image of x, e.g. $f(ab) = ba$. Then, f is involutive, i.e. $f \circ f =$Id (the identity mapping). We consider the powerset residuated semigroup over V^+ with the quasi-Boolean complement, given by f. Let $\mu(B)$ be the set of all b–strings, and let $\mu(A)$ be the set of all strings in V^+ whose last symbol is a. Then, $\mu(C) = \mu(B)$, and $\mu(\neg A)$ equals $M - f[\mu(A)]$, i.e. the set of all strings in V^+ whose first symbol is b. Consequently, μ sends the left side of the sequent to the total set V^+. $\mu((C \rightarrow C) \rightarrow A)$ equals $\mu(A)$, hence μ sends the right side of the sequent to $M - f[\mu(A)]$. Since the latter set is not total, then the sequent is not true.

The decidability of AL with De Morgan negation seems to be an open problem, and similarly for other type logics, mentioned above.

We have discussed frames in which logical constants of type logics are interpreted by means of some operations on sets which are naturally determined by the underlying semigroup (groupoid) structure. A more general approach uses Kripke-style frames for modal logics [78, 80]. A frame is a pair (M, R) such that M is a set of states, and R is a ternary relation on M. For sets $P_1, P_2 \subseteq M$, one defines:

(K1) $P_1 \cdot P_2 = \{z \in M : (\exists x \in P_1, y \in P_2)\, R(x, y, z)\}$,

(K2) $P_1 \backslash P_2 = \{y \in M : (\forall x \in P_1, z \in M)(R(x, y, z) \Rightarrow z \in P_2)\}$,

(K3) $P_1 / P_2 = \{x \in M : (\forall y \in P_2, z \in M)(R(x, y, z) \Rightarrow z \in P_1)\}$.

The structure $(P(M), \subseteq, \cdot, \backslash, /)$ is a residuated groupoid. Every powerset residuated groupoid is a structure of this form, since one can define: $R(x, y, z)$ iff $z = xy$. Consequently, NL is strongly complete with respect to Kripke models. A similar fact holds for AL and Kripke frames, satisfying associativity: (i) if $R(x, y, z)$ and $R(z, u, v)$ then there exists w such that $R(y, u, w)$ and $R(x, w, v)$, (ii) if $R(y, u, w)$ and $R(x, w, v)$ then there exists z such that $R(x, y, z)$ and $R(z, u, v)$.

Kripke frames suggest a generalization of type logics toward multi-modal systems [80, 59]. Besides a ternary relation R, one also regards a binary relation S, and defines unary modalities $\Diamond, \Box$ as follows:

(K4) $\Diamond(P) = \{x \in M : (\exists y \in P) S(x, y)\}$,

(K5) $\Box(P) = \{x \in M : (\forall y \in M)(S(x, y) \Rightarrow y \in P)\}$.

The above are standard modalities of Classical Modal Logic. Moortgat [59] also considers minimal modalities of substructural logics [69]:

(K6) $\Diamond(P) = \{y \in M : (\exists x \in P) S(x, y)\}$,

and $\Box$ is defined as above. Clearly, $\Diamond$ yields the image under S, and $\Box$ yields the coimage under S. The linguistic role of modalities is to control the usage of structural rules in type logics. The basic logic BL is NL with the following rules for minimal modalities:

$$(\Diamond\mathrm{L})\ \frac{X[\bullet A] \vdash B}{X[\Diamond A] \vdash B},\quad (\Diamond\mathrm{R})\ \frac{X \vdash A}{\bullet X \vdash \Diamond A},$$

$$(\Box\mathrm{L})\ \frac{X[A] \vdash B}{X[\bullet\Box A] \vdash B},\quad (\Box\mathrm{R})\ \frac{\bullet X \vdash A}{X \vdash \Box A},$$

where $\bullet$ is a new, unary operation on type structures. Moortgat admits several modality pairs $\Diamond_i, \Box_i$ with the corresponding operations $\bullet_i$, for $i = 1, \ldots, n$. A restricted commutativity can be expressed, say, by the axiom:

$$\Diamond_1 A \otimes B \vdash B \otimes \Diamond_1 A,$$

and similarly for restricted associativity, contraction, etc. One can also employ several residuation triples $\otimes_j, \rightarrow_j, \leftarrow_j$ with the corresponding structure operations $\circ_j$, for $j = 1, \ldots, m$. Associativity (commutativity, contraction) may hold for some of them. Such hybrid systems can easily be modelled by Kripke frames. Some of them admit cut elimination.

An analogous formalism, under the name Generalized Lambek Calculus (GLC), has been proposed in [11]; Dunn [25] studies a closely related system with more general models. One starts from abstract algebras (M, F) such that M is a nonempty set and $F = (f_i)_{i \in I}$ is an indexed family of operations on M. On the powerset $P(M)$, one defines powerset operations:

$$f_i(P_1, \ldots, P_n) = \{f_i(x_1, \ldots, x_n) : x_1 \in P_1, \ldots, x_n \in P_n\},$$

and for each $1 \leq j \leq n$ (n is the arity of f_i), the residuation operation f_i^j is defined as follows: $f_i^j(P_1, \ldots, P_n)$ equals the set of all $x_j \in M$ such that $f_i(x_1, \ldots, x_n) \in P_j$, for all $x_k \in P_k$, $k \neq j$. The resulting structures are obvious generalizations of powerset residuated groupoids. Types of GLC are formed out of primitive types by means of logical constants $\otimes_i, \rightarrow_i^j$ and the structure operation $\circ_i$, for $i \in I$ (their arity is that of f_i). The axioms are (Id), and the inference rules are:

$$(\otimes_i \mathrm{L}) \; \frac{X[\circ_i(A_1, \ldots, A_n)] \vdash B}{X[\otimes_i(A_1, \ldots, A_n] \vdash B},$$

$$(\otimes_i \mathrm{R}) \; \frac{X_1 \vdash A_1; \ldots; X_n \vdash A_n}{\circ_i(X_1, \ldots, X_n) \vdash \otimes_i(A_1, \ldots, A_n)},$$

$$(\rightarrow_i^j \mathrm{L}) \; \frac{X[A_j] \vdash B;\; Y_1 \vdash A_1; \ldots; Y_n \vdash A_n}{x[\circ_i(Y_1, \ldots, \rightarrow_i^j (A_1, \ldots, A_n), \ldots, Y_n)] \vdash B},$$

$$(\rightarrow_i^j \mathrm{R}) \; \frac{\circ_i(A_1, \ldots, X, \ldots, A_n) \vdash A_j}{X \vdash \rightarrow_i^j (A_1, \ldots, A_n)},$$

and (NCUT). In rule $(\rightarrow_i^j \mathrm{L})$, the premise $Y_j \vdash A_j$ is dropped and the type $\rightarrow_i^j (\ldots)$ is the j–th argument of $\circ_i$ in the conclusion. In rule $(\rightarrow_i^j \mathrm{R})$, X is the j–th argument of $\circ_i$ in the premise. GLC admits cut elimination. For $n = 1$, $\otimes_i$, $\rightarrow_i^1$ and $\circ_i$ behave like $\Diamond$, $\Box$ and $\bullet$, respectively. As shown in [46], GLC is strongly complete with respect to powerset frames over abstract algebras. Clearly, $\otimes_i$ is interpreted by f_i and $\rightarrow_i^j$ by f_i^j. By cut elimination, GLC is decidable; actually, all finitely axiomatizable theories over GLC are P-Time decidable [17]. This also holds, obviously, for multi-modal systems without additional structural postulates (rules). Grammars based on GLC generate context-free languages [42, 17].

Systems with many products can be used in linguistics to formalize different modes of composition of expressions. Instead of concatenation, one may use other operations which compose a complex expression from simpler ones. A good candidate is substitution. We enrich the lexicon V with variables $x_1, x_2, \ldots$ and define $f_i(a,b)$ to be equal to the result of substitution of b for x_i in a. In this way, we can analyse discontinuous phrases. For instance, 'if ... then ...' can be identified with 'if x_1 then x_2' and be assigned type $\rightarrow^1_1 (\rightarrow^1_2$ (S,S),S). A similar approach to discontinuity has been proposed by Morrill [61, 62].

Type logics can also be modelled by algebras of relations; see van Benthem [78, 79] for a thorough discussion of information logics formalized in this way. For relations $R, S \subseteq M^2$, one defines:

(RM1) $R \cdot S = \{(x,y) \in M^2 : \exists z((x,z) \in R \& (z,y) \in S)\}$,

(RM2) $R \backslash S = \{(x,y) \in M^2 : (\forall z)((z,x) \in R \Rightarrow (z,y) \in S)\}$,

(RM3) $S/R = \{(x,y) \in M^2 : (\forall z)((y,z) \in R \Rightarrow (x,z) \in S)\}$.

The structure $(P(M^2), \subseteq, \cdot, \backslash, /, I)$, where I is the identity relation, is a residuated monoid. It has been shown in [4] that every residuated monoid is embeddable into a structure of that kind, and consequently, AL0 and AL1 are strongly complete with respect to such models. For AL, one uses restricted frames $P(T)$ such that T is a transitive relation on M (replace M^2 with T in (RM1)-(RM3)). Other completeness and representation theorems for type logics with respect to relational models can be found in [48, 47, 19, 73].

Relational models show a link between type logics and logics of programs (dynamic logics). $R \cdot S$ corresponds to the composition of programs R and S. $R \backslash S$ (resp. S/R) is the weakest post-specification (resp. pre-specification) of program R with respect to program S in the sense of Hoare and Jifeng [35]. If one adds $+$ and $\star$, then the resulting algebras are action algebras in the sense of Pratt [68]. In opposition to Kleene algebras, based on $\cdot, +, \star$, action algebras form a variety with finitely many equational axioms [68]. Many basic properties of action algebras and logics seem unknown, as yet. Since AL and its variants are complete with respect to relational models, proof systems for these logics can be designed in the style of relational proof systems of Orłowska [63, 55]. Labelled deductive systems corresponding to relational models are studied in [48, 47].

Type logics are directly related to linear logics, introduced by Girard [32]. Linear logics are commutative and, besides logical constants $\otimes, \rightarrow, \wedge, \vee$, they admit $\oplus$, $\neg$, 0, 1, $\top$ and $\bot$. $\oplus$ is a dual of $\otimes$; it corresponds to a composition of structures on the right hand of sequents. $\neg$ is a linear negation. 0 is the neutral element for $\oplus$ and 1 for $\otimes$, while $\top$ and $\bot$ are the lattice top and bottom.

The most easy way to explain the meaning of linear constants is to consider algebraic models of linear logic. A bilinear algebra is a residuated monoid in which there exists an element 0, satisfying:

$$0/(a\backslash 0) = a = (0/a)\backslash a,$$

for all elements a. In a bilinear algebra, one defines $a^l = 0/a$, $a^r = a\backslash 0$. One proves $(b^r a^r)^l = (b^l a^l)^r$, for all a, b; this element is denoted by $a \oplus b$.

Bilinear algebras which are bounded lattices are models of Noncommutative Linear Logic of Abrusci [1]. If they additionally satisfy: $0/a = a\backslash 0$, for all a, then they determine Cyclic Linear Logic of Yetter [82]. Commutative bilinear algebras determine Classical Linear Logic of Girard. In commutative bilinear algebras and cyclic bilinear algebras $a^l = a^r$, for all a; this operation corresponds to linear negation. The term 'bilinear algebra' is due to Lambek [51]. Bilinear algebras determine the multiplicative fragment of Noncommutative Linear Logic. As shown in [2], AL1 is a conservative, multiplicative fragment of both Noncommutative Linear Logic and Cyclic Linear Logic. Girard interprets Classical Linear Logic by means of phase spaces. Phase spaces are powerset frames over commutative monoids with a distinguished set $P \subseteq M$. Sets of the form P/Q, for $Q \subseteq M$, are called facts. Linear constants are interpreted as operations on facts. For instance, linear negation correspond to the operation $Q^\bot = P/Q$ and $\otimes$ to the operation $Q_1 \cdot Q_2 = (Q_1 Q_2)^{\bot\bot}$. Linear logics show many profound interactions with logical proof theory and theoretical computer science. Their role for linguistics remains not quite clear, since linear constants lack a natural linguistic sense. However, a graph-theoretic representation of proofs in the form of proof nets in multiplicative fragments of linear logics has attracted some linguists; see e.g. Casadio [21].

At the end of this subsection, we mention a recent innovation in type logics, due to Lambek [52]. Instead of binary operations $\backslash, /$, Lambek uses unary operations l and r, of left and right adjoint, respectively. *A pregroup* is a structure $(M, \leq, \cdot, l, r, 1)$ such that $(M, \leq, \cdot, 1)$ is a p.o. monoid, and the following inequalities:

$$\text{(PRE)}\ a^l a \leq 1 \leq a a^l,\ a a^r \leq 1 \leq a^r a,$$

hold, for every $a \in M$. Equivalently, pregroups are bilinear algebras in which $\cdot = \oplus$ and $0 = 1$. Commutative pregroups coincide with commutative p.o.

groups. A pregroup M is a p.o. group iff $a^l = a^r$, for all $a \in M$. Pregroups which are not p.o. groups are said to be *proper*. As shown in [15], proper pregroups can be neither finite, nor totally ordered, nor bounded. Every pregroup is a residuated monoid with operations $\backslash, /$ defined by: $a\backslash b = a^r b$ and $a/b = ab^l$. A residuated monoid is a pregroup iff $a\backslash(bc) \leq (a\backslash b)c$ and $(ab)/c \leq a(b/c)$, for all elements a, b, c (then, $a^l = 1/a$, $a^r = a\backslash 1$).

Concrete pregroups can be formed out of order preserving functions on a poset $(W, \leq)$ which are downward and upward unbounded:

$$\forall x \exists y (f(y) \leq x),\ \forall x \exists y (f(y) \geq x).$$

One defines $f \leq g$ iff $f(x) \leq g(x)$, for all x, and $(f \cdot g)(x) = f(g(x))$. The so-defined structure is a p.o. monoid with the identity function as the unit. In every p.o. monoid, l and r can be defined as partial operations: a^l is the unique b (if exists) such that $ba \leq 1 \leq ab$, and a^r is the unique b (if exists) such that $ab \leq 1 \leq ba$. Every p.o. monoid contains a greatest pregroup: it consists of all elements a such that $a^{l\ldots l}$ and $a^{r\ldots r}$ exist, for all finite iterations of l and r (in pregroups $a^{lr} = a^{rl} = a$, so mixed iterations need not be considered). In some cases, this greatest pregroup is proper. Lambek's example is the pregroup of all unbounded, order preserving functions on the poset of integers. Other models have been studied in [15].

For linguistics, Lambek offers free pregroups, i.e. pregroups induced by the formal theory (with special axioms of the form $p \vdash q$). For instance, if p_i is the type of personal pronouns in i–th Person, and p is the common type of personal pronouns, then $p_i \vdash p$ is a special axiom. If s is the type of sentences, then 'works' is assigned type $p_3^r s$. Accordingly, 'he works' is parsed as $p_3 p_3^r s$, hence it reduces to s, by (PRE). This method has successfully been used by Lambek and his followers to analyse quite delicate syntactic structures of English, German, Italian and other ethnic languages; see e.g. [22, 53, 45].

Free pregroups can be presented in the form of a formal system; Lambek calls it Compact Bilinear Logic. Primitive types are as above. Atoms are of the form $p^{(n)}$, for primitive p and arbitrary integers n. We interprete atoms as follows: $p^{(0)} = p$, $p^{(n+1)} = (p^{(n)})^r$, $p^{(n-1)} = (p^{(n)})^l$. Types are finite strings of atoms. In pregroups, $(ab)^l = b^l a^l$ and $(ab)^r = b^r a^r$. Thus, for type $a_1 \ldots a_k$, the left adjoint is $a_k^l \ldots a_1^l$, and the right adjoint is $a_k^r \ldots a_1^r$. Special axioms form a finite set of rules $p \vdash q$, for primitive p, q. The system is based on the following rewriting rules:

(P-CON) Contraction: $\Gamma a^{(n)} a^{(n+1)} \Delta \vdash \Gamma \Delta$,

(P-EXP) Expansion: $\Gamma \Delta \vdash \Gamma a^{(n+1)} a^{(n)} \Delta$,

(P-IND) Induced Step: $\Gamma a^{(n)}\Delta \vdash \Gamma b^{(n)}\Delta$ if either n is even and $a \vdash b$ is a special axiom, or n is odd and $b \vdash a$ is a special axiom.

(P-IND) expresses tha fact that in pregroups $a \leq b$ entails $b^l \leq a^l$ and $b^r \leq a^r$. A derivation is a standard rewriting procedure. Lambek [52] shows an important property: if Γ reduces to Γ', then there is a type Δ such that Γ reduces to Δ by (P-CON) and (P-IND) only, and Δ reduces to Γ' by (P-EXP) and (P-IND) only (the switching lemma). As a consequence, if Γ reduces to an atom or ϵ, then the reduction can use (P-CON) and (P-IND) only. Such reductions are derivations in CF-grammars based on (a finite set) of the rewriting rules. Accordingly, grammars based on free pregroups generate context-free languages. The provability is P-TIME decidable, since $\Gamma \vdash \Delta$ is provable iff $\Delta^l\Gamma$ reduces to ϵ [15].

The logic of pregroups can be formalized as a sequent system which admits cut elimination; two versions of such a system are formulated in [16]. The cut elimination theorem for them is closely related to the switching lemma (they are deducible from each other). Let us recall the one-sided version. Atomic formulas are atoms (in the above sense) and the constant 1. Formulas are formed out of atomic formulas by $\otimes$. Adjoints are defined in the metalanguage. For terms, the definition is given above. We set $1^l = 1^r = 1$ and $(A \otimes B)^d = A^d \otimes B^d$, for $d = l, r$. Sequents are finite strings of formulas. The axiom is ϵ, and the inference rules are:

$$\frac{\Gamma AB\Gamma'}{\Gamma(A\otimes B)\Gamma'},\ \frac{\Gamma\Gamma'}{\Gamma 1\Gamma'},$$

$$\frac{\Gamma\Gamma'}{\Gamma AA^l\Gamma'},\ \frac{\Gamma\Gamma'}{\Gamma A^r A\Gamma'}.$$

The cut elimination theorem is the following: (i) if ΓA^l and $A\Delta$ are provable then $\Gamma\Delta$ is provable, (ii) if ΓA and $A^r\Delta$ are provable, then $\Gamma\Delta$ is provable. Γ reduces to Δ in the rewriting system iff $Delta\Gamma^l$ is provable in the sequent system.

We had began with a simple logic AB, went through a variety of type logics, and have finished at a strong system of Compact Bilinear Logic. Relatively weak systems like AB, NL and the strong logic of pregroups are P-TIME decidable, while the intermediate ones are NP-complete. All systems except the last one can easily be interpreted as fragments of Intuitionistic Logic or type logics of lambda calculus. This seems to be impossible for Compact Bilinear Logic, hence its semantical validity remains an open problem.

References

[1] V.M. Abrusci, Phase Semantics and Sequent Calculus for Pure Noncommutative Classical Linear Logic, *Journal of Symbolic Logic* 56 (1991), 1403-1451.

[2] V.M. Abrusci, Classical Conservative Extensions of Lambek Calculus, *Studia Logica* 71.3 (2002), 277-314.

[3] K. Ajdukiewicz, Die syntaktische Konnexität, *Studia Philosophica* 1 (1935), 1-27.

[4] H. Andréka and S. Mikulás, Lambek Calculus and Its Relational Semantics. Completeness and Incompleteness, *Journal of Logic, Language and Information* 3 (1994), 1-37.

[5] Y. Bar-Hillel, C. Gaifman and E. Shamir, On categorial and phrase structure grammars, *Bull. Res Council Israel* F 9 (1960), 155-166.

[6] A. Białynicki-Birula and H. Rasiowa, On the Representation of Quasi-Boolean Algebras, *Bull. Acad. Polonaise Scie.* 5 (1957), 259-261.

[7] W. Buszkowski, Some Decision Problems in the Theory of Syntactic Categories, *Zeitschrift f. math. Logik u. Grundlagen der Mathematik* 28 (1982), 539-548.

[8] W. Buszkowski, Completeness Results for Lambek Syntactic Calculus, *Zeitschrift f. math. Logik u. Grundlagen der Mathematik* 32 (1986), 13-28.

[9] W. Buszkowski, Generative Capacity of Nonassociative Lambek Calculus, *Bull. Polish Academy Scie. Math.* 34 (1986), 507-516.

[10] W. Buszkowski, Gaifman's Theorem on Categorial Grammars Revisited, *Studia Logica* 47 (1988), 23-33.

[11] W. Buszkowski, *Logical Foundations of Ajdukiewicz-Lambek Categorial Grammars* (in Polish), Polish Scientific Publishers, Warsaw, 1989.

[12] W. Buszkowski, Categorial Grammars with Negative Information, in: H. Wansing (ed.), *Negation. A notion in focus*, de Gruyter, Berlin, 1996, 107-126.

[13] W. Buszkowski, The Ajdukiewicz Calculus, Polish Notation and Hilbert Style Proofs, in: K. Kijania-Placek and J. Woleński (eds.), *The Lvov-Warsaw School and Contemporary Philosophy*, Kluwer, Dordrecht, 1998, 241-252.

[14] W. Buszkowski, Mathematical Linguistics and Proof Theory, in: J. van Benthem and A. ter Meulen (eds.), *Handbook of Logic and Language*, Elsevier, Amsterdam, 1997, 683-736.

[15] W. Buszkowski, Lambek Grammars Based on Pregroups, in: P. de Groote, G. Morrill and C. Retoré (eds.), *Logical Aspects of Computational Linguistics*, LNAI 2099, Springer, Berlin, 2001, 95-109.

[16] W. Buszkowski, Sequent Systems for Compact Bilinear Logic, *Mathematical Logic Quarterly* (2003), to appear.

[17] W. Buszkowski, Lambek Calculus with Nonlogical Axioms, in: C. Casadio et al. (eds.), *A Festschrift for Jim Lambek*, CSLI Publications, Stanford, to appear.

[18] W. Buszkowski and B. Dziemidowicz, Restricted Optimal Unification, in: Y. Hamamatsu et al. (eds.), *Formal Methods and Intelligent Techniques in Control, Decision Making, Multimedia and Robotics*, Warsaw, 2000, 1-8.

[19] W. Buszkowski and M. Kołowska-Gawiejnowicz, Representation of Residuated Semigroups in Some Algebras of Relations. The Method of Canonical Models, *Fundamenta Informaticae* 31 (1997), 1-12.

[20] W. Buszkowski and G. Penn, Categorial Grammars Determined from Linguistic Data by Unification, *Studia Logica* 49 (1990), 431-454.

[21] C. Casadio, Non-Commutative Linear Logic in Linguistics, *Grammars* 3/4 (2001), 1-19.

[22] C. Casadio and J. Lambek, An Algebraic Analysis of Clitic Pronouns in Italian, in: P. de Groote, G. Morrill and C. Retoré (eds.), *Logical Aspects of Computational Linguistics*, LNAI 2099, Springer, Berlin, 2001, 110-124.

[23] K. Došen, A Completeness Theorem for the Lambek Calculus of Syntactic Categories, *Zeitschrift f. math. Logik u. Grundlagen der Mathematik* 31 (1985), 235-241.

[24] J.M. Dunn, Perp and Star: Two Treatments of Negation, in: J. Tomberlin (ed.), *Philosophical Perspectives (Philosophy of Language and Logic)* 7 (1993), 331-357.

[25] J.M. Dunn, Partial Gaggles Applied to Logics with Restricted Structural Rules, in: P. Schroeder-Heister and K. Došen (eds.), *Substructural Logics*, Clarendon Press, Oxford, 1993, 63-108.

[26] B. Dziemidowicz, Optimal Unification and Learning Algorithms for Categorial Grammars, *Fundamenta Informaticae* 49 (2002), 297-308.

[27] B. Dziemidowicz, On Learnability of Restricted Classes of Categorial Grammars, *Fundamenta Informaticae*, to appear.

[28] D.M. Gabbay, A General Theory of Structured Consequence Relations, in: P. Schroeder-Heister and K. Došen (eds.), *Substructural Logics*, Clarendon Press, Oxford, 1993, 109-151.

[29] D.M. Gabbay, *Labelled Deductive Systems*, Oxford University Press, Oxford, 1996.

[30] M. Gécség and M. Steinby, *Tree Automata*, Akademiai Kiadó, Budapest, 1984.

[31] E.M. Gold, Language Identification in the Limit, *Information and Control* 10 (1867), 447-474.

[32] J.Y. Girard, Linear logic, *Theoretical Computer Science* 50 (1987), 1-102.

[33] P. de Groote and F. Lamarche, Classical Non-Associative Lambek Calculus, *Studia Logica* 71.3 (2002), 355-388.

[34] H. Hiż, Grammar logicism, *The Monist* 51 (1967), 110-127.

[35] C.A.R. Hoare and H. Jifeng, The weakest prespecification, *Fundamenta Informaticae* 9 (1986), 51-84, 217-252.

[36] E. Husserl, *Logische Untersuchungen*, Halle, 1900-1901.

[37] M. Kanazawa, The Lambek Calculus Enriched with Additional Connectives, *Journal of Logic, Language and Information* 1.2 (1992), 141-171.

[38] M. Kanazawa, Identification in the Limit of Categorial Grammars, *Journal of Logic, Language and Information* 5 (1996), 115-155.

[39] M. Kanazawa, *Learnable Classes of Categorial Grammars*, CSLI Publications, Stanford, 1988.

[40] M. Kandulski, The Equivalence of Nonassociative Lambek Categorial Grammars and Context-Free Grammars, *Zeitschrift f. math. Logik u. Grundlagen der Mathematik* 34 (1988), 41-52.

[41] M. Kandulski, Normal Form of Derivations in the Nonassociative and Commutative Lambek Calculus with Product, *Mathematical Logic Quarterly* 39 (1993), 103-114.

[42] M. Kandulski, On Generalized Ajdukiewicz and Lambek Calculi and Grammars, *Fundamenta Informaticae* 30.2 (1997), 169-181.

[43] M. Kandulski, Derived Tree Languages of Nonassociative Lambek Categorial Grammars with Product, *Fundamenta Informaticae* (2003), to appear.

[44] S. Kapur, *Computational Learning of Languages*, Ph.D. Thesis, Cornell University, 1991.

[45] A. Kiślak, Pregroups versus English and Polish grammar, in: V.M. Abrusci and C. Casadio (eds.), *New Perspectives in Logic and Formal Linguistics*, Papers in Formal Linguistics and Logic, Bulzoni Editore, Roma, 2002, 129-154.

[46] M. Kołowska-Gawiejnowicz, Powerset Residuated Algebras and Generalized Lambek Calculus, *Mathematical Logic Quarterly* 43 (1997), 60-72.

[47] M. Kołowska-Gawiejnowicz, A Labelled Deductive System for Relational Semantics of the Lambek Calculus, *Mathematical Logic Quarterly* 45 (1999), 51-58.

[48] N. Kurtonina, *Frames and Labels. A modal analysis of categorial inference*, Ph.D. Thesis, University of Utrecht, 1995.

[49] J. Lambek, The mathematics of sentence structure, *American Mathematical Monthly* 65 (1958), 154-170.

[50] J. Lambek, On the Calculus of Syntactic Types, in: R. Jakobson (ed.), *Structure of Language and Its Mathematical Aspects*, Proc. Symp. Appl. Math., AMS, Providence, 1961, 166-178.

[51] J. Lambek, Bilinear Logic in Algebra and Linguistics, in: J.Y. Girard, Y. Lafont and L. Regnier (eds.), *Advances in Linear Logic*, Cambridge University Press, Cambridge, 1995, 43-59.

[52] J. Lambek, Type Grammars Revisited, in: A. Lecomte, F. Lamarche and G. Perrier (eds.), *Logical Aspects of Computational Linguistics*, LNAI 1582, Springer, Berlin, 1999, 1-27.

[53] J. Lambek, Type Grammars as Pregroups, *Grammars* 4 (2001), 21-39.

[54] S. Leśniewski, Grundzüge einer neuen System der Grundlagen der Mathematik, *Fundamenta Mathematicae* 14 (1929), 1-81.

[55] W. MacCaull and E. Orłowska, Correspondence Results for Relational Proof Systems with Application to the Lambek Calculus, *Studia Logica* 71.3 (2002), 389-414.

[56] J. Marciniec, Learning Categorial Grammars by Unification with Negative Constraints, *Journal of Applied Non-Classical Logics* 4 (1994), 181-200.

[57] R. Montague, *Formal Philosophy*, Selected papers of R. Montague edited by R. Thomason, Yale University Press, New Haven, 1974.

[58] M. Moortgat, *Categorial Investigations. Logical and Linguistic Aspects of the Lambek Calculus*, Foris, Dordrecht, 1988.

[59] M. Moortgat, Categorial Type Logics, in: J. van Benthem and A. ter Meulen (eds.), *Handbook of Logic and Language*, Elsevier, Amsterdam, 1997, 93-177.

[60] M. Moortgat, Structural Equations in Language Learning, in: P. de Groote, G. Morrill and C. Retoré (eds.), *Logical Aspects of Computational Linguistics*, LNAI 2099, Springer, Berlin, 2001, 1-16.

[61] G. Morrill, *Type Logical Grammar. Categorial Logic of Signs*, Kluwer, Dordrecht, 1994.

[62] G. Morrill, A Generalised Discontinuity, manuscript, LICS, Ottawa, 2003.

[63] E. Orłowska, Relational Proof System for Relevant Logics, *Journal of Symbolic Logic* 57 (1992), 1425-1440.

[64] D. Osherson, D. de Jongh, E. Martin and S. Weinstein, Formal learning theory, in: J. van Benthem and A. ter Meulen (eds.), *Handbook of Logic and Language*, Elsevier, Amsterdam, 1997, 737-775.

[65] M. Pentus, Lambek Grammars are Context-Free, *Proc. of 8th IEEE Symposium on Logic in Computer Science*, 1993, 429-433.

[66] M. Pentus, Models for the Lambek Calculus, *Annals of Pure and Applied Logic* 75 (1995), 179-213.

[67] M. Pentus, Lambek calculus is NP-complete, manuscript, Moscow State University, 2003.

[68] V. Pratt, Action Logic and Pure Induction, in: J. van Eijck (ed.), *Logics in AI*, LNAI 478, Springer, Berlin, 1991, 97-120.

[69] G. Restall, *An Introduction to Substructural Logics*, Routledge, London, 2001.

[70] D. Roorda, *Resource Logics. Proof-Theoretical Investigations*, Ph.D. Thesis, University of Amsterdam, 1991.

[71] T. Shinohara, Inductive Inference of Monotonic Formal Systems from Positive Data, in: S. Arikawa et al. (eds.), *Algorithmic Learning Theory*, Springer, Tokyo, 1990, 339-351.

[72] T.A. Sudkamp, *Languages and Machines. An Introduction to the Theory of Computer Science*, Addison-Wesley, Reading, 1991.

[73] M. Szczerba, Representation Theorems for Residuated Groupoids, in: C. Retoré (ed.), *Logical Aspects of Computational Linguistics*, LNAI 1328, Springer, Berlin, 1997, 426-434.

[74] H-J. Tiede, *Deductive Systems and Grammars*, Ph.D. Thesis, Indiana University, 1999.

[75] H. Uszkoreit, Categorial Unification Grammars, *Proc. 11th International Conference on Computational Linguistics*, Bonn, 1986, 187-194.

[76] J. van Benthem, *Essays in Logical Semantics*, D. Reidel, Dordrecht, 1986.

[77] J. van Benthem, Categorial Equations, in: E. Klein and J. van Benthem (eds.), *Categories, Polymorphism and Unification*, University of Amsterdam, 1987, 1-17.

[78] J. van Benthem, *Language in Action. Categories, Lambdas and Dynamic Logic*, North Holland, Amsterdam, 1991.

[79] J. van Benthem, *Exploring Logical Dynamics*, CSLI, Stanford, 1996.

[80] J. van Benthem, Categorial Grammar at a Cross-Roads, in: C. Casadio et al. (eds.), *A Festschrift for Jim Lambek*, CSLI Publications, to appear.

[81] K. Wright, Identification of unions of languages drawn from an identifiable class, in: *The 1989 Workshop on Computational Learning Theory*, Morgan Kaufmann, San Mateo, 1989, 328-333.

[82] D.N. Yetter, Quantales and (Non-Commutative) Linear Logic, *Journal of Symbolic Logic* 55 (1996), 41-64.

[83] W. Zielonka, Axiomatizability of Ajdukiewicz-Lambek Calculus by means of Cancellation Schemes, *Zeitschrift f. math. Logik u. Grundlagen der Mathematik* 27 (1981), 215-224.

[84] W. Zielonka, On Reduction Systems Equivalent to the Lambek Calculus with the Empty String, *Studia Logica* 71.1 (2002), 31-46.

TRENDS IN LOGIC

1. G. Schurz: *The Is-Ought Problem.* An Investigation in Philosophical Logic. 1997 ISBN 0-7923-4410-3
2. E. Ejerhed and S. Lindström (eds.): *Logic, Action and Cognition.* Essays in Philosophical Logic. 1997 ISBN 0-7923-4560-6
3. H. Wansing: *Displaying Modal Logic.* 1998 ISBN 0-7923-5205-X
4. P. Hájek: *Metamathematics of Fuzzy Logic.* 1998 ISBN 0-7923-5238-6
5. H.J. Ohlbach and U. Reyle (eds.): *Logic, Language and Reasoning.* Essays in Honour of Dov Gabbay. 1999 ISBN 0-7923-5687-X
6. K. Došen: *Cut Elimination in Categories.* 2000 ISBN 0-7923-5720-5
7. R.L.O. Cignoli, I.M.L. D'Ottaviano and D. Mundici: *Algebraic Foundations of many-valued Reasoning.* 2000 ISBN 0-7923-6009-5
8. E.P. Klement, R. Mesiar and E. Pap: *Triangular Norms.* 2000 ISBN 0-7923-6416-3
9. V.F. Hendricks: *The Convergence of Scientific Knowledge.* A View From the Limit. 2001 ISBN 0-7923-6929-7
10. J. Czelakowski: *Protoalgebraic Logics.* 2001 ISBN 0-7923-6940-8
11. G. Gerla: *Fuzzy Logic.* Mathematical Tools for Approximate Reasoning. 2001 ISBN 0-7923-6941-6
12. M. Fitting: *Types, Tableaus, and Gödel's God.* 2002 ISBN 1-4020-0604-7
13. F. Paoli: *Substructural Logics: A Primer.* 2002 ISBN 1-4020-0605-5
14. S. Ghilardi and M. Zawadowki: *Sheaves, Games, and Model Completions.* A Categorical Approach to Nonclassical Propositional Logics. 2002 ISBN 1-4020-0660-8
15. G. Coletti and R. Scozzafava: *Probabilistic Logic in a Coherent Setting.* 2002 ISBN 1-4020-0917-8; Pb: 1-4020-0970-4
16. P. Kawalec: *Structural Reliabilism.* Inductive Logic as a Theory of Justification. 2002 ISBN 1-4020-1013-3
17. B. Löwe, W. Malzkorn and T. Räsch (eds.): *Foundations of the Formal Sciences II.* Applications of Mathematical Logic in Philosophy and Linguistics, Papers of a conference held in Bonn, November 10-13, 2000. 2003 ISBN 1-4020-1154-7
18. R.J.G.B. de Queiroz (ed.): *Logic for Concurrency and Synchronisation.* 2003 ISBN 1-4020-1270-5
19. A. Marcja and C. Toffalori: *A Guide to Classical and Modern Model Theory.* 2003 ISBN 1-4020-1330-2; Pb 1-4020-1331-0
20. S.E. Rodabaugh and E.P. Klement (eds.): *Topological and Algebraic Structures in Fuzzy Sets.* A Handbook of Recent Developments in the Mathematics of Fuzzy Sets. 2003 ISBN 1-4020-1515-1; Pb 1-4020-1516-X

21. V.F. Hendricks and J. Malinowski: *Trends in Logic*. 50 Years Studia Logica. 2003
ISBN 1-4020-1601-8

KLUWER ACADEMIC PUBLISHERS – DORDRECHT / BOSTON / LONDON

FSC
www.fsc.org

MIX
Papier aus verantwortungsvollen Quellen
Paper from responsible sources
FSC® C105338

If you have any concerns about our products,
you can contact us on
ProductSafety@springernature.com

In case Publisher is established outside the EU,
the EU authorized representative is:
Springer Nature Customer Service Center GmbH
Europaplatz 3, 69115 Heidelberg, Germany

Printed by Libri Plureos GmbH
in Hamburg, Germany